Electron Transfer Reactions

Inorganic, Organometallic, and Biological Applications

ADVANCES IN CHEMISTRY SERIES 253

Electron Transfer Reactions

Inorganic, Organometallic, and Biological Applications

Stephan S. Isied, EDITOR

Rutgers, The State University of New Jersey

American Chemical Society, Washington, DC

Library of Congress Cataloging-in-Publication Data
Electron transfer reactions : inorganic, organometallic, and biological applications /
Stephan S. Isied, editor.

p. cm.—(Advances in chemistry series, ISSN 0065-2393 ; 253)

"Developed from a symposium bearing the title, Taube Insights: From Electron
Transfer Reactions to Modern Inorganic Chemistry, organized by the editor and held at
Stanford University, Stanford, California, March 30–31, 1995."

Includes bibliographical references and index.

ISBN 0–8412–3456–6

1. Oxidation-reduction reaction—Congresses. I. Isied, Stephan S., 1946– 1. II.
Series.

QD1.A355   no. 253   [QD63.09]

540 s—dc21 [541.3'93   96–38033

CIP

The paper used in this publication meets the minimum requirements of American National Standard for Information Sciences—Permanence of Paper for Printed Library Materials, ANSI Z39.48-1984.

Foreword

The ADVANCES IN CHEMISTRY SERIES was founded in 1949 by the American Chemical Society as an outlet for symposia and collections of data in special areas of topical interest that could not be accommodated in the Society's journals. It provides a medium for symposia that would otherwise be fragmented because their papers would be distributed among several journals or not published at all.

Papers are reviewed critically according to ACS editorial standards and receive the careful attention and processing characteristic of ACS publications. Volumes in the ADVANCES IN CHEMISTRY SERIES maintain the integrity of the symposia on which they are based; however, verbatim reproductions of previously published papers are not accepted. Papers may include reports of research as well as reviews, because symposia may embrace both types of presentation.

About the Editor

STEPHAN S. ISIED is a professor of chemistry at Rutgers, the State University of New Jersey. He received his Ph.D. in organic chemistry from Stanford University under the mentorship of Professor Henry Taube, and his B.S. and M.S. degrees at the American University in Beirut, Lebanon. He joined the faculty at Rutgers University in 1976 after a postdoctorate at the University of California, Berkeley, with Professor Kenneth Raymond (1974–75). He has held concurrent visiting scientist positions at Rockefeller University, Stanford University, California Institute of Technology, and Brookhaven National Laboratory. He has been honored with an NIH Career Development Award, a Camille and Henry Dreyfus Teacher-Scholar Award, a Johnson & Johnson Discovery Fellowship, and a Rutgers Board of Trustees Excellence in Research Award. His current research interests are in biological electron transfer reactions—especially the relevance of peptide conformation and weak interactions such as H-bonding and hydrophobic effects on charge transport phenomena.

Contents

ELECTRON TRANSFER AND TRANSITION METALS IN BIOLOGY

Henry Taube,
1983 Nobel Laureate

HENRY TAUBE WAS BORN ON NOVEMBER 30, 1915, in Neudorf, Saskatchewan, Canada. He received his education at the University of Saskatchewan (B.S., 1935; M.S., 1937) and at the University of California at Berkeley (Ph.D., 1940). He became a naturalized U.S. citizen in 1941. He served as an instructor at Berkeley from 1940 to 1941; as an instructor and assistant professor at Cornell University from 1941 to 1946; as assistant, associate, and full professor at the University of Chicago from 1946 to 1961; and as professor of chemistry at Stanford University from 1961 to the present. He served as department chairman at Chicago from 1955 to 1959 and as chairman at Stanford from 1972 to 1974. Taube's major awards include the Nobel Prize in chemistry in 1983 and the Priestley Medal, the highest award of the American Chemical Society, in 1985. Taube is a physical inorganic chemist whose contributions have been in the discovery of new chemical reactions and the study of their mechanistic pathways.

Taube's professional career began at Cornell in the years from 1941 to 1946. During that time, his research continued in the same areas as during his graduate student investigations—that is, redox reactions of oxygen- and halogen-containing oxidizing agents. (Interestingly, during this period and up until 1950, Taube had very few publications concerned with transition metal chemistry. However, that would soon change.) In studies of aqueous solutions of chlorate, Taube applied oxygen-atom-labeling studies, using oxygen-18 and radioactive chlorine to study the mechanisms of the redox reactions of oxochlorine species and related molecules in aqueous solutions. In 1955 Taube received the American Chemical Society Award for Nuclear Applications in Chemistry in recognition of his perceptive applications of isotopic labeling techniques.

(Excerpted from *Nobel Laureates in Chemistry 1901–1992*, pp. 660–666, © 1993 American Chemical Society and the Chemical Heritage Foundation. With permission.)

In the area of the inorganic chemistry of metal ions, the contributions of Alfred Werner, leading to an understanding of the composition and structure of coordination complexes, are well known. Although a great deal was known about the reactivity of coordination complexes by 1950, it remained for Henry Taube to place reactivity into a unified concept based on the electronic nature of the metal and on the influence of the ligands around the metal center. In many ways, Taube is the leading figure in modern inorganic chemistry, because he laid the groundwork for many conceptual advancements and probed various aspects of chemical reactivity with new techniques.

The wide-ranging contributions of Henry Taube to the advancement of chemistry are apparent from just a casual survey of the areas in which he has published. In addition to major contributions in classical substitution and redox chemistry of transition metal complexes of Cr, Co, Ru, and Os, Taube's publications include papers on (1) main group nonmetal ions and molecules, particularly oxygen compounds and chlorine oxide compounds; (2) organic acids and amino acids; (3) rare earth compounds; (4) main group metal complexes; (5) sulfur ligands in coordination complexes; (6) organometallic compounds of transition metals. The impressive aspect of these studies is the way in which Taube universally applies the fundamental concepts and techniques that he so masterfully developed and refined.

The greatness of a scientist is not measured only by the number (more than 600) and significance of his publications. An important parameter in the measure of a scientist is the overall impact of his "school" on the scientific community. Taube associates (students, postdocs, visiting faculty, and coauthors) make an impressive list of scientists who have made tremendous contributions in their own right. Of over 250 associates, nearly half have served as faculty on university campuses throughout the United States and the world. The Taube school is represented at places such as Cornell, Iowa State, Georgetown, the State University of New York at Stony Brook, Texas Tech, Georgia Tech, Michigan State, Rutgers, Boston, North Carolina, Indiana, MIT, Rice, and Pittsburgh, to name a few. Over a dozen Taube associates are prominent scientists in foreign countries. A similar number hold positions at national research laboratories such as Los Alamos, Brookhaven, and Argonne. Add to this the number of associates in major industrial laboratories and the influence of Henry Taube on the scientific community is indeed significant.

Henry Taube's work was celebrated and honored in a symposium at a national meeting of the American Chemical Society (in Las Vegas, Nevada, 1982) and in Volume 30 of Progress in Inorganic Chemistry, entitled "An Appreciation of Henry Taube". The impact of Henry Taube is reflected in the prominence of the science and the scientists who contributed to these celebrations.

On a personal level, Henry Taube has been characterized as a modest, unpretentious scientist who generously credits others for inspiration and accomplishments. He is greatly respected by his students and colleagues for his guidance and inspiration to approach chemical problems from a critical but rational viewpoint. The success of the Taube school is due not so much to their training in chemistry as to the infectious enthusiasm and creative approach to chemistry that Taube has passed on to his colleagues. Taube associates transmit a sense of high esteem and appreciation for Taube in all references to their professional and personal associations with him.

A reflection of Taube's generosity and respect for others may be best represented by his choice of topics for his Priestley Medal address. On that occasion, he honored William C. Bray, his Ph.D. mentor, four decades after Bray's death. The warmth, inspiration, and generosity of Bray was obviously a strong influence on Taube, who is surely gratified to have earned similar praise from his students.

Taube is an emeritus professor at Stanford and he continues to investigate and inspire the development of new ideas and new areas of inorganic chemistry that are, to him, an extension of his original ideas on chemical reactivity and mechanisms. He continues to publish on organometallic osmium chemistry, osmium dihydrogen complexes, and a variety of other new reactions with the same enthusiasm that many of his associates have witnessed throughout his career. His immediate family consists of his wife, Mary; two daughters, Linda and Marianna, and two sons, Heinrich and Karl. He enjoys listening to old opera records from his vast collection and is said to have a passion for gardening and sour mash whiskey.

The scientific contributions of Henry Taube have inspired revolutionary advancements in inorganic reaction mechanisms. His direct achievements have laid much of the groundwork, and his associates have built on that to define much of what we know as inorganic reaction chemistry. A simple statement from the nomination papers for the Nobel Prize may be the most appropriate way to end. "Henry Taube founded the modern study of inorganic reaction mechanisms."

JERRY WALSH
University of North Carolina
at Greensboro

Bibliography

"An Appreciation of Henry Taube"; Lippard, S. J., Ed.; Prog. Inorg. Chem. 1983, 30.

Gray, H. B.; Collman, J. P. "The 1983 Nobel Prize in Chemistry"; *Science* 1983, 222, 986–987.

Taube, H. "Electron Transfer between Metal Complexes: Retrospective"; *Science* 1984, 226, 1028–1036.

Taube, H. *Electron Transfer Reactions of Complex Ions in Solution;* Academic Press: New York, 1970.

Taube, H. "Rates and Mechanisms of Substitution in Inorganic Complexes in Solution"; *Chem. Rev.* 1952, 50, 69–126.

The Taube Revolution, 1952–1954

ONE DAY, 40 YEARS AGO, I was sitting in the science library, browsing through recent issues of the *Journal of the American Chemical Society*, searching in vain for something that was both new and interesting in inorganic coordination chemistry. For the *n*th time, I was tempted to abandon my Ph.D. thesis and turn to more promising fields, such as organic chemistry or biochemistry. I sat there, deploring the sad story of classical coordination chemistry: it was created by Alfred Werner in a "Big Bang" in 1893 (*1*) and flourished for a quarter of a century. Although Werner was not the only contributor to the field he created, he was by far the most important and influential one.

The decline of classical coordination chemistry started when Werner's illness forced him to abandon his research around 1915. His premature death in 1919 marked the end of the Classical Age of coordination chemistry and the beginning of its Middle Ages, or rather its Dark Age. Very little important work was accomplished in the three decades that followed. This was in remarkable contrast to organic chemistry and biochemistry, which flourished during the same period: synthetic polymers, kinetics and mechanisms of organic substitution reactions, and the discovery of biochemical cycles, vitamins, and antibiotics were some of the highlights of this period.

Meanwhile, inorganic coordination chemistry was in a state of deep hibernation. Even Pauling's valence bond theory, which furnished new insight and understanding of the nature of chemical bonds of coordination compounds and of their physical properties, failed to awaken it.

Engulfed by these depressing thoughts, my eyes focused on a paper entitled "Evidence for a Bridged Activated Complex for Electron Transfer Reactions" (1954) (*2*). I was not expecting anything unusual, because I had missed the earlier communication to the Editor (1953) (*3*) and the review on "Rates and Mechanism of Substitution Reactions in Inorganic Complexes in Solution" (1952) (*4*). As soon as I started reading, my pulse accelerated with excitement, and my eyes raced from one line to the other, trying to swallow it all in one gulp. When I reached the end, literally out of breath, I rushed to the shelves of

Chemical Reviews to find the 1952 review (4) by Henry Taube. As I finished reading it, late that evening, I knew that the Dark Age of inorganic chemistry had come to an end, and its Renaissance had just begun.

The shift of interest from the traditional topics of coordination chemistry toward the mechanism of its reactions in aqueous solutions was one of the main features of this Renaissance. The first results were, however, disappointing. Attempts to repeat the success story of Ingold's "English Heresy" of the 1930s by applying it to the far more complicated situation existing in aqueous solutions of inorganic reagents had only very limited success (5).

The first step forward was made when Taube, in his definitive review (4), created order in this primordial field by establishing the dichotomy between inert and labile complexes. The confusion between thermodynamic and kinetic stability of coordination compounds had plagued coordination chemistry for decades. No real advance in the field of inorganic reaction mechanism could be achieved until this confusion had been cleared. Although ligand substitution reactions was the main subject of this review, its greatest impact was elsewhere—on the study of the mechanism of redox reactions. It would seem that to investigate the mechanism of substitution reactions of aqua ions was complicated enough—in the early 1950s almost hopelessly complicated. To try and elucidate the mechanism of electron transfer processes between these ill-defined, labile, and elusive species, before they themselves were fully characterized, would seem to be almost foolhardy.

Fortunately, Taube's intuition and vision drove him in this direction. His choice of the prototypical reaction

$$Co(NH_3)_5Cl^{2+} + Cr^{2+} + 5H^+ \rightarrow Co^{2+} + 5NH_4^+ + CrCl^{2+}$$

for gaining insight into the mechanism of a redox reaction between two metal ions was a stroke of genius. It was a carefully planned experiment devised to drive nature into a corner that left it no choice but to confirm the suspicion that electron transfer can take place through a ligand bonded directly to both the reducing and the oxidizing metal ion.

Often, in the history of science, important theories or experiments were proposed simultaneously and independently by two scientists because they were based on new findings that had just been made. In the case of Taube, the opposite happened. The stage was set for his revolutionary work long before it was carried out. The experimental bases, including the crucial rate and equilibrium data, were known for at least 20 years and most of them for more than 30 years. The rate and equilibrium data of the reaction between the chromic aqua ion and the chloride ion were known since the work of Niels Bjerrum, at the beginning of the century (6). The distinction between inert and labile complexes could have been made at any time after that and probably was, subconsciously, in the minds of many coordination chemists. Some even used the term "robust" (instead of "inert") and were obviously quite familiar with this distinction, but no one realized its importance. Some of the techniques used by Taube, such as the use of ^{18}O and of radioactive chloride, were not available in the 1920s or 1930s, but he used them only to reconfirm the results of product analysis of the reaction between the cobaltic and chromous ions. The latter were clear and unequivocal, even without this reconfirmation. In other words, a coordination chemist of the intellectual stature of Henry Taube could have done this work a whole generation earlier. There was none.

MICHAEL ARDON
Hebrew University of Jerusalem

References

1. Werner, A. *Z. Anorg. Chem.* **1893**, *3*, 279.
2. Taube, H.; Myers, H. *J. Am. Chem. Soc.* **1954**, *76*, 2103.
3. Taube, H.; Myers, H.; Rich, R. L. *J. Am. Chem. Soc.* **1953**, *75*, 4118.
4. Taube, H. *Chem. Rev.* **1952**, *50*, 69.
5. Ingold, C. K., et al. *J. Chem. Soc.* **1953**, 2674, 2696.
6. Bjerrum, N. *Z. Phys. Chem.* **1907**, *59*, 339, 581.

Preface

Henry Taube's ideas have had a profound impact on the development of inorganic chemistry in the second half of the 20th century. He began his research studying the chemistry and photochemistry of nonmetallic oxidants such as ozone, hydrogen peroxide, and halogens and their reactions with a variety of inorganic and organic species. Throughout these studies, his early insights into mechanistic questions were very apparent. In 1952 he published a classic series of papers and a review summarizing his ideas on the rates and mechanisms of substitution and electron transfer reactions of complex ions in solution. Taube's definitions of inert and labile metal ions in terms of valence bond theory are of fundamental and historical importance in relating electronic structure theory to the reactivity of metal ions in solution. This relationship constitutes one of the foundations of mechanistic inorganic chemistry and provided inorganic chemists with the concepts and methodologies needed to pursue and advance the field to its present state. His elegant experiments in the use of radioisotopes and NMR methods to assign hydration numbers to water molecules around a metal ion in solution and to determine the rate of water exchange constitute the experimental basis for studying the reactions of these species in solution.

In his research on charge transfer complexes, Taube was able to describe metal–ligand bonds in terms of simple molecular orbital language. As a result of this work the new field of mixed-valence compounds developed into a new area for spectroscopic and kinetic investigations. Ligands with σ-donor, π-donor, and π-acceptor properties are now used to dissect and rationalize the affinity and dynamics of metal–ligand interactions. His recent experiments in the area of organometallic osmium ammines have resulted in reuniting the divergent fields of classical coordination chemistry and organometallic chemistry. This research showed that simple ammine complexes of osmium exhibit many of the reactivity characteristics of organometallic complexes with phosphine, arene, and cyclopentadienyl-type ligands. From his experiments in this area, powerful predictions on the reactivity of organic fragments in the coordination sphere of osmium ammines can be made and put to use in organic synthesis.

Taube's ideas on intramolecular electron transfer in extended binuclear complexes set the stage for the more recent electron transfer studies in peptides, proteins, and other more complex biomolecules. Overall, Taube's integrated approach of experimentation on the kinetics, equilibria, and reactivity of complex ions in solution is a continuing theme throughout his studies and constitutes a foundation for collecting and understanding new facts about inorganic systems.

In March of 1995, the year of his 80th birthday, Taube's students, friends, collaborators, and colleagues from around the world assembled at the Stanford University Chemistry Department for a two-day symposium to honor Taube and to remember the contributions he has made to inorganic chemistry and to their research. Only a limited number of topics influenced by Taube's research could be included in the two-day symposium. However, the participants at the symposium found the experience very rewarding and enjoyable and had a chance to review some of the diverse areas of chemistry that have been affected by Taube's ideas and insights. The symposium included lectures in the area of electron transfer and mixed-valence chemistry, mechanistic inorganic chemistry and photochemistry, organometallic chemistry and catalysis, and electron transfer and the use of transition metal ions in biological systems.

This book presents many of the symposium lectures, emphasizing the common themes of Taube's insights that have led to many advances in these areas. Many readers will find these underlying Taube themes applicable to their research interests. The book begins with a chapter by Taube, a historical perspective of his work on the study of electron transfer reactions and the effects of π-backbonding on the reactivity of metal complexes. In the first section of the book (Chapters 2–7), the theory, mechanisms, and applications of Taube's work in inorganic synthesis organic oxidation, polymerization catalysis, zeolite matrices, and catalytic peroxide reactions are discussed.

In the second section of the book (Chapters 8–18), the various aspects of electron transfer mechanisms of cobalt macrocycles, barriers to atom transfer reactions, nuclear factors in main-group electron transfer reactions, redox reactions of mixed-valence ions, electron delocalization in disulfide ligands, reaction of cysteines with aquo iron(III) species, design of oscillating reactions, and hydrolysis of coordinated nitriles are discussed. The area of inorganic photochemistry is represented by three chapters (11–13) on the calculation of the rate of nonradiative decay of metal-to-ligand charge transfer (MLCT) from spectra, ligand-induced relaxation of excited states in chromium(III) complexes, and time-resolved studies of migratory insertions in manganese carbonyl compounds.

In the third section of the book (Chapters 19–24), Taube's influence on the areas of long-range electron transfer reactions and metal ions in biology is presented. Chapters on electron transfer and the entatic state and electron transfer across protein and peptide networks demonstrate how Taube's ideas have been extended to macromolecular systems. Two chapters discuss the use of ruthenium complexes as DNA probes and potential drugs, showing how the wealth of information on the mechanism and reactions of ruthenium complexes

can lead to rational drug design. The last two chapters discuss the role of inorganic chemistry in cellular mechanisms of host resistance to disease and applications to diagnostic imaging reagents in nuclear medicine.

In this book the authors highlight the influence of Taube insights on their research. Whether in the area of inorganic chemistry, organometallic chemistry, or biology, the underlying themes of kinetics and mechanisms, and binding and affinity, as they affect reactivity provide a foundation for continuing research and discovery. These principles will continue to be the engine that generates new knowledge and opens new areas of research and applications for inorganic reactions.

STEPHAN S. ISIED
Department of Chemistry
Rutgers, The State University of New Jersey
P.O. Box 939
Piscataway, NJ 08550

Acknowledgments

WE GRATEFULLY ACKNOWLEDGE the following organizations: Air Products, BASF, BioMetallics, Catalytica, Mallinckrodt, Miles, and the Stanford University Departments of Chemistry and Chemical Engineering, for their participation and financial contributions to the Henry Taube Symposium and to this book.

From Electron Transfer Reactions to the Effects of Backbonding

Henry Taube

Department of Chemistry, Stanford University, Stanford, CA 94305–5080

The initial motivation for extending the chemistry of ruthenium ammines was to provide new reagents for research on electron transfer reactions. Quite early in pursuing the preparative chemistry, an unexpected capacity of Ru(II) for backbonding was uncovered, and the systematic investigation of its influence on physical and chemical properties became a goal in its own right. The new directions led to the discovery of a number of novel reactions, which in some cases led to novel products. The research has provided documentation of the effect of backbonding on ligand-to-metal charge transfer, ligand–metal distances, the intensities of infrared absorption by π-acceptor ligands, the acid–base properties of such ligands, the properties of co-ligands exerted by the electronic-withdrawing power of π-acceptor ligands, and on Ru(III)/Ru(II) redox potentials and enthalpies of complex formation. Some early results, which demonstrate the superior capacity of the osmium ammines to engage in backbonding, are also described.

ONE OF THE THEMES OF MY EARLY RESEARCH at Stanford, begun late in 1961, was the investigation of the mechanisms of substitution reactions of metal complexes by the application of isotope effects, mainly of oxygen. For example, we carried out kinetic and tracer studies in the alkaline hydrolysis of $[Co(NH_3)_5(O_2CCF_3)]^{2+}$ (*1*) and attempted to generate intermediates such as might be implicated in S_N1 mechanisms (e.g., we hoped to generate $[Co(NH_3)_5]^{3+}$ by the nitrosation of $[Co(NH_3)_5(N_3)]^{2+}$) (*2*).

Mechanisms of Electron Transfer

The major emphasis of the research in my group, however, was the investigation of the mechanisms of electron transfer reactions. In key experiments done at the University of Chicago, the substitution characteristics of the $L(NH_3)_5$-Co(III)/(II) and Cr(III)/Cr(II) couples had been exploited to reveal the participation of atom bridging groups (3, 4) in the electron transfer act, especially for the reaction in eq 1

$$Cr^{2+} + [(NH_3)_5CoCl]^{2+} \rightarrow [(NH_3)_5Co\cdots Cl\cdots Cr(H_2O)_5]^{4+}$$

$$\rightarrow Co(H_2O)_6^{2+} + 5\,NH_4^+ + (H_2O)_5CrCl^{2+} \tag{1}$$

and the work had proceeded to the point that electron transfer through polyatomic bridging groups had been demonstrated (5, 6). Of particular interest is the possibility of electron transfer over extended primary bond systems, for which unambiguous examples, if any, were rare at that time. Our objective was to extend the scope of the investigation to metal centers with different electronic structures. The need to take this direction was recognized in part because the principles of ligand field theory had by then become a part of the scientific culture in the field. The redox active orbitals of both Co(III) and Cr(II) have σ symmetry, and a promising next step was to seek a replacement metal ion capable of accepting an electron in an orbital of π symmetry instead of the Co(III) oxidant. The oxidant is selected to be substitution-inert, and the bridging ligand can be varied systematically, enabling the effect of such changes on reaction rates to be studied. An additional requirement on the replacement is that the metal ion driving force for reduction match that of the cobalt ammines.

Reactivity of Ru(II)/(III) Ammine Complexes

A survey of the literature suggested that the Ru(II) ammines offered the best prospects of meeting all three conditions. The kinetic inertness and the πd vacancy were recognized characteristics of Ru(II) ammines, but information on the redox properties of Ru(II)/(III) ammines was lacking. Preparative work, mainly on Ru(III) tetraammines by Morgan and Burstall (7), and on the Ru(III) pentaammines and the tetraammines by Gleu and co-workers (8), had been done. A major theme of this early work with the Ru(III) tetraammines is the similarity of the Ru(III) isomers to those of the Co(III) tetraammines, which had been described much earlier. In fact, terms such as violeo-, roseo-, purpureo-, and praseo-, descriptive of the colors of the Co(III) ammines, were being used to imply geometrical arrangements, not to describe color. The only Ru(II) ammine complexes described in this early work contain SO_2, HSO_3^-, or SO_3^{2-} as ligands (8), but the significance of the stability of these species in air was not at all obvious to me in my early reading of this literature. Very little relevant additional work on ruthenium ammines had appeared when, ca. 1961, I began to look for πd redox reagents. [An exception is an abstract (9) published in 1959, in which the preparation of $[Ru(NH_3)_4(HSO_3)_2]$ and $[Ru(NH_3)_5(NO)]$-

$Cl_2 \cdot H_2O$ are reported but with no particular reference to their chemical properties.] Considerable work on 1,10-bipyridine and phenanthroline (phen) complexes of ruthenium had also been reported, notably by Dwyer and co-workers (*10*). This work included the study of the redox chemistry and showed that the ruthenium 3+/2+ couples were much more strongly oxidizing than those of the cobalt 3+/2+ complexes, which had been used in the electron transfer research.

Along with the new work that was initiated on ruthenium ammines soon after my arrival at Stanford, we continued our earlier research on cobalt-ammine oxidants with extended bridging groups. This research suffered a severe setback when it became apparent that a number of interesting effects, reported by an unusually enterprising co-worker and me (*11*) in the period 1959–1961, could not be reproduced, and their further exploration was my first priority at Stanford. A good part of the early effort was devoted to setting the record straight, and I take this opportunity to express my appreciation to those who contributed to this essential but, for them, unrewarding, task (*12–14*).

Mechanisms of Electron Transfer with Extended Bridging Groups

The fact that organic bridging groups containing conjugated bond systems can lead to greatly enhanced rates of reduction of their pentaamminecobalt(III) complexes by Cr^{2+}(aq) had been observed in numerous cases. A comparison of electron transfer results for different bridging groups led to the hypothesis that reducibility of the ligand was the key issue (*15*) and that, for the most part, reduction of Co(III) when high rates are observed involves stepwise transfer [i.e., electron hopping rather than the presumed resonance transfer (*5*)]. The revised interpretation of electron hopping as the preferred mechanism of electron transfer through conjugated bond systems for the Co(III)/Cr(II) reactions was bolstered by a considerable number of studies, and several completed somewhat later were particularly persuasive (*16–19*). [This subject is dealt with in some detail in a review article by Meyer et al. (*20*).] The Co(III)–Cr(II) case implied that a stepwise electron transfer mechanism (electron hopping) is enforced by the symmetry mismatch between the carrier ligand π^* orbitals and the metal ion σd orbitals. Resonance transfer would occur if the redox active orbital of at least one of the metal ions is of π symmetry, which in fact proves to be the case when $[Co(NH_3)_5]^{3+}$ is replaced by $[Ru(NH_3)_5]^{3+}$ (*21*).

During this period studies on electron transfer reactions of ruthenium compounds had begun even before the additional incentive of understanding the mechanism of electron transfer (electron hopping vs. resonance transfer) had been defined. The first publication on the redox chemistry of ruthenium ammines deals with the reduction of $[Ru(NH_3)_6]^{3+}$, $[Ru(NH_3)_5Cl]^{2+}$, and $[Ru(NH_3)_5(H_2O)]^{3+}$ by Cr^{2+}(aq) (*22–24*). Several new observations that proved to have important implications for later work are reported in those studies.

First, the reaction of Cr^{2+} with $[Ru(NH_3)_5Cl]^{2+}$ takes place with Cl^- transfer, similar to the analogous reaction with $[Co(NH_3)_5Cl]^{2+}$, whereas that with $[Ru(NH_3)_6]^{3+}$ of necessity involves an outer-sphere-activated complex. The rate advantage of the inner-sphere path, however, is reduced to 80 for the Ru(III) system, compared to the factor of 10^9 observed for Co(III) complexes. Observations made on the properties of the Ru(II) products, such as the inertness of the Ru(II)–NH_3 bond and the lability of the Ru(II)–H_2O bond, are particularly relevant to later work. (The stability of the Ru(II)–NH_3 bond had already been demonstrated in $[Ru(NH_3)_5(SO_2)]^{2+}$, but because SO_2 is a π acid, the inertness in $[Ru(NH_3)_5(H_2O)]^{2+}$ could not be ensured.) Second, catalysis by $[Ru(NH_3)_5(H_2O)]^{2+}$ of the Ru(III) substitution reaction

$$[Ru(NH_3)_5(H_2O)]^{3+} + Cl^- \rightarrow [Ru(NH_3)_5Cl]^{2+} + H_2O \qquad (2)$$

was observed, which implies that the Ru(II)–H_2O bonds are quite labile and implies also that electron transfer between Ru(II) and Ru(III) ammines is facile. Third, and to our astonishment, $[Ru(NH_3)_5(H_2O)]^{2+}$ was found to be incompatible with ClO_4^- (in 1.0 M ClO_4^-, the half-life for Ru(II) at 25 °C is <1 min). At that time the inertness of ClO_4^- in the presence of the more powerful reductant Cr^{2+}(aq) was well-known, so that ClO_4^- was widely and successfully used as the anion of choice when a "noncoordinating counter anion" was needed in aqueous solution. The surprising loss of the reducing titer of $[Ru(NH_3)_5(H_2O)]^{2+}$ (when blanketed under N_2 to exclude O_2) in the presence of ClO_4^- led us to consider N_2 (together with H^+ and ClO_4^-) as the possible "oxidants" of Ru(II).

The rates of reduction of a series of $[Ru^{III}(NH_3)_5L]$ complexes by a variety of reducing agents were studied by Endicott (22–24) and by Stritar (25), and comparisons with data obtained by others for analogous $[Co^{III}(NH_3)_5L]$ complexes were carried out. During the course of this work, the first attempt to prepare a pyrazine (or related heterocyclic) complex of $[Ru(NH_3)_5]^{2+}$ was made. This effort was short-lived because the deep color of the product, a brownish violet, was unexpected and not understood because all the other Ru(II) complexes we had encountered until then were colorless. We concluded that the chemistry had gone awry and turned to other matters. Later Ru(II) and Ru(III) complexes with *N*-heterocyclics were investigated by Gaunder (21), who demonstrated convincingly that the color observed earlier for the purported $[Ru(NH_3)_5(pz)]^{2+}$ (pz = pyrazine) was intrinsic. By then the great capacity of $Ru(NH_3)_5^{2+}$ for backbonding had begun to be appreciated. The observation that N_2 (which had been used to exclude air) reacts with $[Ru(NH_3)_5(H_2O)]^{2+}$ to produce $[Ru(NH_3)_5(N_2)]^{2+}$ and $[(NH_3)_5Ru(N_2)Ru(NH_3)_5]^{4+}$ further demonstrated the great capacity of $[Ru(NH_3)_5(H_2O)]^{2+}$ for π backbonding (26, 27).

As already mentioned, the possibility the N_2 might react with $[Ru(NH_3)_5(H_2O)]^{2+}$ had been entertained, but was not taken seriously either by Endicott or me. The idea that a ruthenium complex of N_2 might be prepared from a high-energy initial state, and that the resulting N_2 complex might prove to be kinetically stable, was featured in my grant application to the Atomic Energy Commission, ca. 1964. The pursuit of the N_2 complex was motivated by the research reported in reference 2, as is indicated by this quotation: "In the reaction of $[Co(NH_3)_5(N_3)]^{2+}$ with NO^+, there is the possibility that the N_2 formed when the coordinated N_4O decomposes remains on the metal ion for a finite length of time." The lifetime of $[Co(NH_3)_5(N_2)]^{3+}$, assuming it is formed, proved to be <5 s. Likewise, the presumed intermediates $[Pt(dien)N_2]^{2+}$ (dien = diethylenetriamine), $[Ru(phen)_2(py)(N_2)]^{2+}$ (phen = phenanthroline, py = pyridine), and $[Co(CN)_5(N_2)]^{2-}$ in the reactions of the corresponding azido complexes with NO^+ are also short-lived (2).

Perhaps because of this history, I was at first skeptical of the claim made by Allen and Senoff (28) that they had prepared a kinetically very stable complex, $[Ru(NH_3)_5(N_2)]^{2+}$, by the reaction of $RuCl_3$ with $[N_2H_5]^+$. (Professor J. P. Collman was a visitor to the Department when the Communication by Allen and Senoff appeared. He happened to overhear a discussion within my research group during which several alternative interpretations of the observations reported by Allen and Senoff were advanced. Although Collman did not comment in the discussion, he looked unusually thoughtful throughout. At the conclusion he made a wager that Allen and Senoff had in fact discovered a dinitrogen complex, a wager I confidently accepted. Only two days later I was obliged to settle. His co-worker, Kang, acting on Collman's telephone suggestion, had convincingly demonstrated that a compound of otherwise puzzling composition was in fact *trans*-$[Ir(P(Ph)_3)_2Cl(N_2)]$, the second dinitrogen complex to be prepared.) However, by the time Harrison joined my research group in 1967 it was for the purpose of developing the Ru(II)–N_2 chemistry further. But the fact that $[Ru(NH_3)_5(N_2)]^{2+}$ would be produced also by the reaction of $[Ru(NH_3)_5(H_2O)]^{2+}$ and N_2 in water nevertheless came as a surprise to me. This discovery was made in an experiment in which Harrison deliberately monitored the spectrum of a solution of $[Ru(NH_3)_5(H_2O)]^{2+}$ (which was prepared by reducing the corresponding Ru(III) complex by Zn/Hg under argon) while it was bubbled with N_2 (26, 27). Without invoking the backbonding capacity of $[Ru(NH_3)_5(H_2O)]^{2+}$, it would be difficult to understand the replacement of a good dipole, H_2O, from the coordination sphere of a cation by the nonpolar N_2, and this discovery was powerfully educational.

Mixed-Valence Complexes of Ruthenium Ammines

The intense color of $[Ru(NH_3)_5(pz)]^{2+}$ and related molecules provided an opportunity that Ford was quick to seize soon after joining my group. A better understanding of the backbonding capacity of $[Ru(NH_3)_5(H_2O)]^{2+}$ and some of its consequences in electronic absorption spectra and in chemical properties resulted from his efforts and those of his co-workers (29) and set the stage for further developments in my laboratory, and in his laboratory, later at the University of California at Santa Barbara.

In the course of studying the backbonding capacity of $[Ru(NH_3)_5(H_2O)]^{2+}$ and its effect on the electronic absorption spectra, what has since been named the Creutz–Taube ion was first synthesized (29). The experiment consisted of adding to a solution containing $[Ru(NH_3)_5(H_2O)]^{2+}$ half the molar amount of pyrazine, and after the violet color had developed, adding the amount of oxidant required to produce the (3+/2+) state. By visual observation Ford and I witnessed a partial fading of color, ascribable to partial oxidation of the Ru(II), and because there was no change in the hue, we were mildly disappointed in concluding that nothing of consequence had transpired. Although the experiment was done deliberately, we had not thought much about the physics behind the color of mixed-valence species and returned to other interests that at that time seemed more compelling.

Fortunately, the subject of binuclear ruthenium pyrazine complexes was taken up again by Creutz, on her own initiative, when she joined my research group somewhat later (30, 31). By this time two important review articles (32, 33) on mixed-valence compounds had appeared, and she found the lessons to be learned there convincing enough so that she restored the near-infrared (NIR) capability of our Cary spectrophotometer, which was at first nonfunctional in the critical region of the spectrum. It was my good fortune to share her feeling of triumph, just a few minutes after she obtained the first spectrophotometric trace of an intervalence band for a complex deliberately synthesized to display this low-energy feature for the binuclear complex shown in structure 1.

Apart from loss in intensity for the 5+ ion, the MLCT (metal-to-ligand charge transfer) absorption differs little from that of the 4+ ion; intervalence absorption appears at 1570 nm (molar absorptivity $\varepsilon = 5.0 \times 10^2$ M^{-1} cm^{-1}).

Other contributions followed from my laboratory, and a productive program involving the synthesis of new mixed-valence molecules, and the study of their properties, was soon initiated by Meyer in his own laboratories (34).

Trifluoromethanesulfonate as a Weakly Coordinating Anion

An advance that was very important to the development of ruthenium ammine chemistry and that proved to be essential to most of the advances made over a

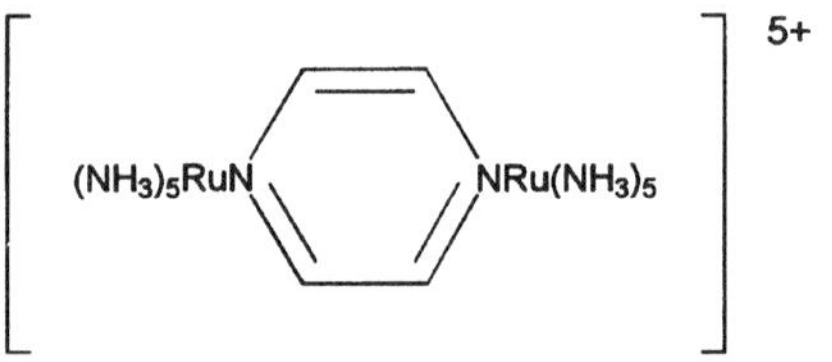

$$\left[(NH_3)_5RuN\bigcirc NRu(NH_3)_5\right]^{5+}$$

1

decade later in the chemistry of osmiumammines was the introduction oɪ new, weakly nucleophilic counterion, $CF_3SO_3^-$. The ClO_4^- anion had been shown to be incompatible with some of the Ru(II) species. In the middle 1960s in a series of lectures at Reed College on basic problems in solution chemistry, I raised the issue of a replacement for ClO_4^-. A representative of Minnesota Mining and Manufacturing Company (3M) responded by describing the virtues of $CF_3SO_3^-$, which, although it was being produced by the company on a small scale, was then not on the market, and was not widely known. Although at the time I was not hopeful that it would prove to be widely useful, I requested research samples from 3M after I returned to Stanford, and we soon learned that it fully met our needs. A single systematic study was done to gauge the nucleophilic power of $CF_3SO_3^-$, and it showed that the rate of aquation of $[Cr(H_2O)_5(O_3SCF_3)]^{2+}$ is not much slower than it is for the perchlorate complex and considerably higher than that of the nitrato complex (35). In no case has reduction of $CF_3SO_3^-$ by Ru(II) [nor, in work done much later, by osmium(II)] been observed. The full potential of the $CF_3SO_3^-$ ion, however, was not realized until a systematic study of "triflate" salts of metal amine complexes was undertaken in Sargeson's laboratory, in which the solubility of salts of these complexes in nonaqueous solvents was demonstrated (36).

Further Reactivity Studies of Ru(II) Ammine Reagents

By now, exploration of the role of backbonding in the chemistry of ruthenium(II) ammines was actively being pursued in my laboratory: as examples, topics of published studies included a bimolecular mechanism for substitution in the reaction of NO with $[Ru(NH_3)_6]^{3+}$ to produce $[Ru(NH_3)_5NO]^{3+}$ (37), evidence of a binuclear nitrous oxide complex of ruthenium (38), disproportionation of ammine (pyridine) ruthenium complexes in alkaline solution (39), influence of backbonding on the hydrate–carbonyl equilibrium for 4-formylpyridine as ligand (40), and nucleophilic attack on cyanoformate induced by coordination to ruthenium ammines (41).

It was probably for the purpose of engaging in research of this kind that Isied joined my research group, and his skill in preparative chemistry was applied to an experiment that provided an unambiguous measurement of the rate of intramolecular electron transfer through an organic ligand in a system of defined geometry (42). An earlier unsuccessful attempt had been made by Professor Kirk Roberts, a faculty visitor to my research group, in which $[Ru(NH_3)_5(H_2O)]^{2+}$ was mixed with *O*-isonicotinatopenta-amminecobalt(III) in the hope that substitution would precede the redox reaction. This kind of approach had been successful in the hands of Gaswick and Haim (43) by the use of $[Fe(CN)_5(H_2O)]^{3-}$ as reductant and Co(III) complexes as oxidants; independently of our work, Gaswick and Haim set out to measure electron transfer rates in the intramolecular mode. Isied, in addition to other research, developed an interest in this important subject. By an ingenious set of reactions, he

$$[(SO_4)(NH_3)_4Ru^{III}N{-}C_6H_4{-}C(=O){-}OCo^{III}(NH_3)_5]^{3+}$$

2

$$[(SO_4)(NH_3)_4Ru^{III}N{-}C_6H_4{-}C_6H_4{-}NCo^{III}(NH_3)_5]^{4+}$$

3

succeeded in preparing a binuclear complex (structure **2**) in which $[Ru(NH_3)_4(SO_4)]^+$ is attached to the nitrogen of a cobalt complex.

On the addition of a suitable reducing agent, Ru(III) is reduced, and the rate of internal electron transfer is monitored by following the decrease in the Ru(II) → *N*-heterocyclic charge transfer absorption. In an important extension of this approach (which also involved refinement of the preparative method), Fischer and Tom (*44*) used derivatives of 4,4′-bipyridine as bridging groups in which the linkage between the rings was varied as to distance (metal-to-metal separation in some cases 13 Å) and as affecting conjugation (*see* structure **3**).

Further studies of mixed-valence molecules and of intramolecular electron transfer ensued, but exploration of the backbonding capacity of Ru(II) ammines was the dominant theme during a period of about 15 years beginning in 1971. In fact this backbonding capacity made possible the advances in mixed-valence molecules and in measuring rates of intramolecular electron transfer that have been outlined. To broaden the scope, the interaction of $[Ru(NH_3)_5(H_2O)]^{2+}$ with a variety of ligands was explored. Notably featured were sulfur donors {e.g., H_2S (to form $[(NH_3)_5RuSH_2]^{2+}$, $K = 1.5 \times 10^3$) and derivatives (*45*) ; trimethylsulfonium ion (*46*), and cyclic disulfur ligands (*47*)}, trimethylphosphite (*48*), and nucleobases (*49, 50*). In none of these studies was synthesis and characterization the sole interest, but chemical behavior, especially the measurement of the affinities of the ligands for the ruthenium center, was usually one of the goals.

Electrochemical measurements provide an important means of characterization of the ruthenium complexes; since the Ru(III)/Ru(II) couple is usually reversible, changes in standard electrode potential $(E)°$ [actually in half-wave potential $(E_{1/2})$ as determined by cyclic voltammetry] provide a measure of the changes in the relative affinities to Ru(III) and Ru(II) as a reference ligand such as H_2O is replaced by a variable ligand, L. Issues pertaining to backbonding that were touched on, in addition to those already mentioned, include the shrinking of the Ru–donor atom distance when the Ru(III)–ligand complex is reduced to the Ru(II) state, observed in some instances when the ligand is a π-acid (*51, 52*); protection of pyrazine to reduction by Cr^{2+}(aq) when it is coordinated to $Ru(NH_3)_5^{2+}$ (*53*); labilization of π-acid on Ru(II) by a π-acid co-ligand, but kinetic stabilization of a σ donor co-ligand (*54*); effect on the acid–base properties of $SO_3^{2-}/HSO_3^-/SO_2$ of binding to Ru(II) (*55*); linkage isomerization coupled to a change in oxidation state when this process is attended by a marked change in backbonding (*56*); and the effects of backbonding on the infrared intensities of the N≡C stretching vibration of nitriles (*57*). Few of the effects exposed in these studies were investigated extensively, although they did serve as examples, often of novel effects.

The most systematic studies were those devoted to determining equilibrium constants for complex formation by ruthenium ammines. Such studies were made in much of the work already cited, and in some additional cases the equilibrium stability was the dominant theme (*58–60*).

The determination of equilibrium constants by measuring concentrations at equilibrium is convenient only when the equilibrium constant is not far from 1 and has proved of limited use in our systems, but a kinetic method has proven to be more widely applicable (*40, 48, 58, 60–62*).

When the equilibrium constants for complex formation are high and the rates of reactions are reasonably rapid, such as in the reactions of $[Ru(NH_3)_5H_2O]^{2+}$ with a large number of interesting ligands, enthalpies can readily be obtained. These enthalpy measurements were made by Wishart (with the cooperation of Breslauer and Isied) at Rutgers University, since after undertaking the project we discovered that the only calorimeter at Stanford was not suited to our purposes. Although data about enthalpy changes ($\Delta H°$) are in themselves of course of interest, for neutral entering ligands the enthalpy data can also be converted to reasonably good values of standard Gibbs free energy of complex formation reaction ($\Delta G°$). The entropy changes ($\Delta S°$) in the reactions of a number of neutral ligands with $[Ru(NH_3)_5(H_2O)]^{2+}$ in water lie in the range –2 to –9 cal deg^{-1} mol^{-1}, and in fact those for the reaction of $[Ni(H_2O)_6]^{2+}$ with neutral ligands also lie in this range. By choosing $-\Delta S° = 5$ cal deg^{-1} mol^{-1} for complex formation, $\Delta H°$ values can be converted to equilibrium constants that are good to somewhat better than an order of magnitude. Because some of the equilibrium constants are as high as 1×10^{30}, such estimates represent a substantial advance in useful knowledge.

In the elementary system of electron bookkeeping adopted by coordination chemists, the ligand regarded as a Lewis base provides an electron pair shared with the metal center, regarded as a Lewis acid. To account for multiple bond formation in complexes such as those being dealt with here, it is necessary to invoke electron donation from metal orbitals to unoccupied orbitals of suitable symmetry on the ligand. Just how important this "correction" term is can be seen from a consideration of the energetics of complex formation obtained. Consider the values of $\Delta H°$ for the replacement of H_2O from $[Ru(NH_3)_5(H_2O)]^{2-}$ by CO and NO^+, an isoelectronic pair, as -38.2 ± 1.4 kcal (60) and -52 ± 2 kcal (62), respectively. Simple electron donation from ligand to metal falls far short of accounting either for the relative values of the enthalpies or for the sign of ΔH.

It is instructive to compare the equilibrium constants, K^{II}, for $[Ru(NH_3)_5(H_2O)]^{2+}$, in which backbonding to a ligand can play an important role, with those for $[Ru(NH_3)_5(H_2O)]^{3+}$, K^{III}, in which this role appears to be negligible. Once K^{II} is measured or estimated from enthalpy data, that for K^{III} can be calculated by making use of the values of $E_{1/2}$ for the $[Ru(NH_3)_5(H_2O)]^{3+/2+}$ couple and for the couple when H_2O is replaced by the ligand of choice. Only one comparison will be made to illustrate the point. The ligand OH^- cannot benefit from the backbonding capacity of Ru(II), and we find the values of K^{II} and K^{III} to be 6×10^2 (45, 63) and 6×10^{11} (64), respectively, with K^{III} higher than K^{II}, as expected. However, when N_2, an indifferent electron pair donor but a π acceptor, is the ligand, the values are 3×10^4 (61) (K^{II}) and $\sim 4 \times 10^{-13}$ (K^{III}), respectively. The ratios of K^{III}/K^{II} change from 10^9 for OH^- to $\sim 10^{-17}$ for N_2, a range of 10^{26}. Were NO^+ rather than N_2 used for the comparison, the role of backbonding would be displayed even more dramatically.

Backbonding and Linkage Isomerization

A number of interesting effects have been observed for ligands that have more than one binding site and that differ in their capacity for backbonding. Imidazole in its reaction with $[Ru(NH_3)_5(H_2O)]^{2+}$ is bound to N1 of the ring. On proton-assisted aquation, the major product is $[Ru(NH_3)_5(H_2O)]^{2+}$ (accompanied by a corresponding amount of imidazolium ion), but in about 10% of the product (*see* structure 4), imidazole remains bound to Ru(II), now at C2 (65).

The oxidation state of C2 is the same as it is in CN^-, and the formation of this minor product is attributable at least in part to the superior π acid character of the C2 site compared to N1. In the reactions of $[Ru(NH_3)_5(H_2O)]^{2+}$ with xanthines, both N- and C-bound complexes are formed (66a), and in later,

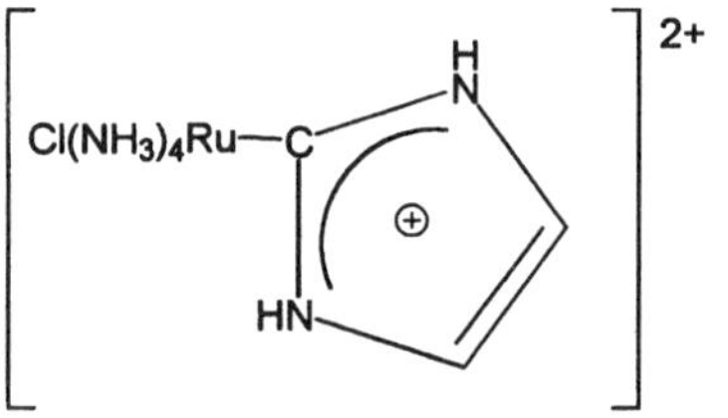

4

independent work by Clarke (*66b*), linkage isomerizations have been described. Effects noted (*67*) in the interaction of $[Ru(NH_3)_5(H_2O)]^{2+}$ with HCN are closely related to those referred to. The solid freshly prepared from strongly acidic solution shows Ru(II) as bound to the N of HCN. On aging of the solid, features appear in the infrared spectrum attributable to C-bound CN^-, as they do also in solid freshly prepared from less acidic solution. In aqueous solution $[(NH_3)_5RuCN]^+$ undergoes rapid loss of NH_3 trans to CN^-; the product of this disruption polymerizes rapidly.

The first example of linkage isomerization accompanying the oxidation of $[Ru^{II}(NH_3)_5L]$, where L is an ambidentate (or multidentate) ligand, was described by Yeh et al. (*68*). When $(CH_3)_2SO$ replaces H_2O in $[Ru(NH_3)_5$-$(H_2O)]^{2+}$, the ligand is S-bound, but on one-electron oxidation ($E_{1/2}$=1.0 V) the immediate product (S-bound) rearranges to the O-bound form ($k = 7 \times 10^{-2}$ s^{-1} at 25 °C). On reduction of the resulting Ru(III) complex ($E_{1/2}$= 0.01 V), the O-bound species rearranges ($k = 30$ s^{-1}) to the original state. Abundant examples of this kind have been encountered in developing the organometallic chemistry of Os(II)/Os(III) ammines, which include the first complex prepared in this program $[Os(NH_3)_5(\eta^2\text{-}(CH_3)_2C=O]^{2+}$ (*69a, 69b*) and the pentaammine complex of aniline (*69c*), in which η^2 binding to the phenyl ring is favored in the 2+ state for osmium, whereas on oxidation to Os(III) rearrangement to the N-bound form occurs.

Use has been made of the sulfoxide function in devising a system that displays important properties of hysteresis in a simple molecule. This was done (*70, 71*) by including in the same molecule a reversible couple with an intermediate $E_{1/2}$ value (between 0.01 and 1 V, the potentials for the two sulfoxide couples referred to previously). This condition is fulfilled by attaching two $Ru(NH_3)_5^{2+}$ to dithiacyclooctane, one sulfur of which has been converted to sulfoxide. Although applications of systems of this kind in "molecular electronics", if indeed possible, lie far in the future, the interconversion of the isomeric states that arise on oxidation and reduction, in this case $Ru^{II}SO/Ru^{III}S$ and $Ru^{III}SO/Ru^{II}S$, respectively, provide lessons in electron transfer coupled to changes in configuration.

Further Amplification of Backbonding in Osmium Ammine Complexes

Important though the consequences of backbonding in the chemistry of ruthenium(II) ammines are, they are very much greater in that of osmium(II) ammines, and affect even the chemistry of osmium(III) ammines. The first use of osmium in my laboratory was a study on the chemistry of $[Os(NH_3)_5(N_2)]^{2+}$, first prepared by Allen and Stevens (*72*).

This study led to the characterization of the first bis-dinitrogen complex, cis-$[Os(NH_3)_4(N_2)_2]^{2+}$, which was prepared by the reaction of $[Os(NH_3)_5$-$(N_2)]^{2+}$ with nitrous acid (*73*)

$$[\text{Os(NH}_3)_5(\text{N}_2)]^{2+} + \text{HNO}_2 \rightarrow \text{cis-}[\text{Os(NH}_3)_4(\text{N}_2)_2]^{2+} + 2\,\text{H}_2\text{O} \qquad (3)$$

The bis-dinitrogen complex proved to be a valuable precursor for the preparation of tetraammines because one of the molecules of N_2 is readily lost and can be replaced by another ligand, and on heating in solution with $[\text{Os(NH}_3)_5\text{N}_2]^{2+}$ it led to the preparation of the very stable mixed-valence molecules $[(\text{H}_2\text{O})(\text{NH}_3)_4\text{Os(N}_2)\text{Os(NH}_3)_5]^{5+}$ and $[\text{Cl(NH}_3)_4\text{OsN}_2\text{Os(NH}_3)_5]^{4+}$ (74). Loss of dinitrogen from $[\text{Os(NH}_3)_5\text{N}_2]^{2+}$ occurs on oxidation by Ce(IV), and opened a convenient route to $[\text{Os(NH}_3)_5(\text{H}_2\text{O})](\text{ClO}_4)_3$ (75), which in principle provides access to pentaammineosmium chemistry. But a violent explosion of fortunately a very small amount of this compound ended the use of ClO_4^- as a counterion, to be replaced by CF_3SO_3^-. The compound $[\text{Os(NH}_3)_5(\text{O}_3\text{SCF}_3)]$-$(\text{O}_3\text{SCF}_3)_2$ was also prepared from $[\text{Os(NH}_3)_5(\text{N}_2)]^{2+}$ as precursor, but with Br_2 as oxidant in neat $\text{CF}_3\text{SO}_3\text{H}$ medium (76). An exploration of some aspects of Os(IV) complexes with bis- and tris-halo ligands was carried out (75, 77); Os(IV) is an oxidation state that is readily accessible when the complex is stabilized by good σ- and π-donor ligands.

The extraordinary stability of $[\text{Os(NH}_3)_5(\text{N}_2)]^{2+}$ to loss of N_2 compared to that of $[\text{Ru(NH}_3)_5(\text{N}_2)]^{2+}$, and the stability of Os(II)/Os(III) mixed-valence species with N_2, as well as pyrazine (78), as bridging group, compared to the ruthenium analogs made it abundantly clear that the capacity of $\text{Os(NH}_3)_5^{2+}$ for backbonding greatly exceeds that of $\text{Ru(NH}_3)_5^{2+}$. Specific quantitative supporting evidence was the finding that pK_a for $[\text{Os(NH}_3)_4\text{Cl(pzH)}]^{2+}$ is 7.6 (79) compared to 2.6 for $[\text{Ru(NH}_3)_5(\text{pzH})]^{3+}$ and 1.9 for the free ligand.

The pentaammine derivative prepared later by Sen, $[\text{Os(NH}_3)_5(\text{pzH})]^{3+}$, which is the exact osmium analog of $[\text{Ru(NH}_3)_5(\text{pzH})]^{3+}$, has a pK_a of 7.4 (80), thus discounting any possibility that Cl^- in place of an ammonia is responsible for the enormous enhancement of the basicity of a nitrogen of pyrazine when the osmium(II) moiety is attached to the other.

Manifestation of the Properties of the Os(II) Center in Metal Ion–Organic Ligands

When Harman joined my research group in 1983, I had already assumed emeritus status. Although I was no longer actively "recruiting" graduate students, his interest in the work of my laboratory and his enthusiasm overcame any reluctance I might have had in making a commitment to a graduate student. The implications of the studies on the chemistry of osmium referred to convinced us that exploring the organometallic chemistry of osmium ammines would prove to be rewarding. He took the initiative in preparing a National Institutes of Health grant proposal for the purchase of a controlled atmosphere box; and in due course a new subarea of chemistry, which bridges traditional coordination chemistry and organometallic chemistry—and which greatly extended the scope of organometallic chemistry in aqueous solution—

emerged. The nature of this chemistry is to some extent revealed in the chapter by Harman that appears in this volume, and I will restrict my account to three early key discoveries.

The compound $Os(NH_3)_5(O_3SCF_3)_3$ has played a crucial role in the work, and a large number of novel species have been prepared by reduction of this compound, using Mg (in the early work) or Zn/Hg, in an appropriate liquid environment. The first complex discovered was the acetone complex of $Os(NH_3)_5^{2+}$ (structure **5**), in which acetone is bound η^2 to the metal center (*69a*). It is very stable, being resistant to the replacement of acetone by H_2O and a number of other good nucleophiles. The description of $[Os(NH_3)_5(\eta^2\text{-}C_6H_6)]^{2+}$ (structure **6**) soon followed (*81*).

The $[Os(NH_3)_5(\eta^2\text{-}C_6H_6)]^{2+}$ complex too is remarkably stable compared to other molecules of its class, but in solution it does disproportionate slowly to yield $[(Os(NH_3)_5)_2(\eta^2\text{-}C_6H_6)]^{4+}$.

$$2\ [\text{complex}]^{2+} \longrightarrow [\text{complex}]^{4+} + C_6H_6 \tag{4}$$

The further systematic development of the organometal chemistry, particularly in exploring the effect of the metal center on the reactivity of organic ligands, has flourished under Harman's direction at the University of Virginia. In my own laboratory Li has pursued the research along more inorganic lines, that is, by turning to tetraammines or bis-ethylenediamine complexes, which, by freeing two sites for occupation by variable ligands, introduces many new opportunities (*82*). The investigation of the chemical and physical properties of these osmium tetraammine and bis-ethylenediamine dihydrogen complexes has been a particularly productive area. The subject of dihydrogen complexes of osmium was in fact opened up by Harman in preparing a dihydrogen complex of very simple composition $[Os(NH_3)_5(\eta^2\text{-}H_2)](O_3SCF_3)_2$, by the novel route of reducing $Os(NH_3)_5(O_3SCF_3)_3$ in methanol as solvent (*83*).

In extending this chemistry to $Os(NH_3)_4^{2+}$ and $Osen_2^{2+}$, a new dimension was introduced into the chemistry of dihydrogen complexes, enabling the systematic study of the effect of a variable co-ligand on the properties of H_2 as a

$$7$$

ligand, and the converse. Moreover, this work introduces an interesting historical note. Already in 1971 Malin and I (*84*) reported that the reduction of *trans*-[Osen$_2$(O)$_2$]$^{2+}$ by Zn/Hg in water leads to the production of a species that we formulated as *cis*-[Osen$_2$(H)$_2$]$^{2+}$, based on the ^{1}H NMR evidence, which revealed that the protons of the organic ligand yielded two signals of equal intensity. Although the species was new, it did not seem particularly interesting to us or to others at that time. Somewhat later this species was used in the preparation of *trans*-[Os^{III}en$_2$Cl$_2$]$^+$ by Coelho and Malin (*85*). In retrospect, that a cis species arises from a trans, and by further chemical action under mild conditions the cis form is converted to a trans, should have piqued our interest, but it must be borne in mind that little was known about the stereorigidity of Os(IV) at that time. In my own laboratories, less than a decade ago, a year of effort was devoted to the photolysis of the supposed [Osen$_2$(H)$_2$]$^{2+}$ in CH$_3$OH in the expectation that [Osen$_2$]$^{2+}$ would be generated, and interesting chemistry would ensue. Although there is photoinduced reaction with the solvent no products were identified. Early in his work on the dihydrogen complexes of Os(II) amines, Li (*86a*) showed that the product of the reduction of [Osen$_2$(O)$_2$]$^{2+}$ in water is in fact *trans*-[Osen$_2$(η^2-H$_2$)(H$_2$O)]$^{2+}$ (structure **7**).

This species (**7**) has proven to be very interesting. It serves as ^{1}H NMR probe for biochemical molecules (*86b*), and as an intermediate in the preparation of new organo-osmium species, including metallocycles (*87, 88*); by oxidation it leads to a class of monohydride complexes of Os(IV) of coordination number VII (*89*); and it provides an opportunity for the systematic study of H$_2$ as a co-ligand affecting the dynamics of substitution and of cis–trans isomerizations (*90*).

Epilogue

In this historical account, a number of episodes that led to advances have been described. In some cases preplanning was rewarded with success. In others, because the mind was inadequately prepared the significance of a chance observation was initially overlooked, but after a time, recovery was made. In the last example mentioned, recovery was made only after two decades had elapsed; that it was made at all has an element of chance, and the happy outcome was a result of the mind being prepared by contributions made by others. All have important elements in common: the successes that my laboratory has witnessed have depended largely on the enthusiasm, persistence, enterprise, and skill of my co-workers (this comment applies also to the many who have contributed to advances not included in this account); all the work has provided me with a sense of excitement throughout my career, and all of it has been educational.

Acknowledgments

This chapter is based on a lecture given during the symposium, the general theme having been suggested by the planning committee. It is an account of how some of the discoveries relating to the announced topic came about. My memory is far from perfect and I apologize for lapses that those directly involved in the work may find.

References

1. Jordan, R. B.; Taube, H. *J. Am. Chem. Soc.* **1964,** *86,* 3890.
2. Jordan, R. B.; Sargeson; A. M.; Taube, H. *Inorg. Chem.* **1966,** 5, 1091.
3. Taube, H.; Myers, H.; Rich, R. *J. Am. Chem. Soc.* **1953,** *74,* 4118.
4. Taube, H.; Myers, H. *J. Am. Chem. Soc.* **1954,** *76,* 2103.
5. Taube, H. *J. Am. Chem. Soc.* **1955,** 77, 4481.
6. Sebera, D. K.; Taube, H. *J. Am. Chem. Soc.* **1961,** *83,* 1785.
7. Morgan, G. T.; Burstall, F. H. *J. Chem. Soc.* **1938,** *1675.*
8. Gleu, K.; Breuel, W. Z. *Anorg. Allg. Chem.* **1988,** *237,* 335.
9. Lever, F. M.; Powell, A. R. *Chem. Soc. Spec. Publ. No. 13* **1959,** 135.
10. Dwyer, F. P.; Sargeson, A. M. *J. Phys. Chem.* **1956,** *60,* 1331.
11. Fraser, R.T.; Taube, H. *J. Am. Chem. Soc.* **1961,** *83,* 2242.
12. Gould, E. S.; Taube, H. *J. Am. Chem. Soc.* **1964,** *86,* 1318.
13. Huchital, D. H.; Taube, H. *J. Am. Chem. Soc.* **1965,** *87,* 5371.
14. Hurst, J. K.; Taube, H. *J. Am. Chem. Soc.* **1968,** *90,* 1178.
15. Taube, H.; Gould, E. S. *Acc. Chem. Res.* **1969,** *2,* 321.
16. Olson, M. V.; Taube, H. *Inorg. Chem.* **1970,** 9, 2072.
17. Gould, E. S.; *J. Am. Chem. Soc.* **1972,** *94,* 4360.
18. Sprecker, H.; Wieghardt, K. *Inorg. Chem.* **1977,** *16,* 1290.
19. Norris, C. R.; Nordmeyer, F. R. *J. Am. Chem. Soc.* **1971,** *93,* 4044.
20. Meyer, T. J.; Taube, H. In *Comprehensive Coordination Chemistry;* Wilkinson, G., Ed.; Pergamon: New York, 1987, p 373.
21. Gaunder, R. G.; Taube, H. *Inorg. Chem.* **1970,** 9, 2627.
22. Endicott, J. F.; Taube, H. *J. Am. Chem. Soc.* **1962,** *84,* 4984.
23. Endicott, J. F.; Taube, H. *J. Am. Chem. Soc.* **1964,** *86,* 1686.
24. Endicott, J. F.; Taube, H. *Inorg. Chem.* **1965,** *4,* 437.
25. Stritar, J.; Taube, H. *Inorg. Chem.* **1969,** *8,* 2281.
26. Harrison, D. E.; Taube, H. *J. Am. Chem. Soc.* **1967,** *89,* 5706.
27. Harrison, D. E.; Weissberger, E.; Taube, H. *Science,* **1968,** *159,* 320.
28. Allen, A. D.; Senoff, C. V. *Chem. Commun.* **1965,** *621.*
29. Ford, P.; Gaunder, R. G.; De Rudd, F. P.; Taube, H. *J. Am. Chem. Soc.* **1968,** *90,* 1187.

30. Creutz, C.; Taube, H. *J. Am. Chem. Soc.* **1969**, *91*, 3988.
31. Creutz, C.; Taube, H. *J. Am. Chem. Soc.* **1973**, *95*, 1086.
32. Hush, N. S. *Prog. Inorg. Chem.* **1967**, *8*, 357.
33. Robin, M. B.; Day, P. *Adv. Inorg. Chem. Radiochem.* **1967**, *10*, 247.
34. Adeyimi, S. A.; Braddock, J. N.; Brown, G. M.; Ferguson, A.; Miller, F. J.; Meyer, T. J. *J. Am. Chem. Soc.* **1972**, *94*, 300.
35. Scott, A; Taube, H. *Inorg. Chem.* **1971**, *10*, 62.
36. Dixon, N. E.; Laurance, G. A.; Lay, P. A.; Sargeson, A. M. *Inorg. Chem.* **1983**, *22*, 846.
37. Scheidegger, H.; Armor, J.; Taube, H. *J. Am. Chem. Soc.* **1968**, *90*, 5928.
38. Armor, J. M.; Taube, H. *Chem. Commun.* **1971**, *7*, 287.
39. Rudd, D. P.; Taube, H. *Inorg. Chem.* **1971**, *10*, 1543.
40. Zanella, A.; Taube, H. *J. Am. Chem. Soc.* **1971**, *93*, 7166.
41. Diamond, S. B.; Taube, H. *J. Chem. Soc. Chem. Commun.* **1974**, 622.
42. Isied, S.; Taube, H. *J. Am. Chem. Soc.* **1973**, *95*, 8198.
43. Gaswick, D.; Haim, A. *J. Am. Chem. Soc.* **1974**, *96*, 7845.
44. Fischer, H.; Tom, G. M. *J. Am. Chem. Soc.* **1976**, *98*, 5512.
45. Kuehn, C. G.; Taube, H. *J. Am. Chem. Soc.* **1976**, *98*, 689.
46. Stein, C. A.; Taube, H. *J. Am. Chem. Soc.* **1978**, *100*, 1635.
47. Stein, C. A.; Taube, H. *J. Am. Chem. Soc.* **1978**, *100*, 336.
48. Franco, D. W.; Taube, H. *Inorg. Chem.* **1978**, *17*, 571.
49. Clarke, M. J.; Taube, H. *J. Am. Chem. Soc.* **1974**, *96*, 5413.
50. Clarke, M. J.; Taube, H. *J. Am. Chem. Soc.* **1975**, *97*, 1397.
51. Gress, M. E.; Creutz, C.; Quicksall, C. O. *Inorg. Chem.* **1981**, *20*, 1522.
52. Wishart, J. F.; Bino, A.; Taube, H. *Inorg. Chem.* **1986**, *25*, 3318.
53. Blesa, M. A.; Taube, H. *Inorg. Chem.* **1976**, *15*, 1454.
54. Isied, S. S.; Taube, H. *Inorg. Chem.* **1976**, *15*, 3070.
55. Isied, S. S.; Taube, H. *Inorg. Chem.* **1974**, *13*, 1545.
56. Yeh, A.; Scott, N.; Taube, H. *Inorg. Chem.* **1982**, *21*, 2542.
57. Johnson, A.; Taube, H. *J. Indian Chem. Soc.* **1989**, *66*, 503.
58. Shepherd, R. E.; Taube, H. *Inorg. Chem.* **1973**, *12*, 1392.
59. Brown, G.; Sutton, J.; Taube, H. *J. Am. Chem. Soc.* **1978**, *100*, 2767.
60. Wishart, J. F.; Breslauer, K. J.; Isied, S. S.; Taube, H. *Inorg. Chem.* **1984**, *23*, 2997.
61. Armor, J.; Taube, H. *J. Am. Chem. Soc.* **1970**, *92*, 6170.
62. Wishart, J. F.; Breslauer, K. G.; Isied, S. S.; Taube, H. *Inorg. Chem.* **1986**, *25*, 1479.
63. Lim, H. S.; Barclay, D. J.; Anson, F. C. *Inorg. Chem.* **1972**, *11*, 1460.
64. Broomhead, J. A.; Basolo, F.; Pearson, R. G. *Inorg. Chem.* **1964**, *3*, 826.
65. Sundberg, R. J.; Shepherd, R.E.; Taube, H. *J. Am. Chem. Soc.* **1972**, *94*, 6558.
66. (a) Clarke, M. J.; Taube, H. *J. Am. Chem. Soc.*, **1975**, *97*, 1397; (b) Clarke, M. J. *Inorg. Chem.* **1977**, *16*, 738.
67. Isied, S.S.; Taube, H. *Inorg. Chem.* **1975**, *14*, 2561.
68. Yeh, A.; Scott, N.; Taube, H. *Inorg. Chem.* **1982**, *21*, 2542–2545.
69. (a) Harman, W. D.; Fairlie, D. P.; Taube, H. *J. Am. Chem. Soc.* **1986**, *108*, 8233–8237; (b) Harman, W. D.; Sekine, M.; Taube, H. *J. Am. Chem. Soc.* **1988**, *110*, 2439; (c) Harman, W. D.; Taube, H. *J. Am. Chem. Soc.* **1988**, *110*, 5403.
70. Sano, M.; Taube, H. *J. Am. Chem. Soc.* **1991**, *113*, 2327–2328.
71. Sano, M.; Taube, H. *Inorg. Chem.* **1994**, *33*, 705–709.
72. Allen, A. D.; Stevens, J. R. *Chem. Comm.* **1967**, *1147*.
73. Scheidegger, H. A.; Taube, H. *J. Am. Chem. Soc.* **1968**, *90*, 3263.
74. Magnuson, R. H.; Taube, H. *J. Am. Chem. Soc.* **1972**, *94*, 7213.
75. Buhr, J. D.; Winkler, J. R.; Taube, H. *Inorg. Chem.* **1980**, *19*, 2416.

76. Lay, P. A.; Magnuson, R. H.; Sen, J. P.; Taube, H. *J. Am. Chem. Soc.* **1982**, *104*, 7658.
77. Buhr, J. D.; Taube, H. *Inorg. Chem.* **1980**, *19*, 2425.
78. Magnuson, R. H.; Lay, P. A.; Taube, H. *J. Am. Chem. Soc.* **1983**, *105*, 2509.
79. Magnuson, R. H.; Taube, H. *J. Am. Chem. Soc.* **1975**, *97*, 5129.
80. Sen, J. P.; Taube, H. *Acta. Chem. Scand.* **1979**, *A33*, 125.
81. Harman, W. D.; Taube, H. *J. Am. Chem. Soc.* **1987**, *109*, 1883.
82. Li, Z.-W.; Harman, W. D.; Lay, P. A.; Taube, H. *Inorg. Chem.* **1994**, *33*, 3635.
83. Harman, W. D.; Taube, H. *J. Am. Chem. Soc.* **1990**, *112*, 2261.
84. Malin, J.; Taube, H. *Inorg. Chem.* **1971**, *10*, 2403.
85. Coelho, A. L.; Malin, J. M. *Inorg. Chim. Acta.* **1975**, *14*, L41.
86. (a) Li, Z.-W.; Taube, H. *J. Am. Chem. Soc.* **1991**, *113*, 8946; (b) Li, Z.-W.; Taube, H. *Science (Washington, D.C.)* **1992**, *256*, 210.
87. Pu, L; Hasegawa, T.; Parkin, S.; Taube, H. *J. Am. Chem. Soc.* **1992**, *114*, 2712.
88. Pu, L: Hasegawa, T.; Parkin, S.; Taube, H. *J. Am. Chem. Soc.* **1992**, *114*, 7609.
89. Li, Z.-W.; Yeh, A.; Taube, H. *J. Am. Chem. Soc.* **1993**, *115*, 10384.
90. Li, Z.-W.; Taube, H. *J. Am. Chem. Soc.* **1994**, *116*, 9506.

PART I:
ELECTRON TRANSFER APPLICATIONS IN ORGANOMETALLIC CHEMISTRY

Structure and Bonding in Molecular Hydrogen Complexes of Osmium(II)

Ian Bytheway[1], J. Simon Craw[1,2], George B. Bacskay[1] and Noel S. Hush[*,1,3]

[1]Department of Physical and Theoretical Chemistry and [3]Department of Biochemistry, University of Sydney, Sydney, New South Wales 2006, Australia

A quantum chemical study of geometries, H_2 binding energies, and HD spin–spin coupling constants using self-consistent field theory, second-order Møller–Plesset theory (MP2), and density functional theory (DFT) techniques for a series of molecular hydrogen complexes $[Os(NH_3)_4 L^z(\eta^2\text{-}H_2)]^{(z+2)+}$ $[L^z = (CH_3)_2CO, H_2O, CH_3COO^-, Cl^-, H^-, C_5H_5N, CH_3CN^-, CN, NH_2OH, and NH_3]$ is described. Electron correlation was found to be of crucial importance in the description of the H–H potential and the equilibrium H–H distance. The MP2 and DFT predictions of the geometries and energetics are in reasonable agreement, but there is noticeable divergence in the predicted H–H distances for weakly bound complexes containing trans ligands with strong π-acceptor properties. The calculated H–H distances range from 0.95 to 1.40 Å and are consistent with stretched molecular hydrogen acting as a ligand rather than dissociating into two atoms bound as hydrides. The H–H distance predicted by both MP2 and DFT methods is in good agreement with that observed in the $[Os(NH_2C_2H_4NH_2)_2(CH_3COO^-)(\eta^2\text{-}H_2)]^+$ complex, the only one for which neutron diffraction data are available.

THE FIRST IDENTIFICATION of an η^2-H_2 complex, $W(CO)_3(P(i\text{-}Pr_3))_2(H_2)$ (i-Pr = isopropyl), by Kubas and co-workers (1) in 1984 marked the beginning of a fascinating and rapidly expanding area of inorganic chemistry. Low-temperature neutron diffraction studies of this complex showed that the hydrogen molecule is bound to the tungsten atom in a sideways manner, with an H–H separation of 0.82 Å (i.e., 10% longer than in H_2), indicating that the H–H

[2]Current address: Department of Chemistry, University of Manchester, Oxford Rd., Manchester, M13 9PL, United Kingdom.

[*]Corresponding author.

bond is somewhat weakened upon complexation (*2, 3*). Since this discovery, the chemistry of dihydrogen complexes has blossomed, with over 150 complexes of this type now known (*4, 5*).

In 1971 Malin and Taube (*6*) synthesized the complex $[Os(en)_2(H_2)]^+$ (en = ethylenediamine), which has since been characterized as a molecular hydrogen complex (*7*). More recently, a series of complexes of the general type $[Os(NH_3)_4L^z(\eta^2\text{-}H_2)]^{(z+2)+}$, where L^z represents a wide variety of ligands, have been synthesized and studied by Li and Taube (*7–9*), and it is these complexes that we shall be concerned with in this chapter. The generic, pseudo-octahedral geometry of these osmium complexes is depicted in Figure 1, which shows clearly the trans relationship between the L^z and H_2 ligands, with the ammonia ligands located in the equatorial sites.

An interesting, well-characterized feature of these species is the marked dependence of hydrogen–deuterium nuclear spin–spin coupling constants, J_{HD}, on the nature of the trans ligand L^z. The value of J_{HD} varies from about 20 Hz when L^z is acetonitrile, to 4 Hz when L^z is acetone, suggesting that J_{HD} correlates with the π-donor characteristics of the trans ligand. By way of comparison, the value of J_{HD} for free HD is 43 Hz (*10*). As the chemical shift of the $\eta^2\text{-}H_2$ signal appears in the spectral window of -20 to 0 ppm, far from other

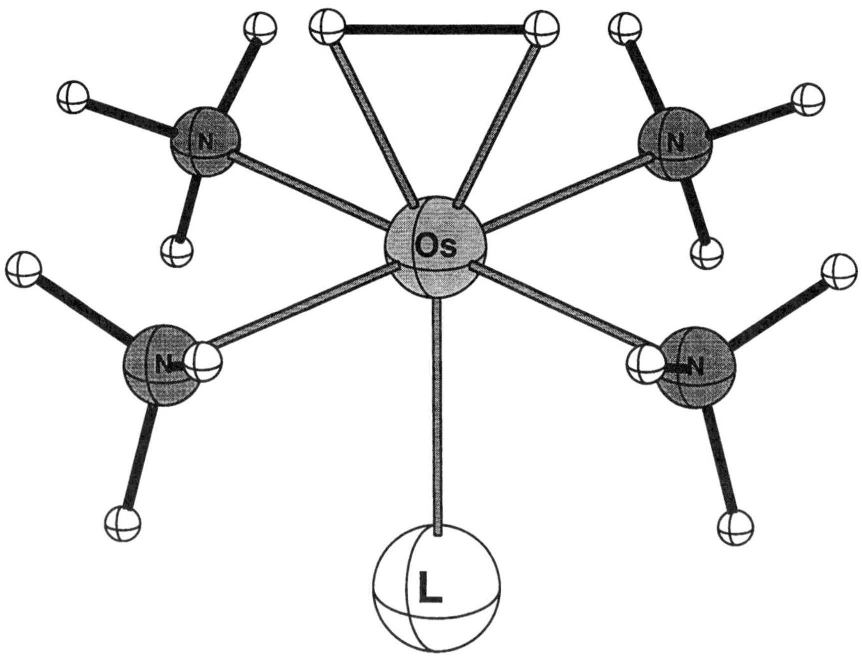

Figure 1. The structure of $[Os(NH_3)_4L^z(\eta^2\text{-}H_2)]^{(z+2)+}$.

resonances, it is a useful diagnostic probe for L^z, which may have important applications in biochemical systems (*11*).

The observed coupling constants in these osmium complexes are thought to be indicative of a substantial increase in H–H distance on complexation, considerably more so than in the aforementioned tungsten complex, in which the observed value of J_{HD} of 34 Hz is consistent with the small 0.1 Å increase in H–H bond length (*1*). There are other complexes with longer H–H bond lengths, for example $ReH_7(pTol_3)$ (pTol = paratoluene), in which an H–H distance is 1.357 Å (*12*), although this finding could be a consequence of steric crowding due to the high coordination number, as in the ReH_7(dppe) [dppe = 1,2-bis(diphenylphosphino)ethane] complex (*13*). The "four-legged piano stool" complexes (*14, 15*) also contain hydrogen ligands separated by distances intermediate between that in free hydrogen and the values observed in "classical" hydrides, although chemically these complexes behave as polyhydrides (*4, 5*).

The aim of the work presented in this chapter has been the characterization of a series of molecular hydrogen complexes of osmium: those synthesized by Li and Taube and some that have yet to be prepared, using quantum chemical methods. In particular, we are concerned with the nature of the Os–H and H–H bonds, the geometries of the complexes, and the influence of the trans ligand on the H–H distance, the binding energy of H_2, and the H–D coupling constant. The role of electron correlation in the description of the properties of the complexes is also examined. These were performed mainly at the second-order Møller–Plesset (MP2) level of theory. However, given the recent developments in density functional theory (DFT) and its obvious computational advantages over conventional methods, we also embarked on a study of these dihydrogen complexes using DFT.

Computational Details

A detailed description of the self-consistent field (SCF)/MP2 computational approaches used has been given elsewhere (*16*); consequently only a brief discussion of the methodology is given here. Effective core potentials (ECP), parameterized so as to account for relativistic corrections, were used in conjunction with double- and triple-ζ quality basis sets. The ECPs and basis sets used are those of Stoll and co-workers (*17, 18*). The Os basis set consists of a [5s4p3d] Gaussian basis set in order to describe the valence 5s, 5p, and 5d electrons, whereas for the C, N, O, and Cl atoms, nonrelativistic ECPs were used along with [2s2p] ([3s3p] for Cl) basis sets to describe the valence electrons. For the hydrogen atoms bound to the osmium, a double-ζ basis set has been used (*19*) extended with a set of 2p polarization functions ($\zeta = 0.80$). The geometries were optimized at the SCF level with respect to the parameters not involving the ligated H_2 moiety, while the Os–H and H–H distances were optimized by pointwise energy calculations using MP2 perturbation theory. In some cases (L^z = CN^-, NH_2OH, or NH_3) all geometrical parameters involving the osmium atom were fully optimized at the MP2 level.

In addition to the SCF and MP2 methods, we have also used techniques based on DFT. These include exchange and correlation effects via a functional, thereby avoiding the lengthy configuration interaction (CI) type expansions of conventional MP and CI methods. The computational advantages of DFT approaches over the standard methods are especially significant when studying large molecules such as transition-metal complexes, making it an attractive alternative. In this work we report results obtained using the BLYP functional (20), that is, a hybrid of Slater's exchange functional and the Lee, Yang, Parr (LYP) correlation functional with Becke's gradient correction.

The H–D coupling constants, J_{HD}, were calculated by the finite perturbation technique of Kowalewski et al. (21), using the unrestricted Hartree–Fock (UHF) and UHF + MP2 (UMP2) method. It was assumed that the Fermi contact term represents the dominant contribution to J_{HD} (22), allowing the spin dipolar and orbital effects to be neglected.

The calculations were performed using a variety of software packages: HONDO (23), MOLECULE (24–26), TURBOMOLE (27, 28) and GAUSSIAN92/DFT (20).

Geometries and Vibrational Frequencies

The optimized key geometrical parameters are given in Table I, along with the available experimental values. Somewhat surprisingly, the MP2 H–H, Os–H, and Os–N distances are quite uniform across the range of complexes studied,

Table I. Calculated Bond Lengths (Å) for the $[Os(NH_3)_4L^z(\eta^2\text{-}H_2)]^{(z+2)+}$ Complexes

	r(H–H)		r(Os–H)	
L^z	MP2	DFT	MP2	DFT
$(CH_3)_2CO$	1.380	1.249	1.596	1.613
H_2O	1.350	1.250	1.590	1.635
CH_3OO^-	1.389	1.316	1.580	1.635
	(1.34)		(1.60)	
Cl^-	1.400	1.314	1.600	1.630
H^-	1.330	0.978	1.630	1.750
C_5H_5N	1.300	0.998	1.616	1.689
CH_3CN	1.330	0.985	1.580	1.691
CN^-	1.293	0.953	1.614	1.746
NH_2OH	1.256	1.031	1.582	1.670
NH_3	1.252	1.057	1.581	1.659

NOTE: L^z means ligand with charge z; r means bond length; MP2 means second-order Møller–Plesset perturbation theory; DFT means density functional theory. Values in parentheses are neutron diffraction distances at 165 K (29).

suggesting that the electronic structures of these molecules are very similar. The qualitative trend in the MP2 distances, for the range of trans ligands considered, is reproduced by DFT, although in same cases, for example, $L^z = C_5H_5N$ and CH_3CN, the predicted distances differ by as much as 0.3 Å. The variation in Os–L^z distances is similarly quite small (ca. 0.1 Å) for the first-row ligands, with an increase of about 0.3 Å when $L^z = Cl^-$. Comparison with experiment is possible for the acetate complex because the crystal structure for the related complex $[Os(en)_2(CH_3COO^-)(\eta^2\text{-}H_2)]^+$ has been determined by both X-ray and neutron diffraction techniques (29). The MP2 and DFT predictions of the H–H separation are 1.39 and 1.32 Å, respectively, which are in good agreement with the observed value of 1.34 Å. The level of consistency between theory and experiment with respect to the other geometrical parameters is similarly quite good.

An unusual feature of these complexes is the crucial role of electron correlation in the description of the geometries, especially the H–H distance. SCF theory predicts a H–H separation of 0.8 Å in the acetate complex, compared with the MP2 value of 1.39 Å (16, 30). Such a large, qualitatively important difference necessitated the examination of the validity of single-reference SCF and MP2 techniques in these calculations. The coefficient of the SCF reference configuration in the MP2 wavefunction was found to be on the order of 0.85 for all of the complexes examined (16). Although this value may seem low at first, it was found to be due to the large number of double excitations that individually make only small contributions (≤ 0.05) to the wavefunction. Thus the SCF configuration is dominant in the wavefunctions, indicating that there are no significant near-degeneracy effects, and hence the use of the MP2 method is valid.

The importance of electron correlation in other dihydrogen complexes such as the polyhydrides of Re, Tc, and Ir has been noted by Haynes et al. (31) and by Lin and Hall (32), for which the energetic ordering of the various classical hydride and $\eta^2\text{-}H_2$ complexes is altered once correlation is accounted for.

In a further test, higher level calculations (i.e., those that provide a more accurate description of electron correlation than MP2) were also carried out for one of the complexes, $L^z = Cl^-$, using the MP3 and averaged coupled pair functional (ACPF) methods (33). The results of these calculations are shown in Figure 2 along with the SCF predictions. Clearly the H–H distance is affected enormously by electron correlation, but refinements beyond MP2 in the level of correlation result in only a slight change in the predicted H–H separation. The MP3 calculation yields a minimum at a H–H separation of 1.38 Å, while the ACPF distance is 1.35 Å, compared with the MP2 value of 1.40 Å. MP4 calculations for the $L^z = NH_3$ complex also confirm these findings (34). It is likely, therefore, that the MP2 H–H distances are too long by about 0.05 Å. Applying such a correction to the H–H separation in the acetate complex does in fact result in almost exact agreement between theory and experiment.

An important feature of the calculations, as shown in Figure 2, is that only a single minimum is obtained, regardless of the level of theory, that is, corre-

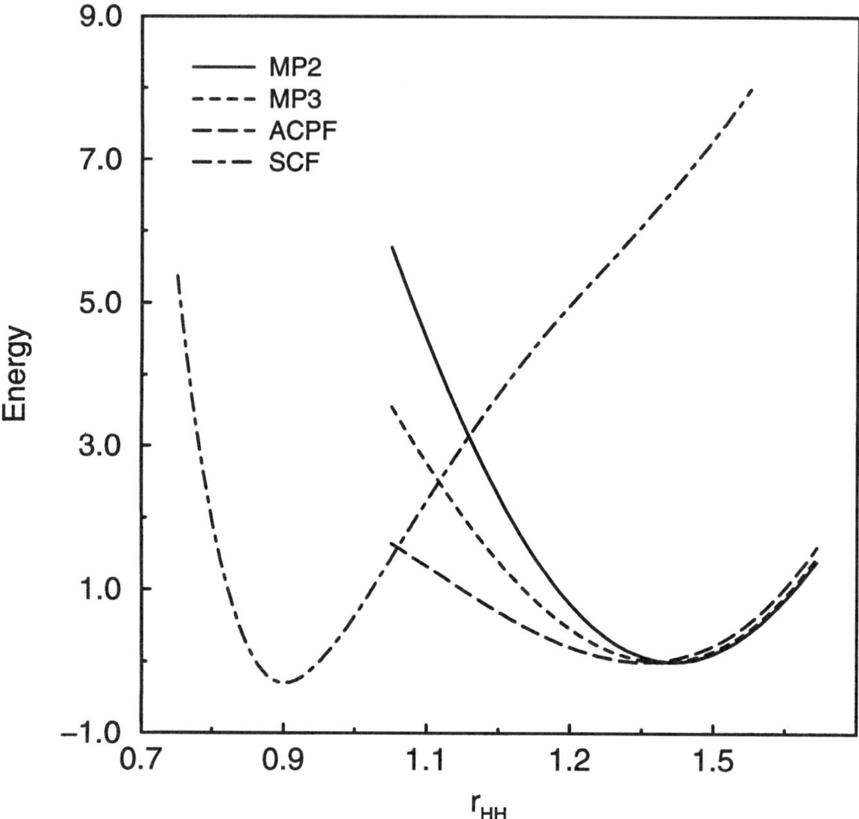

Figure 2. SCF, MP2, MP3, and ACPF energies (kcal/mol) as a function of the H–H distance (Å) for the $[Os(NH_3)_4Cl^-(\eta^2\text{-}H_2)]^+$ complex.

sponding to a stretched η^2-H_2 complex. No *cis*-dihydrides were located on the potential energy surface, in contrast to the results obtained by Hay (35) for the tungsten complexes $[W(CO)_3(PH_3)_2(H_2)]$ and $[W(PH_3)_5(H_2)]$, which showed the existence of stable dihydrogen and dihydride isomers.

The consistency between the MP2 and DFT predictions of the H–H and Os–H distances is reasonable in most cases. However, for certain trans ligands such as pyridine and acetonitrile, the differences are approximately 0.3 Å. It appears that when L^z is a good π-acceptor, DFT tends to predict a longer Os–H separation than MP2, as well as a correspondingly smaller H–H distance. The problem is addressed elsewhere (34).

Finally, we consider the orientation of the H_2 ligand in the complex relative to the NH_3 ligands. As shown in Figure 1, H_2 has been restricted by the symmetry imposed to assume a staggered orientation. This orientation is contrary to the arrangement of ligands in other complexes such as $W(CO)_3(P(i$-

Pr)$_3$)$_2$(H$_2$) (2) and calculated for M(CO)$_5$(H$_2$) (M = Cr, Mo, and W) (36), in which the preferred arrangement is eclipsed. Our MP2 calculations suggest, however, that in the current Os(II) complexes the conformation is staggered, although the barrier to rotation of the H$_2$ ligand is quite low (e.g., 2.5, 1.5, and 0.9 kcal mol^{-1} when L^z = CH$_3$COO$^-$, NH$_3$, and Cl$^-$, respectively). SCF calculations that we carried out for the W(CO)$_3$(PH$_3$)$_2$(H$_2$) complex yielded a value of 1.3 kcal mol^{-1} for the analogous barrier to rotation. (In the tungsten complex SCF theory provides an adequate description of the geometry.)

The Os–H stretching vibrational frequencies have been observed in two ethylenediamine complexes with chloride and pyridine as trans ligands. These may be meaningfully compared with the analogous frequencies for the corresponding tetraammine complexes. Treating the MH$_2$ part of the complex as a pseudotriatomic molecule, where M = [Os(NH$_3$)$_4$L^z]$^{(z+2)+}$, the force constants and hence harmonic frequencies were calculated via pointwise MP2 energy calculations. The results are given in Table II, together with available experimental values (37). The agreement is reasonable once we note that the comparison is between harmonic (theoretical) and anharmonic (experimental) values.

The stretching frequencies are actually typical of hydride stretch modes. For example, the stretching frequency in the $^4\Pi$ ground state of OsH has been calculated to be 2138 cm^{-1} (38). The predicted bend frequencies ν_b are about one-third of the value of the analogous H–H stretch in free H$_2$, which is consistent with a strong Os–H interaction and a correspondingly weaker H–H bond in these complexes.

Bonding and Charge Distribution

The bonding between molecular dihydrogen and a metal atom is usually explained in terms of a simple one-electron picture involving electron donation to the metal from the σ orbitals of the H$_2$ ligand with concomitant back donation from the occupied metal orbitals to the σ^* antibonding orbitals of H$_2$ (32, 35, 39, 40). This explanation is analogous to the Chatt–Dewar–Duncanson model used to interpret the binding in metal–olefin complexes (41, 42). The osmium ion in these molecular hydrogen complexes has the low spin (d^6) con-

Table II. MP2 Harmonic Vibrational Frequencies (cm^{-1}) for the [Os(NH$_3$)$_4$L^z(η^2-H$_2$)]$^{(z+2)+}$ Complexes

L^z	ν_b	ν_a	ν_s	ν_{expt}
Cl$^-$	1318	2335	2358	2155
C$_5$H$_5$N	1283	2310	2305	2285

NOTE: For harmonic vibrational frequencies (ν), subscript s means symmetric stretch, subscript a means asymmetric stretch, and subscript b means bend.
SOURCE: ν_{expt} data are from reference 37.

figuration and has an empty d_σ orbital that is available for σ-donation by the dihydrogen ligand. Back donation is from a filled osmium d_π orbital into the vacant σ^* orbitals of the H_2 ligand, and the overall effect is therefore to weaken and stretch the H–H bond, as indeed is observed.

A molecular orbital (MO) diagram illustrating this bonding scheme is given in Figure 3, which shows that on stretching free H_2 the energies of the σ and σ^* MOs rise and fall, respectively, moving closer to the osmium d_σ and d_π orbitals and thereby allowing for larger highest occupied molecular orbital (HOMO)–lowest unoccupied molecular orbital (LUMO)-type interactions (i.e., larger ligand-to-metal σ donation as well as π back donation). The orbitals not directly involved in bonding to the dihydrogen ligand are designated nonbonding (nb).

The results of population analyses that were carried out can be used to gauge the applicability of this model for the range of complexes studied. Atomic charges, calculated by the Roby–Davidson (*43–46*) and Mulliken methods (*47*), lend support to the proposed bonding model (*16*). In the case of Roby–Davidson charges there is little variation in the net charge on H_2 in all of the complexes studied. When the trans ligand is neutral, H_2 appears to be slightly positive, but in the case of negatively charged trans ligands, H_2 is actually slightly negative. For complexes with a neutral L^z, removal of the H_2 ligand results in a complex in which the osmium atom has a higher positive

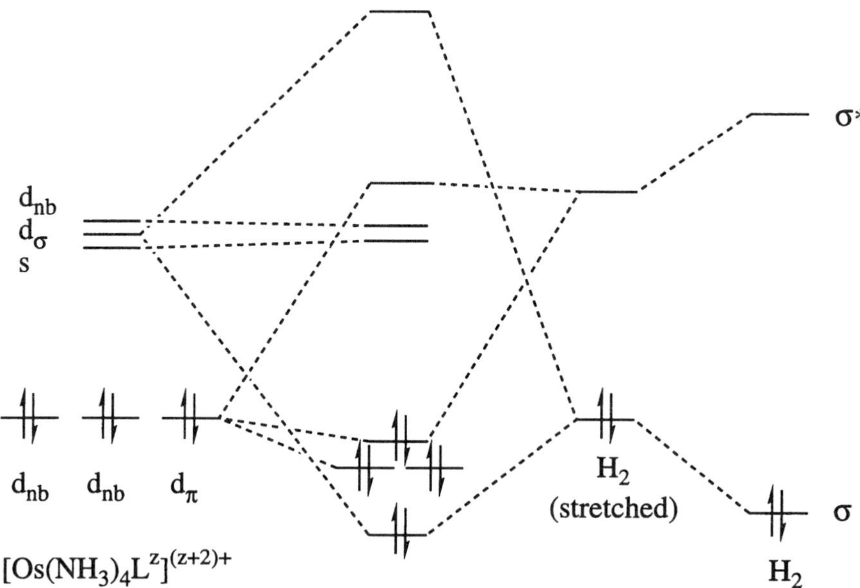

Figure 3. Molecular orbital energy level diagram showing the interaction between the $[Os(NH_3)_4L^z]^{(z+2)+}$ fragment and H_2.

charge, whereas the reverse holds when L^z is an anion. The Mulliken charges are qualitatively similar to those obtained in the Roby–Davidson analyses. The main difference is in the osmium charges, which are predicted to be less positive when calculated by the Mulliken method, while the NH_3 and H_2 ligands are slightly more positive.

To quantify the degree of σ donation and π back donation, we calculated the difference in gross Mulliken populations between the various dihydrogen complexes and the fragments that result by removing H_2. The resulting differences were partitioned into MO contributions involving the σ and σ^* orbitals of H_2 and the d_σ and d_π orbitals of osmium and are displayed in Figure 4, which shows the average amount of charge gained and lost by the osmium atom and the H_2 ligand, respectively, as a result of dihydrogen complex formation. As expected on the basis of the bonding model discussed, the osmium d_σ and H_2 σ^* orbitals gain electron population, while the osmium d_π and H_2 σ orbitals lose electrons. Given the inherent limitations of the Mulliken analysis (*16, 48*), we suggest that the actual amounts of charge transferred should be viewed only as a qualitative to semiquantitative guide to the importance of the charge-transfer mechanism. An alternative approach is the direct study of the energetic effects of charge transfer using the Complete Spatial Orbital Variation (CSOV) method (*49*), as carried out by Craw et al. (*16*), which suggested that charge transfer is indeed the dominant contribution to the Os–H_2 interaction energy.

The binding energies of H_2 in the complexes, ΔE_B, that is, the energy of the reaction

$$[Os(NH_3)_4L^z]^{(z+2)+} + H_2 \rightarrow [Os(NH_3)_4L^z(\eta^2\text{-}H_2)]^{(z+2)+}$$

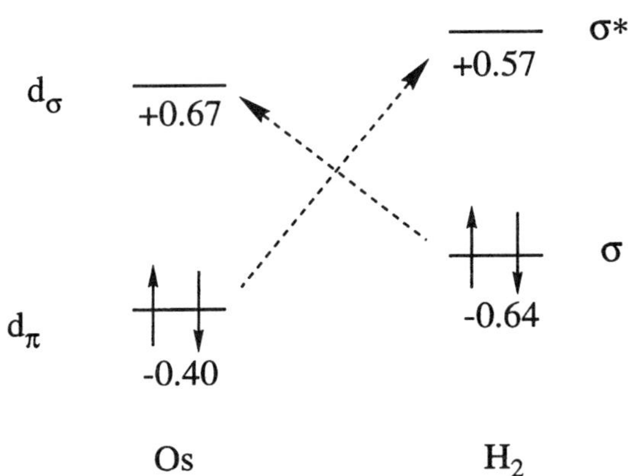

Figure 4. Partitioning of the difference in gross Mulliken populations between $[Os(NH_3)_4L^z(\eta^2\text{-}H_2)]^{(z+2)+}$ complexes and the $[Os(NH_3)_4L^z]^{(z+2)+}$ and H_2 into σ and π contributions (averaged over the series of complexes studied).

provides a quantitative measure of the strength of the Os–H_2 bond. The MP2 and DFT estimates are given in Table III and show quite clearly that the bond between Os and H_2 is quite strong. The trends in binding energy with L^z, as calculated by MP2 and DFT methods, are quite similar, although the DFT predictions are significantly lower, approximately 70% of the MP2 values. The binding energies are expected to correlate with the H–H distance (r_{HH}). The plot of the ΔE_B values against the H–H distance, shown in Figure 5, suggests a reasonable degree of correlation, but with considerable scatter at the low-energy side.

The binding energies in Table III are generally much larger than those obtained for other dihydrogen complexes such as $[W(CO)_5(\eta^2\text{-}H_2)]$ (36), in which the binding energy of H_2 (before zero-point energy corrections) is only -19.8 kcal mol^{-1}. The reason for this difference is most likely the high positive charge of Os (formal charge of 2e), in comparison with W, which appears with zero formal charge in the complex just cited. Indeed, according to our recent work (50) on the $[W(CO)_3(PH_3)_2(\eta^2\text{-}H_2)]$ complex, the binding energy of H_2 is -15.6 kcal mol^{-1}, when calculated at the DFT (BLYP) level, whereas the corresponding MP2 value is -26.7 kcal mol^{-1}. Commensurate with the smaller binding energy in this tungsten complex, the H–H distance is 0.85 Å at the DFT level, indicating a stretch of only 0.09 Å, in stark contrast with those computed for the Os(II) complexes.

H–D Spin–Spin Coupling Constants

Li and Taube (7) have remarked on the sensitivity of the H–D spin–spin coupling constant, J_{HD}, to the trans ligand in a variety of Os(II) dihydrogen com-

Table III. The Calculated Binding Energies of $H_2[Os(NH_3)_4(L^z)(\eta^2\text{-}H_2)]^{(z+2)+}$

	$-\Delta E_B$ (kcal/mol)	
L^z	MP2	DFT
$(CH_3)_2CO$	64.0	45.9
H_2O	57.7	49.7
CH_3COO^-	59.0	44.5
Cl^-	60.8	45.0
H^-	40.2	22.9
C_5H_5N	46.7	32.6
CH_3CN	46.6	33.0
CN^-	40.5	23.7
NH_2OH	48.2	36.5
NH_3	49.1	37.5

NOTE: $-\Delta E_B$ is calculated binding energy of H_2 in $[Os(NH_3)_4L^z(\eta^2\text{-}H_2)]^{(z+2)+}$ for each ligand, L^z.

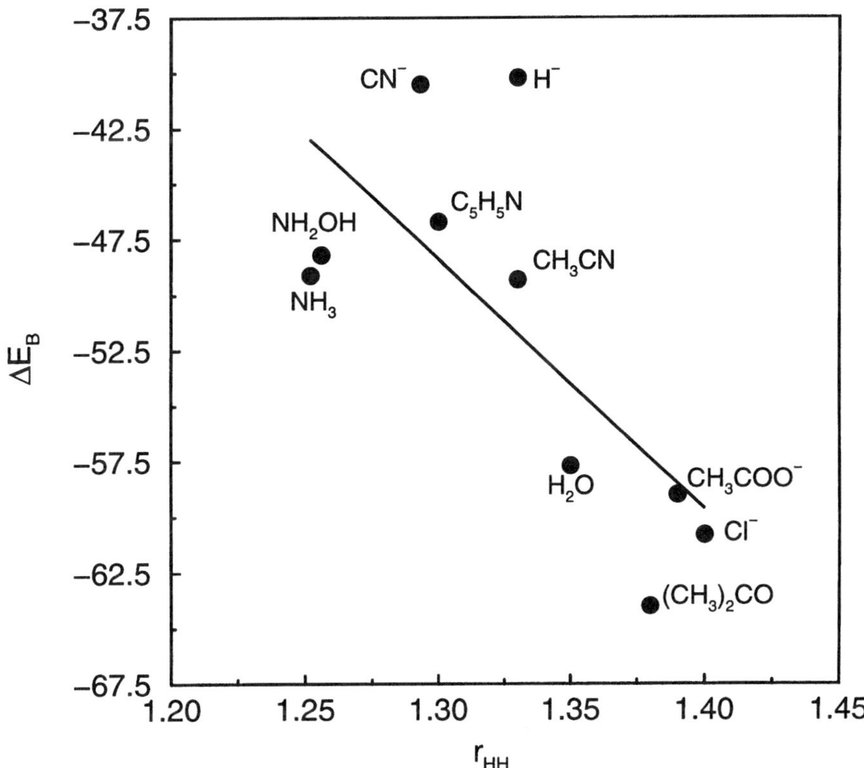

Figure 5. The MP2 calculated binding energies of H_2 in the $[Os(NH_3)_4L^z(\eta^2\text{-}H_2)]^{(z+2)+}$ complexes, ΔE_B (kcal/mol), versus the H–H distance (Å).

plexes and suggested that the H–H bond length may show similar sensitivity. Consequently, J_{HD} is expected to correlate with the H–D distance, that is, to decrease with increasing H–D separation.

We explored the possibility of such relationships by plotting the experimental coupling constants against the H–H separation calculated at both the MP2 and DFT levels of theory (Figure 6). As expected, an inverse relationship between J_{HD} and r_{HH} is apparent. The MP2 calculations predict so little variation in the H–H distances that J_{HD} appears to be perhaps unexpectedly sensitive to r_{HH} although the correlation is somewhat tenuous.

The DFT results predict a greater range of r_{HH} and hence a reduced sensitivity of J_{HD} to distance although qualitatively the same conclusions may be reached: J_{HD} decreases as r_{HH} increases. In free HD the behavior is different from that in the complexes: it monotonically increases throughout the range of H–H distances that span those obtained in the complexes and the smaller free molecule value, which might be thought counterintuitive (51). We must there-

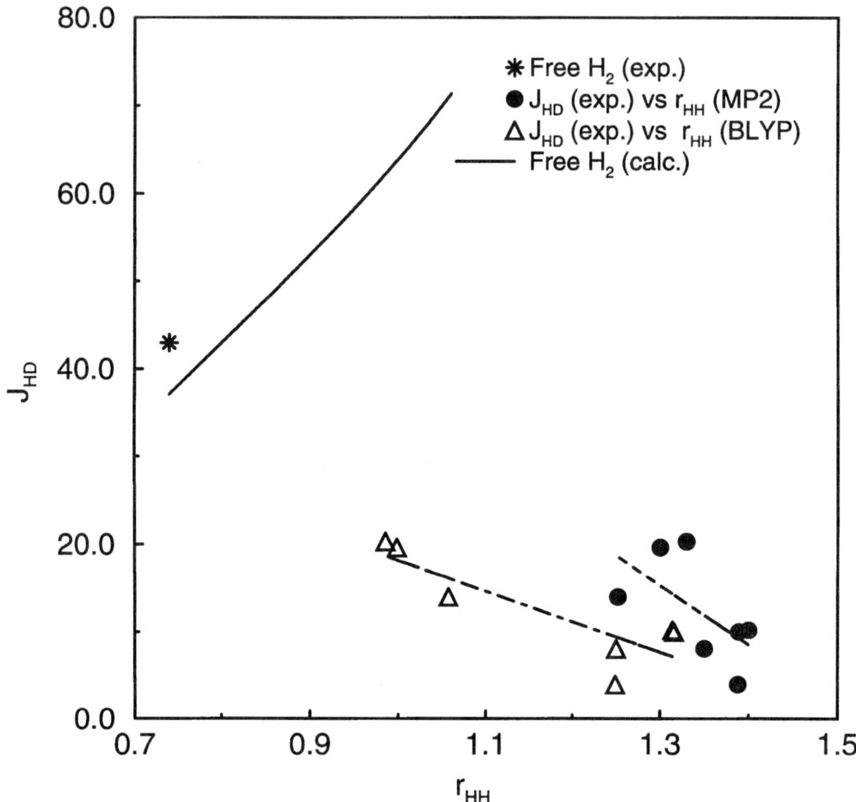

Figure 6. Experimental H–D coupling constant, J_{HD} (Hz), versus the MP2 and DFT H–H distances (Å) in the $[Os(NH_3)_4L^z(\eta^2\text{-}H_2)]^{(z+2)+}$ complexes. A comparison with the calculated value of J_{HD} for free H_2 is also shown.

fore conclude that in the complexes the coupling between the H and D spins is actually suppressed by the osmium atom, that is, by H–Os–D type coupling. Therefore, we should also observe a correlation between J_{HD} and the binding energy ΔE_B of H_2. The plot in Figure 7 shows that this correlation is indeed the case.

We have also computed the Fermi contact contribution to the coupling constants, at both UHF and UMP2 levels of theory (16). A graphical comparison of the UHF and experimental values of J_{HD} is shown in Figure 8. With the obvious exceptions of the systems with charged ligands Cl^- and CH_3COO^-, the calculated UHF coupling constants agree reasonably well with the experimental values. However, the level of agreement between UMP2 and experiment is less satisfactory, suggesting that the electron correlation corrections to J_{HD} are not described adequately by UMP2 theory.

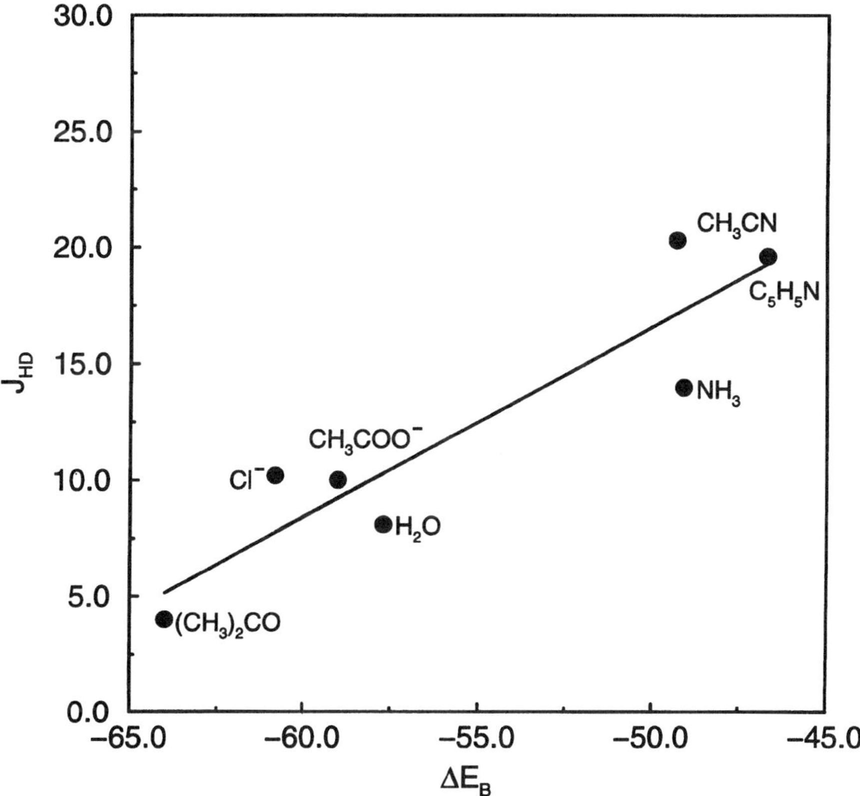

Figure 7. Experimental coupling constant, J_{HD} (Hz), versus the calculated binding energy of H_2, ΔE_B (kcal/mol), for the $[Os(NH_3)_4L^z(\eta^2\text{-}H_2)]^{(z+2)+}$ complexes.

Correlation of Properties with Spectrochemical Parameters of trans Ligands

Given the effect of the trans ligand on the properties of the $[Os(NH_3)_4L^z(\eta^2\text{-}H_2)]^{(z+2)+}$ complexes, it is of great interest to know whether they correlate with some intrinsic property of the trans ligand, such as its spectrochemical constant. If such a correlation were valid, ligands could be chosen so as to "fine tune" the chemical properties of the complex in question. As discussed already, the degree of interaction between the osmium atom and the hydrogen molecule depends on the energetics of the metal d_π and d_σ orbitals relative to those of H_2, in particular on their separation Δ. Consequently, $\eta^2\text{-}H_2$ complexation will be favored by raising the energy of the d_π and/or lowering the energy of the d_σ orbitals, that is, by decreasing Δ. Smaller separation of these orbitals

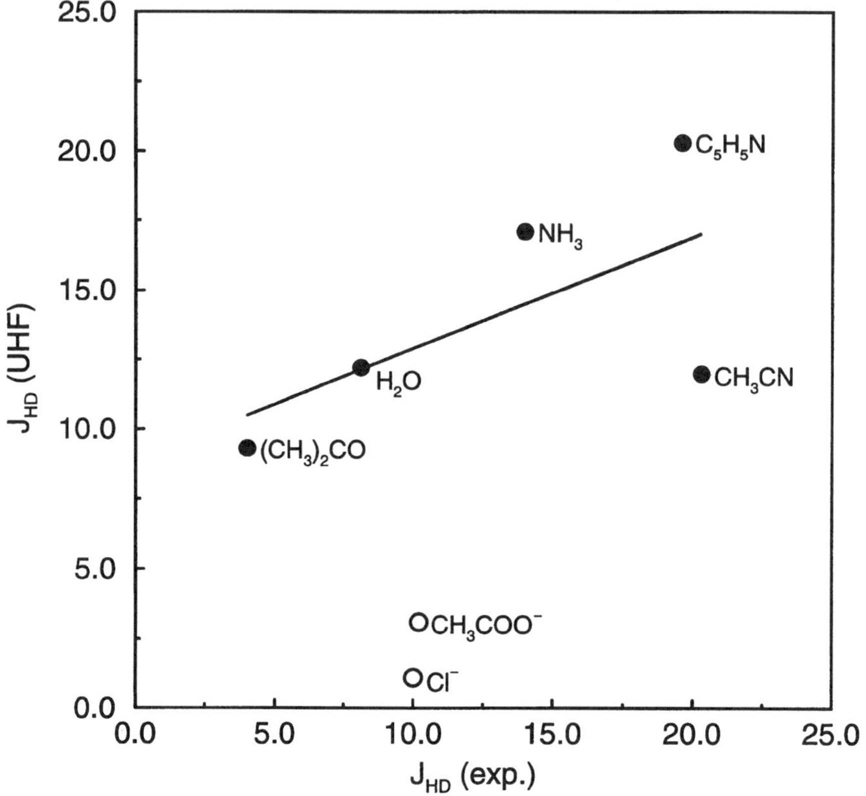

Figure 8. Calculated (UHF) coupling constants, J_{HD} (Hz), versus the experimental values for the $[Os(NH_3)_4L^z(\eta^2\text{-}H_2)]^{(z+2)+}$ complexes.

(i.e., a smaller value of Δ) results in a longer H–H distance and stronger binding of H_2 that would increasingly resemble a classical hydride.

An experimental link between Δ in a given complex and a ligand (L) is the spectrochemical parameter $f(L)$ (52) of the ligand. The angular overlap (AO) model (53) defines Δ as

$$\Delta = 3e_\sigma - 4e_\pi$$

where e_σ and e_π are quantities that describe the σ-donor and π-acceptor properties of the ligand. Within this formalism e_σ is always positive, while e_π is only positive for complexes with moderately large π-donor abilities (e.g., halides) and is negative for strong π acceptors (e.g., CH_3CN).

The relationship between $f(L)$ and the binding energy ΔE_B of H_2 calculated at both the MP2 and DFT levels of theory is shown in Figure 9. Although

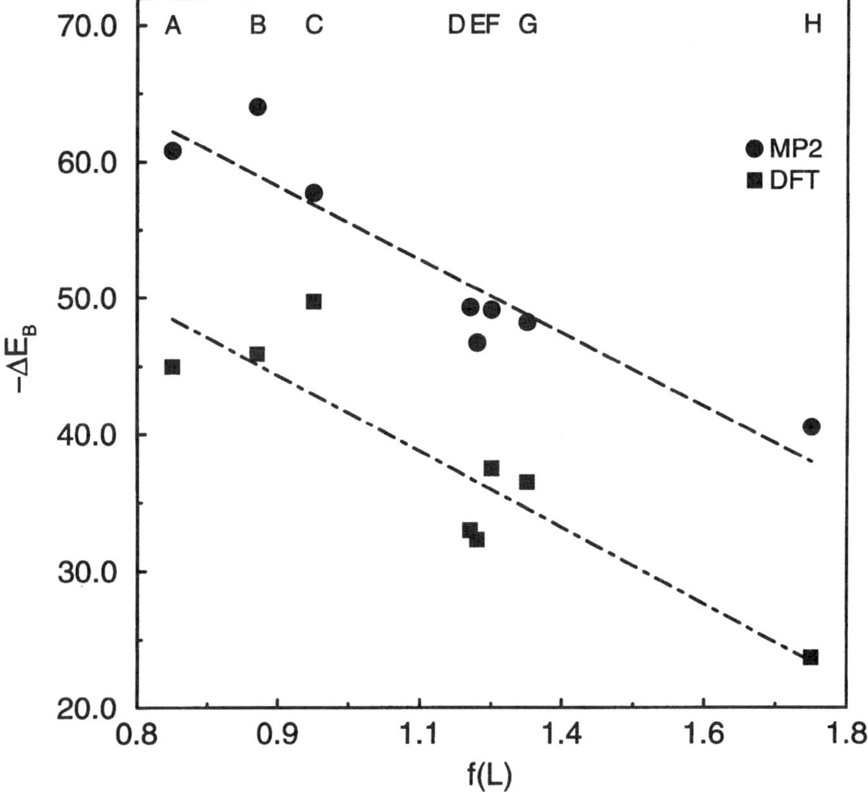

Figure 9. Correlation of the binding energy, ΔE_B (kcal/mol), of H_2 in the $[Os(NH_3)_4L^z(\eta^2\text{-}H_2)]^{(z+2)+}$ complexes with the spectrochemical parameter, f(L). Labels show the f(L) values of the various L^z: $A = Cl^-$, $B = (CH_3)_2CO$, $C = H_2O$, $D = CH_3CN$, $E = C_5H_5N$, $F = NH_3$, $G = NH_2OH$, and $H = CN^-$.

we do not want to place an inordinate amount of emphasis on these relationships, it is worth noting that correlation with these parameters does provide a link between ab initio theory and empirical chemical and spectroscopic properties, and may serve as a possible guide to properties as yet unmeasured.

Summary

We have presented in this chapter the salient features of an extensive study of the geometries, binding energies, H–D coupling constants, and charge distributions for the series of complexes $[Os(NH_3)_4L^z(\eta^2\text{-}H_2)]^{(z+2)+}$ carried out at the MP2 and DFT/BLYP levels of theory. The geometries that were obtained using correlated methods are markedly different from the conventional SCF predictions, especially with regard to the H–H distance. The sensitivity of molecular structure to correlation has been noted for a range of molecular hydrogen as well as hydride complexes (*16, 31, 32, 54*).

The bonding in the η^2-H_2 complexes of osmium may be understood in terms of σ donation by dihydrogen and backbonding that results in electron donation into the σ^* MOs of H_2. The binding energies of the dihydrogen molecule in the complex are consistent with this model, as well as with an angular overlap model that correlates the spectrochemical constant of the trans ligand with the properties of the complexes. Thus, where L^z induces a large splitting of the osmium d_π and d_σ orbitals, H_2 is less strongly bound to Os.

The experimental H–D spin–spin coupling constants, J_{HD}, correlate with the calculated H–H separation as well as with the H_2 binding energy. Because the H_2 binding energy correlates with the spectrochemical constants, $f(L)$, of the trans ligand, it follows that the coupling constants will show a similar dependence on $f(L)$. The variation of J_{HD} with H–H distance is opposite to that in free H_2, showing that the H–Os–H interaction has a strong effect on the spin–spin coupling in complexed dihydrogen.

The preliminary results using DFT suggest that the method is capable of describing the electronic structures of these interesting and unusual osmium complexes. Further DFT studies are currently in progress (*34*).

Note Added in Proof

Since the writing of this chapter, a body of further work on these systems has been completed. In particular, the H–D spin–spin coupling constants have been calculated using DFT with the BLYP functional, and now the agreement with experiment is very good (*55*). The effects of solvation on binding energies and ligand exchange equilibria also have been studied (*56*).

Acknowledgments

The financial support of the Australian Research Council is gratefully acknowledged. For his interest, inspiration, extensive discussions, and unpublished results, we thank Professor Henry Taube.

References

1. Kubas, G. J.; Ryan, R. R.; Swanson, B. I.; Vergamini, P. J.; Wasserman, H. J. *J. Am. Chem. Soc.* **1984**, *106*, 451.
2. Eckert, J.; Kubas, G. J.; Hall, J. H.; Hay, P. J.; Boyle, C. M. *J. Am. Chem. Soc.* **1990**, *112*, 2324.
3. Kubas, G. J.; Unkefer, C. F.; Swanson, B. I.; Fukushima, E. *J. Am. Chem. Soc.* **1986**, *108*, 7000.
4. Jessop, P. G.; Morris, R. H. *Coord. Chem. Rev.* **1992**, *121*, 155.
5. Heinekey, D. M.; Oldham, W. J., Jr. *Chem. Rev. (Washington, D.C.)* **1993**, *93*, 913.
6. Malin, J.; Taube, H. *Inorg. Chem.* **1971**, *10*, 2403.

7.   Li, Z.-W.; Taube, H. *J. Am. Chem. Soc.* **1991**, *113*, 8946.
8.   Li, Z.-W.; Taube, H. *J. Am. Chem. Soc.* **1994**, *116*, 9506.
9.   Li, Z.-W.; Taube, H. *J. Am. Chem. Soc.* **1994**, *116*, 11584.
10. Wimmet, T. F. *Phys. Rev.* **1953**, *91*, 476.
11. Li. Z.-W.; Taube, H. *Science (Washington, D.C.)* **1992**, *256*, 210.
12. Brammer, L.; Howard, J. K.; Johnson, O.; Koetzle, T. F.; Spencer, J. F.; Stringer, A. M. *J. Chem. Soc. Chem. Commun.* **1991**, 241.
13. Howard, J. K.; Mason, S. A.; Johnson, O.; Diamond, I. C.; Crennel, S.; Keller, P. A.; Spencer, J. L. *J. Chem. Soc. Chem. Commun.* **1988**, 1502.
14. Poli, R. *Organometallics* **1990**, *9*, 1892.
15. Lin, Z.; Hall, M. B. *Organometallics* **1993**, *12*, 19.
16. Craw, J. S.; Bacskay, G. B.; Hush, N. S. *J. Am. Chem. Soc.* **1994**, *116*, 5937.
17. Andrae, D.; Häussermann, U.; Dolg, M.; Stoll, H.; Preus, H. *Theor. Chim. Acta* **1990**, *77*, 123.
18. Igel-Mann, G.; Stoll, H.; Preus, H. *Mol. Phys.* **1988**, *65*, 1321.
19. Huzinaga, S. *J. Chem. Phys.* **1965**, *42*, 1293.
20. Frisch, M. J.; Trucks, G. W.; Head-Gordon, M.; Gill, P. W.; Wong, M. W.; Foresman, J. B.; Johnson, B. G.; Schlegel, H. B.; Robb, M. A.; Replogle, E. S.; Gomperts, R.; Andres, J. L.; Raghavachari, K.; Binkley, J. S.; Gonzalez, C.; Martin, R. L.; Fox, D. J.; Defrees, D. J.; Baker, J.; Stewart, J. P.; Pople, J. A. *Gaussian 92/DFT*; Gaussian Inc.: Pittsburgh PA, 1993.
21. Kowalewski, J.; Laaksonen, A.; Roos, B.; Siegbahn, P. *J. Chem. Phys.* **1979**, *71*, 2896.
22. Harris, R. K. *Nuclear Magnetic Resonance Spectroscopy: A Physicochemical View*; Pitman: London, 1983; pp 211–220.
23. Dupuis, M.; Rys, J.; King, H. F. *J. Chem. Phys.* **1976**, *65*, 111.
24. Almlöf, J. Technical Report No. 74-29, 1974; University of Stockholm, Sweden.
25. Roos, B. O.; Siegbahn, P. E. In *Modern Theoretical Chemistry*; Schaefer, H. F. III, Ed.; Plenum: New York, 1977; Vol. 3, 277.
26. Bacskay, G. B. *Chem. Phys.* **1981**, *61*, 385.
27. Ahlrichs, R.; Bär, M.; Horn, H.; Kölmel, C. *Chem. Phys. Lett.* **1989**, *162*, 165.
28. Häser, M.; Ahlrichs, R. *J. Comput. Chem.* **1989**, *10*, 104.
29. Hasegawa, T.; Koetzle, T. J.; Li, Z.; Parkin, S.; McMullan, R.; Taube, H. Presented at the 29th International Conference on Coordination Chemistry, Lausanne, Switzerland, 1992.
30. Craw, J. S.; Bacskay, G. B.; Hush. N. S. *Inorg. Chem.* **1993**, *32*, 2230.
31. Haynes, G. R.; Martin, R. L.; Hay, P. J. *J. Am. Chem. Soc.* **1992**, *114*, 28.
32. Lin, Z.; Hall, M. B. *J. Am. Chem. Soc.* **1992**, *114*, 2928.
33. Gdanitz, R. J.; Ahlrichs, R. *Chem. Phys. Lett.* **1988**, *143*, 413.
34. Bytheway, I.; Bacskay, G. B.; Hush, N. S., *J. Phys. Chem.* **1996**, 100, 6023.
35. Hay, P. J. *J. Am. Chem. Soc.* **1987**, *109*, 705.
36. Dapprich S.; Frenking, G. *Angew. Chem. Int. Ed. Engl.* **1995**, *34*, 354.
37. Taube, H., personal communication, 1993.
38. Benavides-Garcia, M.; Balasubramanian, K. J. *J. Mol. Spectrosc.* **1991**, *150*, 271.
39. Hay, P. J. *Chem. Phys. Lett.* **1984**, *103*, 466.
40. Saillard, J. Y.; Hoffmann, R. *J. Am. Chem. Soc.* **1984**, *106*, 2006.
41. Dewar, M. S. *Bull. Soc. Chim. Fr.* **1951**, *18*, C71.
42. Chatt, J.; Duncanson, L. A. *J. Chem. Soc.* **1953**, 2939.
43. Davidson, E. R. *J. Chem. Phys.* **1967**, *46*, 3320.
44. Roby, K. R. *Mol. Phys.* **1974**, *27*, 81.
45. Heinzmann, R.; Ahlrichs, R. *Theor. Chim. Acta* **1976**, *42*, 33.
46. Ehrhardt, C.; Ahlrichs, R. *Theor. Chim. Acta* **1985**, *68*, 231.

47. Mulliken, R. S. *J. Chem. Phys.* **1955**, *23*, 1833.
48. Maseras, F.; Morokuma, K. *Chem. Phys. Lett.* **1992**, *195*, 500.
49. Bagus, P. S.; Hermann, K.; Bauschlicher, C. W. *J. Chem. Phys.* **1984**, *80*, 4378.
50. Bytheway, I.; Bacskay, G. B.; Hush, N. S., unpublished observations.
51. Bacskay, G. B. *Chem. Phys. Lett.*, **1995**, *242*, 507.
52. Lever, A. P. *Inorganic Electronic Spectroscopy*; Elsevier: Amsterdam, 1984.
53. Gerloch, M.; Slade, R. C. *Ligand Field Parameters*; Cambridge University Press: Cambridge, 1973.
54. Lin, Z.; Hall, M. B. *Coord. Chem. Rev.* **1994**, *135/136*, 845.
55. Bacskay, G. B.; Bytheway, I.; Hush, N. S. *J. Am. Chem Soc.* **1996**, *118*, 3753.
56. Bytheway, I.; Bacskay, G. B.; Hush, N. S. *J. Phys. Chem.* **1996**, *100*, 14899.

3

Osmium(II) Dearomatization Agents in Organic Synthesis

W. Dean Harman

Department of Chemistry, University of Virginia, Charlottesville, VA 22901

The fragment [Os(NH₃)₅]²⁺ forms stable η²-coordinate complexes with a wide variety of aromatic molecules including arenes, pyrroles, and furans. The act of coordination greatly reduces the aromatic character of these ligands and, as a consequence, activates them toward various organic reactions. In particular, the addition of carbon-based electrophiles to arenes, furans, and pyrroles is notably enhanced relative to these free aromatic molecules. The resulting arenium, furanium, and pyrrolium intermediates are stabilized by metal backbonding to the point that they may be isolated and subsequently subjected to a variety of carbon-based nucleophiles. The overall vicinal difunctionalization of two-ring carbons may be accomplished with excellent stereo- and regio-control.

The development of $Os(NH_3)_5(OTf)_3$ $(OTf = CF_3SO_3)$ by Magnuson and co-workers (*1*) was a turning point of sorts for the coordination chemistry of osmium. The publication of a high-yield synthesis for this compound offered a general route not only to osmium(III) analogs of the widely studied ruthenium(III)- and cobalt(III)pentaammine systems, but also to complexes of pentaammineosmium(II). The new entry into this d^6 system was of natural interest to chemists such as Taube (*2*), who had expected the heavy metal center to have a substantially greater tendency to participate in π backbonding than its ruthenium(II) congener. Perhaps even Taube was surprised to discover just how different pentaammineosmium(II) was from ruthenium.

Once a method was established for the reduction of triflatopentaammineosmium(III) in a noncoordinating solvent (*3*), the unusual nature of pentaammineosmium(II) began to emerge. This system, exhibiting chemistry common to both coordination and organometallic disciplines, had a propensity for forming stable π complexes with unsaturated organic ligands that was unparalleled by other transition metals (*4*). Most notably, arenes (*5*), naphthalenes (*6*), pyrroles, furans, thiophenes (*7*), and some pyridines (*8*) all form thermally stable dihapto-coordinate complexes with pentaammineosmium(II) that have been well-characterized (Scheme I).

The chemical nature of aromatic systems is profoundly affected by their coordination to transition metals (*9*). For example, in complexes such as (η^6-arene)Cr(CO)$_3$, and its cationic analogs [e.g., Mn(CO)$_3{}^+$, FeCp$^+$, and RuCp$^+$], the bound arenes are susceptible to nucleophilic addition, ultimately leading to substituted arenes or cyclohexadienes; the application of these systems to organic synthesis has been widely demonstrated (*10*). By contrast, at the onset of our investigation, virtually nothing was known about the chemistry of dihapto-coordinated aromatic systems. Although dihapto-coordinated arene complexes have been reported for several other transition metals including Re(I) (*11, 12*), Rh(I) (*13–15*), Ru(II) (*16*), Mn(I) (*17*), and Ni(0) (*18*), these complexes are generally unstable either to substitution or oxidative addition. In this regard, the pentaammineosmium(II) system shows unprecedented stability in solution. These complexes have solvent substitution half-lives on the order of hours to weeks at 20 °C. Thus, an exciting opportunity emerged to investigate how η^2-coordination alters the chemistry of the *uncoordinated* portion of an aromatic ligand.

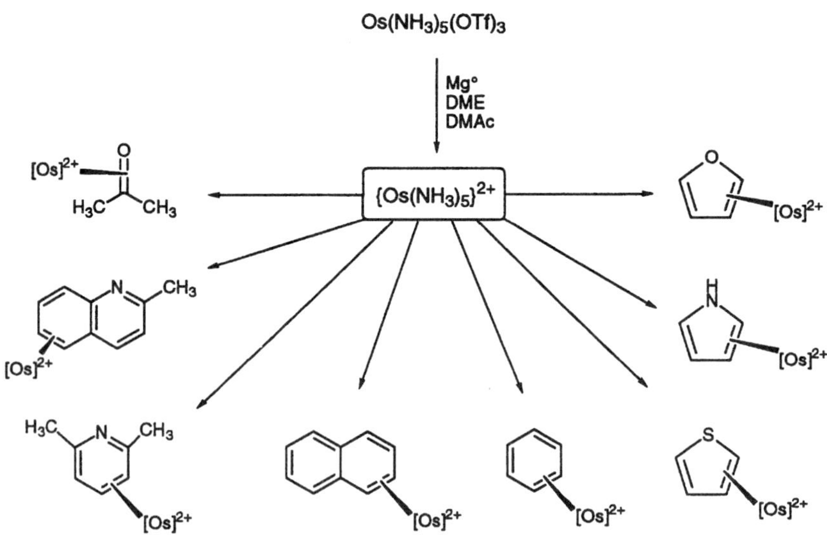

Scheme I. Selected complexes of pentaammineosmium(II) with unsaturated organic ligands ([Os]$^{2+}$ = [Os(NH$_3$)$_5$]$^{2+}$).

Elementary Transformations of Dihapto-Coordinated Arenes

By using dihapto-coordinated arenes in organic transformations, significant advantages were anticipated compared to the η^6-arene methodology. The coordination of only two carbons partially disrupts the aromaticity of the organic ligand and partially localizes the π-electron density within the uncoordinated portion of the ring so that it resembles a conjugated diene. In addition, because the stability of the arene–metal bond is considered to arise largely from a metal-to-ligand backbonding interaction, the electron-rich osmium metal center was anticipated to activate the uncoordinated portion of the bound arene toward electrophilic addition, a function complementary to the nucleophilic activation commonly observed in η^6 systems.

Quantification of the Metal–Arene Interaction. To assess the degree of interaction between the pentaammineosmium(II) moiety and an aromatic system, we embarked on a study concerning the equilibrium between the osmium complex of phenol and its cyclohexadienone isomers (Scheme II). Complexes of phenol (**1**), 2-methyl- (**2**), 3-methyl- (**3**), 4-methyl- (**4**), and 3,4-dimethylphenol (**5**) were prepared and spectroscopically characterized. In all cases, the metal binds the arene in an η^2-fashion across the unsubstituted ortho (C6) and meta (C5) carbons. When any one of these complexes is dissolved in CD_3OD containing a catalytic amount of DOTf, an equilibrium is established between the phenol species and their 2*H*- and 4*H*-phenol isomers. For the parent, the phenol isomer (**1a**) is thermodynamically favored by a ratio of 5:1 over its 4*H*-phenol counterpart. In contrast, when this experiment is repeated with the 2-methyl- (**2a**) and 3-methylphenol (**3a**) derivatives, the 4*H*-phenol isomers **2b** and **3b** are formed as major products ($K_{eq} = 3 \pm 0.5$). For 3,4-dimethylphenol, a 1:1 ratio of the η^2-2*H*-phenol and phenol is observed.

For the free organic system, the equilibrium between 2,4- and 2,5-cyclohexadien-1-ones and phenol heavily favors the aromatic species. Although no data are available in solution, these dienones have been generated by vacuum pyrolysis and partially characterized (*19*). More recently, Shiner et al. (*20*) have determined the gas-phase heats of formation for 2,5- or 2,4-cyclohexadien-1-one as $\Delta H_{298} = -13 \pm 3$ and -17 ± 3 kcal/mol, respectively, compared to the

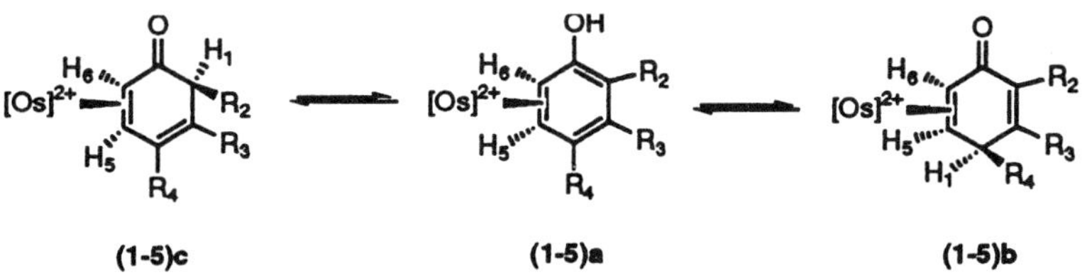

(1-5)c   (1-5)a   (1-5)b

Scheme II. The phenol–dienone equilibrium on pentaammineosmium(II).

corresponding value for phenol (–23 kcal/mol). Thus, ΔH for the tautomerization of phenol to the $2H$-phenol isomer (i.e., 2,4-dienone) is $\sim$6 kcal/mol, and relative to the $4H$-phenol isomer, phenol is favored by about 10 kcal/mol. Assuming that $T\Delta S°$ has a relatively small contribution to $\Delta G°$ at 25 °C (*21*), $\Delta H°$ for the conversion of phenol to 2,5-cyclohexadien-1-one on osmium(II) is close to zero. (Here, T is absolute temperature, and $\Delta S°$, $\Delta G°$, and $\Delta H°$ refer to the standard entropy, free energy, and enthalpy of reaction, respectively.) Thus, η^2-coordination reduces $\Delta H°$ of tautomerization for **1a** by about 10 kcal/mol relative to the free ligand.

In the limit of the formalism, osmium rehybridizes C5 and C6 to form a metallocyclopropane. This rehybridization effectively removes much of the resonance energy that stabilizes the enolic form of the free ligand. It is useful to consider as a model for the phenol complex **1a** the compound 1,3-cyclohexadien-1-ol, in which C–H bonds have replaced those to osmium (Scheme III). For the most part, thermodynamic data for enolization equilibria of dienol systems have been scarce. However, in a study by Pollack et al. (*22*), equilibrium constants for the enolization of 2- and 3-cyclohexenone were determined. From these equilibrium constants, $\Delta G°$ for the enolization of 2- and 3-cyclohexenone was calculated to be –10 and –7 kcal/mol, respectively. Again assuming that $T\Delta S°$ has relatively little contribution to $\Delta G°$ at 25 °C, $\Delta H°$ for the conversion of 1,3-cyclohexadien-1-ol to 2-cyclohexenone is about –10 kcal/mol. Finally, noting that gas-phase values of $\Delta H°$ and those of ΔG_{H_2O} for enolizations vary only by $\sim$2 kcal/mol, we conclude that complexation of pentaammineosmium(II) to phenol effectively reduces the aromatic character of phenol but does not destroy it.

Cyclic voltammetric data indicate that the Os^{III}/Os^{II} reduction potential is highly dependent on the isomeric form of the phenol ligand. Although an exact value for compounds **1a–5a** was not determined, an earlier study (*4*) found typical isomerization or substitution rates for pentaammineosmium (III) com-

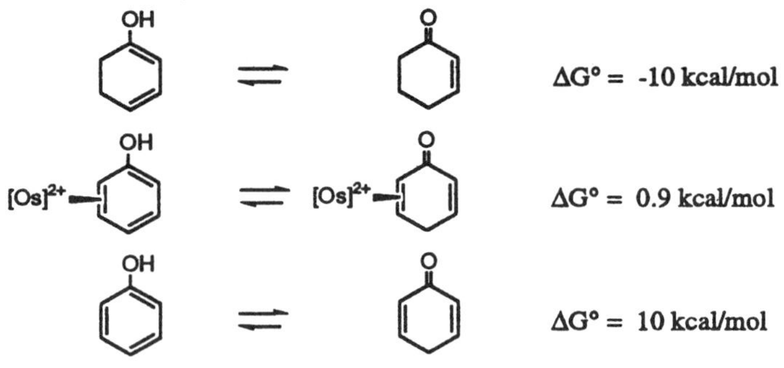

Scheme III. Free energy relationship for the phenol–dienone equilibrium of the η^2-phenol complex (1), uncomplexed phenol, and 1,3-cyclohexadienol.

plexes of η^2-arene complexes to lie from 10^1 to 10^3 s^{-1} . (For the complex [Os(NH$_3$)$_5$(η^2-benzene)]$^{2+}$ in CH$_3$CN, $E_{1/2}$ = 0.31 V (NHE) (200 mV/s), where $E_{1/2}$ is half-wave potential and NHE is normal hydrogen electrode (*4*).]Treating the oxidation of the compounds **1a–5a** as an E_rC_i process (i.e., an electrochemically reversible step followed by an irreversible chemical step) with k = 10^1–10^3 s^{-1} (*23*), a value for the reduction potential $\varepsilon°$ can safely be assigned to that of $E_{p,a}$ ± 100 mV. If we take the phenol complex as a model, the reduction potential of the arene form (**1a**) is observed about 600 mV lower than that of the 4*H*-phenol isomer (**1b**). With these two half-reactions in hand, knowledge of the keto–enol equilibrium constant for Os(II) allows for the determination of the corresponding value for Os(III) (Scheme IV).

Thus, the phenol complex is 13.8 ± 0.8 kcal/mol more stable than its 4*H* isomer at 25 °C. Remarkably, this value represents the same aromatic stabilization as reported for phenol in the gas phase, within experimental error, and indicates that the more electron-deficient Os(III) is completely ineffective in stabilizing the dienone isomer compared to the arene. Thus, the act of η^2-coordination alone does not appear to affect the phenol–dienone equilibrium. Rather, it is the increased π-acidity of the dienone ligand that stabilizes the osmium(II) complex, and this strong π interaction shifts the equilibrium away from the aromatic form upon coordination.

Alkylation and Dearomatization of Electron-Rich Arenes. When an acetonitrile solution of **1** is treated with 1.0 equiv of methyl vinyl ketone (MVK), the solution becomes deep red. Upon addition to ether, a tan salt is obtained (**6**) whose ^{1}H and ^{13}C NMR data show four olefinic doublets over the range of 3.8–6.6 parts per million (ppm) and a ^{13}C carbonyl resonance at 199.9 ppm, features that indicate the formation of a 4*H*-phenol species. In addition,

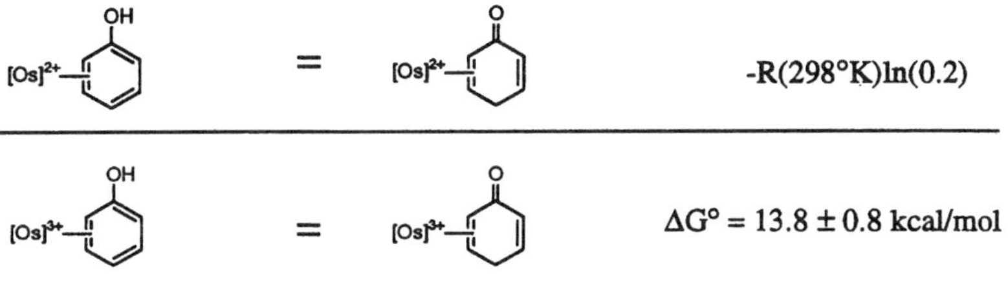

Scheme IV. Determination of $\Delta G°$ for the phenol–dienone isomerization on Os(III).

two sets of methylene resonances plus a methyl and a second carbonyl resonance (209 ppm) are diagnostic for the 3-oxobutyl group. Infrared data show
two carbonyl groups, with corresponding C=O stretching frequencies of 1636
and 1701 cm^{-1}. A cyclic voltammogram of this compound indicates a reversible
one-electron oxidation at 0.93 V (NHE), characteristic of a pentaammine-
osmium(II) complex of an electron-deficient olefin. Taken together these data
are consistent with a single diastereomer (d.e. > 90%) of a 4-alkylated 4H-phe-
nol complex, the C4 conjugate addition product of the electrophile and metallated phenol (Scheme V). The absence of coupling between H5 and H4 and a
substantial nuclear Overhauser effect (NOE) between H4 and the *cis*-ammine
protons indicate that conjugate addition occurs to the α-face of the phenol
ring, opposite to that of metal coordination as shown in Scheme V. (With the
exception of steroid systems, the metal is depicted as β-coordinated.)

Compound **6** is stable in solution, showing no signs of decomposition even
after standing in acidic CH$_3$CN solution for 24 h. However, treatment of **6** with
a moderate base, such as a tertiary amine, results in rearomatization to the 4-
substituted phenol complex (**7**). From a synthetic organic perspective, the reaction outlined in Scheme V is remarkable. Under basic conditions the reaction
of free phenol and MVK leads to alkylation at oxygen exclusively. In the presence of strong Lewis acids, some alkylation of the ring does occur, but problems with multiple alkylation and polymerization limit the value of this
approach. Perhaps the most intriguing and unexpected aspect of the reaction
shown in Scheme V is that in contrast to the 4H-phenol complexes generated
from isomerization of the phenol ligands (**2b** in Scheme II), compound **6**
resists isomerization to the arene in acidic solution. This stability is a direct
result of the stereochemistry of the C4 substituent that, in compounds like **6**,
blocks potential bases from accessing the C4 proton (*24*).

By a method similar to that described for the formation of **6**, the η^2-phe-
nol complex (**1**) undergoes conjugate addition reaction at C4 with a variety of
Michael acceptors (Scheme VI) including those with β-substituents such as in
the formation of compounds **9** and **10**. In most cases, the addition reaction is
most conveniently carried out with an amine base as catalyst. (Many of these

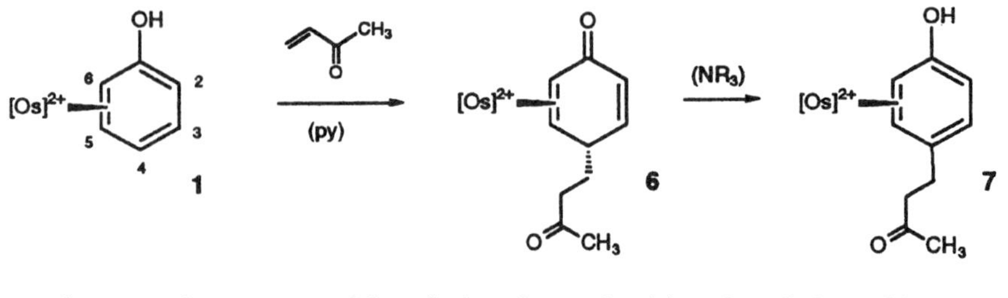

*Scheme V. The reaction of the η^2-phenol complex (1) and methyl vinyl ketone
(MVK).*

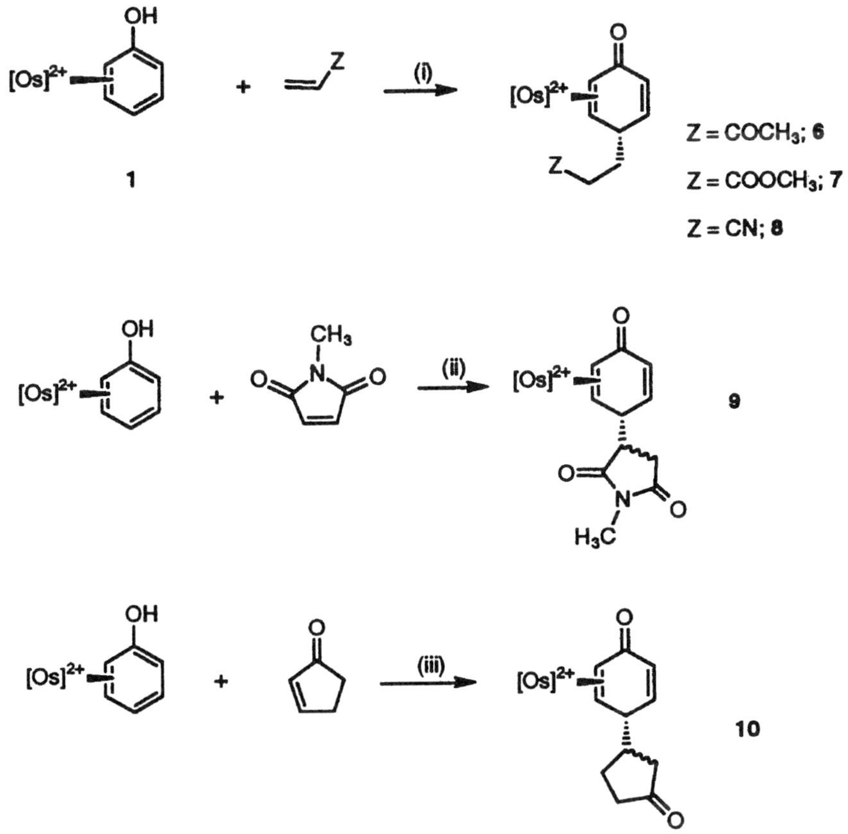

Scheme VI. Conjugate addition reactions with the η^2-phenol complex (1) and various Michael acceptors: (i) 6: MVK/pyridine/CH$_3$CN; 7: methyl acrylate/ Zn(OTf)$_2$/ diisopropylethylamine (DIEA)/CH$_3$CN; 8: acrylonitrile/Zn(OTf)$_2$/DIEA/CH$_3$CN; (ii) 9: N-methylmaleimide/pyridine/CH$_3$CN; (iii) 10: 2-cyclopenten-1-one/pyridine/CH$_3$CN.

reactions may be carried out in the absence of base, but reaction times are considerably longer.) Less reactive electrophiles such as methyl acrylate or acrylonitrile fail to undergo conjugate addition with the phenol complex 1 in the presence of base alone. However, in the presence of a Lewis acid co-catalyst, conjugate addition may be accomplished in good yield (e.g., 7 and 8 in Scheme VI).

A particularly nice example of the versatility of this reaction is shown in Scheme VII, in which the aromatic steroid β-estradiol is complexed to form 11 and subsequently alkylated at C10 (i.e., para) exclusively at −40 °C to give 12 (25). Because the osmium preferentially binds the α-face of the steroid, conju-

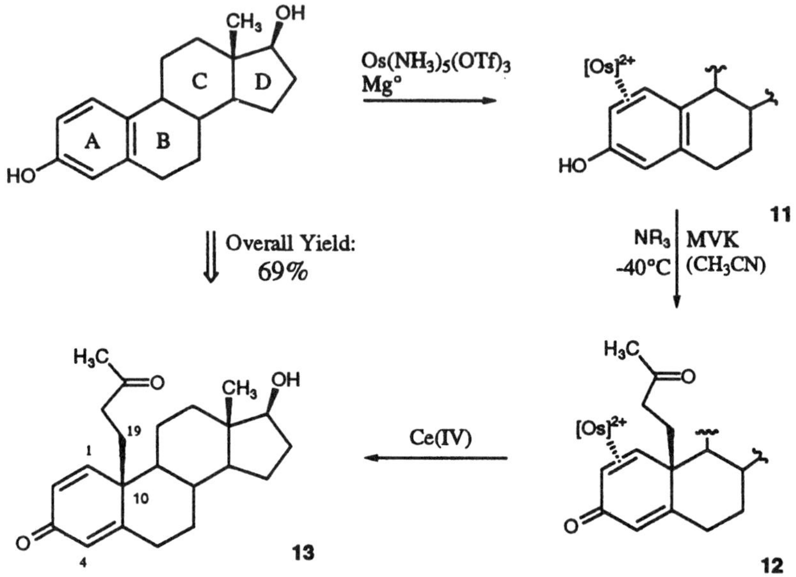

Scheme VII. The C10 alkylation of β-estradiol using osmium(II) to dearomatize the A-ring.

gate addition occurs from the more congested β-face, providing the natural stereochemistry of testosterones. The overall yield of this transformation *after* decomplexation of the dienone product (**13**) is about 70%, for a reaction that has no synthetic counterpart.

As initial deprotonation of the heteroatom with a moderate base is not facile for anilines and impossible for anisoles, the development of electrophilic addition reactions for osmium(II) complexes of these ligands must rely on Lewis acids. We have found that $BF_3 \cdot OEt_2$, $Sn(OTf)_2$, and TBSOTf (TBS = *tert*-butyldimethylsilyl), among others, are tolerated by the metal and may be used to catalyze a wide variety of conjugate addition reactions. Several of these reactions are illustrated in Scheme VIII. For the *N*-ethylaniline complex **14**, the only acid present is the anilinium species itself (**15**), prepared in water.

In the presence of $BF_3 \cdot OEt_2$, anisole complexes such as **19** initially undergo Michael addition at C4 to generate an oxonium–boron enolate (Scheme IX). When the Michael acceptor is *N*-methylmaleimide, this intermediate, characterized by low-temperature NMR, either protonates to give a 4*H*-anisolium species (**21**) or, alternatively, undergoes a ring closure at C1 to provide compound **23**, what is formally a cycloaddition product of the arene and olefin (**26**). Given that the organic ligand can readily be removed from the metal, such a reaction (formally a Diels–Alder reaction) constitutes a potentially valuable new synthetic approach to highly functionalized bicyclo[2.2.2]-

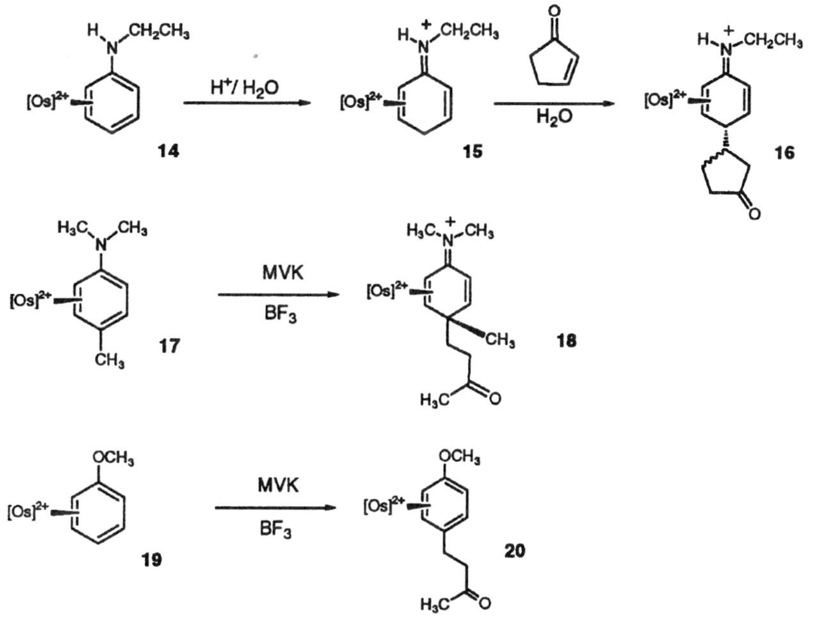

Scheme VIII. Conjugate addition reactions with η^2-aniline and η^2-anisole complexes.

octadienes. If the pentaammineosmium cycloaddition product (**23**) is allowed to stand, loss of ammonia occurs to generate the η^4-coordinate tetraammineosmium analog (**24**).

When the anisole ligand is substituted at C4 (i.e., **25**), Michael addition still occurs para to the methoxy group in Scheme X. Here, because deprotonation on the ring cannot occur, the reaction with MVK generates a 4*H*-anisolium species (**27**) that is stable enough to be isolated as a triflate salt. On the other hand, when 3-butyne-2-one is used as the Michael acceptor, two additional outcomes are observed. Provided that proton sources are carefully excluded, the initially formed boron enolate closes to form an η^2-barrelene complex (**26**) analogous to that shown in Scheme IX. Alternatively, in the presence of a Brønsted acid, the enolate is protonated, and the resulting 4*H*-anisolium species, upon warming to 20 °C, undergoes an alkyl migration of the enone from C4 to C3 to form the disubstituted anisole species **29** in a reaction analogous to the acid-catalyzed cyclohexadienone–phenol rearrangement (Scheme X) (*27a*).

Perhaps the greatest synthetic potential of this dearomatization technology lies in the ability of the metal to stabilize, through metal-to-ligand backbonding, the dienone, anilinium, and anisolium intermediates derived from phenol, aniline, and anisole, respectively. This stabilization sets the stage for subse-

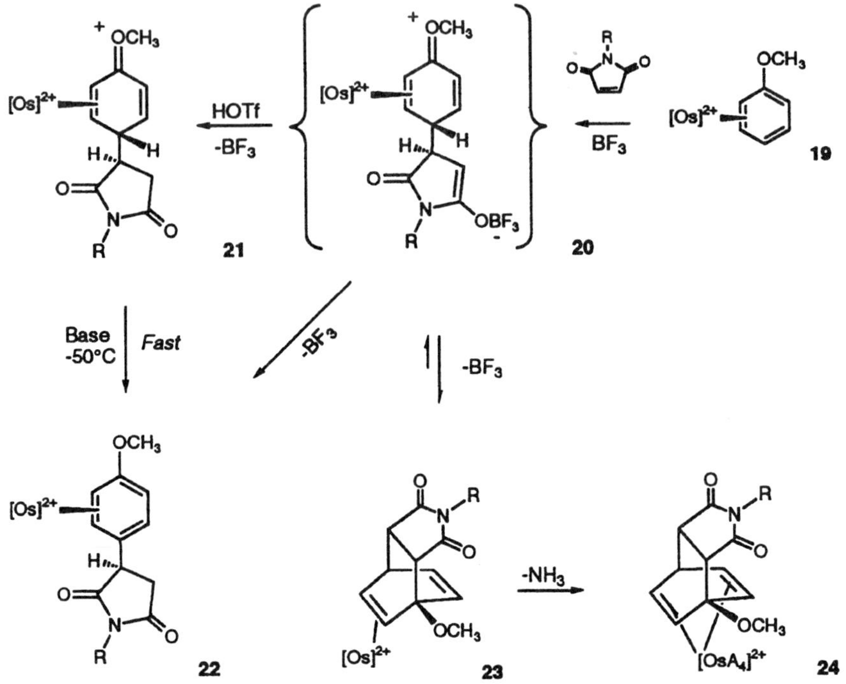

*Scheme IX. A cycloaddition reaction with the η^2-anisole complex (**19**).*

quent nucleophilic addition to the meta position, thereby preventing rearomatization of the ring upon decomplexation of the metal. Examples of both an intra- and intermolecular nucleophilic addition to C3 are shown in Scheme XI, including novel synthetic routes to benzopyran (**31**) and decalin (**32**) ring systems (*27a*).

The key features of the preceding section can be summarized as follows. Coordination of pentaammineosmium(II) to an arene with a single electron-donating substituent activates the arene towards electrophilic addition at C4 and stabilizes the 4*H*-arenium or arene species with respect to rearomatization; consequently, nucleophilic addition at C3 may be accomplished to give highly functionalized dienes. The steric bulk of the metal requires both electrophilic and nucleophilic additions to occur anti to the metal, providing predictable stereocontrol.

Elementary Transformations of Dihapto-Coordinated Heterocycles

Pentaammineosmium(II) forms stable dihapto-coordinated complexes with furans and pyrroles in which the metal binds across C4 and C5. Such coordina-

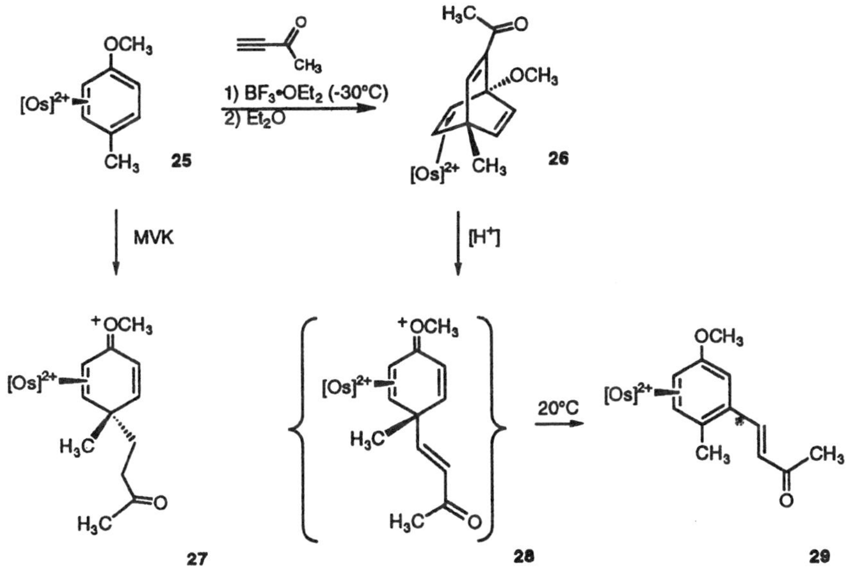

Scheme X. A cycloaddition, retroaldol, and cyclohexadienonium/anisolium rearrangement resulting from the reaction of 3-butynone and the 4-methylanisole complex (25).

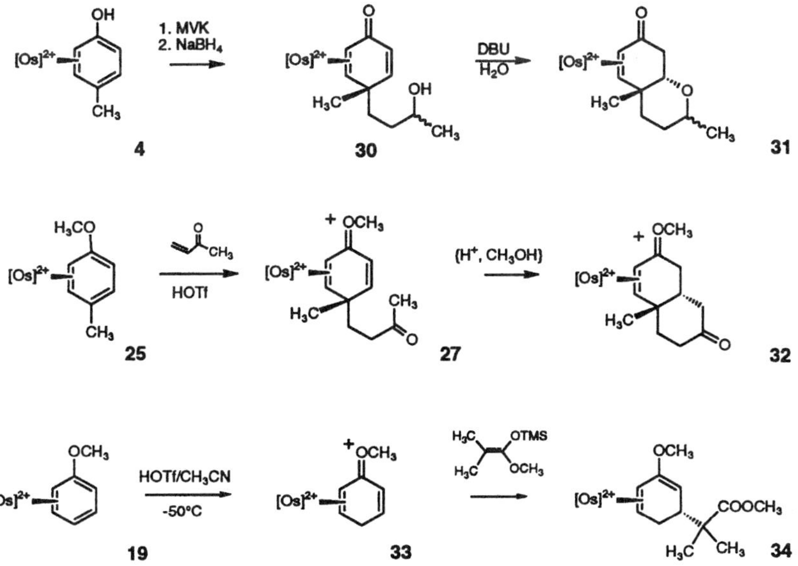

Scheme XI. The osmium(II)-promoted vicinal difunctionalization of the C3–C4 bond in arenes.

tion activates the uncoordinated β-carbon toward electrophilic addition, and, as with the arene chemistry, the resulting addition product, now a 3*H*-pyrrolium or 3*H*-furanium species, is stabilized by the metal backbonding interaction.

η²-Pyrrole Complexes. *Quantification of the Metal–Pyrrole Interaction.* Complexes of pyrrole (**35**), *N*-methylpyrrole (**36**), and 2,5-dimethylpyrrole (**37**) (Chart I) can readily be reversibly protonated at C3 to form stable 3*H*-pyrrolium complex (*27b*). In the case of the 2,5-dimethylpyrrole complex (**38**), the 3*H*-pyrrolium ligand converts over time to its 2*H*-pyrrolium isomer, in which the osmium now coordinates C3 and C4 (**39**). This 2*H*-pyrrolium species in turn undergoes reversible deprotonation at nitrogen to form the neutral 2*H*-pyrrole complex **40** (Scheme XII).

Comparison of the pK_a values shown in Scheme XII with those reported for various enamines and pyrroles demonstrates the extent that the osmium modifies the pyrrolic ligand upon complexation. Thermodynamic protonation of simple enamines occurs at the β-carbon (*28*), analogous to compounds **35–37**. Typically, the conjugate iminium ions have pK_a values ranging from 9 to 12. For comparison, the iminium ion derived from 2,3-dihydro-1,3,4-trimethylpyrrole has a reported pK_a of 9.6 (*28*), a value similar to that of the 2,5-dimethylpyrrolium complex **38**, in which pK_a = 7.5. As the extent of alkylation is reduced, the pK_a for the pyrrolium complexes decreases. The parent species [Os(NH$_3$)$_5$(3*H*-pyrrolium)]$^{3+}$ (**42**) exhibits a pK_a of just over 4, in dramatic contrast to the uncoordinated 3*H*-pyrrolium ion that is about 10 orders of magnitude more acidic (pK_a ~ –5.9) (*29*).

The oxidation waves recorded in a voltammogram of the pyrrole species **35** and that recorded for its β-protonated analog (**42**) differ by more than 1 V, and the protonated form is substantially more difficult to oxidize. Although the irreversible nature of these waves prevents the exact determination of the formal reduction potentials, a crude approximation for the acidity of the osmium(III)-3*H*-pyrrolium species **43** can be obtained (Scheme XIII) by equating the reduction potentials for **35** and **42** with their anodic peak potentials. Combining the two half-reactions corresponding to these redox processes with the pK_a for **42** gives an estimate of –14 for the pK_a of a β-protonated pyr-

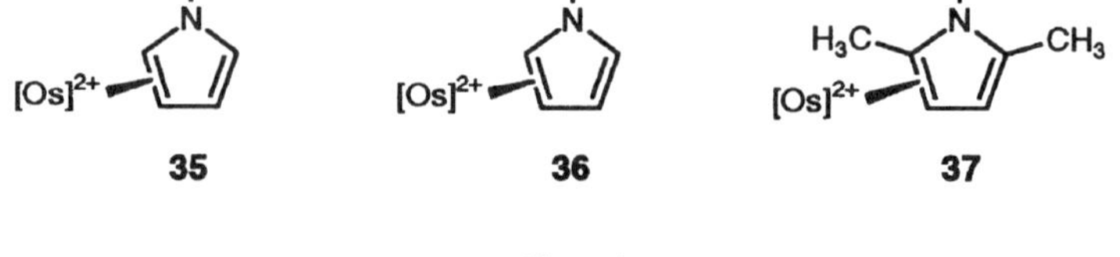

35   **36**   **37**

Chart I.

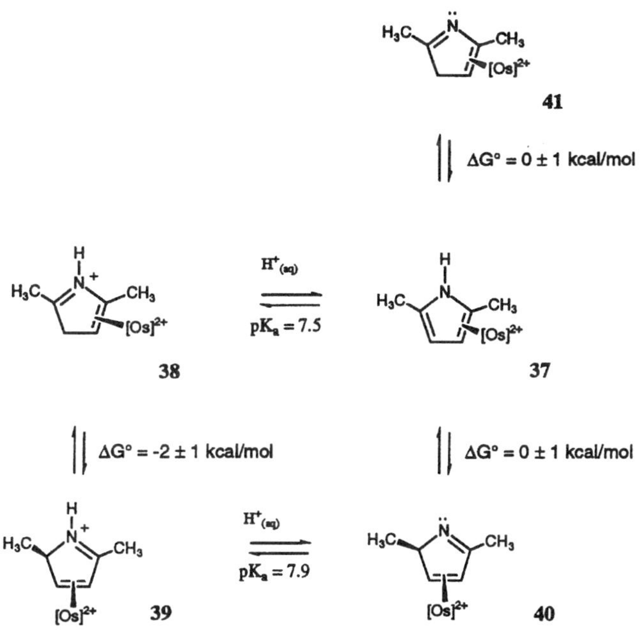

41

$\Delta G° = 0 \pm 1$ kcal/mol

$H^+_{(aq)}$

$pK_a = 7.5$

38

37

$\Delta G° = -2 \pm 1$ kcal/mol

$\Delta G° = 0 \pm 1$ kcal/mol

$H^+_{(aq)}$

$pK_a = 7.9$

39

40

Scheme XII. Thermodynamic relationships between a η^2-1H-pyrrole complex, its 2H- and 3H-isomers, and their corresponding conjugate acids.

role coordinated to osmium(III) (Scheme XIII). Thus, the one-electron oxidation of osmium results in an increase in acidity of close to 18 orders of magnitude. Presumably, the energy associated with rearomatization dominates for the higher oxidation state, in which the interaction of the metal with the heterocycle π system is poor. By contrast, in Scheme XIV a comparison of acidities is made between the osmium(III) and osmium(II) complexes of the 2H-pyrrolium species **39**. In this case, neither acid nor conjugate base are aromatic, and thus the 9-order increase in magnitude of acidity upon oxidation from osmium(II) to osmium(III) is due solely to the decrease in metal-to-ligand π-interactions and electrostatic effects.

Another useful indicator of metal–pyrrole interaction in these η^2-pyrrole complexes is found in the isomerization energies of the osmium(II) complexes (Scheme XII) (*27b*). By utilizing pK_a data, the isomerization energy for the conversion of the 2,5-dimethyl-1H-pyrrole complex (**37**) to the corresponding 2H-pyrrole species (**40**) can be estimated to be $\Delta G° = 0 \pm 1$ kcal/mol. (The details of this calculation are worked out in reference 27b.) Additionally, in a protic solvent such as water, the 1H-pyrrole complex of 2,5-dimethylpyrrole is in measurable equilibrium with its 3H-isomer (*30*). Thus, as indicated in Scheme XII, the 1H-pyrrole, 2H-pyrrole, and 3H-pyrrole isomers are practically isoergic at 25 °C. By contrast, the corresponding (pyrrole → pyrrolenine) isomer-

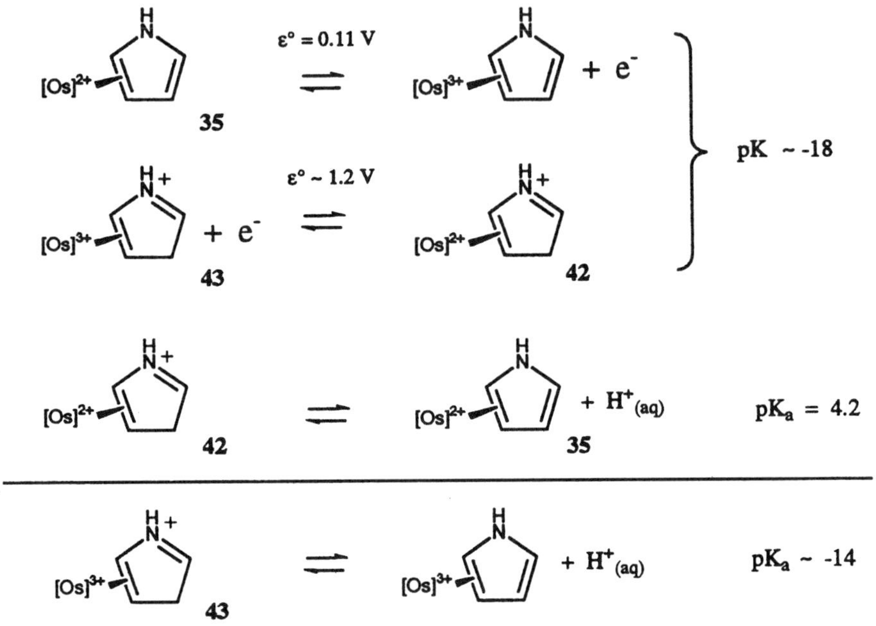

Scheme XIII. Determination of the acidity (pK$_a$) of an η^2-3H-pyrrolium complex of pentaammineosmium(III).

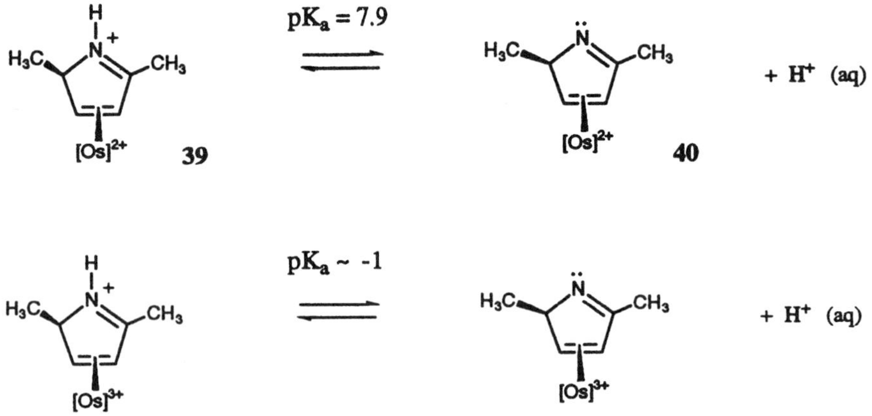

Scheme XIV. Comparison of the acidities for a 2H-pyrrolium ligand bound to osmium(II) and osmium(III).

ization energy has not been experimentally determined, but INDO (*31*) and MINDO/3 (*32*) calculations put this value at 14 and 19 kcal/mol, respectively (average ~ 16 kcal). Thus, the trivalent metal is completely ineffective in stabilizing the 2*H*-pyrrole tautomer over the 1*H*-pyrrole form. However, the highly π-basic osmium(II) metal center causes a dramatic shift in the pyrrole–pyrrolenine equilibrium, effectively erasing 16 of the estimated 20 kcal of resonance stabilization in the 1*H*-pyrrole (*33*).

Electrophilic Additions at the β-Carbon of Pyrrole. The studies just mentioned highlight the degree to which the electron-rich pentaammineosmium(II) system alters the reactivity of the pyrrole ring. In agreement with these chemical observations, crystallographic data indicate that the η²-coordinated pyrrole complexes closely resemble enamines and are expected to undergo electrophilic addition at the β-carbon (*34*). Several examples of this reaction are shown for carbon-based electrophiles in Scheme XV (*35*). This chemistry is in marked contrast to the chemistry of the uncoordinated pyrrole

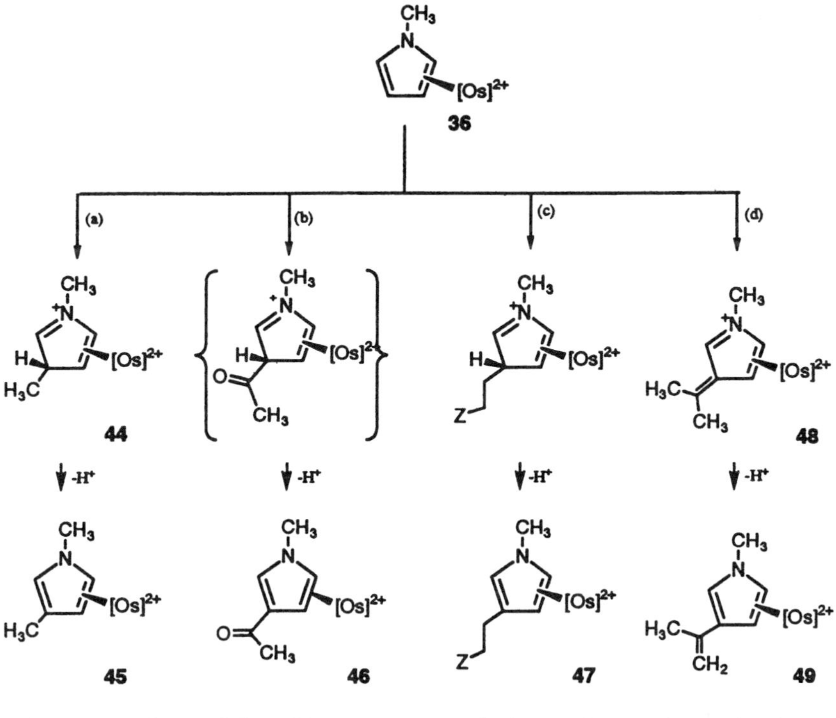

Scheme XV. Electrophilic addition reactions of η²-pyrrole complexes of Os(II). [Os]²⁺ = [Os(NH₃)₅]²⁺. (a) CH₃OTf; (b) Ac₂O [dimethylamino)pyridine (DMAP)]; (c) CH₂=CHZ (Z = C(O)CH₃; COOCH₃; CN); (d) (CH₃)₂CO/TBSOTf.

ligands, in which electrophilic addition and substitution occur preferentially at the α-carbons and multiple alkylated side products are commonly encountered. [For example, when 1-methylpyrrole is treated with 1 equiv of methylacrylate and TBSOTf, five different alkylated pyrrole products are isolated (29).] Several β-substituted pyrrole complexes prepared in a similar fashion to those shown in Scheme XV have proven to be valuable synthons for highly functionalized indoles as shown in Scheme XVI. The β-vinylpyrrole complexes in Scheme XVI are kinetically stable as their pyrrole-bound forms even though the vinyl-bound linkage isomer is heavily favored thermodynamically.

Dipolar Cycloaddition Reactions. Symmetric Os(II)–pyrrole complexes are most stable as the 4,5-η^2 isomer, but these species are in equilibrium with a minor isomer in which the metal is coordinated across C3 and C4. In this configuration, the osmium serves to "isolate" four ring electrons in an azomethineylide configuration, and as a consequence, the 3,4-η^2-pyrrole isomer readily undergoes 1,3-dipolar cycloadditions (20 °C, 1 atm) with electron-deficient olefins to form coordinated 7-azabicyclo[2.2.1]heptenes (36). The cycloaddition reaction is carried out typically in 80–90% yield and is highly stereospecific. In all cases examined, the electrophile adds to the face of the pyrrole ring opposite to metal coordination, and in cases in which the pyrrole nitrogen is not substituted, an exo-addition product dominates, often by >10:1 (Scheme XVII). (These assignments are based on both X-ray crystallographic and NOE

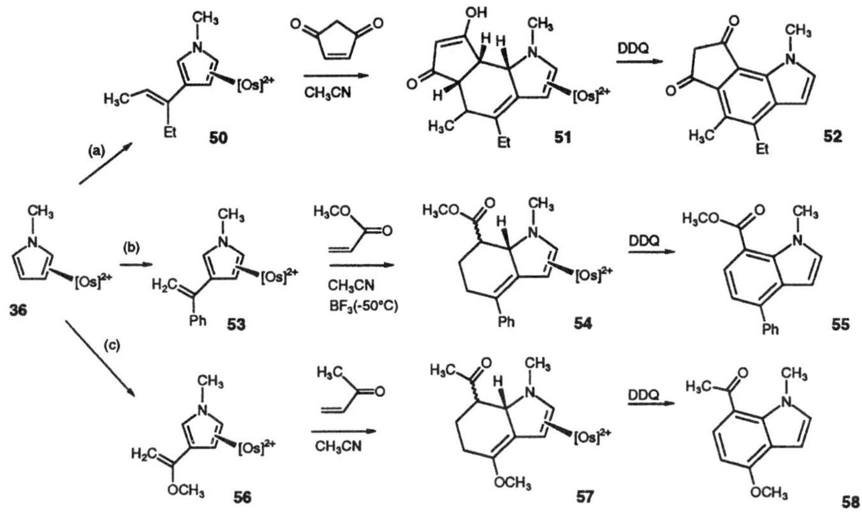

Scheme XVI. Preparation of polysubstituted indoles from pyrrole using pentaammineosmium(II). (a) diethyl ketone/TBSOTf/1,8-diazabicyclo[5.4.0]undec-7-ene (DBU); (b) acetophenone/TBSOTf/DBU; (c) Ac₂O(DMAP)/CH₃OTf/DBU. DDQ is 2,3-dichloro-5,6-dicyanoquinone.

data.). Although these complexes are stable in solution over extended periods of time, the organic ligand itself is highly susceptible to a retro-Diels–Alder reaction to give back pyrrole and olefin. To prevent this decomposition, the azanorbornene ligand is first protonated at nitrogen and then decomplexed and hydrogenated to yield the stable 7-azanorbornane (Scheme XVIII).

Cycloadduct complexes such as **61** may also be used as intermediates in the synthesis of the pyrrolizidine ring system originating from net electrophilic addition at C2. Treatment of **61** with TBSOTf results in a clean retro-Mannich reaction producing the complex **67**, after hydrolysis, in 98% yield (Scheme XIX). This metal-stabilized 2H-pyrrolium ligand may then be reduced stereo-selectively to the corresponding 3-pyrroline complex (**68**) through the use of a hydride reagent (e.g., $NaBH_4$; d.e. > 90%). NOE studies are consistent with both cycloaddition and hydride addition occurring from the face of the ring anti to the metal moiety. Thus, hydride is added stereoselectively to the more congested face of the ligand. Heating the 3-pyrroline complex **68** effects decomplexation, and subsequent aqueous workup induces ring closure to the γ-lactam to give a single diastereomer of the pyrrolizidine nucleus **69** (overall isolated yield from **61** = 65%). Here the unnatural angular substituent at C8 is conveniently established from α-substitution of the pyrrole precursor.

η^2**-Furan Complexes.** As with pyrroles, η^2-furan complexes of penta-ammineosmium(II) undergo accelerated electrophilic addition at the β-carbon. Although far less stable than the 3H-pyrrolium complexes previously

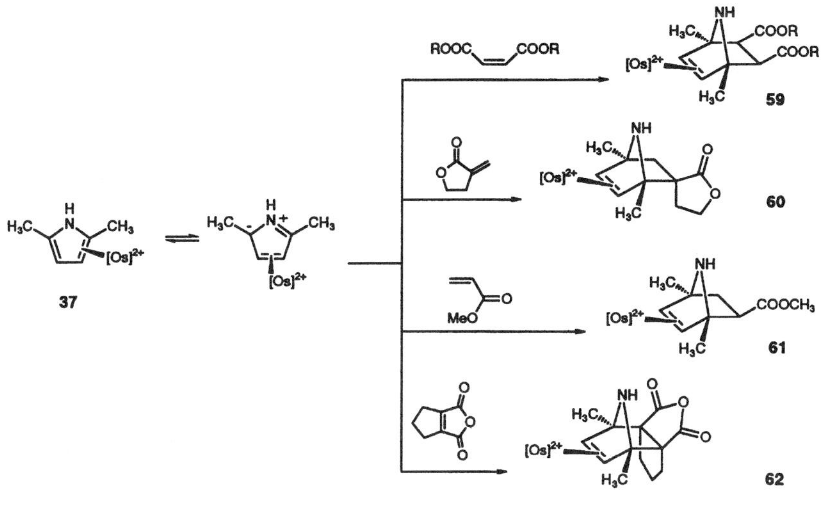

Scheme XVII. Assorted dipolar cycloaddition reactions with η^2-pyrrole complexes. [Os]$^{2+}$ = [Os(NH$_3$)$_5$]$^{2+}$.

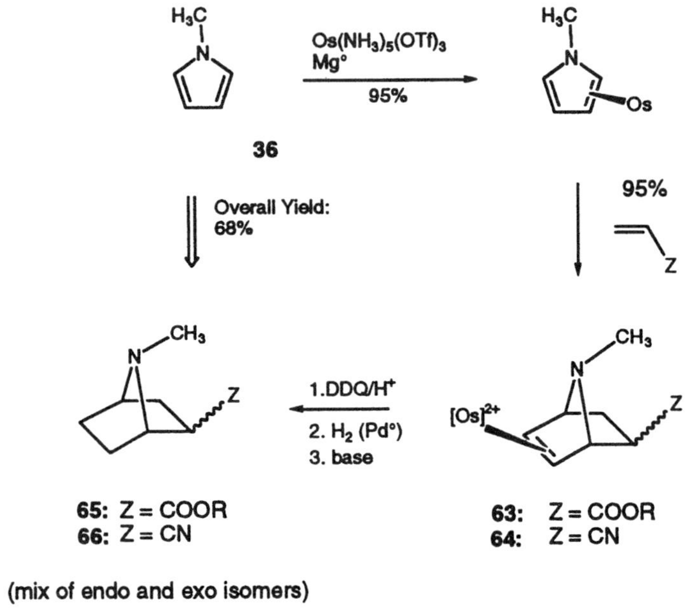

Scheme XVIII. Formation of 2-substituted azabicycloheptanes from N-methyl-pyrrole.

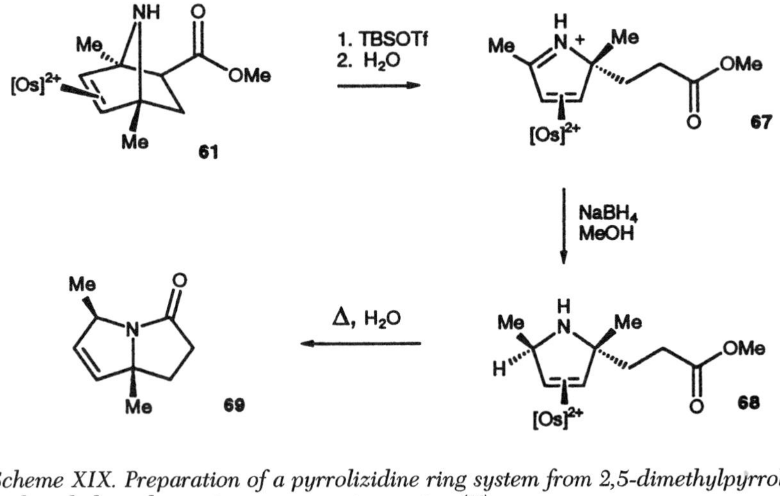

Scheme XIX. Preparation of a pyrrolizidine ring system from 2,5-dimethylpyrrole and methyl acrylate using pentaammineosmium(II).

described, the 3*H*-furanium ligand is stabilized by metal-to-ligand backbonding to a degree that allows nucleophiles to be added to the α-carbon without the complication of deprotonation at the β-carbon. In Scheme XX, several examples of the vicinal difunctionalization of η^2-furans are shown (37, 38).

Beyond Pentaammineosmium(II)

A long-range goal of our program is to develop other π-basic metal systems that mimic and ultimately improve on the ability of pentaammineosmium(II) to reversibly "dearomatize" aromatic molecules through η^2-coordination. By varying the electronic and steric qualities of the metal center, the affinity for aromatic ligands may be optimized, a chiral coordination environment could

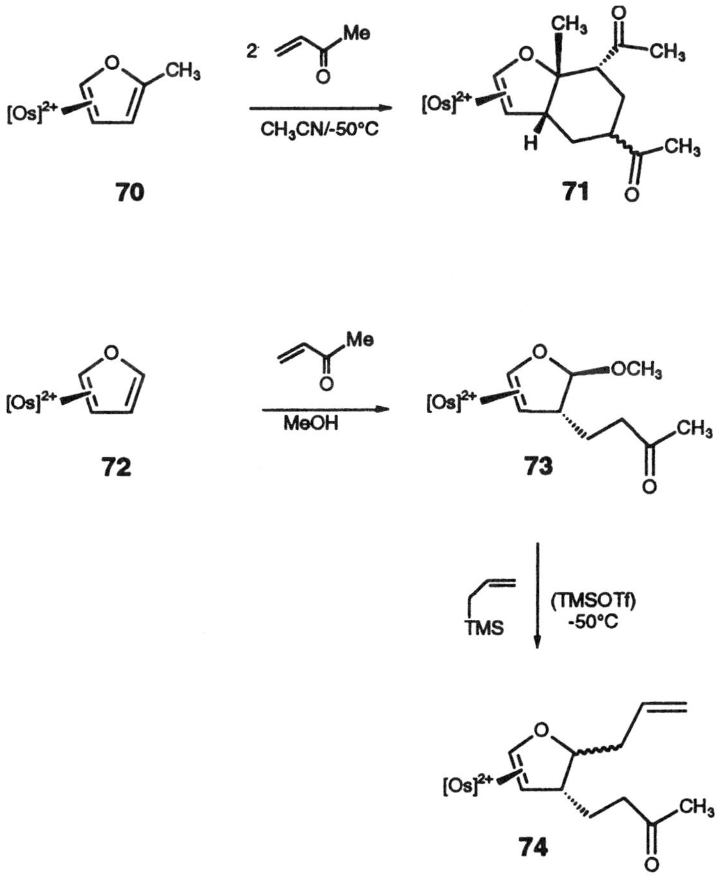

Scheme XX. Intra- and intermolecular examples of the vicinal difunctionalization of an η^2-furan ligand.

be prepared for the added feature of asymmetric induction, and metal–aromatic bond strengths could be adjusted to facilitate the development of a catalytic process.

After five years of exploring this issue, it has become clear that replicating the activity of the pentaammineosmium(II) system is not a trivial task. Pentaammineruthenium(II) triflate (39) and pentaammineiridium(III) triflate (40), species that are isoelectronic to the osmium(II), fail to form stable η^2-coordination complexes with aromatic systems. Presumably, the lower orbital energies in these systems do not provide sufficient π-interactions with an arene to overcome the loss of resonance energy that would accompany complexation. Furthermore, our continuing investigation into the coordination chemistry of rhenium indicates that chloropentaaminerhenium(I) systems have reducing potentials that are less than –2.0 V (NHE) and are unstable in solution, with or without arene present (41). Yet, examples of η^2-complexation to arenes have been reported for several other transition metals with organometallic ancillary ligands (11–16), although none appears to be stable enough to be used as dearomatization agents in organic synthesis. In all of these cases, however, the ancillary ligand set is considerably more bulky than in the pentaammineosmium(II) system, and this steric factor is likely to play a major role in the stability of the desired η^2-arene systems.

To better understand to what extent steric effects alter the stability of an η^2-arene system, we prepared a series of *cis*- tetraammineosmium(II) complexes of the form $[\text{Os(NH}_3)_4(\text{N}_2)(\text{L})]^{2+}$, where L = NH$_3$, methylamine, propylamine, *tert*-butylamine, pyrroline, valine methyl ether, quinuclidine, pyridine, trimethylphosphine, and acetonitrile (42). From these dinitrogen species, the corresponding osmium(III) triflate analogs {i.e., $[\text{Os(NH}_3)_4(\text{L})(\text{OTf})](\text{OTf})_2$} were prepared and evaluated as potential synthons to dearomatization agents. In the cases of the weak π acids pyridine, trimethylphosphine, and acetonitrile, reduction in the presence of benzene or anisole failed to generate detectable amounts of any η^2-arene species. Reduction in the presence of excess arene, where L = a bulky amine (e.g., *tert*-butylamine or quinuclidine), produced binuclear complexes of the form $\{[\text{Os(NH}_3)_4(\text{L})]_2(\eta^2{:}\eta^2\text{-arene})\}^{4+}$ as the only products. Even the presence of a single *tert*-butyl group attached to a *cis*-amine is enough to increase the rate of arene substitution to the point that mononuclear arene complexes cannot be isolated. Even when L = a small amine (i.e., methylamine or propylamine), η^2-complexation is stifled but for a different reason. In this case, the osmium abstracts a β-hydrogen from the aliphatic amine, and an η^2-iminium hydride species is formed (Scheme XXI) (43). Although this species is in equilibrium with its pentaamine isomer in methanol, the equilibrium lies far enough toward the iminium hydride that η^2-complexation is thermodynamically disfavored for simple arenes. Thus, the stability of η^2-arene complexes is much more sensitive to small electronic and steric perturbations than we originally envisioned.

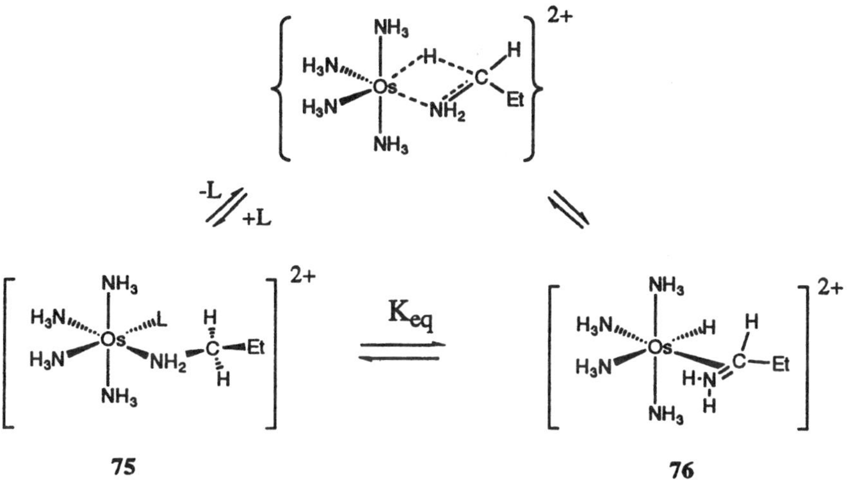

75   **76**

L = MeOH, CH$_3$CN, benzene

Scheme XXI. The reversible β-hydride elimination of an aliphatic amine bound to osmium(II).

Conclusions

Henry Taube's fascination with π-backbonding in transition-metal complexes has impacted a number of different areas in chemistry, and, in the present case, ultimately evolved into an investigation of pentaammineosmium(II) as a potential tool for organic synthesis. The series of studies highlighted in this chapter demonstrates that the electron-rich nature of the pentaammineosmium(II) system coupled with its low steric profile make it ideally suited for use as an η^2-complexation agent for aromatic systems. These complexes of arenes and aromatic heterocycles undergo dramatically enhanced ligand-centered reactions with carbon electrophiles and in this regard represent a fundamentally new method for the activation of aromatic systems.

References

1. Lay, P. A.; Magnuson, R. H.; Taube, H. *Inorg. Synth.* **1986**, *24*, 269.
2. Taube, H. *Pure Appl. Chem.* **1979**, *51*, 901.
3. Harman, W. D.; Taube, H. *Inorg. Chem.* **1987**, *26*, 2917.
4. Harman, W. D. Ph. D. Dissertation, Stanford University, 1987.
5. Harman, W. D.; Sekine, M.; Taube, H. *J. Am. Chem. Soc.* **1988**, *110*, 5725.
6. Harman, W. D.; Taube, H. *J. Am. Chem. Soc.* **1988**, *110*, 7555.
7. Cordone, R.; Harman, W. D.; Taube, H. *J. Am. Chem. Soc.* **1989**, *111*, 5969.
8. Cordone, R.; Taube, H. *J. Am. Chem. Soc.* **1987**, *109*, 8101.

9. Kane-Maguire, L. P. *Chem. Rev.* **1984,** *84,* 525.

10. Semmelhack, M. F. In *Comprehensive Organic Synthesis;* Trost, B. M., Ed.; Pergamon Press: Oxford, United Kingdom, 1991; Vol. 2, p 517.

11. Sweet, J. R.; Graham, W. G. *J. Am. Chem. Soc.* **1983,** *105,* 305.

12. Heijden, H.; Orpen, A. G.; Pasman, P. *J. Chem. Soc. Chem. Commun.* **1985,** 1576.

13. Jones, W. D.; Feher, F. J. *J. Am. Chem. Soc.* **1984,** *106,* 1650.

14. Jones, W. D.; Feher, F. J. *J. Am. Chem. Soc.* **1989,** *111,* 8722.

15. Chin, R. M.; Dong, L.; Duckett, S. B.; Partridge, M. G.; Jones, W. D.; Perutz, R. N. *J. Am. Chem. Soc.* **1993,** *115,* 7685.

16. Harman, W. D.; Taube, H. *J. Am. Chem. Soc.* **1988,** *110,* 7555.

17. Wang, C.; Lang, M. G.; Sheridan, J. B. *J. Am. Chem. Soc.* **1990,** *112,* 3236.

18. Brauer, D. J.; Krüger, C. *Inorg. Chem.* **1977,** *16,* 884.

19. Lasne, M. C.; Ripoll, J. L. *Tetrahedron Lett.* **1980,** *21,* 463.

20. Shiner, C. S.; Vorndam, P. E.; Kass, S. R. *J. Am. Chem. Soc.* **1986,** *108,* 5699.

21. Rappaport, Z.; *The Chemistry of Enols*; Wiley: Chichester, United Kingdom, 1990; p 341.

22. Pollack, R. M.; Blotny, G.; Dzingeleski, G. *J. Org. Chem.* **1990,** *55,* 1019.

23. Bard, A. J.; Faulkner, L. R. *Electrochemical Methods;* Wiley: New York, 1980; p 429.

24. Kopach, M. E.; Harman, W. D. *J. Am Chem. Soc.* **1994,** *116,* 6581.

25. Kopach, M. E.; Kelsh, L. P.; Stork, K. C.; Harman, W. D. *J. Am. Chem. Soc.* **1993,** *115,* 5322.

26. Kopach, M. E.; Harman, W. D. *J. Org. Chem.* **1994,** *59,* 6506.

27. (a) Kopach, M. E., Ph.D. dissertation, University of Virginia, 1995; (b) Myers, W. H.; Koontz, J. I.; Harman, W. D. *J. Am. Chem. Soc.* **1992,** *114,* 5684.

28. Cook, A. G. *Enamines, Synthesis, Structure and Reactions;* Marcel Dekker: New York, 1988; p 77.

29. Chadwick, D. J. In *Pyrroles: Part One*; Jones, R. A., Ed.; Wiley: New York, 1990.

30. Hodges, L. M.; Gonzalez, J.; Koontz, J. I.; Myers, W. H.; Harman, W. D. *J. Org. Chem.* **1995,** *60,* 2125. (In our original report, we misassigned the 3*H*-pyrrole isomer as a hydrated form of pyrrole; *see* reference 35.)

31. Catalán, J.; de Paz, J. G.; Sánchez-Cabezudo, M.; Elguero, J. *Bull. Soc. Chim. Fr.* **1966,** *103,* 429.

32. Karpfen, A.; Schuster, P.; Berner, H. *J. Org. Chem.* **1979,** *44,* 374.

33. Lloyd. D.; Marshall, D. R. *Chem. Ind. (London)* **1972,** 335.

34. Myers, W. H.; Sabat, M.; Harman, W. D. *J. Am. Chem. Soc.* **1991,** *113,* 6682.

35. Hodges, L. M.; Moody, M. W.; Harman, W. D. *J. Am. Chem. Soc.* **1994,** *116,* 7931. (*See* reference 30.)

36. Koontz, J. I.; Gonzalez, J.; Myers, W. H.; Hodges, L. M.; Sabat, M.; Nilsson, K. R.; Neely, L. K.; Harman, W. D. *J. Am Chem Soc.* **1995,** *117,* 3405.

37. Chen, H.; Hodges, L. M.; Liu, R.; Stevens, W. C.; Sabat, M.; Harman, W. D. *J. Am. Chem. Soc.* **1994,** *116,* 5499.

38. Liu, R.; Chen, H.; Harman, W. D. *Organometallics* **1995,** *14,* 2861.

39. Harman, W. D.; Taube, H. *J. Am. Chem. Soc.* **1988,** *110,* 7555.

40. Harman, W. D.; Barrera, J. unpublished results.

41. Orth, S. D.; Barrera, J.; Sabat, M.; Harman, W. D. *Inorg. Chem.* **1993,** *32,* 594.

42. Orth, S. D. Ph. D. Dissertation, University of Virginia, 1994.

43. Barrera, J.; Orth, S. D.; Harman, W. D. *J. Am. Chem. Soc.* **1992,** *114,* 7316.

A Novel, High-Yield System for the Oxidation of Methane to Methanol

Roy A. Periana

Catalytica, Inc., 430 Ferguson Drive, Mountain View, CA 94043

A novel, homogeneous system for the selective, low-temperature, catalytic oxidation of methane to methanol is reported. The net reaction catalyzed by mercuric ions, Hg(II), is the oxidation of methane by concentrated sulfuric acid to produce methyl bisulfate, water, and sulfur dioxide. The reaction is efficient. At a methane conversion of 50%, 85% selectivity to methyl bisulfate (~43% yield, the major side product is carbon dioxide) was achieved at a molar productivity of 10^{-7} mol/cm^3 s and Hg(II) turnover frequency of 10^{-3} s^{-1}. Separate hydrolysis of methyl bisulfate and reoxidation of the sulfur dioxide with air provides a potentially practical scheme for the oxidation of methane to methanol with molecular oxygen. This yield is the highest single-pass yield of methanol so far reported for a catalytic methane oxidation. The primary steps of the Hg(II)-catalyzed reaction were individually examined, and the essential elements of the mechanism were identified. The Hg(II) ion reacts with methane by an electrophilic displacement mechanism to produce an observable species, CH_3HgOSO_3H. Under the reaction conditions, CH_3HgOSO_3H readily decomposes to CH_3OSO_3H and the reduced mercurous species, Hg_2^{2+}. The catalytic cycle is completed by the reoxidation of Hg_2^{2+} with H_2SO_4 to regenerate Hg(II) and by-products SO_2 and H_2O. Thallium(III), palladium(II), and the cations of platinum and gold also oxidize methane to methyl bisulfate in sulfuric acid.

THIS CHAPTER HONORS HENRY TAUBE'S CONTRIBUTION to an important area of industrial research. The selective oxidation of methane to methanol is an important scientific and commercial objective and one that Catalytica has long sought to address through innovative chemistry. When I arrived at Catalytica, Taube was already there in the capacity of Scientific Advisor. He held this position while still maintaining an active academic role at Stanford University. This unusual relationship prompted me to ask him why he felt the need to interact with Catalytica at a time in his life when he could be enjoying the accolades of his career. His answer typifies the man and scientist: "I wanted to do something useful."

When I arrived Taube had already helped to lay the foundation for the approach to selective oxidation. His approach, as is characteristic of Taube's style, was simple and unambiguous. Our engineers told him that an economically competitive methane to methanol process would have to be carried out at high, one-pass methane conversion and selectivity. He realized that this would only be possible if a process could be found in which the methanol was less reactive than methane during the oxidation step. Realizing that designing a catalyst that exhibited such discrimination was an enormous task, Taube wondered instead if the methanol could be "protected" from the catalyst by chemical modification. The concept of protection is well-established in organic chemistry. For example, the nitro group slows or "protects" nitrobenzene from further electrophilic nitration relative to benzene. However, it was not clear whether this concept could be applied to a saturated system in which the bonds do not involve π-electrons.

Taube's fundamental approach now came to bear. He collected data (Figure 1) from the literature that showed unambiguously that electron-withdrawing groups on the methyl group would be expected to reduce the rate of the reaction of the substituted substrate relative to methane in reactions involving electrophiles. Thus, in reaction with hydroxy radicals, while methanol was four times more reactive than methane, acetonitrile was approximately two orders of magnitude less reactive! Taube noted that these effects should be much greater for two-electron oxidations. This insight led us to focus on the identification of catalysts for methane oxidation that operated via two-electron processes and for protecting groups that could slow the reaction of methanol relative to methane. This clear logic, and Taube's championing of the program, was pivotal in raising the funds to implement the research program. The results that I am about to discuss are some of the fruits of this program.

Catalytica has been involved in programs aimed at the development of new chemistry for the selective oxidation of methane. Our efforts are directed at replacing the current capital-intensive methane-to-methanol technology based on syngas with a more efficient, direct oxidation process that is less capital intensive. The most economic alternative to the syngas technology would be a hypothetical process for the direct, high-yield, one-step oxidation of methane to methanol. Economic evaluations indicate that even for such an idealized process, single-pass conversions in excess of 30% at greater than 80% selectivity are required for an economical process. Here, %conversion $\equiv \{([CH_4]_{initial} - [CH_4]_{final})/[CH_4]_{initial}\} \times 100$; %selectivity to $CH_3OH \equiv [CH_3OH]/([CH_4]_{initial} - [CH_4]_{final})\} \times 100$; and %yield $\equiv$ (%conversion) $\times$ (%selectivity). (High selectivity is essential because low selectivity results in the formation of CO_2 and the generation of heat. Removal of heat is a large part of the process costs. High con-

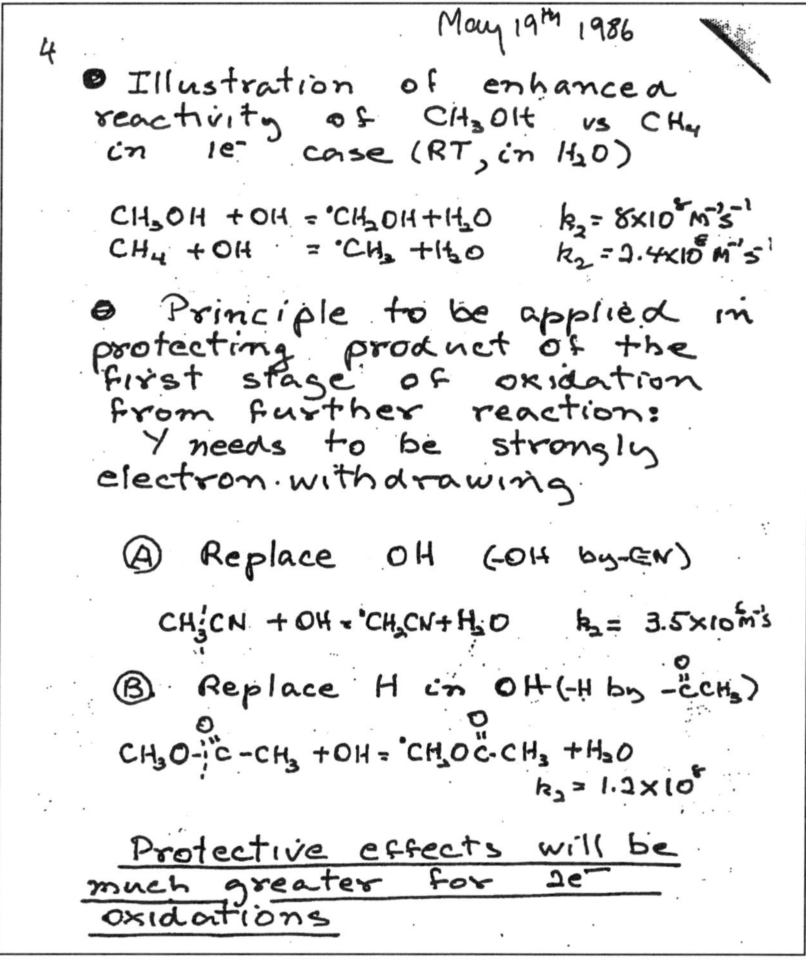

Figure 1. Henry Taube's original notes showing that methanol "protection" should be possible.

version minimizes methane recycling and increases plant efficiency.) Direct oxidation processes have been investigated, and in all attempts only low yields (~2%) have been reported (*1–4*). In this chapter we describe the development of a novel homogeneous catalytic system for the direct, selective oxidation of methane to methanol in ~43% yield per pass.

Novel Homogeneous Catalytic Systems

In the 1970s, new reactions were reported that allow reaction of methane at quite low temperatures without the intervention of free radicals (Figure 2).

Broadly classified, these reactions all involve two-electron redox changes that could be divided into reactions of methane with protons/carbonium ions developed by Olah (5, 6), and reactions of methane with metal complexes (7, 8) first illustrated by the work of Shilov (M = Pt; 7). The essential characteristics of these reactions are (1) the reactions occurred at low temperature (<100 °C); (2) electrophilic processes are important; (3) the reactions are homogeneous, and (4) efficient catalysis is lacking. We, as well as others, were very interested in the work with metal complexes because the low temperatures of reactions would favor high selectivity. However, it was clear that a major challenge would be to develop true catalytic systems that lead to functionalized methane products rather than exhibiting only C–H activation. ["Activation" is used to describe a process in which substitution of a stronger C–H bond (375–440 kJ/mol) occurs to produce a weaker metal–carbon bond (210–335 kJ/mol). "Functionalization" is used to refer to a process in which the metal–carbon bond is replaced by any bond except a C–H bond.] In many of the reactions reported, no catalysis was possible as the reactive species deactivated in the presence of oxidants. In other cases only a few turnovers were reported.

In our work we focused on the Shilov-type, electrophilic chemistry with Pt(II), Co(III), and Pd(II) (Figure 3), as these reactions with methane exhibited catalysis as well as produced "protected" products. With this chemistry, a conceptual scheme that allows the catalytic oxidation of methane could be envisioned, as shown in Scheme I. In this scheme, methane is oxidized in step 1 by a metal oxidant, a redox electrophile, to produce an ester of methanol, CH_3X, along with the reduced form of the catalyst MX_{n-2}. In step 2, the catalyst is regenerated by use of oxygen or another inexpensive, regenerable oxidant. In step 3, the methyl ester is hydrolyzed to produce methanol, and the acid is recycled. The net transformation is the oxidation of methane to methanol with oxygen.

Other researchers have expanded on this chemistry to include other metals, and the essential details of the claims from these studies are summarized in

- Superacid : $CH_4 + H^+ \longrightarrow CH_5^+$

- Carbonium : $CH_4 + CH_3^+ \longrightarrow CH_3\text{-}CH_3 + H^+$

- Transition metal : $L_nM + CH_4 \longrightarrow L_nM(CH_3)(H)$
  $MX_n + CH_4 \longrightarrow MX_{n-2} + CH_3X + HX$

- Rare earth : $MR + CH_4 \longrightarrow MCH_3 + RH$

- Metal-oxo : $_nM{=}O + CH_4 \longrightarrow L_nM(CH_3)(OH)$

Figure 2. Facile non-free-radical reaction with methane.

Figure 4. The Shilov chemistry (7) generates a mixture of methyl chloride and methanol from the reaction of Pt(II)/(IV) chlorides with methane in water.

Catalysis using heteropolyacids have been reported, but only a few turnovers are possible before the system deactivates. Sen (8) reported the stoichiometric oxidation of methane to methyl trifluoroacetate with Pd(II) and claims catalysis with the use of peracids. Moiseev and co-workers (9) have reported the Co(III)-catalyzed oxidation of methane to methyl trifluoroacetate using oxygen. Although the Pt(II)/(IV) work has been confirmed by several workers, we have been unable to reproduce the Sen and Moiseev work or significantly improve on these or the Shilov systems. Our results are shown in Figure 5. Although we found that we could increase the stoichiometric yields by changing the acid solvents, the reactions were very inefficient because no effective catalysis could be developed. These reactions are all presumed to operate by the simplified conceptual scheme shown in Scheme I. The main challenge in these systems seems to be the reoxidation of the reduced forms of the metal catalyst as shown in step 2 of Scheme I. On the basis of the known difficulties of reoxidizing the noble metals Pd and Pt as well as the powerful oxidizing abilities of Co(III), these results are perhaps not surprising.

$$MX_n + CH_4 \xrightarrow{< 160\ ^\circ C} CH_3X + HX + MX_{n-2}$$

M	X	% yield[a]	Reference
Pt(II)	Cl, OH	~15	Shilov, et al., *Kinet. Katal.* **24**, 486 (1983)
Pd(II)	CF_3CO_2	60	Sen, et al., *J. Am. Chem. Soc.* **109**, 8109 (1987)
Co(III)	CF_3CO_2	90	Moiseev, et al. *J. Chem. Soc. Chem. Commun.* 1049 (1990)

[a] based on metal added

- Catalysis reported with peracids (Pd) and O_2 (Co)

- Pd(II) results questioned by Moiseev

Figure 3. Shilov-type, electrophilic redox reactions with methane.

$$MX_n + CH_4 \longrightarrow CH_3X + HX + MX_{n-2}$$

$$MX_{n-2} + 1/2\ O_2 + 2\ HX \longrightarrow MX_n + H_2O$$

$$CH_3X + H_2O \longrightarrow CH_3OH + HX$$

$$\text{Net} \quad CH_4 + 1/2\ O_2 \longrightarrow CH_3OH$$

Scheme I. Possible scheme for a catalytic mechanism for the oxidation of methane to methanol.

$$MX_n + CH_4 \xrightarrow{< 160\ ^\circ C} CH_3X + HX + MX_{n-2}$$

M	X	% yield[a]	Reference
Pt(II)	Cl, OH	~15	Shilov, et al., *Kinet. Katal.* **24**, 486 (1983)
Pd(II)	CF_3CO_2	60	Sen, et al., *J. Am. Chem. Soc.* **109**, 8109 (1987)
Co(III)	CF_3CO_2	90	Moiseev, et al. *J. Chem. Soc. Chem. Commun.* 1049 (1990)

[a] based on metal added

- Catalysis reported with peracids (Pd) and O_2 (Co)

- Pd(II) results questioned by Moiseev

Figure 4. Yields of Shilov-type reactions of transition metal complexes with methane.

- Co(III) did not react with methane

- The Pd(II) system oxidized methane:

$$PdX_2 + CH_4 \xrightarrow[HX]{180\ ^\circ C} CH_3X + HX + Pd(0)$$

X	% yield[a]
CF_3CO_2	< 5 (60% reported by Sen)
HSO_4	30
CF_3SO_3	75

[a] based on metal added

- Yield and catalytic efficiency too low

Figure 5. Yields from the Pd(II) and Co(III) systems obtained in our laboratory.

Attempting to expand on this chemistry, we focused on an important common characteristic of the reactions shown in Figure 2. An emerging body of theoretical and experimental work suggested that all of the reactions involve the reactive species acting as an electrophile toward methane early on the reaction coordinate (*10–13*). This mode of interaction could be considered on a simplified level as a Lewis base (methane)–Lewis acid (proton or transition-metal electrophile) adduct as shown in Figure 6. Such interactions can be treated conceptually by Hard–Soft Acid–Base (HSAB) theory or by simplified perturbation analyses such as Frontier molecular orbital theory. These simple, conceptual considerations suggest that the characteristics of metal complexes that would efficiently react with methane are "soft" electrophiles characterized by (1) large ionic radii, (2) high-density of low-lying states, (3) low-lying

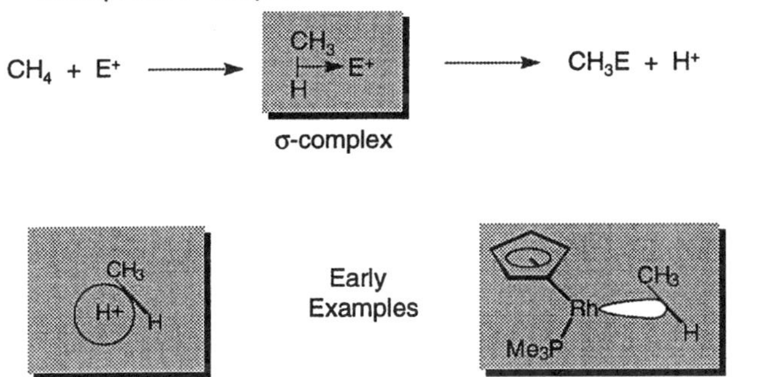

Figure 6. Simplified Lewis acid–base interactions of methane with metal electrophiles.

LUMOs (high ionization potential), and (4) a LUMO with σ symmetry. Additionally, as shown in Scheme I, if methane oxidation is required, a two-electron redox-active metal with the appropriate oxidation potential is required.

These characteristics are exhibited by the isoelectronic, late-third-row metal ions Au(I), Hg(II), and Tl(III) with the filled d shell configuration, $[Xe]5d^{10}5s^0$. Tl(III) is a powerful two-electron oxidant that is known to undergo thallation reaction with benzene (Figure 7).

This system was examined first for reaction with methane (Scheme II). The reaction of thallic triflate with methane was found to result in almost quantitative yield of methyl triflate, based on added thallic triflate. This post-transition metal was more efficient than the transition metals for CH activation!

Although the reaction of thallium triflate with methane was efficient, the reaction was stoichiometric and no attempts to make it catalytic were successful. This result was undoubtedly due to the high redox potential of the Tl(III)/(I) couple. We then turned to Hg(II). The soft electrophilic properties of Hg(II) are exemplary of this group (*14*) (Figure 8). It is the quintessential "soft" cation based on HSAB theory. The ionic radius of Hg(II) is large (1.1 Å), and the LUMO is a low-lying s-orbital (ionization potential = 18.7 eV) with σ-symmetry. The ion is a moderate oxidizing agent ($E^\circ = 0.9$ V). Consistent with these properties, Hg(II) is perhaps best known to the organometallic chemist for the "mercuration" reaction with arenes, one of the first well-established metal–carbon bond-forming reactions (*15, 16*). In this reaction, shown in Figure 8, the Hg(II) acts as an electrophile, displaying reactivity that is consistent with the characteristics discussed previously.

It would be misleading to suggest that Hg(II) and Tl(III) were our first choices for a new methane oxidation catalyst based on the principles just considered; they were not. However, these considerations did give us the required

- Tl(III) is a powerful electrophile with known 2-e$^-$ oxidation chemistry:

$$Tl(III) + 2e^- \longrightarrow Tl(I) \qquad E^\circ = 1.25 \text{ V}$$

- Tl(III) is "soft" and known for thallation of benzene

$$C_6H_6 + Tl(III) \longrightarrow C_6H_5\text{-}Tl(III) + H+$$

- Compare:

$$CH_4 + Tl(III) \longrightarrow CH_3\text{-}Tl(III) + H+$$

Figure 7. Known properties of Tl(III) as a "soft" two-electron redox electrophile.

$$Tl_2O_3 + 6\,CF_3SO_3H \longrightarrow 2\,Tl(CF_3SO_3)_3 + 3\,H_2O$$

$$Tl(CF_3SO_3)_3 + CH_4 \longrightarrow CF_3OSO_2OCH_3 + CF_3SO_3H + Tl(CF_3SO_3)$$
$$\sim 85\% \text{ yield based on Tl(III)}$$

- Reaction clean!

- First example of post-transition metal oxidation of methane

- No reaction in weaker acids

- No catalysis observed with sulfuric acid or oxygen

Scheme II. Remarkably efficient oxidation reaction of thallic triflate with methane.

- The electronic configuration of Hg(II) is [Xe] $5d^{10}\,5s^0$:
  - IP = 18.7 eV

  - LUMO is largely s character (σ symmetry)

  - Small ΔE between s and p orbitals (polarizable)

  - Ionic radius ~ 1.1 Å

  - E° [Hg(II)/Hg(I)] = 0.9 V (easier to oxidize than Tl(I))

- Hg(II) is also well known to activate arenes:

$$ArH + HgX_2 \longrightarrow Ar\text{-}HgX + HX$$

Figure 8. "Soft" redox properties of Hg(II).

motivation to examine this class of third-row, d^{10} metal ions. Such motivation was important because Hg and Tl are considered "posttransition metals" and, as such, have not been of much interest to the present-day catalytic community. This lack of interest results from the filled-shell configuration (that tends to limit the coordination number to two) and the extensive exploration of the chemistries of these elements in the early 1960s.

The reaction of Hg(II) salts with methane was examined in various acid media. The reaction proceeded quite efficiently in triflic acid (CF_3SO_3H), and ~50% yield of methyl triflate is produced, based on added mercuric triflate. Reactions were carried out under 34.5 bar methane (containing 3% Ne as internal standard), at 180 °C for 3 h with 10 mL of triflic acid containing 2.0 mmol of mercury(II) triflate (generated by prior in situ reaction of mercuric oxide with the triflic acid solvent) in a 50-mL high-pressure Autoclave Engineers Hastaloy-C reactor equipped with Desperi-Max gas–liquid mixer. Routine analyses were carried out using HPLC to quantify the methanol produced by hydrolysis of a reaction aliquot. In selected cases, qualitative and quantitative ^{13}C NMR analyses of the crude reaction mixtures with acetic acid as an internal standard (added after reaction) were used to confirm the results. In reported cases, the mass balance on methane was >90%. This mass balance was obtained by accounting for unreacted methane, methyl bisulfate (as methanol after hydrolysis), and carbon dioxide. To obtain good mass balance, the methane/Ne mixture was dispensed from a known-volume, known-pressure reservoir. This procedure allowed the total moles of methane delivered to the reactor to be determined. The moles of methane remaining after reaction and carbon dioxide produced were determined by GC analysis of the gas phase using Ne as an internal standard.

In this reaction, the mercuric triflate is reduced to mercurous triflate. Moreover, no metallic mercury is observed, suggesting that the reaction occurs via a two-electron process as shown in Figure 9. The reaction is quite selective in triflic acid, as no carbon dioxide or other overoxidation products are observed. The rate of reaction with methane was found to correlate with the acidity of the acid solvent. Thus, the reaction occurs with decreasing rates with the acids ($CF_3SO_3H > CF_3CO_2H$) and does not occur at all in acetic acid under typical reaction conditions.

A more important reaction occurs in 100% sulfuric acid. In this solvent, the reaction was found to be catalytic in Hg(II) for the selective oxidation of methane to methyl bisulfate as shown by the 2000% yield in Figure 9. Consistent with the stoichiometry shown in Figure 10, the coproduction of sulfur dioxide was observed. ^{13}C-enriched methane resulted in ^{13}C-labeled methanol with the same level of enrichment. This finding confirmed that the methanol was not produced from a carbon-containing contaminant.

The reaction is quite efficient and, to our knowledge, unprecedented. In 1-L, batch reactions, methane conversions as high as 50% with 85% selectivity to methyl bisulfate (~43% yield, the major side product is carbon dioxide) and

$$2\,HgX_2 + CH_4 \xrightarrow[180\,°C]{HX\,(100\%)} CH_3X + HX + Hg_2X_2$$

X	% yield[a]
CF_3SO_3	~ 50
HSO4	~ 2000

[a] based on metal added

- Oxidation occurs via a two-electron change

- No metallic Hg observed

- Reactions are selective; very little CO_2 observed

- Reactions in Hg(II)-H_2SO_4 system are catalytic !!!!

Figure 9. Reactions of Hg(II) with methane.

- H_2SO_4 is the oxidant:

$$CH_4 + 2\,H_2SO_4 \xrightarrow{Hg(II)} CH_3OSO_3H + 2\,H_2O + SO_2$$

- At 1-liter scale:
  - Methane conversion = 50%
  - Selectivity to methyl bisulfate = 85%
  - Volume productivity = 10^{-7} mol/cm^3.s
  - Turnover frequency of Hg(II) = 10^{-3} s^{-1}

- Use of $^{13}CH_4$ produced $^{13}CH_3OH$ (after hydrolysis)

Figure 10. High-yield oxidation of methane by sulfuric acid catalyzed by Hg(II).

molar productivity rates of 10^{-7} mol/cm^3 s with Hg(II) turnover frequencies of 10^{-3} s^{-1} have been observed. Reactions were carried out as described for the preceding reaction but using a 1-L Hastaloy-C reactor and 300 mL of a 0.1 M solution of $Hg(OSO_3H)_2$ in 100% sulfuric acid. The selectivity was observed to decrease at higher conversions.

On the basis of these results and assuming (1) an exaggerated molar solubility of methane (500 psig) in sulfuric acid at 180 °C of ~0.02 M, in a reactor of equal gas–liquid volume, and (2) a simplified kinetic scheme as shown in Figure 11, the ratio of k_{obs1}/k_{obs2} can be calculated to be ~100. Thus, in the Hg(II)/H_2SO_4 system, methane is significantly more reactive than methyl bisulfate toward reaction with the Hg(II)/H_2SO_4 system! This finding was confirmed by control experiments comparing the rate of overoxidation of methyl bisulfate by the Hg(II)/H_2SO_4 system to the rate of methane oxidation. These results show that the concept of "protection" can result in substantial increases

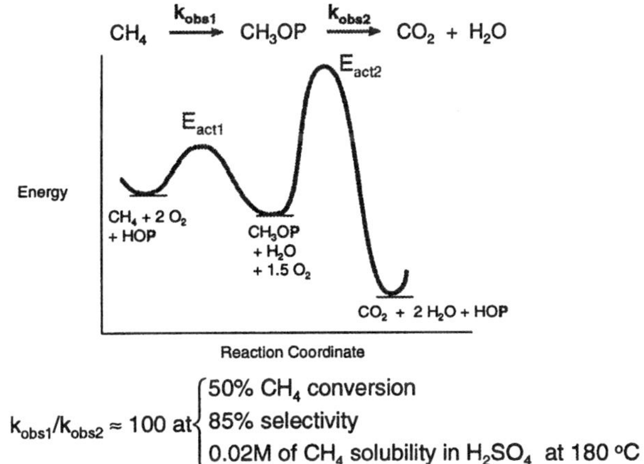

Figure 11. Protection strategy does allow high one-pass yields of protected methanol.

in one-pass yield for methane oxidation. The advantages of the protection strategy become quite apparent when the maximum one-pass yields of nonprotection schemes are compared to this result (Figure 12).

The chemistries of the other $[Xe]5d^{10}5s^0$, isoelectronic cations, Au(I) and Tl(III), were briefly examined in sulfuric acid, and the work is continuing. These species react quite efficiently with methane at 180 °C to produce methyl bisulfate but only in less than stoichiometric reactions based on added metal ion; no efficient catalysis was observed. Both of these species are powerful oxidants ($E° > 1.0$ V) and are not readily oxidized by hot sulfuric acid. Similar results were obtained with Pd(II) and Pt(II)/(IV).

Process Scheme

The Hg(II)-catalyzed conversion of methane to methanol with the concomitant reduction of sulfuric acid to sulfur dioxide, as shown in Figure 10, cannot as such be the basis for an economical synthesis of methanol. However, sulfuric acid is the single largest commodity chemical produced in the world today and is prepared from the oxidation of sulfur dioxide. Thus, the technology for the reoxidation of sulfur dioxide to sulfuric acid (via sulfur trioxide) with air (20% oxygen) is practiced on a large scale and is relatively inexpensive. By combining this step with reactions shown in Scheme III, a potentially practical process can be described. In this scheme, sulfuric acid functions as an oxygen-atom transfer reagent. To our knowledge, this scheme is the first reported example of the use of sulfuric acid in this manner. The net transformation of the established steps is the selective oxidation of methane to methanol with molecular oxygen. As shown in Figure 12, the ~43% methanol yield disclosed

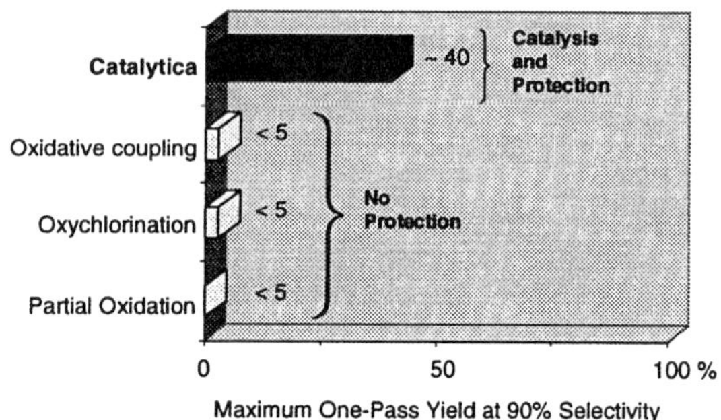

Figure 12. Comparison of "protection" to "nonprotection" methane oxidation schemes.

$$CH_4 + H_2SO_4 + HOP \xrightarrow[\text{catalyst}]{\text{fast}} CH_3OP + 2 H_2O + SO_2 \xrightarrow{\text{slow}} CO_2 + H_2O$$

$$SO_2 + 1/2 O_2 + H_2O \longrightarrow H_2SO_4$$

$$CH_3OP + H_2O \longrightarrow CH_3OH + HOP$$

Net Reaction: $\quad CH_4 + 1/2 O_2 \longrightarrow CH_3OH$

Scheme III. Possible scheme for a high, one-pass yield methane to methanol process.

here represents a significant advance compared to the typical 2% yields obtained for the gas-phase oxidation of methane to methanol.

The comparison made in Figure 12 is not an entirely fair one. The process described in Scheme III is not a direct, one-step process (methyl bisulfate is produced in situ) to a useful methane-derived product as are the partial oxidation processes plotted in Figure 12. However, the comparisons can be made on the basis that both the Catalytica reaction and the partial oxidation reactions result in the selective oxidation of one of the C–H bonds of methane. This mechanism is in contrast to the syngas process, in which the methane is first converted to carbon monoxide and then reduced to methanol.

Reaction Mechanism

The catalyzed oxidation of methane by mercuric ions is a remarkably efficient reaction. Understanding the molecular basis for such a system would be valuable to the continuing search for even more efficient methane oxidation catalysts. The mechanistic work is in progress, and the three key steps in the catalytic sequence have been established. The working model, based on the

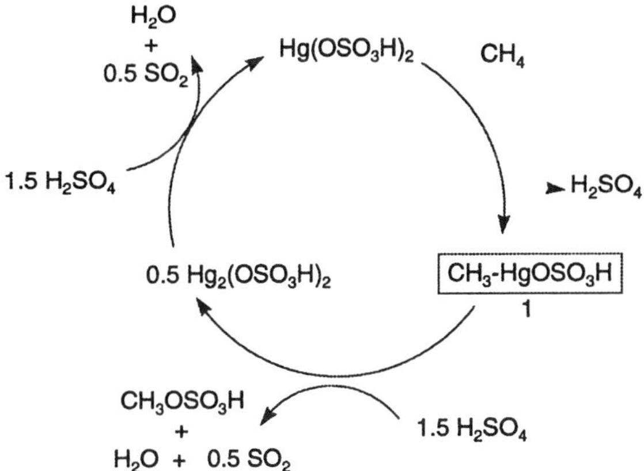

Figure 13. Proposed mechanism for the oxidation of methane to methyl bisulfate by the Hg(II)/H$_2$SO$_4$ system

available data, is shown in Figure 13. The activation of methane is proposed to occur via a net electrophilic displacement reaction with mercuric bisulfate to produce methyl mercuric bisulfate, **1**. This species then decomposes to the product and the reduced species, mercurous bisulfate, in the functionalization step. In the reoxidation step, the mercurous bisulfate is oxidized by sulfuric acid, regenerating mercuric bisulfate.

In the activation step, an electrophilic displacement mechanism is assumed because of the similarity of this reaction to other electrophilic reactions involving hydrocarbons and metal electrophiles, such as the reactions of Pt(II) with alkanes (Figure 4) and the well-established electrophilic substitution reaction of arenes with Hg(II) (Figure 8) (*15, 16*). Processes involving free radicals are not considered likely on the basis of the high selectivity of the reaction. Another possible mechanism, oxidative addition to produce an intermediate Hg(IV) methyl hydride species, has been considered but is deemed unlikely because it would require participation and disruption of the stable $5d^{10}$ filled shell configuration of Hg(II).

In the Pt(II), Pd(II), and Hg(II) electrophilic substitution reactions of hydrocarbons (*9, 15, 16*) the reaction rates are reported to increase with the decreasing basicity of the counter anion and increasing acidity of the solvent. This same trend is observed for the HgX$_2$/HX/CH$_4$ system with the reaction rate decreasing in the order X = CF$_3$SO$_3$H > CF$_3$CO$_2$H >> CH$_3$CO$_2$H as noted previously. The higher reactivity of methane relative to methyl bisulfate ($k_{obs1}/k_{obs2} \approx 100$, *see* Figure 11) is also consistent with an electrophilic reaction. The bisulfate group is electron-withdrawing, and substitution of this group for a C–H bond of methane would be expected to retard participation of the C–H bonds of these substituted species in electrophilic reactions.

Several pieces of evidence have been obtained for the involvement of the activation step as shown in Figure 13 to produce methyl mercuric bisulfate, **1**. Direct observations of crude reaction mixtures by ^{13}C and ^{199}Hg NMR spectroscopy show that **1** is present at a low steady-state concentration during the reaction with methane. The ^{13}C and ^{199}Hg NMR spectra are shown in Figures 14 and 15, respectively. The use of ^{13}C-enriched methane confirms that this species is produced by reaction with methane and not by some contaminant. The identity of intermediate **1** was confirmed by comparison to a sample prepared independently by treatment of dimethyl mercury, $(CH_3)_2Hg$, with 100% sulfuric acid. This reaction occurs readily at room temperature and is quantitative for the formation of **1**. The intermediate, **1**, can also be synthesized by the reaction of methyl mercury hydroxide, $[(CH_3-Hg)_3O]OH$, with sulfuric acid.

Further evidence that **1** is produced by methane activation was provided by the reaction of CH_4 with $^2H_2SO_4$ in the presence of Hg(II). Under these conditions, one deuterium was incorporated into the CH_4 as observed by GC–mass spectroscopy. This finding can be explained by the formation of **1** by methane activation with Hg(II), followed by deuterolysis of **1** to produce $CH_3{}^2H$ (Scheme IV). Independent reactions of synthesized **1** with H_2SO_4 confirmed that protolysis, the microscopic reverse of the activation step, can occur at 180 °C to produce $Hg(OSO_3H)_2$ and methane.

In addition to the protolysis reaction of synthesized **1** to produce methane, methyl bisulfate is also produced when synthesized **1** is heated at 180 °C in H_2SO_4 (Figure 16). This result confirms that the functionalization step proposed in Figure 13 is plausible. On the basis of the known chemistry of metal alkyls in protic media it was expected that the rate of protolysis would be significantly faster than the rate of functionalization. However, we unexpectedly found that the formation of methyl bisulfate from synthesized **1** was quite efficient and proceeded in ~50% yield upon treatment of **1** at 180 °C in 100%

^{13}C-NMR spectra of the reaction mixture

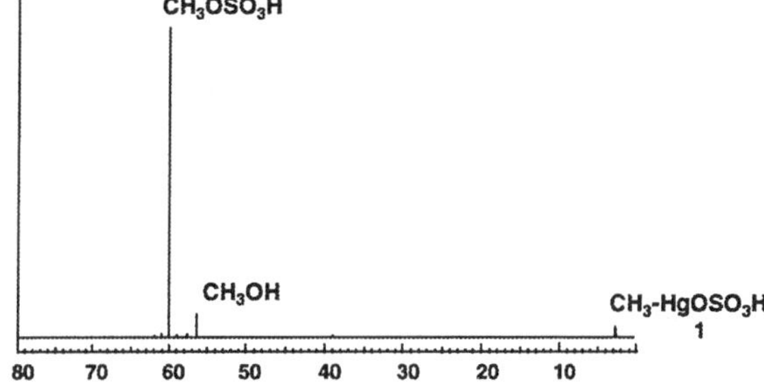

Figure 14. ^{13}C NMR spectrum of the crude reaction mixture resulting from the oxidation of $^{13}CH_4$ (100% enriched) with $Hg(OSO_3H)_2$ in H_2SO_4.

¹⁹⁹Hg-NMR spectra of the reaction mixture

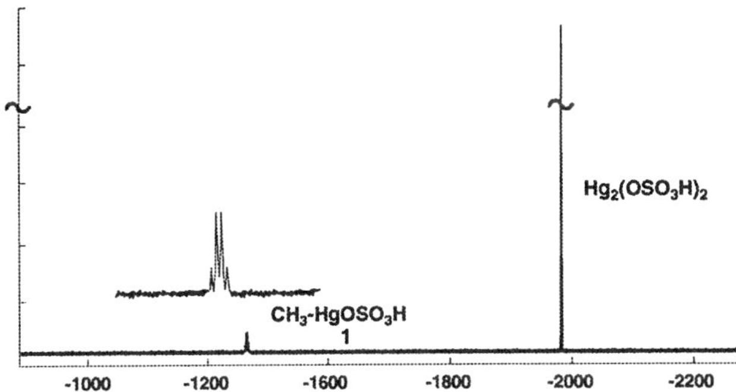

Figure 15. ^{199}Hg NMR spectrum of the crude reaction mixture resulting from the oxidation of CH_4 with $Hg(OSO_3H)_2$ in H_2SO_4.

$$CH_4 + D_2SO_4 \xrightarrow{Hg(II)} CH_3D + CH_3OSO_3H + D_2O + SO_2$$

~ 8% conv.

$$[CH_3OSO_3H] / [CH_3D] = 2 \pm 1$$

$$CH_4 + Hg(OSO_3H)_2 \xrightarrow{- H_2SO_4} CH_3\text{-}HgOSO_3H \xrightarrow{+ D_2SO_4} CH_3D + Hg(OSO_3H)_2$$

1

$D = {}^2H$

Scheme IV. Deuterium incorporation into methane occurs via intermediate, 1.

Synthesis

$$(CH_3)_2Hg + H_2SO_4 \xrightarrow{25\,°C} CH_3\text{-}HgOSO_3H + CH_4$$

1

$$[(CH_3\text{-}Hg)_3O]OH + 3\,H_2SO_4 \xrightarrow{25\,°C} 3\,CH_3\text{-}HgOSO_3H + 2\,H_2O$$

1

Protolysis

$$CH_3\text{-}HgOSO_3H + H_2SO_4 \xrightarrow{180\,°C} CH_4 + Hg(OSO_3H)_2$$

1

Figure 16. Synthesized 1 reacts with sulfuric acid to produce methyl bisulfate.

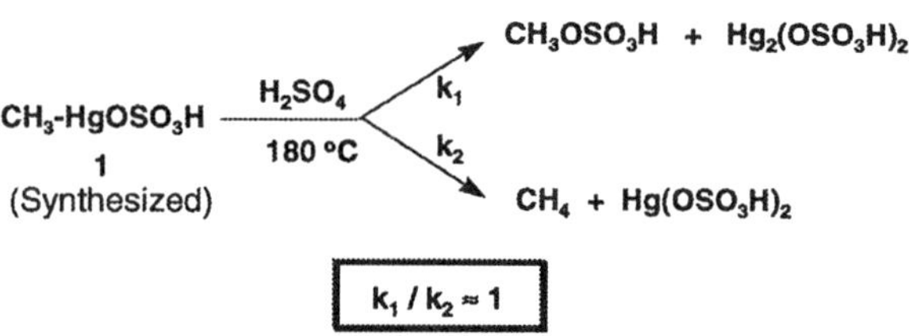

Scheme V. Decomposition of 1 to methane and methyl bisulfate.

H_2SO_4. The only other observed product was methane resulting from protolysis. Consistent with these yields, the rates of formation of methane and methyl bisulfate from synthesized 1 were examined, and both reactions were found to be first order in 1 with $k_1 \approx k_2$ (Scheme V).

On the basis of the mechanism proposed in Figure 13 and the similarities in the rate of functionalization and the rate of protolysis from synthesized 1, it could be predicted that, if 1 were a key intermediate in the Hg(II)-catalyzed oxidation of methane, methane oxidations carried out in $^2H_2SO_4$ should result in approximately equal yields of methyl bisulfate and deuterium-exchanged methane. Just such a correlation was observed. As shown in Scheme IV, approximately equal yields of $CH_3{}^2H$ and methyl bisulfate were obtained from the Hg(II)-catalyzed reaction of methane with $^2H_2SO_4$ at low methane conversion (<8%, to ensure kinetic control) under typical catalytic oxidation conditions. Under these conditions only $CH_3{}^2H$ was detected; no polydeuterium-incorporated methane was observed. Interestingly, no deuterium incorporation was observed in the methyl bisulfate.

These data provide strong evidence for the formation of 1 by the activation step proposed in Figure 13. However, our observations do not imply that $Hg(OSO_3H)_2$ is the species that reacts with methane. At this time, the identity of this species is unknown. However, a likely candidate is the solvated cation $[Hg(OSO_3H)]^+$, which is produced by ionization of $Hg(OSO_3H)_2$. An intriguing possibility is that a complex involving coordinated methane and such a species might be an intermediate on the reaction pathway to 1. Such coordinated complexes have been proposed (*see* references 10–13).

The evidence for the reoxidation step shown in Figure 13 is straightforward. Hot, concentrated sulfuric acid is a powerful oxidant and was observed to rapidly oxidize mercurous bisulfate to mercuric bisulfate with the formation of sulfur dioxide. The reaction kinetics were examined and the oxidation was found to be first order in mercurous bisulfate. The decomposition of 1 to produce methyl bisulfate warranted further investigation because it is a rare example of metal alkyl functionalization in a C–H activation system. Although many systems are known that activate C–H bonds, only a small number both activate and functionalize alkane substrates. The kinetics of this process was briefly examined to distinguish between reaction occurring by a bimolecular process

- **Bimolecular Reaction**

$$CH_3\text{-}HgOSO_3H + Hg(OSO_3H)_2 \longrightarrow CH_3OSO_3H + Hg_2(OSO_3H)_2$$

- **Unimolecular Reaction (solvolysis)**

$$CH_3\text{-}HgOSO_3H \longrightarrow [CH_3]^+ + Hg(0) + HSO_4^- \longrightarrow CH_3OSO_3H + Hg(0)$$

Hg(0) rapidly oxidized by H_2SO_4 and Hg(II)

Figure 17. The lack of effect of added Hg(II) suggests that 1 decomposes via a unimolecular pathway.

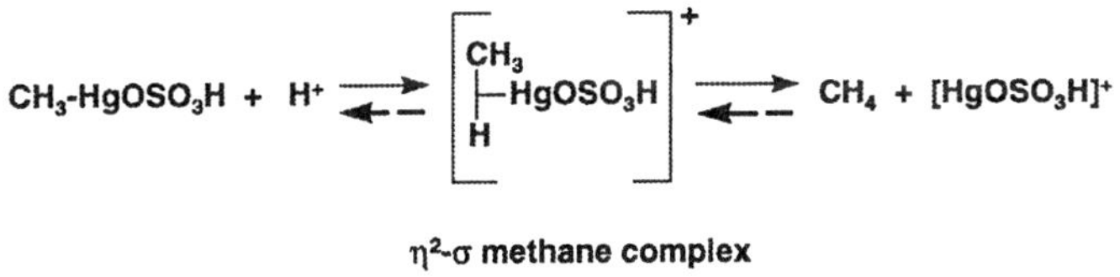

$$CH_3\text{-}HgOSO_3H + H^+ \rightleftharpoons \left[\begin{array}{c} CH_3 \\ |\text{-}HgOSO_3H \\ H \end{array} \right]^+ \rightleftharpoons CH_4 + [HgOSO_3H]^+$$

$\eta^2\text{-}\sigma$ **methane complex**

- Protolysis is the microscopic reverse of methane activation

Figure 18. A key detail that remains to be addressed is whether methane complexes are intermediates during the CH activation reaction.

of **1** and mercuric bisulfate, and a unimolecular solvolysis process of **1**. Interestingly, the data support a unimolecular process (Figure 17). Although we presume that the reaction occurs by a unimolecular solvolytic (S_N1) process, the data could also be interpreted by assuming a pseudo-first-order, S_N2-type reaction between 1 and H_2SO_4 or HSO_4^-.

Conclusions

The mercury(II)–sulfuric acid system described in this chapter provides important precedent that true catalytic systems can be developed for the selective, low-temperature, overall oxidation of methane to methanol. The catalytic mercury(II)–sulfuric acid system is well-suited for study because the key intermediates can be synthesized and examined under controlled conditions. Few low-temperature, truly catalytic systems for methane oxidation can be examined in such detail. Thus far, the strong evidence for intermediacy of methyl–mercury species in the activation of methane adds credibility to the general supposition that the electrophilic, metal-mediated oxidation of alkanes can occur via intermediate alkyl metal species (7, 8). Some key details that remain to be addressed are the identity of the Hg(II) species that interacts productively with methane and whether σ-complexes (Figure 18) are intermediates or transition states in the CH activation process. The information gained from the discovery and study of this system should be useful in guiding the development of more efficient systems.

Acknowledgments

We are deeply indebted to Henry Taube, Marguerite Blake Wilbur Professor Emeritus at Stanford University and scientific advisor to Catalytica, Inc., for the insight and guidance provided during all phases of our investigation of selective methane oxidation. The results described here were obtained in the course of a joint research and development program among Petro-Canada, Catalytica, Inc., and Techmocisco, Inc., a wholly owned subsidiary of Mitsubishi Oil Co., Ltd.

References

1. Hunter, N. R.; Gesser, H. D.; Morton, L. A.; Yarlogadda, P. S. *Appl. Catal.* **1990**, *57*, 45.
2. *Methane Conversion by Oxidative Processes;* Wolf., E. E., Ed.; Van Nostrand Reinhold: New York, 1991; p 403.
3. Gesser, H. D.; Hunter, N. R.; Prakash, C. B. *Chem. Rev.* **1985**, *85*, 235.
4. Foster, N. R. *Appl. Catal.* **1985**, *19*, 1.
5. Olah, G. A.; Surya Prakash, G. K.; Sommer, J. *Superacids;* John Wiley & Sons: New York, 1985.
6. Olah, G.A.; Parker, D. G.; Yoneda, N. *Angew. Chem. Int. Ed. Engl.* **1978**, *17*, 909.
7. Shilov, A. E. *Activation of Saturated Hydrocarbons by Transition Metal Complexes;* D. A. Reidel: Dordrecht, Netherlands, 1984.
8. Sen, A. *Acc. Chem. Res.* **1988**, *21*, 421.
9. Varagaftik, M. N.; Stolarov, I. P.; Moiseev, I. I. *J. Chem. Soc. Chem. Commun.* **1990**, *15*, 1049.
10. Periana, R. A.; Bergman, R. G. *J. Am. Chem. Soc.* **1986**, *108*, 7332.
11. Wasserman, E. P.; Moore, C. B.; Bergman, R. G. *Science (Washington, D.C.)* **1986**, *55*, 315.
12. Brown, C. E.; Ijshikawa, Y.; Hackett, P. A.; Rayner, D. M. *J. Am. Chem. Soc.* **1990**, *112*, 2530.
13. Koga, N.; Morokuma, K. *J. Phys. Chem.* **1990**, *94*, 5454.
14. McAuliffe, C. A. *The Chemistry of Mercury;* MacMillan: London, 1977.
15. Parshall, G. W. *Homogeneous Catalysis: The Application and Chemistry of Catalysis by Homogeneous Transition Metal Complexes;* Wiley-Interscience: New York, 1980.
16. Henry, P. M. *Catalysis by Metal Complexes: Palladium Catalyzed Oxidations of Hydrocarbons;* D. A. Reidel: Dordecht, Netherlands, 1980; Vol. 2.

5

Intrinsic Ancillary Ligand Effects in Cationic Zirconium Polymerization Catalysts

David E. Richardson

Department of Chemistry, University of Florida, Gainesville, FL 32611–7200

The intrinsic electrophilicity of zirconocenium polymerization catalysts has been obtained by determining the gas-phase rates of reaction of various catalyst ions with dihydrogen and unsaturated hydrocarbons. The expected decrease in electrophilicity when cyclopentadienyl (Cp) is replaced with more electron-donating Cp derivatives is observed. Thus, the increased polymerization activity often observed for complexes with more electron-donating ligands is probably a result of increased rate of initiation due to lower ion pair binding energy, decreased termination rates due to inhibition of β-elimination, or both.

Electrophilic Group 4 Metallocene Polymerization Catalysts

Homogeneous and supported alkene polymerization catalysts based on group 4 metallocenes have recently moved from the basic research laboratory to practical operation in the production of polymer in large-scale plants (*1*). The fundamental active site in these catalysts is now widely accepted to be a cationic complex of the general type $[L_2MR]^+$, where L is typically a cyclopentadienyl (Cp) derivative and R represents methyl or the growing polymer chain (*2*). The cation is formed by reaction of a catalyst precursor with a strong Lewis acid such as an aluminum alkyl or $B(C_6F_5)_3$ (*2*). Polymerization of an alkene is based on the repetitive insertion of monomer into the M–R bond to form a long-chain saturated polyalkene. An example of a typical titanium metallocene/trimethylaluminum catalyst is shown in Scheme I.

Scheme I greatly simplifies the true nature of the catalyst systems, which in practice can be a complex mixture of a metallocene precursor, activating

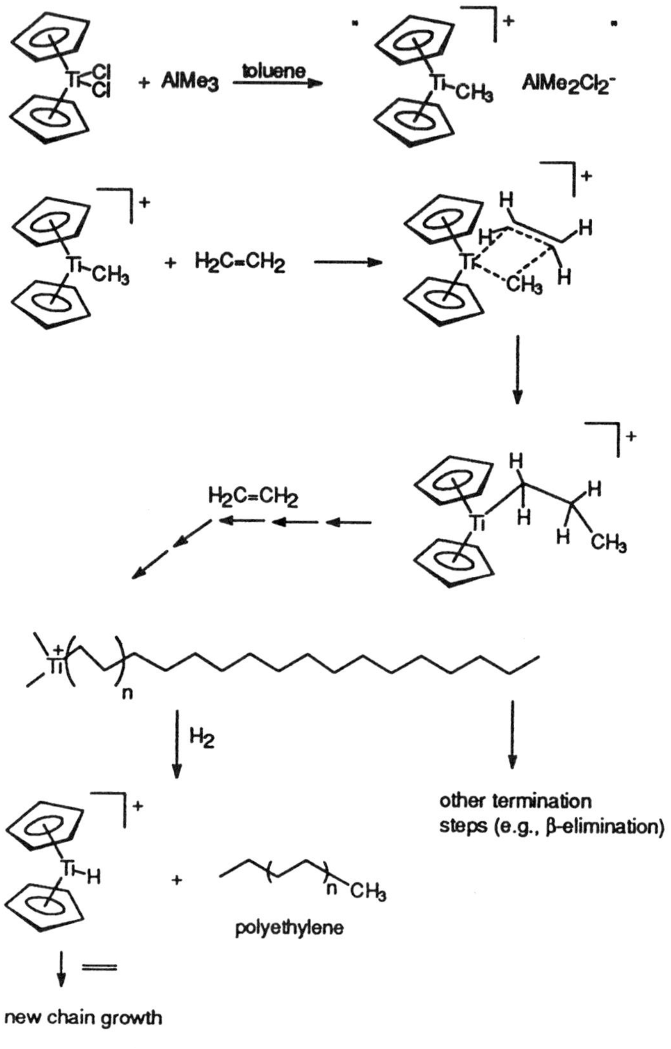

Scheme I.

Lewis acid, alkylating agents, supports, and other agents that promote efficient polymerization or control molecular weight (e.g., H_2). The cation itself is not "bare", but instead is solvated and ion-paired to varying extents depending on the solvent used, the ancillary ligands on the metallocene, and the type of counterions present. The co-catalysts, such as methylalumoxane (MAO) (3), are not always well-characterized. Thus, the activity of a given metallocene can be varied extensively by altering many aspects of the catalyst system. For example, a Lewis basic solvent such as acetonitrile can effectively eliminate polymerization activity in an otherwise active catalyst (*4*). Therefore, typical poly-merizations are run in a nonpolar aprotic solvent such as toluene (or the mono-mer itself for higher alkenes), and the anionic counterion is chosen to be as unreactive and noncoordinating as possible (*5*). However, as shown by recent work (*6*), even under these conditions, substantial ion-pairing must be over-come before monomer can be activated and inserted.

One of the attractive features of these catalysts is the possibility for "tun-ing" the catalyst for production of polymers with a variety of desirable proper-ties, which can include molecular weight, stereochemistry, and copolymeriza-tion with other monomers. This tuning is often accomplished by variation of the ancillary ligand set in the metallocene catalyst precursor. The catalytic activity of the 14-electron complex cation $[L_2MR]^+$ arises from its electron-deficient nature, which encourages binding and activation of the monomer toward insertion. Thus, the incorporation of electron-donating ligands would be expected to decrease reactivity of the catalyst in a reaction in which it acts as an electrophile (as in Scheme I). However, just the opposite is observed in many instances. Both permethylcyclopentadienyl (Cp*) and indenyl (Ind) are well-known to be more electron releasing than Cp itself (*7*), yet they can lead to more active catalysts in some cases (*8, 9*). Systematic studies of alkene poly-merization by substituted bis(indenyl) zirconium(IV) catalysts showed that electron-withdrawing substituents (e.g., X = Cl, F) lead to reduced activity compared to X = H (*10, 11*). In these latter examples (*10, 11*), electronic effects are assumed to be important factors because the site of substitution is remote from the metal center.

These observations illustrate one of the more confusing aspects of predict-ing changes in activity of the catalysts as the ancillary ligands are varied. The explanations in the literature for relative catalyst activity focus on the rates of initiation, propagation, and termination steps as well as the strength of the ion pairing of the cation with the counterion, all of which can be affected in differ-ent ways by ancillary ligand substitution. A method that examines *intrinsic* electrophilicity of the active species in these catalysts could clarify the origins of the activity trends, and we turned to gas-phase methods for this reason.

We have previously examined the gas-phase ion–molecule reactions of the 14-electron cation $Cp_2ZrCH_3^+$ by using Fourier transform mass spectrometric methods (*12–14*). In these studies, the intrinsic reactivity of the cation could be probed without interference due to ion pairing or solvation. In general, all of the observed gas-phase reaction pathways can be explained by using a known mech-anism in solution (either insertion, deinsertion, β-H shift, or C–H activation).

To assess the intrinsic electrophilicity of zirconocenium cations as a func-tion of ancillary ligands, we decided to determine the kinetic influence of dif-ferent Cp substituents on the rates of some of the previously observed reac-tions (eq 1–3).

$$[L_2Zr–CH_3]^+ + H_2 \rightarrow [L_2Zr–H]^+ + CH_4 \qquad k_1 \qquad (1)$$

$$[L_2Zr{-}CH_3]^+ + C_2H_4 \rightarrow [L_2Zr{-}C_3H_5]^+ + H_2 \quad k_2 \qquad (2)$$

$$[L_2Zr{-}CH_3]^+ + H_2C{=}C(CH_3)_2 \rightarrow$$

$$[L_2Zr{-}H_2C{-}C(CH_3){=}CH_2]^+ + CH_4 \quad k_3 \qquad (3)$$

Equation 1 represents the most straightforward and sterically undemanding reaction in the group. Hydrogenolysis (eq 1) and allylic C–H activation of isobutene (eq 3) at the Zr–Me bond proceed via 4-center, 4-electron transition states (Scheme II). Reaction with ethylene is initiated by insertion followed by β-hydride shift and C–H activation of the allylic hydrogen to eliminate dihydrogen (Scheme II).

Parameterization of Electronic Effects for Cyclopentadienyl Ligands

Ancillary ligand effects on reactivity at a metal center can be modeled as a combination of electronic and steric effects, and this approach has been widely used to rationalize the variations in reactivity in group 4 polymerization catalysts (2). Attempts to probe electronic effects by analysis of polymerization activity have been instructive but indirect approaches to separating these effects. We believe that the rates of gas-phase reactions (eqs 1–3) are more direct measures of electronic effects.

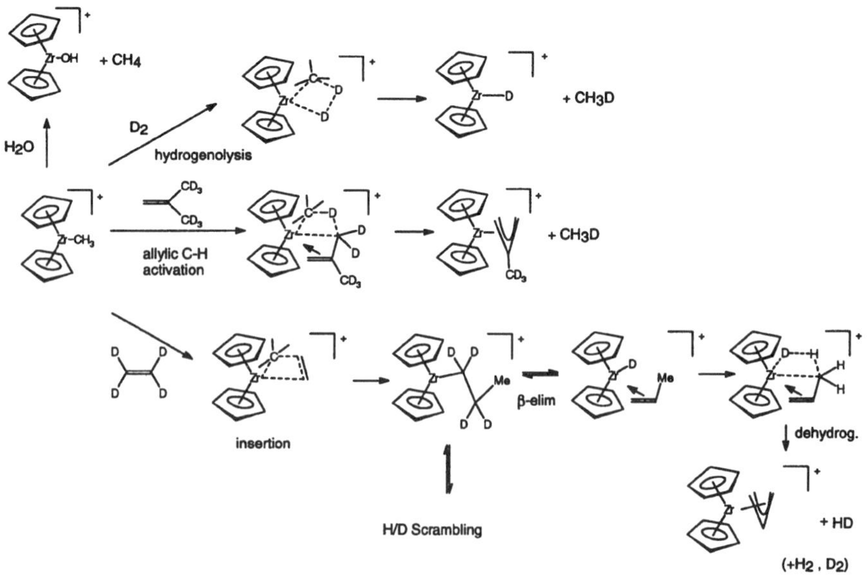

Scheme II.

At this point, it is useful to review our understanding of gas-phase substituent effects in Cp derivatives. We have applied the technique of gas-phase electron-transfer equilibria (ETE) to determine the ionization free energies for the reaction $LL'M \rightarrow LL'M^+ + e^-$, where L and L' are Cp derivatives and M is either Fe, Ru, or Ni (*15–19*). By assigning ligand parameters γ_L to Cp ($\gamma_L = 0$) and Cp* ($\gamma_L = -1$), it is possible to derive a set of parameters for a variety of Cp derivatives (Table I). It is assumed in the derivation of the parameters that ligand effects are additive for L and L'. Negative values of γ_L indicate a tendency to decrease the metallocene ionization energy relative to Cp, whereas positive values increase the ionization energy. The general trends are not unexpected and largely coincide with other measures of electron-donating and electron-withdrawing tendencies of substituents, but the values of γ_L are specific for metallocenes in the gas phase and are based on thermal equilibrium reactions. In addition, fused ring ligands such as Ind and fluorenyl (Flu) are readily parameterized.

Table I. Cyclopentadienyl Derivative γ Parameters

Positive		*Negative*	
Ligand	γ	*Ligand*	γ
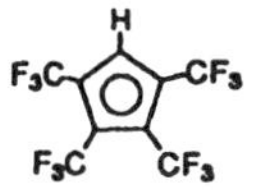 1,2,3,4-tetra(CF$_3$), 5-H Cp	3.1	 Me-Cp	−0.20
1-OSi(Et)$_3$, 2,3,4,5-tetra(CF$_3$) Cp	1.6	Et-Cp	−0.22
pentafluoro Cp (C$_5$F$_5$)	1.5	CH$_2$NMe$_2$-Cp	−0.22
pentachloro Cp (C$_5$Cl$_5$)	1.1	SiMe$_3$-Cp	−0.24
pentabromo Cp (C$_5$Br$_5$)	1.0	CH$_2$CH$_2$CH$_2$CH$_3$-Cp	−0.30
NO$_2$-Cp	0.79	t-Bu-Cp	−0.32
Cp	0[a]	indenyl	−0.41
		fluorenyl	−0.65
		pentamethyl Cp (Cp*)	−1[a]

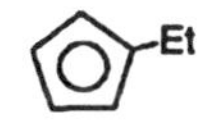

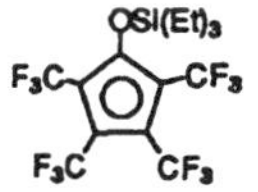

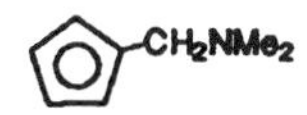

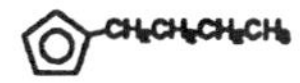

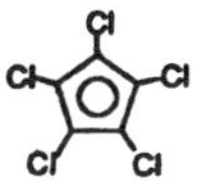

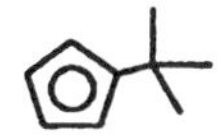

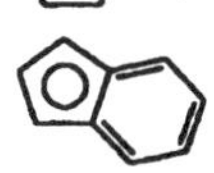

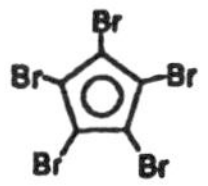

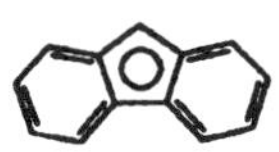

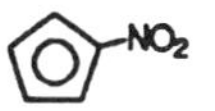

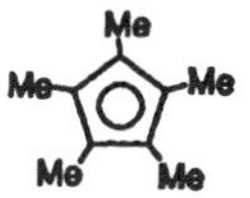

[a]By definition.

Rates and Mechanisms of Reactions of $L_2ZrCH_3^+$ *with Dihydrogen and Alkenes*

Kinetics for the reactions in eqs 1–3 were determined for the cations (**1–5**) shown in Chart I (*20a, 20b*). Besides the parent cation **1**, Ind- and Flu-substituted complexes (**2**, **3**, and **5**) were examined along with complex **4**, which features a silyl linker between the Cp ligands.

Unfortunately, it has not been possible to determine the reaction kinetics for complexes bearing alkyl substituents because the alkyl group is rapidly C–H activated intramolecularly as shown in eq 4 for the permethylated cation.

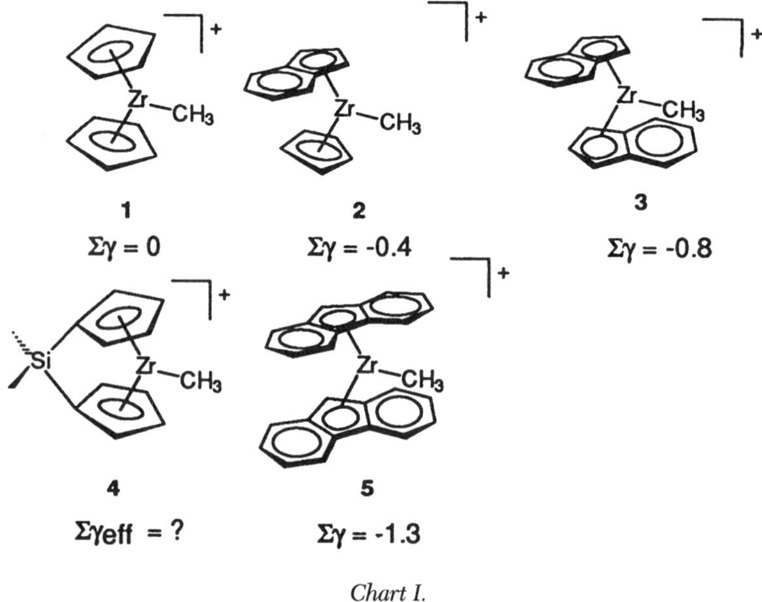

$$\Sigma\gamma = -2.0 \qquad + \ CH_4 \qquad (4)$$

The experimental methods for obtaining kinetic data by using Fourier transform ion cyclotron resonance mass spectrometry were similar to those described previously (*20c*). The observed ion–molecule reaction pathways were modeled as a series of pseudo-first-order elementary steps. In addition to

Chart I.

the reaction with the substrate to produce the desired product ion, reaction of the methyl cation with background water ($\sim 10^{-9}$ Torr) forms the metallocene hydroxide ion ($[L_2ZrOH]^+$, *see* Scheme II), and reaction of various cations with the neutral dimethyl parent compound produces dimer ions (i.e., binuclear Zr complex ions). These alternate pathways were incorporated into the full kinetic model used to fit the data. The resulting differential equations were solved to yield an analytical solution describing the time dependence of the intensity for each product ion and reactant ion, and the time dependence of ion intensities was fit to the model by optimizing the rate constants simultaneously. The second-order rate constants for the reactions in eqs 1–3 are plotted vs. summed γ parameters in Figure 1. An effective value of $\Sigma\gamma$ for **4** was chosen ($+0.16$) to give the best fit to the lines derived from fits to the Ind and Cp complexes. Rates for the reactions of **5** were immeasureably slow, and the best fit lines for the measured rate constants were used to predict the rate constants for **5** shown in Figure 1.

From Figure 1 it is clear that more electron-donating ligands substantially retard the reactions in eqs 1–3. In addition, the general trends are the same for

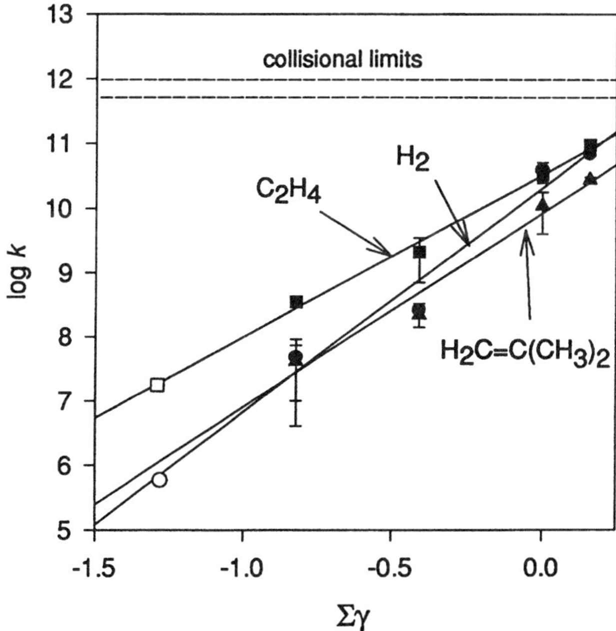

*Figure 1. Plot of log(k) vs. $\Sigma\gamma$ for k_1 (filled circles), k_2 (filled squares), and k_3 (filled triangles). The solid lines represent the best fit to the available parameters. The dashed lines are the Langevin collisional limits for the second-order rate constants (upper line for C_2H_4, lower for H_2). Open symbols are the predicted values of log(k) for the reaction of **5** (see Chart I) with H_2 (open square) and C_2H_4 (open circle).*

all three substrates, that is, the electrophilicities of the cations decrease in the order 4 > 1 > 2 > 3 > 5. The hydrogenolysis reaction is the most sensitive to the value of $\Sigma\gamma$.

Relationship of Intrinsic Reactivity to Solution Kinetics

A simplified mechanism is shown in Scheme III for the solution polymerization of a 1-alkene by a cationic zirconocene catalyst. Actual initiation of the chain reaction can only occur once the solvent-separated ion pair is trapped by a monomer insertion. Propagation is terminated in this scheme by a β-elimination step, and the zirconocene hydride is assumed to be rapidly deactivated. In actual catalytic systems, hydride intermediates can contribute to propagation by reacting further with monomer, but dimerization of the hydride complexes may be one mechanism of catalyst inactivation (21).

In the limiting case where $k_{-1}[\mathrm{X}^-] >> k_{\mathrm{ini}}[\text{monomer}]$, the rate of overall propagation (R_{p}, which is monomer consumption rate) is proportional to $(K_{\mathrm{ip}}k_{\mathrm{ini}}k_{\mathrm{prop}})/k_{\mathrm{term}}$, where $K_{\mathrm{ip}} = k_1/k_{-1}$ and K_{ip}, k_{ini}, k_{prop}, and k_{term} are the ion pairing equilibrium constant and rate constants for initiation, propagation, and termination, respectively. Thus, when ion pairing is faster than reaction with monomer, the overall rate of polymerization depends on the equilibrium constant for ion-pair separation and the rates of initiation, propagation, and termination. Any attempt to ascertain the separate effects of modification of ancillary ligands on each of these steps by measuring overall catalyst activity will be difficult at best.

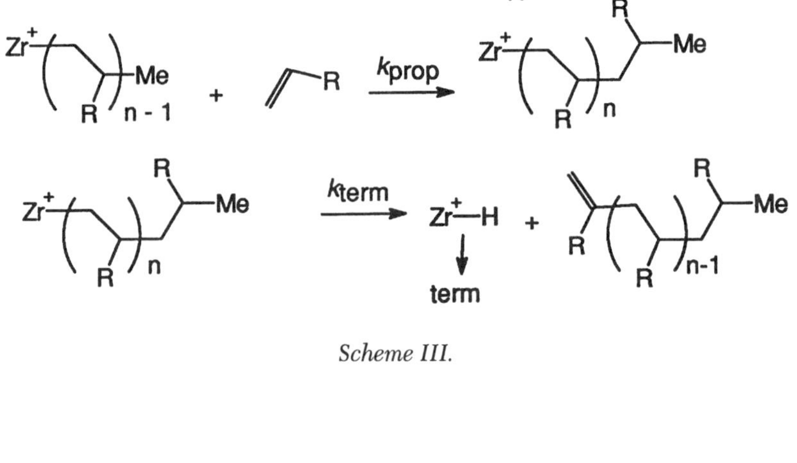

Scheme III.

The gas-phase data provide an intrinsic order of electrophilicity as a function of the ancillary ligand set. It is not surprising that all three reactions studied follow the same trend **4 > 1 > 2 > 3 > 5**, since each reaction has a rate-determining step that is controlled by a 4-center, 4-electron transition state (i.e., either insertion, deinsertion, or C–H activation).

Scheme IV illustrates the relationship between the potential energy surfaces for the gas-phase reactions and the condensed-phase reactions using the ethylene as an example substrate. The Cp ancillary ligands are left off the structures for simplicity. The energies where shown are based on a combination of theory (22) and known thermodynamic quantities for hydrocarbons. The remaining details are only qualitative. The observed gas-phase reaction pro-

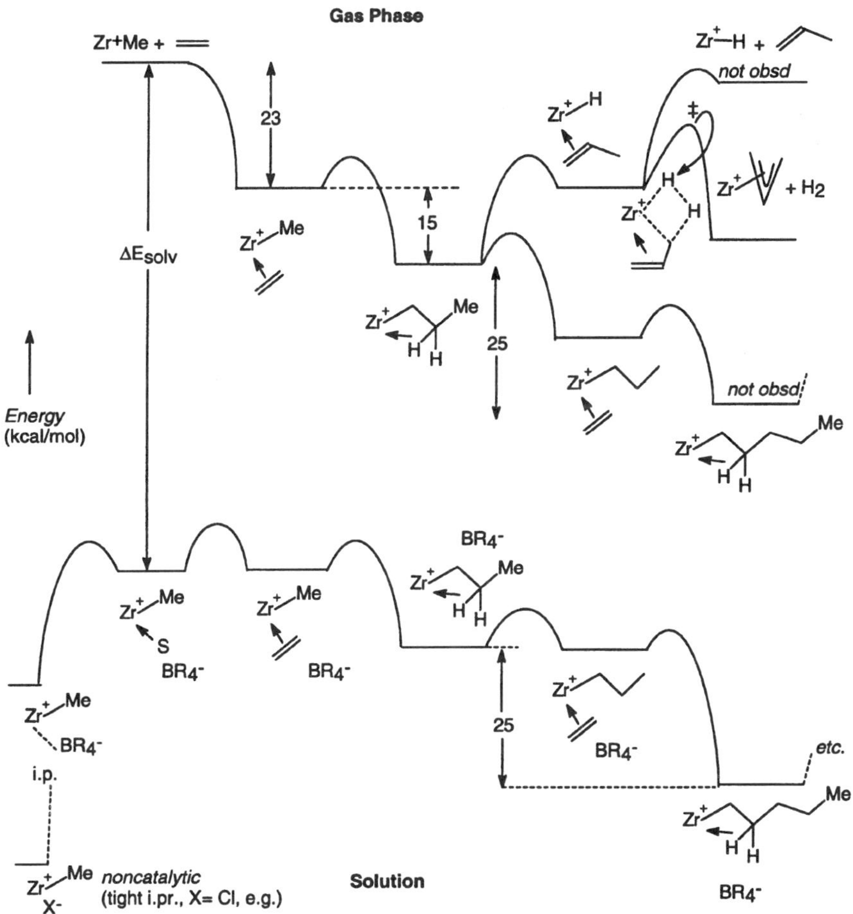

Scheme IV.

duces the allyl complex and dihydrogen (eq 2). Other gas-phase pathways are shown that are not observed. One of these, the elimination of propylene (i.e., β-elimination), is not observed because the transition state for that near-thermoneutral reaction is higher than that of the observed path. Although the insertion of a second ethylene into the Zr–propyl bond is substantially downhill, the bimolecular nature of the reaction means that it will not compete with the unimolecular elimination of dihydrogen.

To connect the gas-phase potential energy with that for solution, we introduce the solvation energies for each species along the various pathways. A borate anion is shown as the weakly coordinating counterion, but the ion-pair binding energy even in the case of borate ions is substantial (>10 kcal/mol). Solvation energies for coordinatively saturated metallocenium ions are on the order of 20–40 kcal/mol in polar solvents depending on the Cp substituents (*14–17*). For typical nonpolar solvents used in polymerization studies (i.e., benzene or toluene) the range would be expected to be $\sim$15–30 kcal/mol. To that must be added the binding energy for specific inner-sphere solvation (even the "nonbasic" toluene probably coordinates via a C–H bond to the 14-electron cation with a significant binding energy). Thus, although ethylene is predicted to bind to the cation with an exoergicity of –23 kcal/mol, in solution it is assumed that the solvent and ethylene are roughly equal in donor strength and the energies of the solvent (S) and ethylene adducts in Scheme IV are shown as equal.

On the basis of the order of electrophilicity observed in the gas-phase reactions of eqs 1–3, we suggest that the order of k_{prop} is most likely **4 > 1 > 2 > 3 > 5**. Assuming that the solvation energies of all intermediates and transition states are constant for a given ancillary ligand set except for the 14-electron methylzirconium cation, the trends observed in the gas-phase kinetics will also be observed in solution. For example, consider the insertion of ethylene into the Zr–Me bond (i.e., initiation). The transition state energy for this reaction can influence the rate of insertion in the gas phase even though it lies below the energy of the initial reactants. Significant kinetic barriers due to negative activation energies are commonly encountered in ion–molecule reactions. Although seemingly irrational to solution chemists, a negative barrier can slow a gas-phase reaction, and the effect can be modeled for metal complex ion–molecule reactions by statistical models such as RRKM (*23*).

To convert to the solution phase, all species (including activated complexes) are solvated. For the 14-electron Zr–Me$^+$, some additional solvation energy results from inner-sphere solvation, but the other species are assumed to have equal solvation energies. Therefore, although the initiation reaction in solution is less exoergic than the gas-phase reaction, the trend in the rate as a function of ancillary ligands should be the same in both phases.

If indeed the propagation reaction rate (controlled by k_{prop} in Scheme III) decreases with more electron-donating ligands such as Ind, then the observed increase in the activity for the Ind catalyst must arise from the ion-pair equilibrium constant, initiation rate, or termination rate in Scheme III. The most likely candidate is the ion-pair binding constant. Bulkier ligands may reduce the binding constant for the counterion to the cationic catalyst and thereby accelerate the polymerization by increasing the equilibrium amount of the solvent-separated cation.

Summary

The intrinsic electrophilicity of zirconocenium polymerization catalysts has been obtained by determining the gas-phase rates of reaction of various catalyst ions with dihydrogen and unsaturated hydrocarbons. The expected decrease in electrophilicity when Cp is replaced with more electron-donating Cp derivatives is observed. Thus, the increased polymerization activity often observed for complexes with more electron-donating ligands may result from an increase in the rate of initiation (due to lower ion-pair binding energy), a decrease in the termination rate (inhibition of β-elimination), or both.

Acknowledgments

I express gratitude to Henry Taube, my doctoral advisor (Stanford University, 1976–1980). It is hoped that the work described here illustrates the insights that can be achieved through a Taube-inspired application of thermodynamics and kinetics to fundamental problems in chemistry.

I also thank co-workers and collaborators who have made this research possible, especially N. George Alameddin, Matthew F. Ryan, Allen R. Siedle (3M), and John R. Eyler. Other co-workers who have contributed to the work in gas-phase organometallic chemistry are listed in the references.

This work was supported by a grant from the National Science Foundation (CHE9311614).

References

1.   Haggin, J. *Chem. Eng. News* **1995**, *73(18)*, 7.
2.   Mohring, P. C.; Coville, N. J. *J. Organomet. Chem.* **1994**, *479*, 1.
3.   Sinn, H.; Kaminsky, W.; Vollmer, H.; Woldt, R. *Angew. Chem.* **1980**, *92*, 396.
4.   Jordan, R. F. *Adv. Organomet. Chem.* **1991**, *32*, 325.
5.   Yang, X.; Stern, C. L.; Marks, T. J. *J. Am. Chem. Soc.* **1994**, *116*, 10015 and references therein.
6.   Deck, P. A.; Marks, T. J. *J. Am. Chem. Soc.* **1995**, *117*, 6128.
7.   Gassman, P. G.; Winter, C. H. *J. Am. Chem. Soc.* **1988**, *110*, 6130.
8.   Resconi, L.; Piemontesi, F.; Franciscono, G.; Abis, L.; Fiorani, T. *J. Am. Chem. Soc.* **1992**, *114*, 1025.
9.   Giannetti, E.; Nicoletti, G. M.; Mazzocchi, R. *J. Polym. Sci., Polym. Chem. Ed.* **1985**, *23*, 2117.
10. Lee, I.-K. ; Gauthier, W. J.; Ball, J. M.; Iyengar, B.; Collins, S. *Organometallics* **1992**, *11*, 2115.

11. Piccolrovazzi, N.; Pino, P.; Consiglio, G.; Sironi, A.; Moert, M. *Organometallics* **1990**, *9*, 3098.
12. Christ, C. S.; Eyler, J. R.; Richardson, D. E. *J. Am. Chem. Soc.* **1988**, *110*, 4038.
13. Christ, C. S.; Eyler, J. R.; Richardson, D. E. *J. Am. Chem. Soc.* **1990**, *112*, 596.
14. Christ, C. S.; Eyler, J. R.; Richardson, D. E. *J. Am. Chem. Soc.* **1990**, *112*, 4778.
15. Richardson, D. E. In *Organometallic Ion Chemistry*; Freiser, B. S., Ed.; Kluwer Academic Publishers: Dordrecht, 1996; Ch. 8.
16. Richardson, D. E.; Ryan, M. F.; Khan, Md. N.I.; Maxwell, K. A. *J. Am. Chem. Soc.* **1992**, *114*, 10482.
17. Ryan, M. F.; Eyler, J. R.; Richardson, D. E. *J. Am. Chem. Soc.* **1992**, 114, 8611.
18. Ryan, M. F.; Richardson, D. E.; Lichtenberger, D. L.; Gruhn, N. *Organometallics* **1994**, *13*, 1190.
19. Ryan, M. F.; Siedle, A. R.; Burk, M. J.; Richardson, D. E. *Organometallics* **1992**, *11*, 4231.
20. (a) Alameddin, N. G.; Ryan, M. F.; Eyler, J. R.; Siedle, A. R.; Richardson, D. E. *Organometallics* **1995**, *14*, 5005; (b) Richardson, D. E.; Alameddin, N. G.; Ryan, M. F.; Eyler, J. R.; Hayes, T.; Siedle, A. R. *J. Am. Chem. Soc.* in press; (c) Sharpe, P.; Richardson, D. E. *Coord. Chem. Rev.* **1989**, *93*, 59.
21. Stehling, U.; Diebold, J.; Kirsten, R.; Röll, W. Brintzinger, H.-H.; Jüngling, S.; Mülhaupt, R.; Langhauser, F. *Organometallics* **1994**, *13*, 964.
22. Woo, T. K.; Fan, L.; Ziegler, T. *Organometallics* **1994**, *13*, 2252.
23. Richardson, D. E.; Eyler, J. R. *Chem. Phys.* **1993**, *176*, 457.

6

Zeolites Offer Variety as Ligands

J. N. Armor* and Brian H. Toby[1]

Air Products and Chemicals, Inc., 7201 Hamilton Blvd., Allentown, PA 18195

This chapter presents some general observations that extend the concepts that govern the use of zeolites as catalysts. Traditionally, zeolites are used for the production of petrochemicals, and their reactivity is associated with either their acidity or their shape selectivity. Zeolites also offer unusual chemical environments for transition metal ions. Cations exchanged into zeolites often site at defects or at positions of high coordinative unsaturation. Such effects have been previously demonstrated with type A, X, and Y zeolites. Each zeolite structure type offers a different ligating environment to the cation. This variation is illustrated by the use of computer graphics to show the various cation environments of cobalt, copper, and nickel ions in ZSM-5 vs. faujasite and type A zeolites.

SCIENTISTS HAVE MARVELED for more than 20 years over the structure, utility, and importance of zeolites as catalysts and adsorbents. A great deal has been written about the strong acidity of zeolites and their shape selectivity (*1*). Silicate molecular sieves are neutral, but the replacement of Al atoms for some Si atoms creates an overall negative charge on the aluminosilicate lattice, charge which must be balanced with protons, alkali cations, or transition metal ions. These cations add new dimension to the utility of zeolites with a majority of past work focused upon the Na or H cations.

Past work has shown that transition metal cations can impart remarkable and dramatic reactivity to zeolites as catalysts (*2–12*). The role of cation sites and their coordination has been discussed, especially for A, MOR, X, and Y type zeolites. In 1975, Williamson et al. (*13*) reported that transition metal

---

[1]Current address: Reactor Radiation Division, National Institute of Standards and Technology, Gaithersburg, MD, 20899.

*Corresponding author.

complexes can be formed around the cationic sites within Y zeolite. In their reviews Klier (*14*) and Schoonheydt et al. (*15*) describe the multiple types of sites that exist for transition metal ions in zeolite X, Y, or A. Jablonski et al. (*16*) showed that the occupation and distribution of Ni^{2+} and Co^{2+} ions in zeolites depends strongly on the temperature, dehydration level, and time, with X, Y and A type zeolites all behaving quite differently. Dedecek and Wichterlova (*17*) argue that photoluminescence studies show that two different Cu sites exist in Cu–ZSM-5, which differ in population and their reducibility (*17*). Miyamoto et al. (*18*) recently described the use of molecular dynamics calculations and computer graphics to describe the location of Cu(I) within ZSM-5. Thomas and colleagues (*19–21*) have reported the results of in situ structure determinations on NiY. Dramatic shifts in the Ni siting occur upon temperature programming to 400 °C, hydrothermal treatment, or reduction at high temperatures. Upon dehydration the zeolite undergoes a dramatic rearrangement trapping the nickel cations within the smaller sodalite cavities (*20*).

Ozin and co-workers (*10, 22, 23*) use the term "zeolate" to described the zeolite as a ligand in Y zeolites. Ozin and Ozkar (*22*) describe the typical "crown ether like" coordination of oxygens around cations in Y zeolites, in which the cations are coordinatively unsaturated having gigantic electrostatic fields associated with these cations. These Y zeolites are viewed as exhibiting novel coordination chemistry within the cavity of the supercage.

In a recent article, Armor (*2*) suggested that different zeolites can offer a variety of different and unusual ligand effects. This chapter seeks to illustrate the concept that zeolites offer a multitude of ligand fields (not only within the same structure, but also across all the different topologies of zeolites available) by using computer graphics as a tool to visualize the multiple ligand environments around the cations within zeolites.

Experimental Procedures

We examined the published structures of Ni–faujasite [X-type], Ni–A, Ni–ZSM-5, and CuX using a computer graphics package developed by Biosym Technologies, Inc. The approach used was to look only at the structure and "peel away" the framework not directly involved with the cations. Particular focus was placed on cations that would be more highly exposed and thus more reactive. These cations would be coordinatively saturated and might even have alternative sites in the presence of water; therefore, we selected anhydrous structures where possible. Cations that were buried within sodalite cages, other small pockets, or highly coordinated were not considered in our simplification. We wanted to see what the environment was like around highly exposed cations; therefore, we tried to select structures with limited amounts of hydrated cations.

The single crystal structure of dehydrated copper-exchanged [via multiple exposure of single crystals of natural faujasite to 1 M solutions of Cu(II) sulfate or acetate] faujasite with Si/Al = 2.42 by Maxwell and deBoer (*24*) was refined to an R factor of 0.050 with Cu^{2+} located in sites I, I', II, II', and III. Multiple exposure to Cu(II) sulfate or acetate encouraged the full exchange of Cu^{2+} cations into the faujasite. The data of Olson (25) for nickel faujasite showed that the nickel cations prefer site I, but are distributed among I', II, and II' as well. The nickel was batch exchanged into the faujasite crystal at 90 °C over 60 days. Olson also commented that a trace of residual water may play a major role in determining the distribution of nickel ions. The data of Zhenji et al. (26) were collected on a polycrystalline sample at 500 °C prepared by Ni exchange of H–ZSM-5 with Si/Al = 30 to 1.8 wt% Ni. The Ni cations were located by using the Rietveld method. The single crystal structure data of Firor and Seff (27) for nickel-exchanged A zeolite were based on a single crystal of NaA exchanged by exposure to a continuous stream of 0.1 M Ni acetate at 25 °C. The R^2 value was 0.115, and residual water was present in the crystal. With the water present, the nickel cations do not coordinate to the anionic lattice, but rather to the water molecules.

Location of Cations

Using one cation in a variety of topologies, we sought to examine differences in the coordination around each cation. Unfortunately we were limited in the number of transition metal ions in structures other than Y- and A-type zeolites. We could only find structures with a highly exchanged level of cations for one common cation, Ni^{2+}; hence we chose to compare NiA, Ni–faujasite, and Ni–ZSM-5 with Cu–faujasite. Using the structures described in the experimental section already and the general approach with the Biosym software, we generated a number of images that were quite revealing, not for their similarities but for their differences.

Figure 1 shows the Ni cation sitting on a trigonal array of oxygens from the lattice with three water molecules providing additional oxide ligands. Two images are provided: one looking from the center of the supercage out toward the walls where the Ni is located, and the other from the side showing the position of the Ni vs. the supercage and the sodalite cage. The Ni–O (lattice) distance is quite long (2.5–2.7 Å), and the Ni is not formally coordinated to the anionic lattice. The same authors found these structural similarities in Co^{2+} and Fe^{2+} A zeolites as well.

Ni and Cu faujasite are compared in Figure 2. Ni–O distances vary from 2.0 to 3.02 Å, depending on whether one focuses on Ni in the sodalite cage, supercage, or water of hydration. Because we sought to examine exposed and coordinatively unsaturated cations, we focused on site II, where the Ni–O distance varied from 2.260–3.028 Å. (The Ni in site I was a near-perfect octahedron with the six lattice oxygen atoms of the sodalite cage.)

For Cu–faujasite, looking only at the differences, one sees a number of intriguing features. Cu–O in I and I' sites vary from 2.12 to 3.53 Å and from 2.29 to 3.05 Å in sites II and II'. In contrast the Cu–faujasite shows the Cu cation in site II sits closer to the wall than the Ni in the structure at the left in Figure 2, and there is also some Cu in the site III position [Cu–O = 2.16–2.77 Å]. Comparisons here must be made with caution because the structures will be affected by the Si/Al ratio, by the dehydration level, and by how the

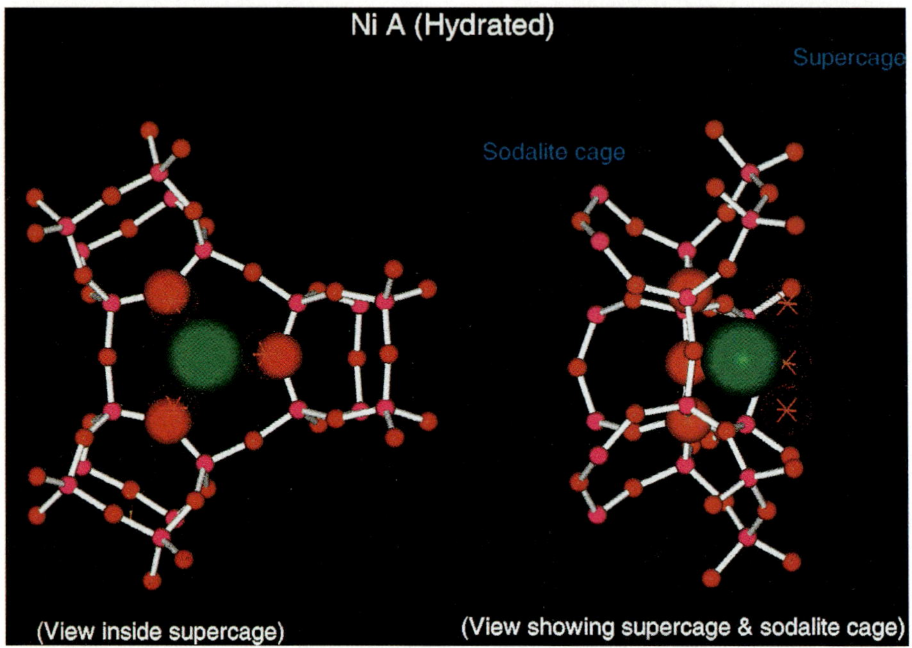

Figure 1. Image of Ni in A zeolite. Green represents Ni cations, large red balls represent nearest oxygen atoms from Si–O–Al lattice, and X represents other oxygens from water molecules

exchange is made, as well as their thermal history. Nonetheless, the Cu in site III is clearly very exposed to incoming molecules.

In Figure 3, Ni–ZSM-5 is compared with Ni–faujasite of Figure 2 to show the dramatic differences in coordination around the Ni cation. In this ZSM-5 structure the Ni cation is not located at the intersection of the two channels. If one looks down the straight channel no Ni cations are seen; however, looking down the zigzag channel one sees the Ni cations sited on a V-shaped array of lattice oxide ions. The Ni cations are only weakly bonded to the framework with Ni–O distance varying from 2.23 to 3.25 Å.

Discussion

The images on Ni and Cu in A, faujasite, and ZSM-5 show the wide variation in siting, coordination number, and degree of exposure of cations within zeolites. The Ni cation can exist in a variety of coordination geometries not only within the same zeolite, but across different families of zeolites. Clearly zeolites themselves offer a multitude of ligand environments that are very sensi-

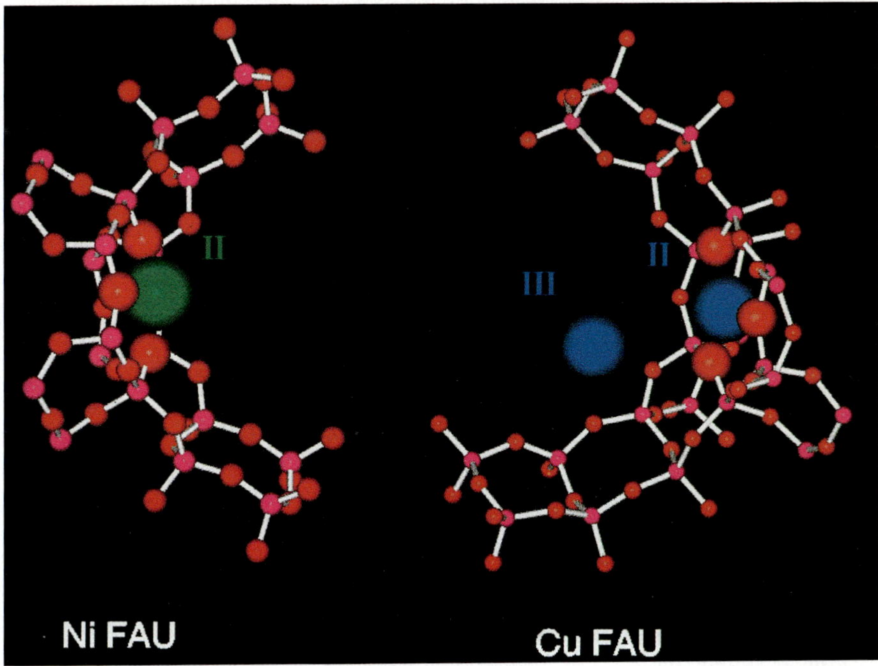

Figure 2. Image of Ni in faujasite (FAU) compared with Cu in faujasite. Green represents Ni cations, blue spheres represent Cu cations, and large red balls represent nearest oxygen atoms from Si–O–Al lattice.

tive to the preparation, degree of hydration, thermal history, and degree of exchange. In addition one can also expect variations with changes in the Si/Al ratio and the oxidation number of the cation.

In comparison the chemistry of transition metal cations adds another facet to zeolite science. Protons or alkali ions would seem to be a bit simpler, because of the dissimilarity of one transition metal to another, the multiple oxidation states possible on the cation, and the rich chemistry (e.g., hydrolysis and condensation) that is possible. There is a wealth of possibilities beyond the relative simplicity of protons and alkali cations. The multiple charge on transition metal ions and the charge balance bring out other issues that still are not well-understood. On the other hand, the localized, charged sites within the zeolites offer an opportunity for maximum dispersion of the cations within the zeolite. A number of opportunities to use zeolites for novel coordination chemistry and catalysis were discussed in a recent article (2).

Clearly, there is no reason to believe that different cation sites within a zeolite will exhibit the same catalytic reactivity. All these possible locations and the sensitivity of the sites to the conditions of reaction demand that investiga-

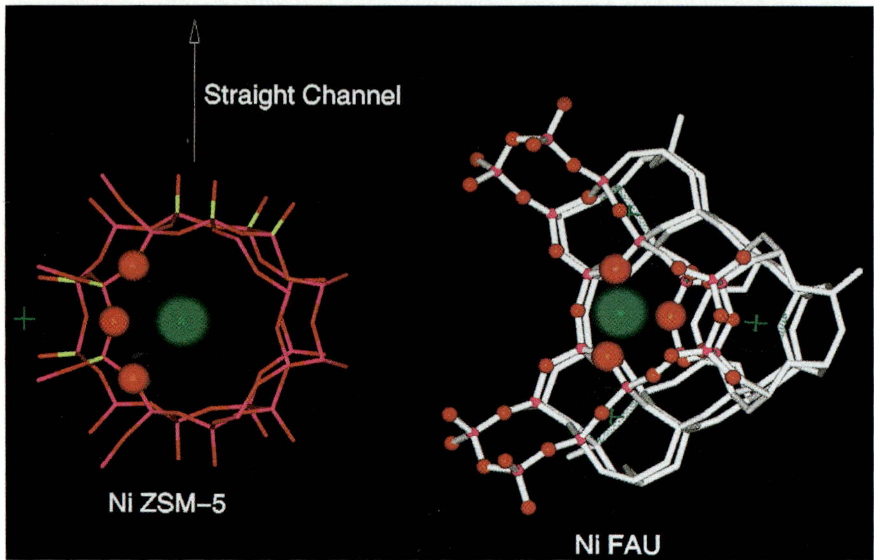

Figure 3. Image of Ni in ZSM-5 compared with Ni in faujasite (from Figure 2). Green represents Ni cations, and large red balls represent nearest oxygen atoms from Si–O–Al lattice.

tors must be very specific in describing the preparation and experimental conditions as well as the history of their samples. Dramatic differences can exist between zeolites of comparable composition if they are handled differently. More importantly, unless we understand what happens to the cations in the zeolite with time under the desired reaction conditions, we cannot hope to understand or control the catalytic chemistry.

Furthermore, because cations may find greater coordination numbers available at deflect sites in the structures, variation in the preparation conditions for the zeolite may produce compositionally similar materials with very different catalytic reactivities. There is also more evidence (28) to show that intergrowth structures also can occur during the synthesis of some zeolites. This process results in the formation of both hexagonal and cubic arrays in the structure. Where the boundary layers for these arrays occur, one can expect very different types of coordination sites. This concept remains unexplored and untested with reactive transition metal cations.

In their in situ structure studies, Dooryhee et al. (20) conclude that a great deal of mobility and structure rearrangement occurs with NiY upon dehydration. Unlike Ca^{2+} and Cu^{2+}, the Ni^{2+} cations do not locate preferably at one or two specific sites. There is a rich and complex chemistry possible upon substituting transition metal ions in zeolites. Much more data needs to be accumulated that compares the structure with the catalysis under the same set of conditions. Although computational approaches may hold promise for prediction of the location of the transition metal cation sites, at present accurate predictions cannot be achieved and are difficult to confirm.

Our observations using the Biosym software are limited by the quality and quantity of data available on cation-exchanged zeolites. Although much is known about FAU, LTA, and MOR, little has been established about the newer topologies such as ZSM-5 (MFI), and FER. The difficulty in obtaining and then uniformly exchanging single crystals magnifies the problem of getting good structures on fully exchanged material. Furthermore, cations can move around upon dehydration or thermal treatment. If cations do move around, which ones are moving and are these the cations responsible for catalysis? These are all important aspects that must be addressed. We need to understand where the cations are that are responsible for unique adsorption or catalysis.

Recent work by Cohen de Lara et al. (*29*) and Coulomb et al. (*30*) describe the interaction of gases such as Ar, Kr, H_2, and hydrocarbons within zeolites. A simulation study of copper ions in ZSM-5 has just appeared by Sayle et al. (*31*) that sites the location of the extra lattice copper species, but we still do not know why these sites are unique for the important reaction of NO decomposition (*32*). The in situ studies of Thomas and his colleagues (*19*) on structures of zeolites under a variety of atmospheres and at a variety of temperatures represent the type of leading edge science that must continue and be embraced by others. Considering the commercial importance of ZSM-5 as a catalyst and all the intriguing reactions that have been reported recently with this material, it is amazing how little structural data is available to firmly establish the structure with regard to cation positions. This problem is further complicated by the difficulty in distinguishing between Si and Al, even with neutron diffraction.

Summary

By using published structures on transition metal ions in zeolites, we have shown that dramatic differences in coordination number are possible even with a common cation in a variety of zeolite topologies. Not only is there the possibility of multiple coordination sites, but these sites differ dramatically from one zeolite topology to another. Thus one can envision zeolites offering an entire spectrum of ligand field effects; that is, each zeolite offers a complex and unique ligand field with which cations can interact and coordinate. Additional data and comparisons must be made under conditions of temperature and degree of hydration that simulate the conditions under which these materials are used as catalysts.

Acknowledgments

We thank Air Products and Chemicals, Inc. for the permission to publish this work and for internal review of this manuscript by C. G. Coe and H. Cheng.

References

1. Chen, N. Y. *Shape Selective Catalysis in Industrial Applications;* Dekker: New York, 1989.
2. Armor, J. N. In *Science and Technology in Catalysis in 1994;* Izumi, Y.; Arai, H.; Iwamoto, M., Eds.; Studies in Surface Science and Catalysis Series; Kodansha Press: Tokyo, Japan, 1995; Vol. 92, pp 51–62.
3. Li, Y.; Armor, J. N. U.S. Patent 5,149,512, 1992.
4. Li, Y.; Armor, J. N. *Appl. Catal.* **1993,** *2,* 239–256.
5. Li, Y.; Armor, J. N. *J. Catal.* **1994,** *150,* 376–387.
6. Li, Y.; Armor, J. N. *Appl. Catal.* **1992,** *1,* L21–L29.
7. Faraj, M. K. U.S. Patent 5,312,995, 1994.
8. Boreskov, G. K. *Proc. Int. Congr. Catal. 5th, 1972* **1973,** 68/981–68/996.
9. Karmakar, S.; Green, H. L. *J. Catal.* **1994,** *148,* 524–533.
10. Ozin, G. *Chem. Mater.* **1992,** *4,* 511–521.
11. Gut, G; Auf Deregen, K. *Helv. Chim. Acta* **1974,** *57,* 441.
12. Denkewicz, R. P.; Weiss, A. H.; Kranich, W. L. *J. Washington Acad. Sci.* **1984,** *74,* 19–26.
13. Williamson, W. B.; Flentge, D. R.; Lunsford, J. H. *J. Catal.* **1975,** *37,* 258–266.
14. Klier, K. *Langmuir* **1988,** *4,* 13–25.
15. Schoonheydt, R. A.; Vandamme, L. J.; Jacobs, P. A.; Uytterhoeven, J. B. *J. Catal.* **1976,** *43,* 292.
16. Jablonski, J. M.; Mulak, J.; Romanowski, W. *J. Catal.* **1977,** *47,* 147–158.
17. Dedecek, J.; Wichterlova, B. *J. Phys. Chem.* **1994,** *98,* 5721–5727.
18. Miyamoto, A.; Himei, H.; Oka, Y.; Maruya, E.; Katagiri, M.; Vetrivel, R.; Kubo, M. *Catal. Today* **1994,** *22,* 87–96.
19. Couves, J. W.; Jones, R. H.; Thomas, J. M.; Smith, B. J. *Adv. Mater.* **1990,** *2,* 181–183.
20. Dooryhee, E.; Catlow, C. A.; Couves, J. W.; Maddox, P. J.; Thomas, J. M.; Greaves, G. N.; Steel, A. T.; Townsend, R. P. *J. Phys. Chem.* **1991,** *95,* 4514–4521.
21. Thomas, J. M.; Williams C; Rayment, T. *J. Chem. Soc. Faraday Trans. 1* **1988,** *84,* 2915–2931.
22. Ozin, G. A.; Ozkar, S. *Chem. Mater.* **1992,** *4,* 511–521.
23. Ozin, G; Steele, M. *Macromol. Symp.* **1994,** *80,* 45–61.
24. Maxwell, I. E.; de Boer, J. J. *J. Phys. Chem.* **1975,** *79,* 1874–1879.
25. Olson, D. H. *J. Phys. Chem.* **1968,** *72,* 4366–4373.
26. Zhenyi, L; Wangjin, Z; Qin, Y; Guanglie, L; Wangrong, L; Shuju, W.; Youshi, Z; Bingxiong, L. *Stud. Surf. Sci. Catal.* **1986,** *28,* 415–422.
27. Firor, R. L.; Seff, K. *J. Phys. Chem.* **1976,** *82,* 1650–1655.
28. Espeel, P. H.; De Peuter, G.; Tielen, M. C.; Jacobs, P. A. *J. Phys. Chem.* **1994,** *98,* 11588–11596.
29. Cohen de Lara, E.; Kahn, R.; Tarek, M. *Mol. Cryst. Liq. Cryst.* **1994,** *248,* 5–11.
30. Coulomb, J.; Llewellyn, P.; Martin, C.; Grillet, Y.; Rouquerol, J. poster from the Proceedings of the 5th International Adsorption Conference, Assilomar, CA, May, 1995.
31. Sayle, D. C.; Perrin, M. A.; Nortier, P.; Catlow, R. A. *J. Chem. Soc. Chem. Commun.* **1995,** *9,* 945–947.
32. Armor, J. N. *Appl. Catal.* **1992,** *1,* 221.

Reactions Catalyzed by Methylrhenium Trioxide

James H. Espenson and Mahdi M. Abu-Omar

Ames Laboratory and Department of Chemistry, Iowa State University, Ames, IA 50011

*Methylrhenium trioxide (CH$_3$ReO$_3$ or MTO) reacts with H$_2$O$_2$ to form two peroxo complexes, CH$_3$ReO$_2$(η^2-O$_2$), **A**, and CH$_3$ReO(η^2-O$_2$)$_2$(H$_2$O), **B**. Peroxide binding to MTO is an equilibrium process, rapid but not instantaneous, characterized by the equilibrium constants K$_1$ = 16 L mol^{-1} and K$_2$ = 132 L mol^{-1} at pH 0, μ = 2.0, and 25 °C. MTO catalyzes the oxidation of many organic and inorganic substrates by H$_2$O$_2$. The evaluation of the catalytic kinetics showed that both **A** and **B** react with a given substrate at comparable rates. The various steps of peroxide activation consist of nucleophilic attack of substrate on peroxide ions that have become electrophilically activated by binding to MTO. The versatility of MTO as a catalyst is demonstrated by its ability to catalyze the oxidation even of electron-deficient substrates such as β-dicarbonyl compounds. Catalyst deactivation occurs in the presence of hydrogen peroxide, much more rapidly than in its absence, and produces methanol and perrhenate ions as final products. The rate of decomposition is dependent on both acid and hydrogen peroxide concentrations.*

HENRY TAUBE'S RESEARCH CONTRIBUTIONS over the years have often dealt with the chemistry of oxygen in the broadest sense. On the occasion of the symposium honoring his 80th birthday, it seemed only fitting to take note of some recent advances in one area that build upon this theme. Our primary subject is the reactivity of hydrogen peroxide, and secondarily other reactions that followed from the principles that the peroxide studies revealed. From a maze of reactions and possible reactions that peroxide can undergo, we wish to extract only one that seems, on its face, to be quite simple. It consists of the transfer of a single oxygen atom from peroxide to a substrate, leaving behind a molecule of water:

$$H_2O_2 \rightarrow \{O\} + H_2O \tag{1}$$

If this reaction could be made to dominate, useful laboratory and commercial oxidations might follow. In particular, electrophilic reactions of hydrogen peroxide are free of wastes and by-products (*1, 2*), eliminating problems and costs associated with environmental cleanup.

However, reaction 1 occurs too slowly for practicality because peroxide reactions have high activation energies (*3*), and the free-radical pathways for peroxides often lead to undesirable mixtures of products. By devising and applying a catalyst for the electrophilic pathway, not only will intolerably slow reactions be accelerated, but also the radical and other side reactions will be made unimportant.

Thus we shall review here the chemistry of hydrogen peroxide as catalyzed by methylrhenium trioxide (CH_3ReO_3 or MTO), to present the reactions that MTO is capable of catalyzing, and the steps and intermediates believed to be active in the catalytic chemistry. MTO is not the only catalyst for the electrophilic reactivity of hydrogen peroxide. High-valent compounds such as molybdates, tungstates, chromates, and vanadates, among others, do so as well; their capabilities have been described elsewhere (*1*). The MTO catalyst does offer the advantage of being suitable for aqueous and nonaqueous environments, and it does not feature pH equilibria or oligomer formation processes that tend to complicate quantitative mechanistic work with many other catalysts. Our research has focused only on MTO itself. Cp^*ReO_3 is not an effective catalyst, and few systematic studies have been carried out on other alkyls. MTO is a remarkably stable substance both as the pure solid and in dilute solutions, in which neither oxygen nor acid nor water has deleterious effects.

The Rhenium Catalyst

MTO is a colorless compound, first prepared in 1979 (*4*) as needlelike crystals from residues of $(CH_3)_3ReO_2$ preparations left open to the atmosphere. An improved route was devised from dirhenium heptoxide and tetramethyl tin (eq 2) (*5*), later refined by the incorporation of perfluoroglutaric anhydride (eq 3) (*6*). Recently, the less costly trichloroacetic anhydride has been used in its place (*7*).

$$Re_2O_7 + Sn(CH_3)_4 \rightarrow CH_3ReO_3 + (CH_3)_3SnOReO_3 \tag{2}$$

$$Re_2O_7 + 2\,SnMe_4 + \text{(perfluoroglutaric anhydride)} \longrightarrow 2\,CH_3ReO_3 + \text{(bis(trimethylstannyl) perfluoroglutarate)} \tag{3}$$

MTO can easily be purified by vacuum sublimation. These are some spectroscopic characteristics: IR in CH_2Cl_2, 1000 (w), 967 cm^{-1} (vs) (*8*); ^{1}H-NMR,

δ(CH$_3$) = 2.63 in CDCl$_3$ (*4*); ^{13}C-NMR, 19.03 ppm [quartet, ^2J(C,H) = 138 Hz, which is typical of sp^3-hybridized carbon (*8*)]; UV–vis in H$_2$O: 239 nm (ε 1900 L mol^{-1} cm^{-1}), 270 nm (sh, ε 1300 L mol^{-1} cm^{-1}) (*9*).

Because MTO decomposes under X-ray irradiation even at low temperature, its structure was determined in the gas phase by electron diffraction (*8*). It possesses C$_{3v}$ symmetry with a Re–C bond distance of 2.060(9) Å and a Re=O bond length of 1.709(3) Å. The C–Re=O bond angle of 106.0(2)° reflects the steric demand of the three oxygen atoms. The short Re–C bond distance deserves a comment here. A contribution from a carbene-type structure [i.e., the enol CH$_2$=ReO$_2$(OH)] was considered by us and others (*8, 10*). Based on the ^2J(C,H) from the ^{13}C-NMR and the lack of hydrogen exchange for deuterium in D$_2$O (Scheme I), such a tautomer of MTO was found to be insignificant. Therefore, the recorded d(Re–C) of 2.06 Å corresponds to a standard Re(VII)–C(sp^3) distance for coordination number four.

With only 14 valence electrons, at least in a formal sense, it comes as no surprise that MTO acts as a Lewis acid toward monodentate and bidentate nitrogen bases, such as aniline, ammonia, pyridine, bipyridine, and ethylenediamine (*11*). With monodentate ligands, MTO forms a trigonal bipyramidal adduct with the nitrogen base trans to the methyl group. Bidentate ligands coordinate to MTO, resulting in an octahedral complex with the bidentate ligand trans to two oxo ligands leaving the third oxo ligand trans to methyl. Bond lengths for a few of these CH$_3$ReO$_3$(L) adducts are compared with those from the parent MTO complex in Table I.

The Re–C distances in the adducts are comparable to each other and slightly longer than those in MTO. This slight increase in the Re–C bond

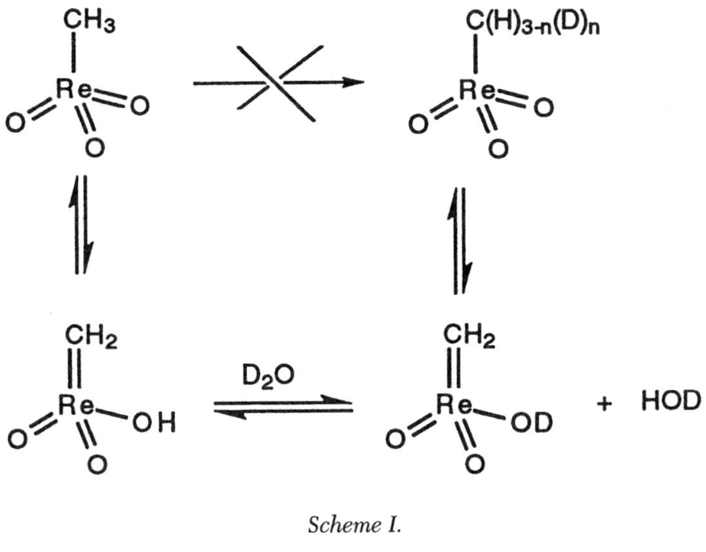

Scheme I.

Table I. Bond Lengths for Some MeReO$_3$(L) Adducts Compared to Those in MTO

Complex	Re–C (Å)	Re=O (Å)	Others (Å)	Ref.
MTO (CH$_3$ReO$_3$)	2.060 (9)	av 1.709 (3)		8
[CH$_3$ReO$_3$Cl]$^{\ominus}$	2.128 (9)	av 1.697 (6)	Re–Cl 2.629 (2)	84
CH$_3$ReO$_3$(NH$_2$Ph)	2.095 (5)	av 1.699 (3)	Re–N 2.469 (4)	11
CH$_3$ReO$_3$(NH$_2$Ph)	2.119 (5)	Re=O^1 1.736 (3) Re=O^2 1.687 (3) Re=O^3 1.698 (3)	Re–N 2.333 (5)	11

NOTE: MTO means methylrhenium trioxide.

length might simply arise from the increase in coordination number from four to five. On the other hand, the Re=O distances do not increase upon ligand addition to the coordination sphere of the rhenium, supporting the notion that the Re=O bond order is mainly two in MTO without much contribution from a bond order of three (Re$^-$ $\equiv$ O$^+$). The aniline adduct, CH$_3$ReO$_3$(NH$_2$Ph), has two isomers in the solid state: one with the ligand trans to the methyl group, and the other with the aniline trans to an apical oxo ligand. The isomer with the aniline trans to an apical oxo ligand exhibits a unique Re=O bond, in which the bond length is greater than that of the other two Re=O bonds, reflecting the trans influence of the aniline ligand.

Hartree–Fock calculations on the MO$_3$ (M = Mn, Tc, and Re) and CH$_3$ fragments (12) demonstrated that ReO$_3$ and TcO$_3$ are isolobal to CH$_3$, allowing the comparison between CH$_3$ReO$_3$ and CH$_3$CH$_3$. The a$_1$ highest occupied molecular orbitals (HOMOs) of MO$_3$ (M = Tc or Re) are very comparable in energy (0.37 Hartree for TcO_3 and 0.39 Hartree for ReO_3) to the a_1 HOMO of CH_3 (0.43 Hartree) in C_{3V} symmetry. Therefore, like MTO, CH_3TcO_3 is expected to be stable according to this theoretical study. On the other hand, orbital energies of MnO_3 do not match those of CH_3; indeed CH_3MnO_3 is unknown.

MTO exchanges its oxo ligands with ^{17}O-labeled water, as shown with ^{17}O-NMR (*13*). The exchange was also verified by dissolving MTO in ^{18}O-labeled water and then measuring the mass spectrum of MTO. The mass spectrum displayed signals at 249 (CH_3ReO_3) and 251 ($CH_3ReO_2^{18}O$), demonstrating the incorporation of one ^{18}O ligand into MTO (*14*). A reasonable intermediate by which ligand exchange may be accomplished is shown in structure **1**. Supporting the proposed intermediate is the fact that MTO in different solvents displays different chemical shifts in ^1H-NMR and ^{17}O-NMR (*15*). For example, $\delta(^{17}O)$ in $CHCl_3$ is 829, but 861 in CH_3OH.

Over the course of days, concentrated aqueous solutions of MTO form a polymeric gold-colored precipitate in 70% yield (*16–19*). The reaction can be accelerated by heating to ~70 °C. This polymer is insoluble in all solvents, except hydrogen peroxide and in solutions of nitrogen bases such as pyridine. Under these circumstances the polymer forms the peroxide complex $CH_3Re(O)(O_2)_2(H_2O)$ (known independently and referred to hereinafter as **B**) and the CH_3ReO_3L adducts, respectively. Elemental analyses, IR spectroscopy, and powder diffraction established the formula of "poly-MTO" as $[C_{0.91}H_{3.3}ReO_{3.0}]_n$. Thus the polymer is basically MTO with ~10% of the methyl groups missing and some additional hydrogen, probably from molecules of water substituting for the missing methyl groups. Poly-MTO conducts electricity and exhibits weak paramagnetism. When pressure is applied to residues of poly-MTO, crystalline monomeric MTO is formed. The proposed structural model for the polymer bears similarity to the perovskite structure of ReO_3: a three dimensional lattice consisting of octahedral Re atoms and ⋯Re–O–Re–O⋯ chains. Consistent with that, at high-temperature poly-MTO eliminates CH_4 mainly to give violet ReO_3. The process by which MTO forms a polymer opens a new avenue to oxide ceramics and possibly heterometallic intercalation compounds. The proposed structure of poly-MTO is shown in structure **2** (*16*).

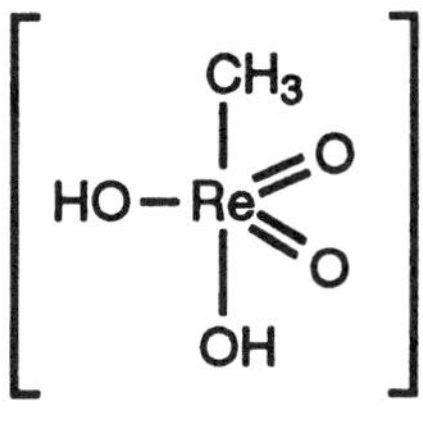

$$\left[\begin{array}{c} CH_3 \\ | \\ HO-Re{\Large\lessgtr}^O_O \\ | \\ OH \end{array} \right]$$

1

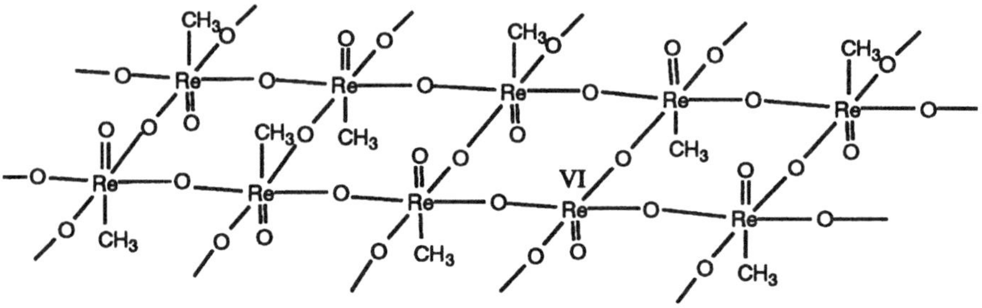

2

Rhenium Peroxides

Composition and Structure of Peroxides.   Aqueous or organic solutions of MTO, treated with excess H_2O_2, change from clear to yellow accompanying the formation of peroxo complexes. There are two rhenium peroxides, which we have called **A** and **B**. They have been characterized in solution by NMR, UV–vis, and IR. Additionally, the structure of the diglyme adduct of the diperoxide, **B**, was determined by X-ray analysis (*20*). The compound contains two η^2-coordinated peroxo ligands and a methyl group in the equatorial plane, and an apical oxo ligand positioned trans to a coordinated water molecule; the coordinated water is hydrogen-bonded to the triether diglyme. Hence, the structure of **B** can be described as a pentagonal bipyramid. The Re–OH$_2$ distance (2.25 Å) is relatively long, indicating that the water is loosely bound to the rhenium. In contrast, the surprisingly short Re–CH$_3$ distance (2.13 Å) illustrates the stability of the methyl group even though it is adjacent to the oxidizing peroxo ligands.

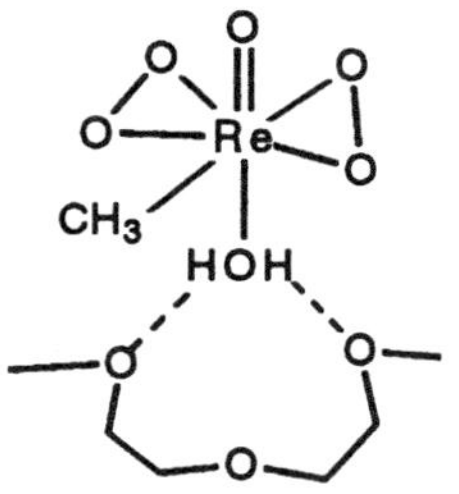

B

The diperoxide **B** has also been characterized in solution by spectroscopic methods. In THF and water an absorption maximum occurs at 360 nm (ε 1100 L mol^{-1} cm^{-1}) (*10, 21, 22*). The IR spectrum in CH_2Cl_2 or Nujol displays the following stretching frequencies that support the assigned structure: $\nu(\mathrm{Re}{=}\mathrm{O})$ = 1020 cm^{-1} (vs), $\nu(\mathrm{Re}{-}\eta^2{-}\mathrm{O}_2)$ = 330 (w), $\nu(\mathrm{O}_2)$ = 877 (w), and $\nu(\mathrm{Re}{-}\mathrm{CH}_3)$ = 571 (s) (*20*). The methyl group is observed in both ^{1}H– and ^{13}C–NMR: ^{1}H in D_2O at 25 °C, $\delta(\mathrm{CH}_3)$ = 3.0 ppm (s); THF-d^8 at –40 °C $\delta(\mathrm{CH}_3)$ = 2.7 ppm, $\delta(\mathrm{Re}{-}\mathrm{OH}_2)$ = 6.4 ppm (br) (*10*); ^{13}C[^{1}H]–NMR in D_2O at 20 °C, $\delta(\mathrm{CH}_3)$ = 31

(*20*); ^{13}C–NMR in THF-d^8 at –10 °C $\delta(CH_3)$ = 30.5 (q, $^2J(^{13}C,^1H)$ = 135 Hz) (*20*). The ^{17}O–NMR spectrum of **B** was reported (*20*) in Et_2O at –20 °C: $\delta(Re{=}O)$ = 762, $\delta(Re–\eta^2\text{-}O_2)$ = 422 and 363, and $\delta(H_2O)$ = –26.

The monoperoxide **A** has proved to be more difficult to characterize than **B** because **A** is usually a minor species (*see* Peroxide-Binding Equilibria) that irreversibly decomposes to MeOH and ReO_4^- (*see* Catalyst Integrity). Because **A** and **B** are formed more slowly in THF than in water, the UV–vis and ^1H–NMR spectra of **A** in THF are available (*10*). The UV spectrum of **A** shows λ_{max} = 305 nm (ε = 730 L mol^{-1} cm^{-1}), and a ^1H–NMR singlet at 2.4 ppm. The spectrum of **A** has been resolved in aqueous solution from a global fitting routine (*23*) that uses the continuous spectra at different $[H_2O_2]$ (*10*). The results of this global fitting are the equilibrium constants for the formation of **A** and **B**, as well as the spectra of **A** and **B**. Although the spectrum of **A** in water from the global fitting is not an exact match to that measured in THF, they are in reasonable agreement given the procedures and the fitting errors. Although the kinetics for the formation of **A** and **B** are much slower in THF than in water, this kinetic relationship is not necessarily true for other organic solvents; for example, the formation of the 1:1 rhenium–peroxide in MeOH is much faster than in water (*see* Kinetics of Peroxide-Forming Reactions).

Water coordination to **A** and **B** was examined by variable-temperature ^1H–NMR over the range of +20 to –50 °C (*10*). Solutions of **B** (generated in situ with high $[H_2O_2]$) in THF-d^8 at $T \leq$ –25 °C showed two water peaks, free water at 5.9 ppm and coordinated water at 6.4 ppm. A low temperature was needed to resolve the two because the water is reasonably labile. Ready exchange of coordinated and solvent water is expected in view of the relatively long $Re–OH_2$ bond distance (2.25 Å) (*20*). On the other hand, the ^1H-NMR spectrum of **A** (generated in situ at low $[H_2O_2]$) in THF-d_8 was recorded from 20 °C to –55 °C. No peak for water coordinated to **A** was observed. Given that the water peak for the otherwise similar **B** was easily detected it seems that either **A** lacks a coordinated water or its water is much more labile than that of **B**.

Peroxide-Binding Equilibria.   The equilibrium constants for the processes depicted in Scheme II were evaluated from the absorbances in the range 360–420 nm over a wide range of hydrogen peroxide concentrations. The absorbance changes at every wavelength were consistent with the two-reaction equilibrium model. Because H_2O_2 contributes negligibly at $\lambda \geq 360$ nm, we obtain

$$\frac{(Abs)_\lambda}{[Re]_T} = \frac{\varepsilon_0 + \varepsilon_A K_1[H_2O_2] + \varepsilon_B K_1 K_2[H_2O_2]^2}{1 + K_1[H_2O_2] + K_1 K_2[H_2O_2]^2} \tag{4}$$

The data at each of several wavelengths were fitted to this equation. The equilibrium constants ($\pm$ ~10%) in water, 1:1 water/acetonitrile, and methanol at 25 °C are given in Table II.

$$CH_3ReO_3 \underset{H_2O}{\overset{K_1\ H_2O_2}{\rightleftharpoons}} \mathbf{A} \underset{}{\overset{K_2\ H_2O_2}{\rightleftharpoons}} \mathbf{B}$$

Scheme II.

Table II. Equilibrium Constants for Peroxide-Binding of MTO in Different Solvents

Solvent	K_1 (L mol^{-1})	K_2 (L mol^{-1})	Ref.
H$_2$O			
$\mu = 0.10$ M , pH 1	7.7	145	22
$\mu = 2.0$ M, pH 1	16	132	10
H$_2$O–CH$_3$CN (1:1)			
$\mu = 0.10$ M, pH 1	13	136	39
MeOH	261	814	27

The second molecule of hydrogen peroxide is bound substantially more tightly than the first indicating cooperativity in the coordination equilibria. This binding order is unusual when compared to the usual stepwise metal–ligand binding, in which successive Lewis acid–base interactions weaken with the binding of the next base.

Kinetics of Peroxide-Forming Reactions.   The rhenium peroxides **A** and **B** form fairly rapidly, attaining their equilibrium concentrations in about 1 min. Despite this finding, the analysis of the kinetic data for the catalytic reactions cannot be carried out by assuming an equilibrium model. Many of the catalytic reactions occur rapidly enough that such an assumption would be invalid. More important, however, the equilibria between MTO, **A**, and **B** are continually being upset as one form or the other of the catalyst is consumed. The concentration of the reactive species will momentarily be lower than its equilibrium value. To characterize the kinetic steps in the catalytic sequence, which are of inherent interest, the kinetics of some of the peroxide-forming reactions have been studied.

Scheme III gives the notation for the reactions. The "extra" water molecule shown in the first step is not taken into account explicitly in the rate law because the activity of water generally is constant.

The kinetics of this reaction were carried out such that [H$_2$O$_2$] was, in effect, constant in each experiment. Keeping [H$_2$O$_2$] constant was achieved either by setting [H$_2$O$_2$] >> [Re]$_T$ or, in experiments at lower ratios, by selecting [H$_2$O$_2$]$_0$ such that the reactions did not proceed far to the right. Despite the

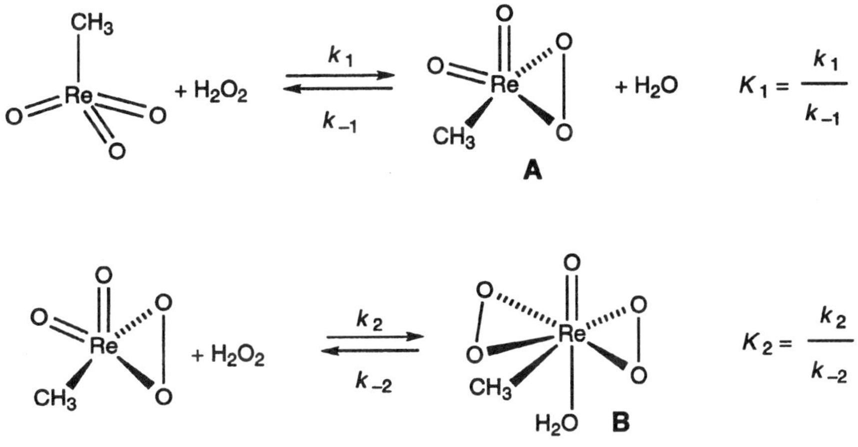

Scheme III.

simplification of $[H_2O_2]$ being constant, this model represents a complex kinetic scheme. Moreover, the two reactions could not, for the most part, be separated for individual study, although k_1 can be independently evaluated from one feature of the catalytic kinetics.

If we postulate that the equilibrium reactions producing compounds **A** and **B** also describe the kinetics, the rate equations are:

$$\frac{d[\mathbf{A}]}{dt} = k_1[CH_3ReO_3][H_2O_2] - k_{-1}[\mathbf{A}] - k_2[\mathbf{A}][H_2O_2] + k_{-2}[\mathbf{B}] \tag{5}$$

$$\frac{d[\mathbf{B}]}{dt} = k_2[\mathbf{A}][H_2O_2] - k_{-2}[\mathbf{B}] \tag{6}$$

The resulting expressions for [**A**] and [**B**] for the case $[H_2O_2] >> [Re]_T$ are the sums of two exponentials. The two relaxation times are complex functions of the four separate rate constants (*24–26*). The buildup of [**B**] is given by:

$$[\mathbf{B}]_t = [Re]_T \left\{ 1 + \frac{\lambda_3}{\lambda_2 - \lambda_3} \exp(-\lambda_2 t) - \frac{\lambda_2}{\lambda_2 - \lambda_3} \exp(-\lambda_3 t) \right\} \tag{7}$$

Parameters λ_2 and λ_3 are related to the parameters of the kinetic scheme. Useful combinations of the two are the sum and the product:

$$\lambda_2 + \lambda_3 = (k_1 + k_2)[H_2O_2] + k_{-1} + k_{-2} \tag{8}$$

$$\lambda_2 \times \lambda_3 = k_1 k_2[H_2O_2]^2 + k_1 k_{-2}[H_2O_2] + k_{-1}k_{-2} \tag{9}$$

The data fit gave the values of two of the forward rate constants, and those for the reverse reactions were then calculated from the equilibrium constants. The results were obtained in methanol (27) and water (22).

The rate constants are pH-independent. There is, however, a mild dependence on ionic strength. This dependence is quite unusual, because ions are not involved, and appears to reflect changes in the activity of water. The values are $k_1 = 77$ L mol^{-1} s^{-1}(μ 0.1), 55.3 (μ 1.0), and 42.2 (μ 2.0); and $k_{-1} = 9.0$ s^{-1} (μ 0.10) and 2.62 (μ 2.0). The steps given in Scheme III make it clear why the activity of water might be involved as a variable. On the other hand, this claim should be taken advisedly, since no data are available other than those given here (21).

On the other hand, the use of catalytic chemistry affords independent values of k_1. The reaction shows an appreciable solvent kinetic isotope effect, $k_1^H/k_1^D = 2.8$, and the activation parameters (28) are $\Delta H^\ddagger = 29.0$ kJ mol^{-1}, $\Delta S^\ddagger = -116$ J mol^{-1} K^{-1}, and $\Delta V^\ddagger = -10.6$ cm^3 mol^{-1}.

Quite clearly, in the formation of **A** from CH_3ReO_3, two protons must move, despite which the rate is pH-independent in the range 0–4.5. The transformation might happen via an intermediate σ-hydroperoxo complex, which never attains a high concentration and evidently undergoes no reaction other than conversion to the η^2–O_2^{2-} species. This possibility is depicted in Scheme IV.

η^1-Hydroperoxo complexes are not numerous. Those isolated as solids include $[MoO(O_2)_2(OOH)]\cdot2C_5H_5N$ (29), $[Pd(O_2CR)(OOH)]_2$ (R = CH_3 and CF_3) (30), and $[Pt(L–L)(CF_3)(OOH)]$ (L–L = $Ph_2P(CH_2)_2PPh_2$) (31–33). Other η^1–hydroperoxo complexes are known in solution, including $(H_2O)_5CrOOH^{2+}$ (34–36) and $LCo(OH_2)OOH^{2+}$ (L = cyclam and hexamethylcyclam) (37, 38).

The formation of **A** is characterized by a low activation enthalpy, a substantial and negative activation entropy, and a negative activation volume. The three activation parameters consistently point to a rate-controlling step that features bond making more than bond breaking. Thus only a small enthalpy barrier presents itself. The structure of the transition state appears to be a highly ordered one, not far advanced toward the product. The activation parameters suggest to us that the first step of Scheme IV is rate-controlling. To

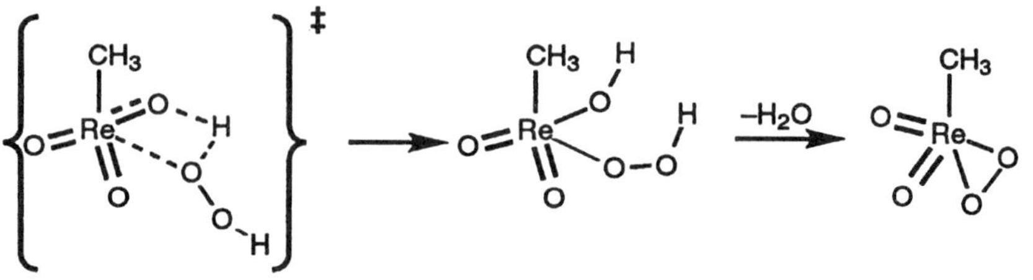

Scheme IV.

allow for incomplete O–H bond making/breaking in light of the kinetic isotope effect, proton transfer from hydrogen peroxide to an oxo group is involved in the transition state, despite there being no kinetic dependence on $[H_3O^+]$ (28).

Catalytic Oxidations

Oxidations of Ar₃E (E = P, As, or Sb).   The oxidation of organic phosphines, Ar_3P mostly, as well as Ph_3As and Ph_3Sb, was studied in 1:1 CH_3CN/H_2O (39). The reactivity of these substrates toward **A**, the catalytically active 1:1 peroxide, was evaluated, and in two instances the 2:1 adduct **B** was studied as well. Systematic changes in the substituents on the phosphorus were made to vary the nucleophilicity of the phosphine and its cone angles; the kinetic effects are discernible, although they lie in a narrow range.

The reactions with MTO appeared to follow zeroth-order kinetics during the initial part of the reaction, but an approximate single exponential toward the end. The catalyzed reaction showed first-order dependences on $[H_2O_2]$ and $[Re]_T$.

At low $[H_2O_2]$, $[B]$ is negligible, and the reactive form of the catalyst is **A**, and eqs 10 and 11 apply. With the steady-state approximation for **A**, eq 12 is obtained, in which $[Re]_T = [MTO] + [A]$. It applies when $[B]$ is negligible.

$$CH_3ReO_3 + H_2O_2 \underset{k_{-1}}{\overset{k_1}{\rightleftharpoons}} CH_3Re(O)_2(\eta^2\text{-}O_2) + H_2O \tag{10}$$

$$Ar_3E + CH_3Re(O)_2(\eta^2\text{-}O_2) \overset{k_3}{\longrightarrow} O = Ar_3E + CH_3ReO_3 \tag{11}$$

$$v_c = \left(\frac{-d[Ar_3E]}{dt}\right)_c = \frac{k_1 k_3 [Re]_T [H_2O_2][Ar_3E]}{k_{-1} + k_3[Ar_3E] + k_1[H_2O_2]} \tag{12}$$

At high concentrations of substrate, such that $k_3[Ar_3E] >> k_{-1}$, the limiting form of the rate law is

$$v_c = \left(\frac{-d[Ar_3E]}{dt}\right)_c = k_1[Re]_T[H_2O_2] \tag{13}$$

When $[H_2O_2]$ is limiting, the kinetic traces follow a single exponential decay with $k_\psi = k_1[Re]_T$. A set of experiments with PPh_3 over a wide range of $[Re]_T$ yielded a value of $k_1 = 32.5$ L mol⁻¹ s⁻¹ at 25 °C in 1:1 CH_3CN–H_2O at pH 1. From K_1 (*see* Peroxide-Binding Equilibria), we have $k_{-1} = 2.5$ s⁻¹.

As can be seen from eq 12, the conditions needed for evaluation of k_3 and k_{-1} are low $[H_2O_2]$ with $[Ar_3E]$ varied in the region in which k_{-1} and $k_3[Ar_3E]$ are comparable. The values of k_3 for the catalyzed reactions are tabulated, together with those for the uncatalyzed reactions, in Table III. The activating power of MTO, defined as k_3/k_u, is also presented and shows a remarkable

Table III. Summary of Rate Constants for Reactions of A and B with Ar₃E

Substrate	k_u (L mol⁻¹ s⁻¹) (uncatalyzed)	k_3 (L mol⁻¹ s⁻¹) (catalyst A)	k_4 (L mol⁻¹ s⁻¹) (catalyst B)	Activation (k_3/k_u)
P(p-CH$_3$C$_6$H$_4$)$_3$	4.1	9.4×10^5		2.3×10^5
P(C$_6$H$_5$)$_3$	3.0	7.3×10^5	22×10^5	2.4×10^5
P(p-ClC$_6$H$_4$)$_3$	1.87	4.8×10^5		2.6×10^5
P(p-CF$_3$C$_6$H$_4$)$_3$	0.71	3.4×10^5		4.8×10^5
P(o-CH$_3$C$_6$H$_4$)$_3$	0.119	1.9×10^5		1.6×10^6
P(C$_6$H$_5$)$_2$(CH$_3$)	7.5	3.9×10^6		5.2×10^5
PTA		3.2×10^3		
P(C$_6$F$_5$)$_3$	1.6×10^{-3}	1.3×10^2	3.47×10^2	8×10^4
As(C$_6$H$_5$)$_3$	0.084	3.7×10^5		4.4×10^6
Sb(C$_6$H$_5$)$_3$	1.33	5.3×10^5		4×10^5

NOTE: 1,3,5-Triaza-7-phospha-adamantane (PTA) has the following structure:

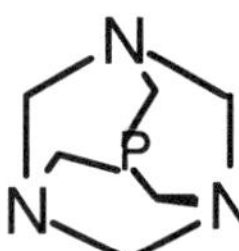

SOURCE: Data are from reference 39.

acceleration of 10^5–10^6. Such a large activation indicates major electronic changes in the peroxide upon coordination.

Because **B**, like **A**, contains peroxide ligands, it is reasonable to anticipate that **B** is catalytically active as well (eq 14).

$$\text{Ar}_3\text{E} + \textbf{B} \xrightarrow{\ k_4\ } \text{O} = \text{EAr}_3 + \textbf{A} + \text{H}_2\text{O} \tag{14}$$

The catalytic kinetics of **B** were studied for two different substrate, PPh$_3$ and P(C$_6$F$_5$)$_3$. Pseudo-first-order conditions were attained by using $[\textbf{B}] \geq 10[\text{Ar}_3\text{P}]$ and $[\text{H}_2\text{O}_2]_0 \geq 100[\text{Re}]_\text{T}$, so that $[\textbf{B}]$ remained constant throughout the reaction. The high $[\text{H}_2\text{O}_2]$ results in conversion of almost all the starting MTO into **B**. Kinetic simulations by KinSim (40) showed that, under these conditions, the reaction should be first-order with respect to substrate, as observed. Under these conditions k_ψ becomes

$$k_\psi = k_4[\textbf{B}] + \frac{k_3[\textbf{B}]}{K_2[\text{H}_2\text{O}_2]} \tag{15}$$

Both PPh$_3$ and P(C$_6$F$_5$)$_3$ substrates have $k_4 \sim 3k_3$ (Table III), which establishes that **A** and **B** have similar reactivities toward the phosphines.

The triarylphosphines with different para substituents exhibited a linear Hammett correlation (41, 42), supporting a common mechanism. The negative sign of the reaction constant ($\rho = -0.63$) is in agreement with nucleophilic

**Table IV. Values of k_3 for Selected Phosphines Compared
with Tolman's Electronic Parameter (v)**

	k_3 ($L\ mol^{-1}\ s^{-1}$)	v (cm^{-1})
Ph_2MeP	39×10^5	2065
Ph_3P	7.3×10^5	2069
$(C_6F_5)_3P$	1.3×10^2	2091

attack of the substrate on the electrophilic oxygen. The electronic trend exhibited by the k_3 values of $MePh_2P$, Ph_3P, and $(C_6F_5)_3P$ also supports the proposed mechanism. The rate constants are compared with Tolman's electronic parameter (v) (*43*) in Table IV.

The kinetic trends for Ph_3E follow the order P > Sb > As, although the rate differences are within a factor of 2. This order is not unprecedented; aryl stibines are oxidized by alkaline hydrogen peroxide (*44*) more readily than aryl arsines. The same trend was also observed for the oxidations of Ar_3E by peroxodisulfate (*45*), peroxodiphosphate (*46*), and *tert*-butyl hydroperoxide (*47*).

The resemblance in reactivities between Ph_3P and its arsine and stibine analogs is not unprecedented. In fact some metal–peroxo catalysts have been used for arsines as well as phosphines (*48*, *49*). For example, $Ru(\eta^2\text{-}O_2)(NO)(PPh_3)_2(CO)(SCN)$ oxidizes Ph_3As 3–4 times faster than Ph_3P under the same conditions. Thus we conclude that **A** shows relatively little discrimination among P, As, and Sb. This result is possibly a consequence of the high electrophilic activation of the η^2-coordinated peroxide ion $(O_2)^{2-}$ upon binding to Re(VII), since hydrogen peroxide alone as an oxidant shows more discrimination between P and As than **A** does.

Because the oxidation of phosphines by H_2O_2 is assumed (*50*) to entail nucleophilic attack by the phosphine on the O–O bond (eq 16), it is reasonable to assume that the same type of nucleophilic attack by Ar_3E occurs in the MTO-catalyzed reaction. That is, the P, As, or Sb attacks the oxygen of the coordinated peroxide ion of **A**.

$$R_3P + \ \ H\text{O-OH} \ \longrightarrow \ R_3P\cdots O\text{-}O \ \longrightarrow \ R_3P{=}O + H_2O \quad (16)$$

The possibility of substrate coordination prior to oxygen transfer was considered. As a general rule, it is related to the kinetic effect of steric hindrance (*51*). Whereas steric factors are rather unimportant in typical electrophilic reactions, such as the acid-catalyzed oxidation of sulfides with hydrogen peroxide (*52*), they play a major role in processes that involve the coordination of substrate. For example, in the metal-catalyzed homogeneous hydrogenations of olefins, in which the olefin is coordinated to the metal complex, the retardation

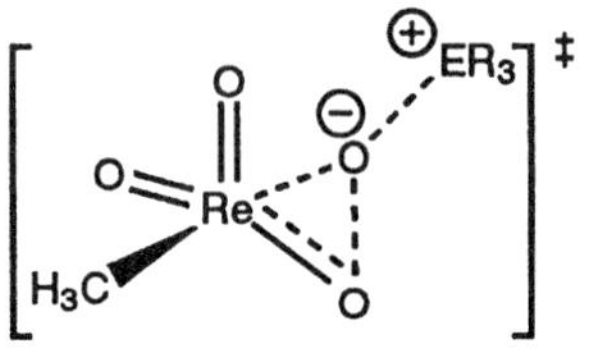

3

is remarkable (53). The reactivity ratio of cyclohexene/1–methylcyclohexene with certain Rh complexes is >40 (53).

Because (p-tolyl)$_3$P and (o-tolyl)$_3$P are electronically the same [Tolman's electronic parameter, ν, for both is 2067 cm^{-1} (43)] but sterically very different [cone angles (43) of 145° for the *para*-methyl and 194° for the *ortho*-methyl], we compared their reaction rates with **A** (Table III). The small difference in rate lends support to the electrophilic mechanism in which attack of the nucleophilic substrate on the peroxide oxygen is the major pathway. Therefore, we suggest the structure **3** for the transition state, and a similar one could be drawn for **B**.

Catalyzed Oxidations of Organic Sulfides.   The following catalyzed reaction (14) is the subject of this section:

$$R_2S + H_2O_2 \rightarrow R_2SO + H_2O \tag{17}$$

where the groups R may be aryl or alkyl (or H, although H leads to further reactions, resulting in the production of RSSR and RSO$_3$H). The general scheme presented previously is valid here as well. That is to say, the rhenium peroxides **A** and **B** transfer oxygen atoms upon nucleophilic attack by the sulfide as evidenced by (1) the functional dependences of the reaction rate upon concentrations, (2) the kinetic effects of changing the R groups or a substituent on them, and (3) isotopic labeling experiments with oxygen-18. We shall examine each in turn.

The kinetic data follow Michaelis–Menten kinetics, in that the rate approaches zeroth-order kinetics at the beginning of each run, and drifts toward first-order kinetics near the end. This finding also accords with the hyperbolic dependences of the initial reaction rate on both [H$_2$O$_2$] and [R$_2$S]. At low concentrations of hydrogen peroxide, only a minimal amount of **B** is formed. When these conditions are realized, **A** will be the predominant peroxide and the principal reactive species, and the rate law is

$$\frac{d[R_2SO]}{dt} = \frac{k_3[Re]_T[H_2O_2][R_2S]}{\dfrac{k_{-1} + k_3[R_2S]}{k_1} + [H_2O_2]} \tag{18}$$

This treatment allowed the evaluation of k_3 for the oxygen-transfer reaction between **A** and R_2S, the other rate constants being known independently. At very high concentrations of hydrogen peroxide, however, the absorbance-time traces adhered more closely to first-order kinetics. Equation 18 was no longer applicable because the higher concentrations of **B** then needed to be considered in the kinetic analysis. The steady-state approximation was applied to both [**A**] and [**B**]. The rate of the reaction, with the full set of chemical equations, is given by Equation 19

$$\frac{d[R_2SO]}{dt} = \frac{k_1 k_3 [Re]_T [H_2O_2][R_2S] + \dfrac{k_1 k_2 k_4 [Re]_T [R_2S][H_2O_2]^2}{k_4 [R_2S] + k_{-2}}}{k_{-1} + k_3 [R_2S] + k_1 [H_2O_2] + \dfrac{k_1 k_2 [H_2O_2]^2}{k_4 [R_2S] + k_{-2}}} \tag{19}$$

in which k_{-2} may be replaced by k_2/K_2. This approach reduces the unknowns in eq 19 to two, k_4 and k_2, the other constants being known independently. The data adhere to the functional forms required by eqs 18 and 19, and the catalytic rate constants are thus obtained reliably and reproducibly. Lest readers be daunted by the complexity of these equations, a reminder should be given that the equilibrium constants and the other rate constants are known independently, other than the imprecise k_2, allowing considerable simplification. Moreover, a great many of the experiments were carried out by the method of initial rates. This method simplifies the differential equations to algebraic equations and allows a ready examination of the functional dependences of rate upon concentrations. Aside from this, the conditions for the kinetics experiments can in certain circumstances be set such that the rate constants of greatest interest in the catalytic processes, k_3 and k_4, can be determined more straightforwardly. These methods were developed in conjunction with epoxide and diol-forming reactions of styrenes (54) (*see* Oxidation of Styrenes, Alkenes, and Alkynes).

As to the mechanism of the oxygen-transfer step in the catalytic process, consider the rate constants given in Table V. Values are shown for k_3 and (where determined) for k_4. The sulfur atom attacks the peroxo group of **A** or **B**. From such a transition state the sulfoxide product would be formed directly; at the same time, CH_3ReO_3 is regenerated from **A**, and **A** from **B**. Hydrogen peroxide is activated by the electron-poor, high-valent rhenium center, increasing the electrophilicity of the peroxide group; nucleophilic attack by the sulfur atom is thus made more facile. A diagram of the proposed transition state, which is supported by the kinetic effects of the substituents and by a Hammett linear-free-energy analysis, is (*14*) shown in structure 4.

To ascertain which oxygen atom of **A** transfers to the sulfide, we first asked whether both of the oxygen atoms in the peroxide unit of **A** originate from the hydrogen peroxide. That is, does the O–O bond remain unbroken upon forma-

Table V. Rate Constants for the Oxidation of Arylalkyl Sulfides and Dialkyl Sulfides (25 °C, in 1:1 acetonitrile-water at pH 1)

Substrate	A k_3 (L mol^{-1} s^{-1})	B k_4 (L mol^{-1} s^{-1})
Ph–S–CH$_3$	2.65×10^3	9.65×10^2
p-CH$_3$C$_4$H$_4$–S–CH$_3$	4.3×10^3	
p-CH$_3$O-o-CH$_3$C$_4$H$_3$–S–CH$_3$	8.5×10^3	
p-ClC$_4$H$_4$–S–CH$_3$	1.6×10^3	
p-NH$_3^+$C$_4$H$_4$–S–CH$_3$	5.7×10^2	7.0×10^1
Ph–S–CH=CH$_2$	1.49×10^2	
Ph–CH$_2$–S–CH$_3$	5.4×10^3	
Ph$_2$S	1.18×10^2	3.2×10^1
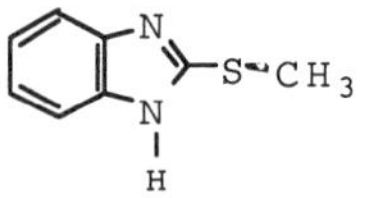 (benzimidazol-2-yl)–S–CH$_3$ (N–H)	4×10^{-1}	4×10^{-1}
(CH$_3$CH$_2$)$_2$S	2.0×10^4	
Pentamethylenesulfide	2.0×10^4	
(Me$_2$CH)$_2$S	1.6×10^4	

SOURCE: Data are from reference 14.

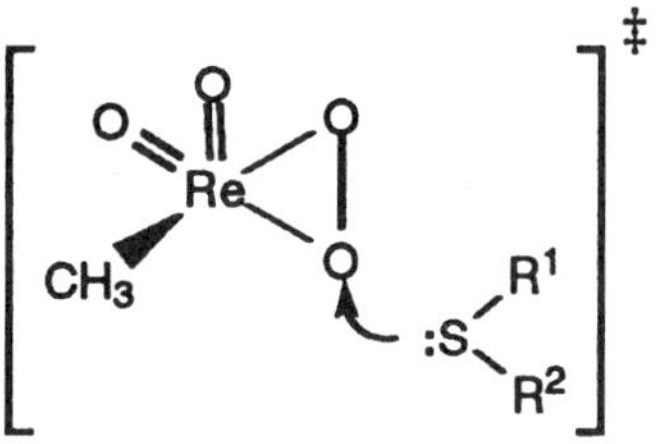

4

tion of **A** from MTO, or does the rhenium peroxide contain one oxygen atom from the hydrogen peroxide and another from the CH$_3$ReO$_3$? Oxygen atoms exchange between H$_2$O and CH$_3$ReO$_3$ (20); in our work we verified by ^{18}O–labeling that the exchange occurs readily in neutral and acidic solutions.

The oxidation of methyl phenyl sulfide in ^{18}O-labeled water, and thereby with ^{18}O-labeled CH$_3$ReO$_3$, yields only the ^{16}O-sulfoxide. In the formation of **A** the peroxo group remains intact. The rhenium peroxide therefore transfers to methyl phenyl sulfide an oxygen atom originating with the original hydrogen peroxide, consistent with the previous depiction.

The same kinetic pattern accounts for the catalytic oxidation of a cobalt(III) thiolate complex (55):

$$CoSR^{2+} \xrightarrow{\ A\ } CoS(O)R^{2+} \xrightarrow{\ A\ } CoS(O)_2R^{2+} \qquad (20)$$

The rate constants are $k_3 = 4.3 \times 10^5$ L mol^{-1} s^{-1} for the first stage of oxidation and $k_3 = 2.65 \times 10^2$ L mol^{-1} s^{-1} for the second in aqueous solution at 25 °C. The uncatalyzed reaction between $(en)_2CoSR^{2+}$ and H_2O_2 is very much slower. The catalytic enhancements can be approximated by comparing the catalytic rate with the uncatalyzed reaction of H_2O_2. The rate constants are summarized in Table VI for several catalysts. The enhancement is particularly striking, and it shows that MTO accelerates these reactions by factors of ~10^5. The powerful rate enhancement arising from the rhenium catalyst is much like the large kinetic accelerations found for the phosphines. This finding is also true for the second-stage oxidation of $(en)_2Co(S(O)CH_2CH_2NH_2-)^{3+}$ (55).

Aniline Oxidations (27).   The catalyzed oxidation of *N,N*-dimethylanilines proceeds quantitatively to the amine oxide:

$$X-\!\!\!\bigcirc\!\!\!-NMe_2 + H_2O_2 \xrightarrow{cat.\ MTO} X-\!\!\!\bigcirc\!\!\!-\overset{\underset{\uparrow}{O}}{N}Me_2 + H_2O \qquad (21)$$

Because the peroxide catalyst **A** is unstable at high pH (*see* Catalyst Integrity), and the amine is of course unreactive in acidic solution, the data were obtained

Table VI. Rate Constants for the Oxidation of a Cobalt Thiolate Complex

Oxidant	k_3 *(L mol^{-1} s^{-1})*	Activation (k_3/k_u)	Ref.
A	4.2×10^5	3.1×10^5	55
$WO(O_2)_2$	3.7×10^5	2.7×10^5	85
$WO(OH)(O_2)_2^-$	3.4×10^4	2.5×10^5	85
$MoO(O_2)_2$	1.91×10^4	1.4×10^4	85
$MoO(OH)(O_2)_2^-$	2.4×10^3	1.8×10^3	85
H_2O_2	$k_u = 1.36$	—	86

NOTE: Reaction conditions are 25 °C in H_2O at pH 1.

Table VII. Rate Constants for the Oxidation of para-Substituted Dimethylanilines by Compound A

X	k_3 (L mol^{-1} s^{-1})
CH$_3$	24.5
H	18.4
F	12.7
Br	8.7
NO$_2$	1.9

NOTE: X column entries are para-substituted groups.

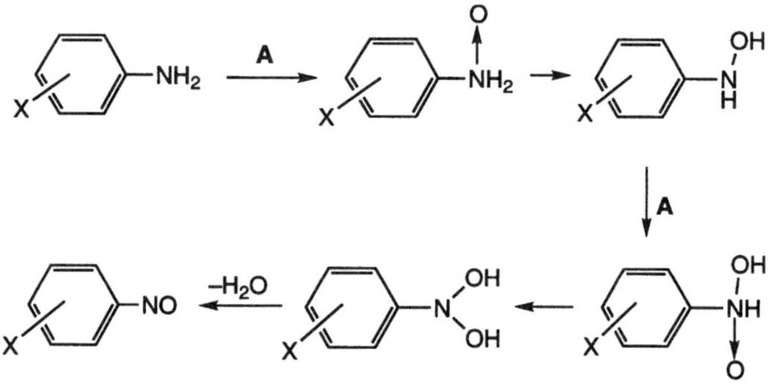

Scheme V.

in nearly dry methanol. For *para*-substituted compounds the k_3 values at 25 °C in methanol are given in Table VII.

The data support the analogous mechanism: nucleophilic attack of the amine on a peroxidic oxygen of **A**. The rate constants fit a Hammett correlation, with $\rho = -1.19$, suggesting attack by nitrogen lone-pair electrons of the aniline on a peroxidic oxygen of **A**.

Aniline and aryl-substituted anilines are oxidized to nitrosobenzenes. The initial step is the same, O-atom transfer, following which tautomerization yields the aryl hydroxylamine, which is subject to further oxidation and then dehydration (Scheme V).

In support of this model, independent studies with phenyl hydroxylamine afforded $k_3 = 1.8 \times 10^2$ L mol^{-1} s^{-1}, compared to 2.0 L mol^{-1} s^{-1} for PhNH$_2$. This difference accounts for the failure of the hydroxylamine to accumulate during the oxidation of aniline. The small amount of azobenzene (= diphenyl diazene) arises from a noncatalytic reaction between nitrosobenzene and aniline in methanol:

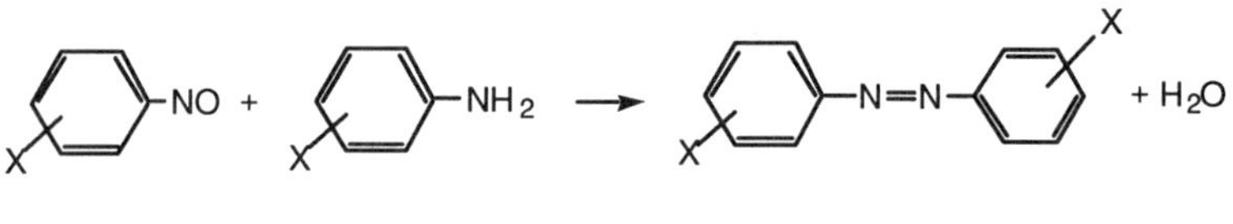

$$\text{ArNO} + \text{ArNH}_2 \longrightarrow \text{Ar-N=N-Ar} + \text{H}_2\text{O} \qquad (22)$$

Catalytic Oxidations of Halide Ions.   Bromide ions are oxidized by hydrogen peroxide, although the spontaneous reaction (eqs 23 and 24) occurs rather slowly without a catalyst (*56, 57*).

$$H_2O_2 + 2Br^- + 2H^+ = Br_2 + 2H_2O \tag{23}$$

$$v_u = \left\{ 3.8 \times 10^{-7} + 2.3 \times 10^{-4}[H^+] \right\} [H_2O_2][Br^-] \text{ mol L}^{-1} \text{ s}^{-1} \tag{24}$$

MTO is an effective catalyst (*21*), and the reaction starts with the same general scheme presented previously. The rate constants (at 25 °C and 2.0 M ionic strength) are $k_3 = 3.35 \times 10^2$ L mol^{-1} s^{-1} for **A** and $k_4 = 1.9 \times 10^2$ L mol^{-1} s^{-1} for **B**. These rate constants refer to the reactions

$$Br^- + CH_3Re(O)_2(O_2) \, (= A) \xrightarrow{\ k_3\ } BrO^- + CH_3ReO_3 \tag{25}$$

$$Br^- + CH_3Re(O)(O_2)_2(H_2O) \, (= B) \xrightarrow{\ k_4\ } BrO^- + CH_3Re(O)_2(O_2) + H_2O \tag{26}$$

The initial product is indeed hypobromite ions, which (as HOBr) partitions between Br$_2$ formation (upon reaction with Br$^-$ and H$^+$) or for O$_2$ formation (by reaction with H$_2$O$_2$). Thus the course of the reaction depends upon the relative concentrations of Br$^-$ and H$_2$O$_2$ as well as the pH. The net stoichiometry reflects, in one limit, the oxidation of bromide ions to bromine and tribromide ions, and in the other the disproportionation of hydrogen peroxide, jointly catalyzed by MTO and bromide ions. The balance is governed by the competitive rates. In principle, at least, the ultimate event would be only peroxide disproportionation because Br$_2$ hydrolyzes to regenerate the bromide catalyst. In actuality, however, the decomposition of some rhenium peroxides sets in at long times, and some bromine will usually remain. The original article should be consulted for the details of the multiple reactions occurring. In any event, the kinetic step in these transformations is oxygen transfer to Br$^-$ from **A** or **B**.

The oxidation of chloride ions proceeds in a similar manner, although the thermodynamics of the oxygen transfer step are more closely balanced (*58*) and dependent on the pH. The kinetic study (*59*) gave these rate constants (at 25 °C and 2.0 M ionic strength): $k_3 = 0.06$ L mol^{-1} s^{-1} for **A** and $k_4 = 0.13$ L mol^{-1} s^{-1} for **B**. Comparing the rate constants for Br$^-$ and Cl$^-$, we find a kinetic factor of $\sim 10^3$ in favor of bromide, which is more easily oxidized.

Oxidations of Styrenes, Alkenes, and Alkynes.   Styrenes are converted to epoxides with hydrogen peroxide–MTO; negligible reaction occurs without a catalyst. The epoxides are obtained in acetonitrile; in aqueous or semiaqueous media the epoxide is an intermediate, and the 1,2-diol is isolated as the final product, consistent with the acid-catalyzed ring-opening reaction that epoxides are known to undergo (*54*).

The slower reactions of the styrenes required the refinement of the techniques for the catalytic reactions. In one approach, with $[H_2O_2] > 0.5$ M, the absorbance-time values followed first-order kinetics. Indeed, at high $[H_2O_2]$, the amount of **A** present is negligible, and the formation of **B** during the reaction is much faster than the reaction of **B** with styrene; that is, $k_2[H_2O_2] >> k_4[\text{styrene}]$. The concentration of **B** was constant during the reaction and essentially equal to $[Re]_T$. Under these conditions the rate of the reaction becomes:

$$v = k_4[\mathbf{B}][\text{styrene}] = k_4[Re]_T[\text{styrene}] = k_\psi[\text{styrene}] \tag{27}$$

Another approach is based on the fact that the equilibrium concentrations of **A** and **B** will not change during the reaction if the following conditions are maintained throughout the reaction time:

$$k_1[H_2O_2] + k_{-1} >> k_3[\text{styrene}] \quad \text{and} \quad k_2[H_2O_2] + k_{-2} >> k_4[\text{styrene}] \tag{28}$$

which allows the rate law to be written as

$$v = (k_3[\mathbf{A}] + k_4[\mathbf{B}])\,[\text{styrene}] = k_\psi[\text{styrene}] \tag{29}$$

In this method, kinetic measurements were carried out with constant $[Re]_T$ and constant [styrene]. To ensure a significant contribution from **A**, $[H_2O_2]$ was not used in very large excess, so that **B** was not entirely dominant in that $K_2 > K_1$. The data fit pseudo-first-order kinetics, with $k_\psi = k_3[\mathbf{A}]_{eq} + k_4[\mathbf{B}]_{eq}$. The values of k_3 and k_4 were calculated using the nonlinear least-squares fitting; an equation with single independent variable can also be used by dividing k_ψ by $[\mathbf{A}]_{eq}$:

$$\frac{k_\psi}{[\mathbf{A}]_{eq}} = k_3 + k_4 \frac{[\mathbf{B}]_{eq}}{[\mathbf{A}]_{eq}} \tag{30}$$

Plots of $k_\psi/[\mathbf{A}]_{eq}$ against $[\mathbf{B}]_{eq}/[\mathbf{A}]_{eq}$ give straight lines with slopes of k_4 and intercepts of k_3. The rate constants are relatively insensitive to steric hindrance, but increase with the nucleophilicity of the styrene, or electron-donating groups on the olefinic carbons or on the aromatic ring, enhancing the rate. The rate constants for *meta*- and *para*-substituted styrenes follow a linear Hammett relationship; correlation with σ^+ gave $\rho = -0.93$. Table VIII presents a number of the rate constants, which were obtained from kinetic determinations that employed optical, NMR, and thermometric detection of the reaction progress (54).

Electron-donating groups on the aromatic ring or on the olefinic carbons enhance the C–C bond cleavage. For example, no C–C bond cleavage was observed with styrene, whereas *trans*-4-propenylanisole produced, in addition,

Table VIII. Rate Constants for the Reactions of Styrenes with A(k_3) and B (k_4)

Styrene	k_3 *(L mol^{-1} s^{-1})*	k_4 *(L mol^{-1} s^{-1})*
$PhCH=CH_2$	—	0.11
$3\text{-}MeC_6H_4\text{--}CH=CH_2$	—	0.12
$4\text{-}MeC_6H_4\text{--}CH=CH_2$	0.38	0.18
$2,4\text{-}Me_2C_6H_3\text{--}CH=CH_2$	—	0.38
$2,4,6\text{-}Me_3C_6H_2\text{--}CH=CH_2$	0.13	0.12
$PhC(Me)=CH_2$	—	0.47
trans-$PhCH=CHMe$	0.51	0.22
cis-$PhCH=CHMe$	0.74	0.28
$PhCH=CMe_2$	1.00	0.70
$PhCH=CH(CH_2OH)$	—	0.14
trans-$PhCH=C(Me)CH_2OH$	0.73	0.43
$4\text{-}MeOC_6H_4\text{--}CH=CH_2$	0.7	0.60
trans-$MeO\text{-}4\text{-}C_6H_4\text{--}CH=CHMe$	2.79	1.0
$PhCH=CH(OMe)$	14.1	16

NOTE: Reaction conditions are 1:1 CH_3CN/H_2O, pH 1, 25 °C.

10–15% of 4-methoxybenzaldehyde and 4-methoxybenzyl alcohol. β-Methoxystyrene, on the other hand, underwent complete C–C bond cleavage. The products were benzaldehyde and formaldehyde. The same products also resulted from the reaction of β-methoxystyrene with *m*-chloroperoxybenzoic acid in CD_3CN/D_2O at pD 1.

During olefin epoxidation by CH_3ReO_3–H_2O_2 in *tert*-butanol, catalytic ring opening of epoxides to 1,2-diols also occurs (*60*). In dry organic solvents, such as CH_2Cl_2 and THF, the 1,2-diols react with CH_3ReO_3 to form Re–diol chelate complexes (*61*) (eq 31).

$$\tag{31}$$

Formation of the diolate may reduce the catalytic turnover. Epoxides bind more strongly to Mo^{VI}–peroxo complexes than the corresponding olefins do (*62*), inhibiting epoxidation and leading to further oxidation of the metal-coordinated epoxide (*63, 64*). The addition of 1,2-diols did not alter the rate of epoxidation of styrene by CH_3ReO_3–H_2O_2. The epoxide ring opening is catalyzed by the acid (0.10 M $HClO_4$) not by CH_3ReO_3. The reaction of 1,2-diols with Re(VII) is inhibited by water since d^0 early transition metals (Lewis acids) have high affinity for bases such as water, present in high concentration relative to the 1,2-diols.

The rate of reaction of β-methoxystyrene with CH_3ReO_3–H_2O_2 increases with the amount of water in the solution, consistent with the higher values of k_1, k_{-1}, k_2, and k_{-2}, in H_2O (22, 27). There is also an effect on the catalytic steps: the rate constants for *trans*-4-propenylanisole are $k_4 = 1.1$ L mol^{-1} s^{-1} in 1:1 CH_3CN/H_2O, and 0.45 L mol^{-1} s^{-1} in a 4:1 solvent mixture. Styrene has k_4 0.021 L mol^{-1} s^{-1} in methanol, compared to 0.11 in 1:1 CH_3CN/H_2O.

Clearly **A** and **B** epoxidize styrenes at similar rates, at variance with a report that only **B** does so (20). Olefins are epoxidized by $M(O)(O_2)_2$HMPA (HMPA = hexamethylphosphoric triamide; M = Mo and W). This diperoxide is more reactive than the monoperoxide $M(O)_2(O_2)$HMPA when M = W. For Mo the opposite was found (65). **A** is less sensitive to steric factors than **B**. For example, *trans*-4-propenylanisole with **A** reacts about three times faster than **B** ($k_3/k_4 = 3.4$), whereas for the less sterically hindered 4-vinylanisole, $k_3/k_4 = 1.3$.

The very similar reactivities of **A** and **B** should be noted, not only toward styrenes, but also with respect to several of the species cited earlier: Br$^-$, ArNMe$_2$, R$_2$S, and Ar$_3$P, for example. This might at first seem surprising, in that CH_3ReO_3 is the product of **A**, whereas **A** itself is the product of **B**; clearly the two products, CH_3ReO_3 and **A**, are quite different. Nonetheless, **A** and **B** hardly differ in free energy with respect to the extra molecule of H_2O_2/H_2O by which they differ; that is, K_1 is only 10.6 L mol^{-1}, or $\Delta G°_1 = -6$ kJ mol^{-1}. Both **A** and **B** contain $Re(\eta^2\text{-}O_2^{2-})$, the active chemical unit.

The oxidation of alkynes with hydrogen peroxide is catalyzed by MTO. Internal alkynes are oxidized to α-diketones and carboxylic acids, the latter from the complete cleavage of the triple bonds. The major product, the α-diketone, can be rationalized by postulating an oxirene intermediate, likely proceeding by way of a "double epoxide" to the final α-diketone (66, 67).

$$^1R-C\equiv C-R^2 \xrightarrow{\textbf{A or B}} \left\{ R-C\overset{\overset{\displaystyle O}{\triangle}}{=}C-R^2 \right\} \xrightarrow{\textbf{A or B}} \left\{ {}^1R-C\underset{O}{\overset{O}{\diamondsuit}}C-R^2 \right\} \longrightarrow {}^1R-\overset{\overset{\displaystyle O}{\|}}{C}-\underset{\underset{\displaystyle O}{\|}}{C}-R^2 \quad (32)$$

Rearrangement products were also observed for aliphatic alkynes. Terminal alkynes gave primarily carboxylic acids and their derivatives and α-ketoacids. By-products are formed as well, depending on the solvent and the alkyne. They can be accounted for by side reactions that are plausible for the intermediates depicted (66).

Rhenium(VII) Diolates

Rhenium(VII) complexes containing a chelated bis(diolate) ligand have been prepared by refluxing MTO with 2,3-dimethyl-2,3-diol (61, 68). MTO and an epoxide also lead to the diolates (69). Five epoxides were used: tetramethyl epoxide, styrene epoxide, *cis*-cyclododecane epoxide, *cis*-stilbene oxide, and *trans*-stilbene oxide. All except the last reacted with MTO to give the corresponding bis(alkoxy)rhenium(VII) compounds in nearly quantitative yield.

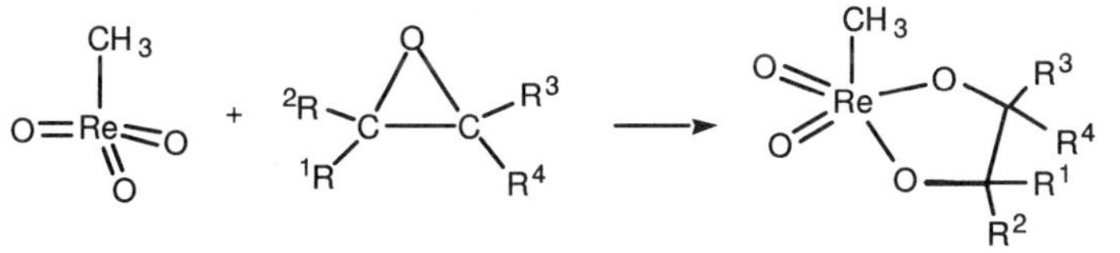

$$\text{(33)}$$

We propose that the oxygen atom of the epoxide approaches the rhenium atom at a site opposite to the Re–C bond, because *trans*-stilbene oxide fails to react with MTO (Scheme VI).

The bis(alkoxy)rhenium(VII) complexes react with triphenyl phosphine in dry benzene at room temperature:

$$CH_3Re(O)_2(OCR_2CR_2O) + PPh_3 \rightarrow CH_3ReO_3 + Ph_3P=O + R_2C=CR_2 \quad \text{(34)}$$

This release of the alkene is strongly enhanced by the phosphine, and the reaction occurs essentially upon mixing. Without the phosphine, alkenes are released only slowly, if at all. The mechanism of the unimolecular reactions that lead to alkene release has not been studied, whereas the unimolecular release of alkene from rhenium(V) diolates has been (*68, 70*).

The Re(V) and Re(VII) diolates differ not simply in oxidation state, but in the compounds they form. The Re(V) compound produces R–ReO$_2$, (*71*) whereas Re(VII) gives the stable R–ReO$_3$.

Three processes appear closely interrelated: (1) alkene epoxidation via **A**; (2) diolate formation from an epoxide; and (3) alkene release from the diolate, which has been characterized for Re(V) (*71*). The interrelationship is diagrammed in Scheme VII. Epoxidation occurs via the peroxide **A**, proceeding through one or more intermediates to either epoxide or diolate (*69*). In practice, epoxidation is rapid, as is the conversion of MTO to **A** with hydrogen per-

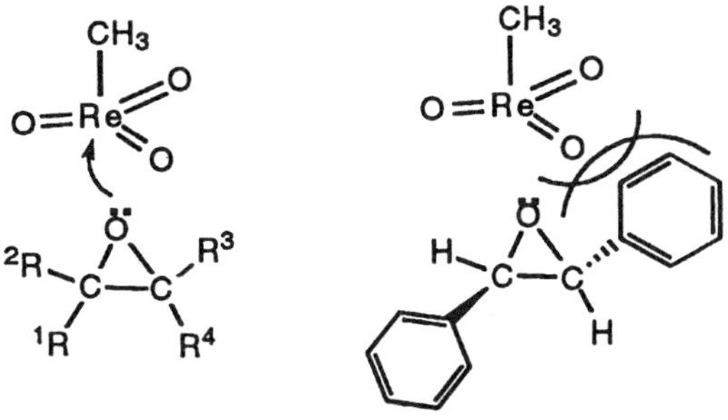

Scheme VI.

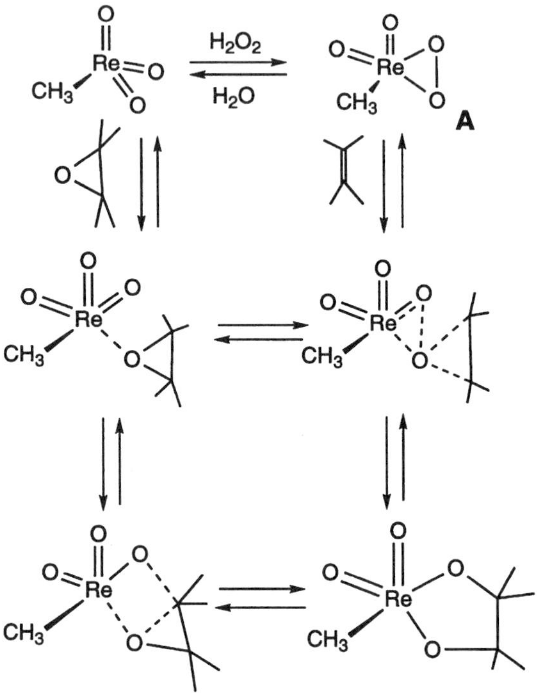

Scheme VII.

oxide. Thus diolate formation is of minimal importance until the supply of peroxide is exhausted, assuming the alkene was taken in excess. Alternatively, without peroxide, the epoxide converts to the diolate.

In light of this, we offer a new interpretation of the results on Re(V) diolates (70), in which the possibilities shown in Scheme VIII were put forth. Gable suggests a CH_2 migration step. Rather than O-to-Re migration, we prefer an O-to-O migration, as in Scheme VII.

This set of interconversions may be related to the interesting report that the conversion of the newly-formed 1,2-epoxy-cyclohexane (from cyclohexene + H_2O_2 + MTO catalyst) to 1,2-cyclohexanediol is inhibited by the addition of coordinating ligands, for example, bipyridine ($7, 60$). This effect was suggested to arise from the blocking of coordination positions, consistent with the model shown.

Oxygen and Oxygen Transfer

Molecular oxygen oxidizes PAr_3 very slowly at room temperature, and MTO has proved to be an efficient catalyst (72). Neither **A** nor **B** could be detected;

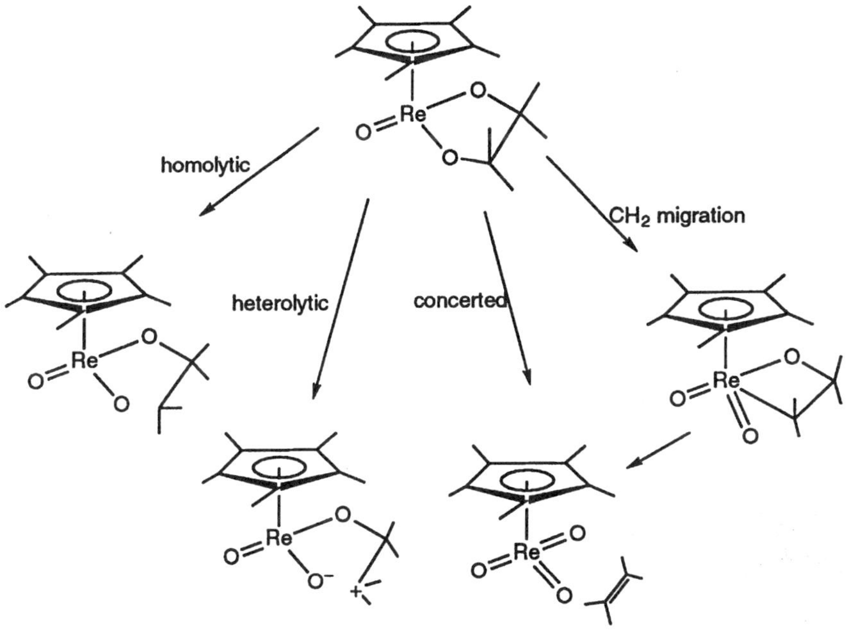

Scheme VIII.

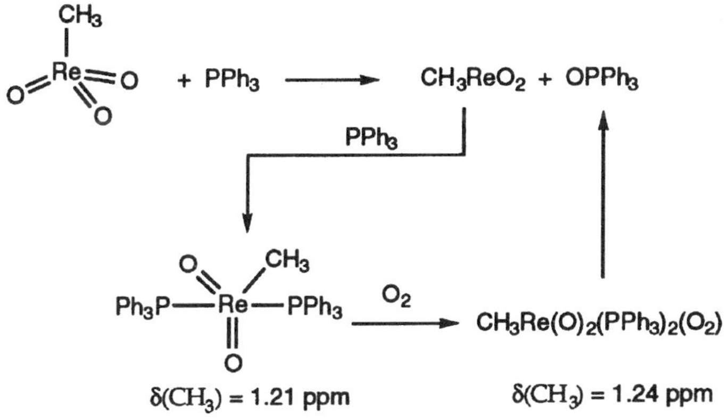

Scheme IX.

they either are not involved or are highly dilute. This reaction may involve an oxygen-containing intermediate (Scheme IX).

The rhenium(V) intermediate $CH_3Re(O)_2$ [or $CH_3Re(O)_2\cdot OPPh_3$], partitions between alternative reactions, governed by the presence of oxygen or of the substrate–oxide, shown in Scheme X.

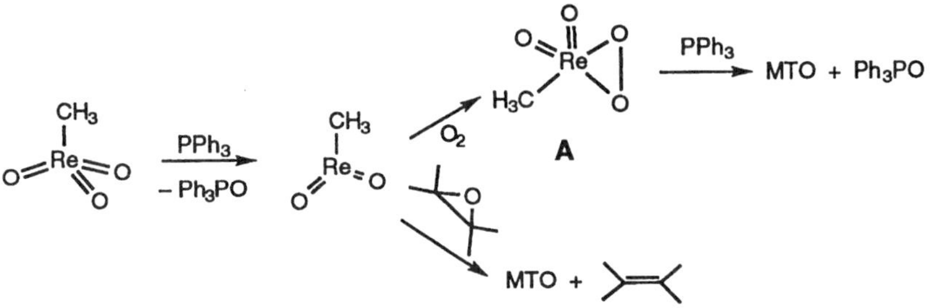

Scheme X.

Scheme X also accounts for certain oxygen transfer reactions. For example, a substrate such as an epoxide, when treated anaerobically with triphenylphosphine in the presence of MTO at room temperature, forms the olefin (72). Another class of substrate is sulfoxides, which give sulfides as the only product. MTO also catalyzes oxygen transfer from tertiary amine oxides to triphenylphosphine, forming the amine and triphenylphosphine oxide (72).

The deoxygenations of triphenylarsine oxide and triphenylstibine oxide by Ph_3P, catalyzed by MTO, were finished within 30 min at room temperature under argon; triphenylarsine and triphenylstibine were formed almost quantitatively:

$$Ph_3E{=}O + PPh_3 \xrightarrow{\text{cat. MTO}} Ph_3E + O{=}PPh_3 \qquad (35)$$

Catalyst Integrity

Alone in water and organic solvents, dilute solutions of MTO persist for days with <2% decomposition. When H_2O_2 is added to MTO, however, decomposition ensues at rates dependent on the peroxide concentration and the pH (10). In organic solvents, such as THF, acetonitrile, and acetone, **B** is stable at ambient temperature, especially with excess H_2O_2. NMR experiments in THF-d^8 revealed that at low $[H_2O_2]$ (generally $\leq Re_T$, such that $[\mathbf{A}] >> [\mathbf{B}]$) methanol is the decomposition product. In addition, perrhenate ion was detected in quantitative yield. This reaction occurs more slowly in acidic aqueous solution. The two pathways that account for the decomposition of the catalyst in the presence of H_2O_2 are presented in eqs 36 and 37. The two pathways are kinetically indistinguishable because they both involve the same transition state, namely, $[CH_3ReO_5H^-]^{\ddagger}$.

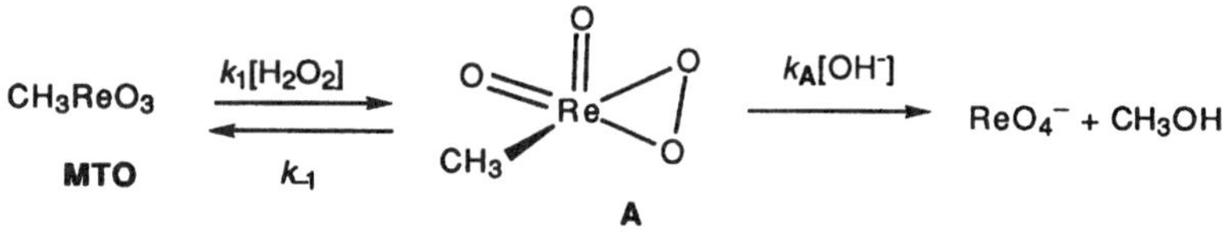

$$CH_3ReO_3 \;+\; HOO^- \;\xrightarrow{\;k_{MTO}\;}\; ReO_4^- + CH_3OH \qquad (37)$$

$$\mathbf{MTO}$$

The peroxide steps to **A** and **B** are sufficiently rapid to be treated as prior equilibria, which was verified independently. The decomposition according to eq 36 yields

$$-\frac{d[\mathbf{A}]}{dt} = f_{\mathbf{A}}k_{\mathbf{A}}[OH^-][Re]_T \qquad (38)$$

where

$$[Re]_T = [MTO] + [\mathbf{A}] + [\mathbf{B}]$$

$$f_{\mathbf{A}} = \frac{K_1[H_2O_2]}{1+K_1[H_2O_2]+K_1K_2[H_2O_2]^2}$$

On the other hand, the rate law according to the reaction of MTO with HOO_2^- as represented by eq 37 would give the following rate law:

$$-\frac{d[MTO]}{dt} = f_{MTO}k_{MTO}[HO_2^-][Re]_T \qquad (39)$$

where

$$f_{MTO} = \frac{1}{1+K_1[H_2O_2]+K_1K_2[H_2O_2]^2}$$

$$[HO_2^-] = \frac{K^a_{H_2O_2}[H_2O_2]}{[H^+]} = \frac{K^a_{H_2O_2}[OH^-][H_2O_2]}{K_w}$$

The rate constant rises at low $[H_2O_2]$ and then falls according to eq 38 or eq 39. The kinetic dependence of the rate of decomposition on $[H_3O^+]$ was determined in the pH range 1.77–6.50. The pH profile (log k_ψ versus pH) is linear with a slope of 0.95 ± 0.01. The value of $k_{\mathbf{A}}$ at $\mu = 0.01$ M is 6.2×10^9 L mol^{-1} s^{-1} and $k_{MTO} = 4.1 \times 10^8$ L mol^{-1} s^{-1}.

If methanol forms from **A**, the mechanism would most likely involve a nucleophilic attack of OH⁻ at the methyl group; we envisage that this step leads either directly to the products (eq 40) or to a rhenium(V) peroxo intermediate, which then rearranges to form perrhenate ion (eq 41). Methanol and perrhenate ions could also be formed by a kinetically indistinguishable mechanism in which MTO is attacked by HOO^- instead of OH^- attacking **A**. In the later possibility, we speculate an attack of the HOO^- ion on the rhenium center to form a peroxo intermediate that is analogous to **A**; this intermediate rearranges to yield perrhenate and methanol as final products (eq 42).

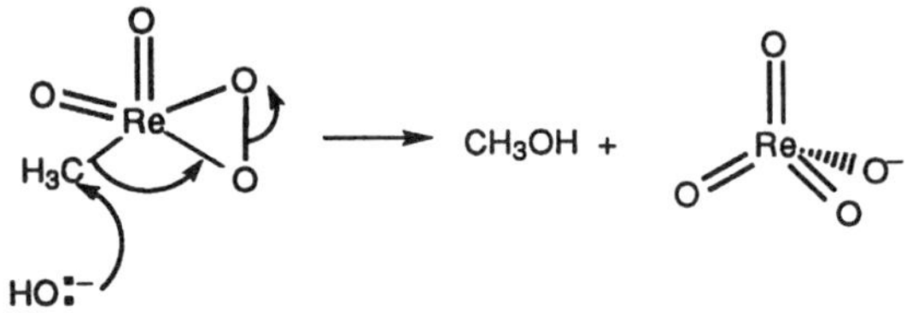

(40)

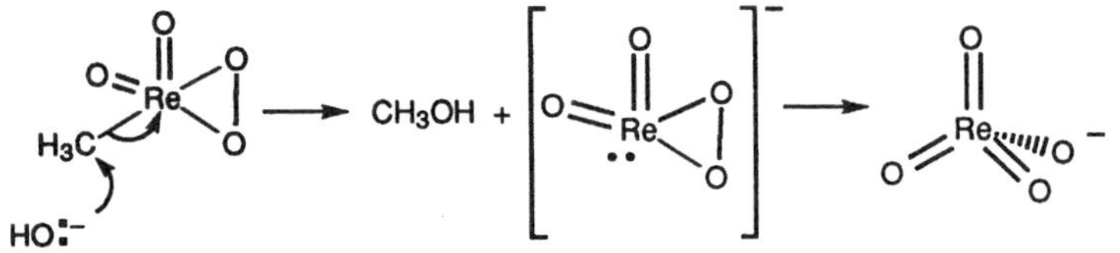

(41)

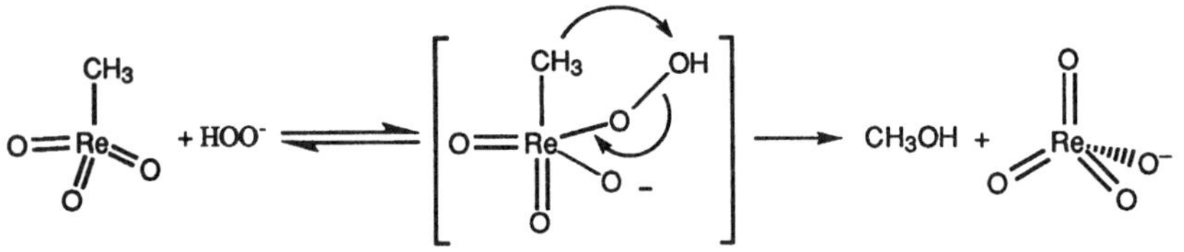

(42)

Attempts to distinguish between the two viable pathways (eqs 40 or 41 and 37) through labeling experiments that utilize ^{18}O-enriched H_2O_2 have not been limited because of the oxygen exchange of perrhenate ion itself. Hence the mass spectrum of the perrhenate produced from the decomposition of $CH_3Re^{16}O_3$ with $H_2^{18}O_2$ yields masses that correspond to $Re^{16}O_4^-$ rather than $Re^{16}O_2{}^{18}O_2^-$ (eqs 40 and 41) or $Re^{16}O_3^{18}O^-$ (eq 42). Nevertheless, because the values of k_A and k_{MTO} are close to the diffusion limit, the reaction of MTO with HO_2^- may be more plausible than that of OH^- with **A**. The hydroperoxide anion is notably more nucleophilic than hydroxide (*73*). Therefore, one expects peroxide-induced decomposition should proceed more rapidly than hydroxide-induced decomposition, particularly because MTO itself persists for long times in aqueous solutions under these concentrations. In summary, the mechanism expressed in eq 42 appears to be the pathway best able to account for the experimental observations and the kinetic data; it has a second-order rate constant (k_{MTO}) of 4.1×10^8 L mol^{-1} s^{-1}.

A precedent relevant to the decomposition of this catalytic system is the formation of $(Bu^tO)_2Mo(O)_2$ from the reaction of $Mo(OBu^t)_4$ with molecular oxygen in dilute solutions (*74*). In the experiments we reported, the evolution of oxygen from **B** might also need to be considered, although that aspect of the situation remains unresolved at this writing.

The Decomposition of MTO Itself.   MTO is known to decompose to CH_4 and ReO_4^- in alkaline solutions, presumably via a hydroxide ion adduct (*11, 13*). The hydroxide adduct of MTO was observed by ^{1}H-NMR (δ 1.9, s) and UV–vis at 340 nm (ε 1500 L mol^{-1} cm^{-1}). This observation lends support to the existence of an intermediate, as shown in Scheme XI.

The dependence of the rate upon [OH$^-$] is in agreement with the Scheme XI and the following rate law:

$$v = \frac{(k_5 K_w + k_6 K_a)[\text{OH}^-]}{K_w + K_a[\text{OH}^-]} \times [\text{Re}]_T \tag{43}$$

The fitting of the data yields $K_a = 1.2 \times 10^{-12}$ mol L^{-1} and $k_5 K_w$ or $k_6 K_a = 2.7 \times 10^{-16}$ mol L^{-1} s^{-1}. The matter cannot be resolved any further on the basis of kinetic data. The value of K_a for MTO was confirmed by spectrophotometric titration of MTO and NaOH; measured absorbances at 360 and 390 nm yield an average K_a of 2.2×10^{-12} mol L^{-1}.

Baeyer–Villiger Oxidations: The Case of β-Diketones

The catalytic reactions of the CH_3ReO_3–H_2O_2 system have been extended to cyclic β-dicarbonyl compounds (*75a*). The electrophilic peroxides **A** and **B** readily react with the electron-rich systems cited previously. Enolic tautomers of carbonyl compounds are electron-deficient owing to the conjugation of the double bond with the electron-withdrawing CO group. Therefore, β-diketones can be used as a test for the ability of **A** and **B** to oxidize electron-deficient substrates. Another interest in β-diketones stems from their use in natural product syntheses (*75b*), such as those that are useful in the syntheses of naturally occurring antibiotics (*76*).

Cyclic β-diketones are slowly cleaved with MTO–H_2O_2. The cyclic β-diketones investigated in this study are present mainly ($\geq$95%) in the enolic form in solution (*77*). The initial oxidation step affords a 2-hydroxy-1,3-dicarbonyl intermediate, which is observed in the ^{1}H-NMR. This hydroxy intermediate is then oxidized further to yield the corresponding organic acids. These general observations are summarized in Scheme XII for 2-methyl-1,3-cyclopentanedione.

Because these oxidations were conducted in 1:1 acetonitrile–water solvent with 0.10 or 1.0 M HClO$_4$ (the acid is needed for catalytic stability in aqueous

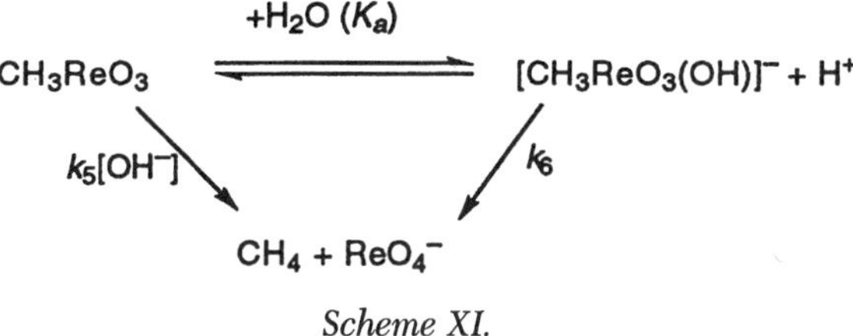

Scheme XI.

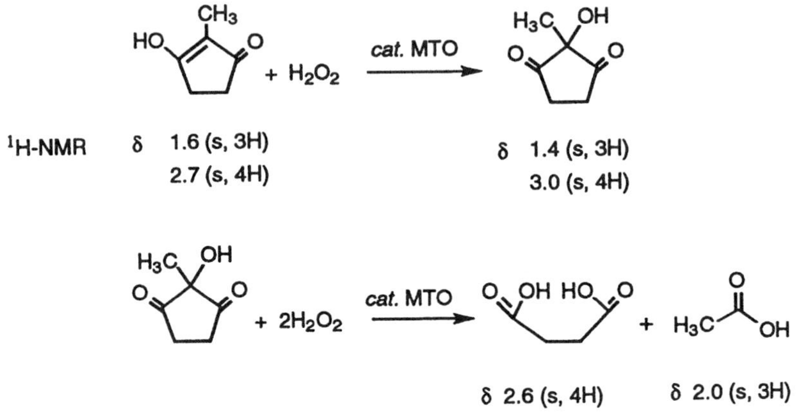

^{1}H-NMR   δ   1.6 (s, 3H)   δ   1.4 (s, 3H)

                                 2.7 (s, 4H)   3.0 (s, 4H)

                                         δ 2.6 (s, 4H)   δ 2.0 (s, 3H)

Scheme XII.

solution; *see* Catalyst Integrity), the 2-hydroxy-1,3-diketone intermediate arises from the ring opening of the epoxide, catalyzed by acid as well as MTO (*20, 54, 60*). The next step in the oxidation involves C–C bond cleavage, possibly via a Baeyer–Villiger mechanism, which involves oxygen insertion into the C–C bond. MTO has been reported as an effective catalyst for the Baeyer–Villiger oxidations of cyclic ketones to give the corresponding lactones (*60, 78*). The lactone, in our case, is then hydrolyzed by water to afford an α-diketone intermediate that is easily cleaved by peroxide to yield the final products, organic acids. We propose the epoxidation of the enolic tautomer initially, followed by oxygen insertion into C–C bond through a Baeyer–Villiger oxidation, and finally rupture of the α-diketo intermediate to afford products. This mechanism is illustrated for the case of 2-methyl-1,3-cyclopentanedione in Scheme XIII.

    Oxygen insertion into a C–C bond (Baeyer–Villiger oxidation step) may proceed through a cyclic Re–peroxo intermediate (structure **5**), illustrated for the reaction of 2-methyl-2-hydroxy-1,3-cyclopentanedione with **A** (a similar intermediate could be drawn for **B**). Such an intermediate has been proposed for the rhenium-catalyzed Baeyer–Villiger oxidations of cyclic ketones (*78*), as well as for the enzyme catechol dioxygenase (*79, 80*), which catalyzes the oxygenation of catechol to 2, 4-hexadienedioic acid.

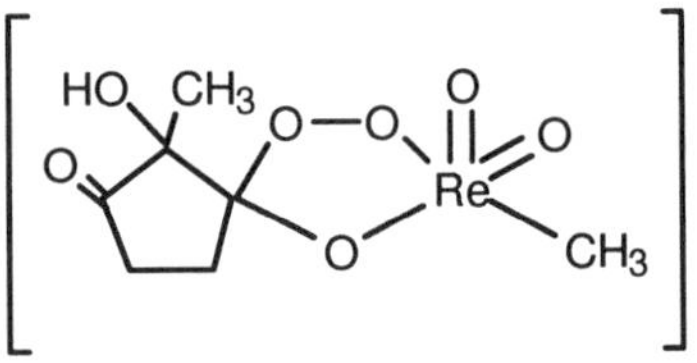

5

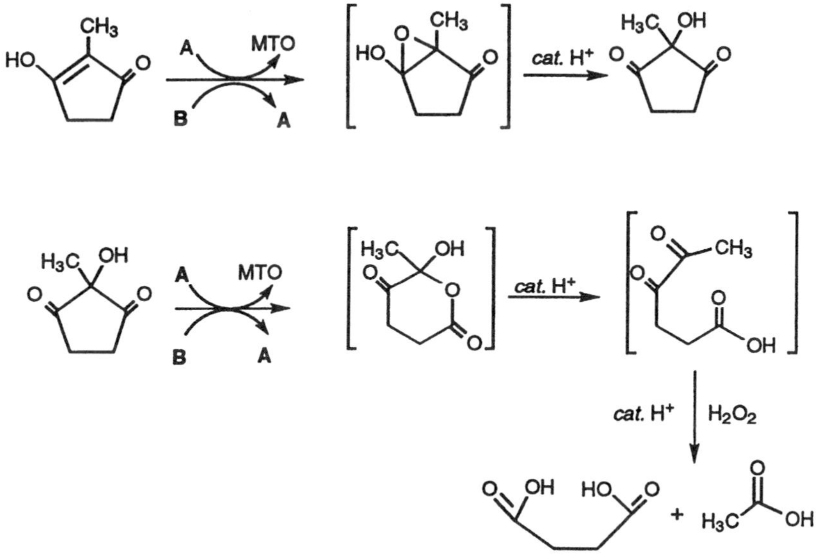

Scheme XIII.

α-Diketones are easily oxidized by hydrogen peroxide to give organic acids, catalyzed by both acid and base (*81–83*). Therefore, the lactone obtained from oxygen insertion into the C–C bond hydrolyzes to afford an α-diketone intermediate, which is then cleaved by H_2O_2 to yield organic acids.

A list of the different cyclic β-diketones investigated thus far is presented in Table IX alongside the final products and yields obtained from catalytic oxidations with MTO–H_2O_2 system.

The kinetics were studied under conditions in which the peroxide reactions of MTO can be regarded as rapid prior equilibria (*see* Oxidations of Styrenes, Alkenes, and Alkynes). This assumption is possible because the oxidation steps of β-diketones as presented by eqs 44 and 45 are much slower than the formation of **A** and **B**.

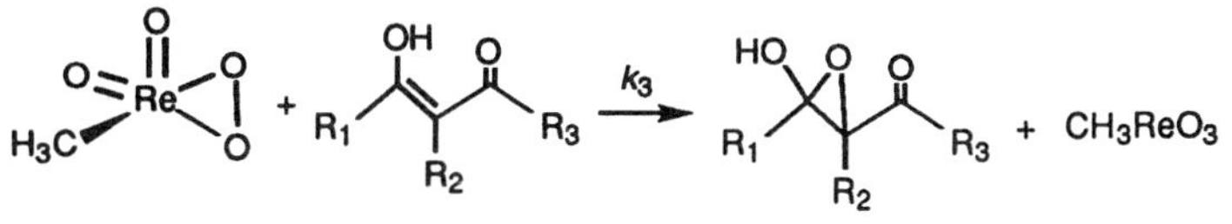

$$\text{(44)}$$

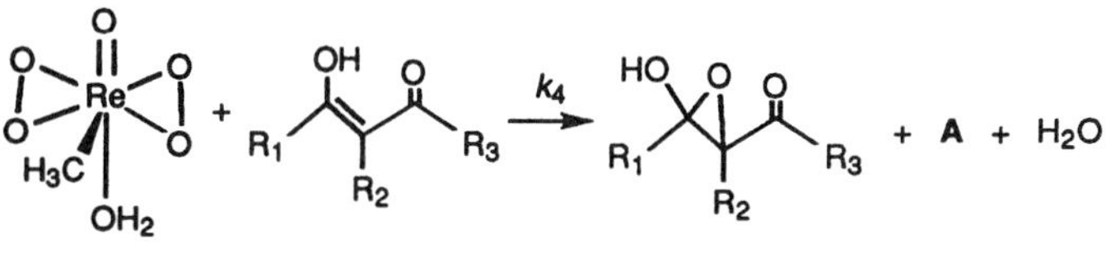

$$\text{(45)}$$

Table IX. Products and Yields from the Oxidations of Cyclic β-Diketones by Hydrogen Peroxide, as Catalyzed by MTO

Compound	Products	%Conversion	%Yield
5,5-dimethyl-1,3-cyclohexanedione (enol form)	HOOC–CH$_2$–C(CH$_3$)$_2$–CH$_2$–COOH + HCOOH	66	100
2-chloro-5,5-dimethyl-1,3-cyclohexanedione (enol form)	HOOC–CH$_2$–C(CH$_3$)$_2$–CH$_2$–COOH + 2,2-dichloro-5,5-dimethyl-1,3-cyclohexanedione	95	74, 26
1,3-cyclohexanedione (enol form)	HOOC–(CH$_2$)$_3$–COOH + HCOOH	85	100
2-methyl-1,3-cyclohexanedione (enol form)	HOOC–(CH$_2$)$_3$–COOH + CH$_3$COOH	95	100
2-methyl-1,3-cyclopentanedione (enol form)	HOOC–CH$_2$–COOH + CH$_3$COOH	100	100
2-hydroxy-3-methyl-2-cyclopentenone	HOOC–C(CH$_3$)(OH)–CH$_2$–CH$_2$–COOH	95	100
2-acetylcyclopentanone [a]	HOOC–(CH$_2$)$_3$–COOH + CH$_3$COOH	90	100
1,3-diphenyl-1,3-propanedione (enol form) [b]	Ph–CO–CH(OH)–CO–Ph	62	85

NOTE: General reaction conditions are MTO:β-diketone:H$_2$O$_2$ = 1:12:70 at 25 °C, in 1:1 CD$_3$CN/D$_2$O and pD 1. %Conversion values are based on starting material. %Yield values are calculated by NMR based on conversion.

[a] In CD$_3$CN at 25 °C.

[b] In CDCl$_3$ at 25 °C, MTO:β-diketone:H$_2$O$_2$ = 1:10:50.

Table X. Rate Constants for the Reactions of A and B with a Couple of Cyclic β-Diketones

Compound	k_3 *(L mol^{-1} s^{-1})*	k_4 *(L mol^{-1} s^{-1})*
1,3-Cyclohexanedione	0.28	0.11
2-Methyl-1,3-cyclohexanedione	0.86	0.18

The rate law under these conditions becomes

$$v = k_3[\beta - \text{diketone}][\mathbf{A}]_{\text{eq}} + k_4[\beta - \text{diketone}][\mathbf{B}]_{\text{eq}} \qquad (46)$$

Because $[\text{H}_2\text{O}_2]$ is in excess over [β-diketone], the equilibrium concentrations of **A** and **B** remain unchanged throughout the reaction, and therefore,

$$k_\psi = k_3[\mathbf{A}]_{\text{eq}} + k_4[\mathbf{B}]_{\text{eq}} \qquad (47)$$

The equilibrium concentrations of **A** and **B** were calculated using the independently determined K_1 and K_2 for the peroxide reactions (*see* Kinetics of Peroxide-Forming Reactions). The rate constants for the reactions of **A** and **B** with 1,3-cyclohexanedione and 2-methyl-1,3-cyclohexanedione are given in Table X.

We envisage a nucleophilic attack by the substrate on the electrophilic bound-peroxide oxygen. The electronic trend displayed by dimedone (structure **6**) and monochlorodimedone (MCD) (structure **7**) agrees with this mechanism. For dimedone, k_4 is 0.11 and for MCD 0.018 L mol^{-1} s^{-1}. The retardation by Cl supports nucleophilic attack by the enol.

Conclusions

MTO is an amazingly versatile reagent. The major aspect of its reactivity that we have reviewed herein relates to its ability to transfer a single oxygen atom to an appropriate substrate. Many nucleophilic substrates can be oxidized, provided an electron-rich center that is able to be oxidized by oxygen-atom addition, and that oxygen transfer is thermodynamically allowed. The last point does not usually pose a problem, given the relevant bond strengths (58).

Looking to the future, it has already become clear that MTO will be able to catalyze an even wider series of transformations than those already characterized. One of the promising areas is the transfer of NR and CR$_2$ groups instead of the isoelectronic oxygen atom. These alternative forms of the catalyst

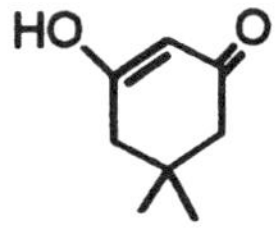

6

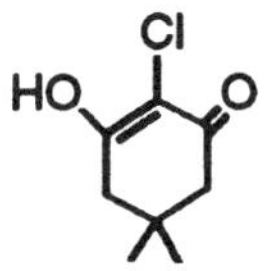

7

will function as nitrenoid and carbenoid equivalents, in the sense that **A** and **B** are oxene equivalents.

Acknowledgments

This research was supported by the U.S. Department of Energy, Office of Basic Energy Sciences, Division of Chemical Sciences under contract W-7405-Eng-82. J. H. Espenson particularly wishes to express his gratitude to his co-workers in this area, who are the co-authors of the referenced publications.

References

1. Strukul, G. *Catalytic Oxidations with Hydrogen Peroxide as Oxidant*; Kluwer Academic: Dordrecht, The Netherlands, 1992.
2. Sheldon, R. A. *Top. Curr. Chem.* **1993**, *164*, 23–28.
3. Kahr, K.; Beetha, C. *Chem. Ber.* **1960**, *93*, 132.
4. Beattie, I. R.; Jones, P. J. *Inorg. Chem.* **1979**, *18*, 2318.
5. Herrmann, W. A.; Romão, C. C.; Kiprof, P.; Behm, J.; Cook, M. R.; Taillefer, J. *J. Organomet. Chem.* **1991**, *413*, 11.
6. Herrmann, W. A.; Kühn, F. E.; Fischer, R. W.; Thiel, W. R.; Romão, C. C. *Inorg. Chem.* **1992**, *31*, 4431.
7. Herrmann, W. A.; Fischer, R. W.; Rauch, M. U.; Scherer, W. *J. Mol. Catal.* **1994**, *86*, 243.
8. Herrmann, W. A.; Kiprof, P.; Rydpal, K.; Tremmel, J.; Blom, R.; Alberto, R.; Behm, J.; Albach, R. W.; Bock, H.; Solouki, B.; Mink, J.; Lichtenberger, D.; Gruhn, N. E. *J. Am. Chem. Soc.* **1991**, *113*, 6527.
9. Kunkely, H.; Türk, T.; Teixeira, C.; de Meric de Bellefon, C.; Herrmann, W. A.; Vogler, A. *Organometallics* **1991**, *10*, 2090.
10. Abu-Omar, M. M.; Hansen, P. J.; Espenson, J. H. *J. Am. Chem. Soc.* **1996**, *118*, 4966.
11. Herrmann, W. A.; Kuchler, J. G.; Weichselbaumer, G.; Herdtweck, E.; Kiprof, P. *J. Organomet. Chem.* **1989**, *372*, 351.
12. Szyperski, T.; Schwerdtfeger, P. *Angew. Chem., Int. Ed. Engl.* **1989**, *28*, 1228.
13. Herrmann, W. A. *Angew. Chem., Int. Ed. Engl.* **1988**, *27*, 1297.
14. Vassell, K. A.; Espenson, J. H. *Inorg. Chem.* **1994**, *33*, 5491.
15. Herrmann, W. A.; Kühn, F. E.; Roesky, P. W. *J. Organomet. Chem.* **1995**, *485*, 243.
16. Herrmann, W. A.; Fischer, R. W.; Scherer, W. *Adv. Mater.* **1992**, *4*, 653.
17. Herrmann, W. A.; Fischer, R. W. *J. Am. Chem. Soc.* **1995**, *117*, 3223.
18. Herrmann, W. A.; Scherer, W.; Fischer, R. W.; Blümel, J.; Kleine, M.; Mertin, W.; Gruehn, R.; Mink, J.; Boysen, H.; Wilson, C. C.; Ibberson, R. M.; Bachmann, L.; Mattner, M. *J. Am. Chem. Soc.* **1995**, *117*, 3231.
19. Genin, H. S.; Lawler, K. A.; Herrmann, W. A.; Fischer, R. W.; Scherer, W. *J. Am. Chem. Soc.* **1995**, *117*, 3244–3252.
20. Herrmann, W. A.; Fischer, R. W.; Scherer, W.; Rauch, M. U. *Angew. Chem., Int. Ed. Engl.* **1993**, *32*, 1157.
21. Espenson, J. H.; Pestovsky, O.; Huston, P.; Staudt, S. *J. Am. Chem. Soc.* **1994**, *116*, 2869.
22. Yamazaki, S.; Espenson, J. H.; Huston, P. *Inorg. Chem.* **1993**, *32*, 4683.
23. Binstead, R. A.; Zuberbühler, A. D. *Specfit*; Spectrum Software Associates, P.O. Box 4494, Chapel Hill, NC 27515:

24. Bernasconi, C. F. *Chemical Relaxation;* Academic Press: New York, 1976; pp 23–29.

25. Espenson, J. H. *Chemical Kinetics and Reaction Mechanisms 2nd ed.;* McGraw-Hill: New York, 1995; p 77.

26. Bernasconi, C. F. *Techniques of Chemistry;* Wiley-Interscience, John Wiley & Sons: New York, 1986; Vol. 6; Part 1, pp 435–443.

27. Zhu, Z.; Espenson, J. H. *J. Org. Chem.* **1995,** *60,* 1326.

28. Pestovsky, O.; van Eldik, R.; Huston, P.; Espenson, J. H. *J. Chem. Soc. Dalton Trans.* **1995,** 133.

29. Carpentier, J. L.; Mitschler, A.; Weiss, R. *Acta. Crystallogr. Sect. B : Struct. Sci.* **1972,** *28,* 1288.

30. Roussel, M.; Mimoun, H. *J. Org. Chem.* **1980,** *45,* 5381.

31. Michelin, R. A.; Ros, R.; Strukul, G. *Inorg. Chim. Acta* **1979,** *37,* L491.

32. Strukul, G.; Ros, R.; Michelin, R. A. *Inorg. Chem.* **1982,** *21,* 495.

33. Zanardo, A.; Michelin, R. A.; Pinna, F.; Strukul, G. *Inorg. Chem* **1989,** *28,* 1648.

34. Brynildson, M. E.; Bakac, A.; Espenson, J. H. *Inorg. Chem.* **1987,** *27,* 2592.

35. Scott, S. L.; Bakac, A.; Espenson, J. H. *Inorg. Chem.* **1991,** *30,* 4112.

36. Wang, W. D.; Bakac, A.; Espenson, J. H. *Inorg. Chem.* **1993,** *32,* 5134.

37. Kumar, K.; Endicott, J. F. *Inorg. Chem.* **1984,** *23,* 2447.

38. Geiger, T.; Anson, F. C. *J. Am. Chem. Soc.* **1981,** *103,* 7489.

39. Abu-Omar, M. M.; Espenson, J. H. *J. Am. Chem. Soc.* **1995,** *117,* 272.

40. Barshop, B. A.; Wrenn, C. F.; Frieden, C. *Anal. Biochem.* **1983,** *130,* 134.

41. Hammett, L. P. *Physical Organic Chemistry;* McGraw-Hill Book Co.: New York, 1970.

42. Hammett, L. P. *Chem. Rev.* **1935,** *17,* 125.

43. Tolman, C. A. *Chem. Rev.* **1977,** *77,* 313.

44. Sidgwick, N. V. *The Chemical Elements and Their Compounds;* Oxford University Press: London, 1962; Vol. 1, p 776.

45. Srinivasan, C.; Pitchumani, K. *Int. J. Chem. Kinet.* **1982,** *14,* 1315.

46. Srinivasan, C.; Pitchumani, K. *Can. J. Chem.* **1985,** *63,* 2285.

47. Hiatt, R.; McColeman, C.; Howe, G. R. *Can. J. Chem.* **1975,** *53,* 559.

48. Cenini, S.; Fusi, A.; Capparella, G. L. *J. Inorg. Nucl. Chem.* **1971,** *33,* 3576.

49. Sutin, N.; Yandell, J. K. *J. Am. Chem. Soc.* **1973,** *95,* 4847.

50. Denney, D. B.; Goodyear, W. F.; Godstein, B. *J. Am. Chem. Soc.* **1960,** *82,* 1393.

51. Bortolini, O.; Campello, E.; DiFuria, F.; Modena, G. *J. Mol. Catal.* **1982,** *14,* 63.

52. Modena, G. *Gazz. Chim. Ital.* **1959,** *89,* 834.

53. James, B. R. *Homogeneous Hydrogenation;* Wiley-Interscience: New York, 1973; pp 204–287.

54. Al-Ajlouni, A.; Espenson, J. H. *J. Am. Chem. Soc.* **1995,** *117,* 9243.

55. Huston, P.; Espenson, J. H.; Bakac, A. *Inorg. Chem.* **1993,** *32,* 4517.

56. Mohammed, A.; Liebhafsky, H. A. *J. Am. Chem. Soc.* **1934,** *56,* 1680.

57. Bray, W. C. *Chem. Rev.* **1932,** *10,* 161.

58. Holm, R. H.; Donahue, J. P. *Polyhedron* **1993,** *12,* 571.

59. Hansen, P. J.; Espenson, J. H. *Inorg. Chem.* **1995,** *34,* 5839.

60. Herrmann, W. A.; Fischer, R. W.; Marz, D. W. *Angew. Chem., Int. Ed. Engl.* **1991,** *30,* 1638.

61. Herrmann, W. A.; Watzlowik, P.; Kiprof, P. *Chem. Ber.* **1991,** *124,* 1101–1106.

62. Zahonyi-Budo, E.; Simandi, L. I. *The Role of Oxygen in Chemistry and Biochemistry;* Elsevier Science Publishers: Amsterdam, The Netherlands, 1988; Vol. 33, p 219.

63. Chaumette, P.; Mimoun, H.; Saussine, L.; Fischer, J.; Mistchler, A. *J. Organomet. Chem.* **1983,** *250,* 291.

64. Simandi, L. I.; Zahonyi-budo, E.; Bodnar, J. *Inorg. Chim. Acta.* **1982,** *65,* L181.
65. Simandi, L. I. *Catalytic Activation of Dioxygen by Metal Complexes;* Kluwer Academic Publishers: Dordrecht, The Netherlands, 1992; p 109.
66. Zhu, Z.; Espenson, J. H. *J. Org. Chem.* **1995,** *60,* 7727.
67. Lewars, E. G. *Chem. Rev.* **1983,** *83,* 519.
68. Gable, K. P. *Organometallics* **1994,** *13,* 2486.
69. Zhu, Z.; Al-Ajlouni, A.; Espenson, J. H. *Inorg. Chem.* **1996,** *35,* 1408.
70. Gable, K. P.; Juliette, J. J. *J. Am. Chem. Soc.* **1995,** *117,* 955.
71. Gable, K. P.; Phan, T. N. *J. Am. Chem. Soc.* **1994,** *116,* 833.
72. Zhu, Z.; Espenson, J. H. *J. Mol. Catal* **1995,** *103,* 87.
73. Carey, F. A.; Sundberg, R. J. *Advanced Organic Chemistry, Part A, 3rd Ed.;* Plenum Press: New York, 1990; p 268.
74. Chisholm, M. H.; Folting, K.; Huffman, J. C.; Kirkpatrick, C. C. *Inorg. Chem.* **1984,** *23,* 1021.
75. (a) Abu-Omar, M.; Espenson, J. H. *Organometallics* **1996,** *15,* 3543; (b) Salzmann, T. N.; Ratcliffe, R. W.; Christensen, B. G. *Tetrahedron Lett.* **1980,** *21,* 1193.
76. Wasserman, H. H.; Han, W. T. *Tetrahedron Lett.* **1984,** *25,* 3743.
77. Toullec, J. *The Chemistry of Enols;*John Wiley & Sons: New York, 1990; Chapter 6.
78. Herrmann, W. A.; Fischer, R. W.; Correia, J. G. *J. Mol. Catal.* **1994,** *94,* 213.
79. Que, L. *J. Chem. Educ.* **1985,** *62,* 938.
80. Lippard, S. J.; Berg, J. M.. *Principles of Bioinorganic Chemistry;* University Science Books: Mill Valley, CA, 1994; pp 315–318.
81. March, J. *Advanced Organic Chemistry;* 4th ed.; Wiley: New York, 1992; Chapter 19.
82. Ogata, Y.; Sawaki, Y.; Shiroyama, M. *J. Org. Chem.* **1977,** *42,* 4061.
83. Cocker, W.; Grayson, D. H. *J. Chem. Soc. Perkin Trans. 1* **1975,** 1347.
84. Herrmann, W. A.; Kuchler, J. G.; Kiprof, P.; Riede, J. *J. Organomet. Chem.* **1990,** *395,* 55.
85. Elder, R. C.; Kennard, G. J.; Payne, M. D.; Deutsch, E. *Inorg. Chem.* **1978,** *17,* 1296.
86. Adzamli, I. K.; Deutsch, E. *Inorg. Chem.* **1980,** *19,* 1366.

PART II:
ELECTRON TRANSFER AND MECHANISTIC INORGANIC CHEMISTRY

8

Steric Effects in Redox Reactions and Electron Transfer Rates

Rodney J. Geue, John V. Hanna, Arthur Höhn, C. Jin Qin, Stephen F. Ralph, Alan M. Sargeson, and Anthony C. Willis

Research School of Chemistry, The Australian National University, Canberra, Australian Capital Territory 0200, Australia, and Commonwealth Scientific and Industrial Research Organisation, Division of Coal and Energy Technology, North Ryde, New South Wales 2113, Australia

Several key cobalt(II/III) hexaamine systems are discussed in which steric factors enhance electron transfer rates by a factor of more than 10^6 and effect redox potential changes in excess of 1.2 V. In some instances conformational isomers alone show dramatic differences. Electron transfer rates between ob_3 and lel_3 conformations of metal ion cages arising from $\Lambda[Co(R\text{-}pn)_3]^{2+/3+}$ and $\Lambda[Co(S\text{-}pn)_3]^{2+/3+}$ ions (pn = 1, 2-propanediamine) vary by more than 10^3 fold. The larger cavity $[Co(Me_8\text{-}tricosaneN_6)]^{3+}$ ions ($Me_8\text{-}tricosaneN_6$ = 1,5,5,9,13,13,20,20-octamethyl-3,7,11,15,18,22-hexaazabicyclo[7.7.7]tricosane) have orange and blue conformations that differ in redox potential by >0.5 V. X-ray crystal structure views of the orange form and of the unusually air-stable cobalt(II) complex as their nitrate salts are presented. Also, solid-state ^{59}Co NMR and reflectance spectra imply a very weak ligand field at the cobalt site of the blue form.

STERIC EFFECTS HAVE ALWAYS BEEN INTERESTING especially in their ability to influence isomer distributions and stereochemistry (1). In more recent time, however, they have been impressive in their ability to influence other properties. It is that aspect that this chapter will address and especially in relation to the metal ion cage redox potentials and electron transfer reactions.

To begin with, the self-exchange electron transfer of the [Co(sepulchrate)]$^{3+/2+}$ (Figure 1) couple is a good example. Structural studies and molecular mechanics calculations of the two complex ions indicate that both

Steric Effects on Electron Transfer

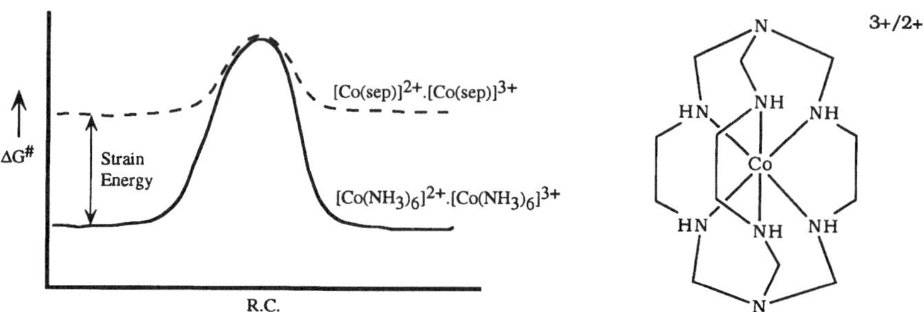

Figure 1. The [Co(sepulchrate)] structure and proposed energetics for the electron transfer reactions of the cage and [Co(NH$_3$)$_6$]$^{3+/2+}$.

oxidation states are strained relative to the [Co(NH$_3$)$_6$]$^{3+/2+}$ or [Co(en)$_3$]$^{3+/2+}$ (en = 1,2-ethanediamine) couples (2, 3). In both instances the metal ion and the organic cage compromise their needs to give a minimum energy structure, but the strain is evident in both parts. Crudely, it can be argued that the cavity of the cage is a little too large for Co(III) and too small for Co(II). This condition means that the strain in both ground states helps the reactant pair toward the activated complex, and the free energy barrier for electron transfer is thereby effectively lowered (Figure 1). The process largely contributes to a 10^6 fold enhancement in rate relative to the parent couple [Co(NH$_3$)$_6$]$^{3+/2+}$. It is a very good example of an "entatic state" (4). Also, it is an important demonstration of how influential these steric forces can be in affecting electron transfer rates and the reversibility of such redox couples.

Another example of some published results (5) involving cages (Figure 2) arising from Λ[Co(S-pn)$_3$]$^{3+}$ and Λ[Co(R-pn)$_3$]$^{3+}$ is displayed in Table I (pn = 1,2-propanediamine). These two molecules stabilize two conformations of the cage ligand, namely the *lel*$_3$ and *ob*$_3$ forms. Their structures have been established by X-ray crystallographic analysis (Gainsford, G. J., Department of Scientific and Industrial Research, New Zealand; Robinson, W. T., University of Canterbury, New Zealand, unpublished data), and the respective conformations are stabilized because of the demand that the methyl groups adopt an equatorial orientation for the five-membered Co–pn-type ring systems. Thus, the two diastereoisomers shown are not interconvertible and are readily separated by chromatography.

Several important features arose from this study. The *ob*$_3$ and *lel*$_3$ isomers had vastly different redox potentials for the Co(III)/(II) couple, $\Delta E \sim 0.3$ V ($\sim 10^5$ in equilibrium constant), and the self exchange electron transfer rate constants differed by ~ 30 fold. The two isomers shown in Figure 2 have a facial arrangement of the methyl groups (*fac*), and there are equivalent isomers where the methyl groups are arranged meridionally (*mer*). The regiospecificity

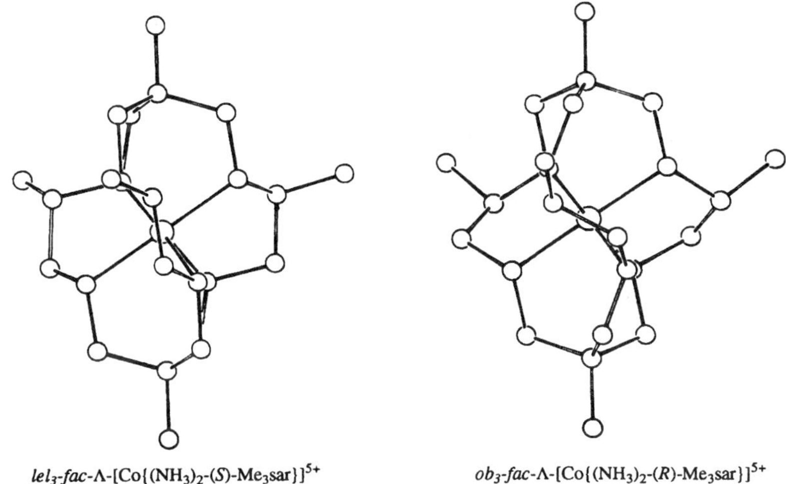

Figure 2. Stable conformers of the $[Co((NH_3)_2, Me_3–sar)]^{5+}$ cages. (Reproduced with permission from reference 5. Copyright 1989.)

Table I. Rate Constants and Redox Potentials for Diastereoisomers of $[Co\{(NH_3)_2,Me_3\text{-}sar\}]^{5+/4+}$ ions at 25 °C

Isomers	Rate Constants $(M^{-1} s^{-1})$	E′ (V vs. SHE)
Enantiomeric pairs	$k_{1,1}$	
fac-lel₃	0.031[a]	0.015
mer-lel₃	0.033[a]	0.015
fac-ob₃	0.97[a]	−0.295
mer-ob₃	1.00[a]	−0.325
$[Co\{(NH_3)_2sar\}]^{5+/4+}$	0.025[a]	0.02
Diastereoisomeric pairs	$k_{1,2}$	
mer-Λ-lel₃ + fac-Δ-ob₃	54[a], 89[b]	0.34
mer-Δ-lel₃ + fac-Δ-ob₃	45[a]	
fac-Δ-lel₃ + fac-Δ-ob₃	40[a]	0.34
fac-Λ-lel₃ + fac-Δ-ob₃	32[a]	
mer-Δ-lel₃ + mer-Δ-ob₃	17[a]	0.31
mer-Λ-lel₃ + mer-Δ-ob₃	14[a]	
fac-Λ-lel₃ + mer-Δ-ob₃	13[a]	0.31
fac-Δ-lel₃ + mer-Δ-ob₃	10[a], 53[b]	

NOTE: SHE means standard hydrogen electrode.

[a]Observed rate constant at conditions of 25 ± 0.1 °C in 0.1 M CF_3SO_3H/0.1 M CF_3SO_3Na.

[b]Calculated rate constant from the Marcus cross-reaction relationship: $k_{12} = (k_{11}.k_{22}.K_{12}.f_{12})^{0.5}$.

SOURCE: Reproduced with permission from reference 5. Copyright 1989.

of these substituents did not affect the redox properties much at all, but this stereochemical factor, the combinations of chirality about the metal and the chiral C center, generate twelve pairs of redox couples to give electron transfer rate constants (Table I). Their values show that $\Delta\Delta$ and $\Delta\Lambda$ combinations differ little. The variations arising from the *mer* and *fac* orientations of the methyl group vary by as much as a factor of 5, but the lel_3 and ob_3 combinations vary by ~2000-fold. What is also evident is a substantial change in λ_{max} for the Co(III) lel_3 and ob_3 forms from ~480 to ~450 nm, respectively, for the absorption band of $^1A_{1g} \rightarrow {}^1T_{1g}$ parentage. Clearly, the ob_3 conformer has a stronger ligand field than the lel_3 form, and these effects are carried through to the redox properties. However, they all emanate from the demand of the methyl group to remain in an equatorial position in all the isomers. The other inference one can gather from these data is that $\Delta\Delta$ and $\Lambda\Delta$ combinations and *mer–mer*, *fac–fac*, and *mer–fac* combinations do not influence the docking of the ions in the activated complex in any profound way. So for very similar molecules, the steric effects influenced the conformations and thereby the redox and electron transfer rates substantially.

Then it became interesting to see what happened when the cage was expanded in size. This cage expansion clearly would enhance its ability to stabilize larger metal ions, and the result (6) is displayed in Figure 3b. Compared with a comparable sarcophagine-type cage, Figure 3a, the redox potential is now much more positive, clearly indicating the Co²⁺ ion is relatively stabilized in the larger cavity. For the Co(III) complex, the λ_{max} (band origin $^1A_{1g} \rightarrow {}^1T_{1g}$) occurs at a longer wavelength, and the complex is crimson instead of the yel-

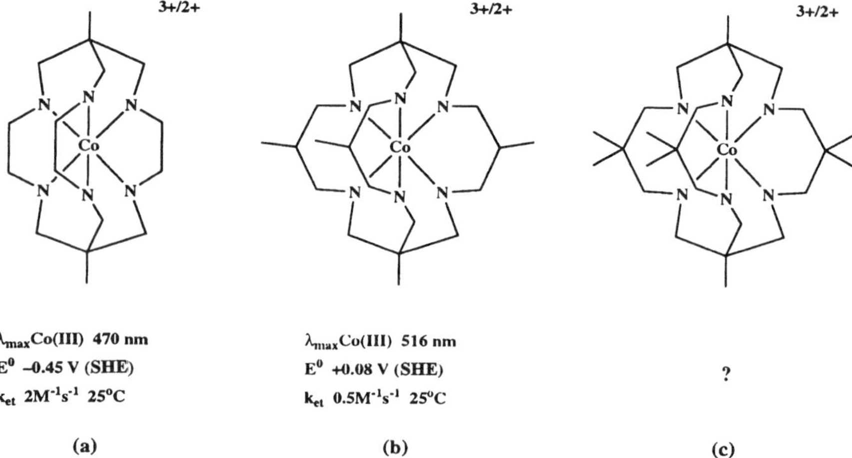

Figure 3. Spectra maxima, redox potentials and electron transfer self-exchange rate constants for sar and tricosane cages; a is [Co(Me₂-sar)]³⁺/²⁺, b is [Co(fac-Me₅-D₃ₕtricosaneN₆)]³⁺/²⁺, and c is [Co(Me₈-D₃ₕtricosaneN₆)]³⁺/²⁺.

low-orange characteristic of Co(III) hexaamine compounds in general. These data imply a weaker ligand field and an expanded cavity, and that shows up as longer Co(III)–N bonds (>2.02 Å). The electron transfer self-exchange rate, however, does not alter very much. Now the cage is a slightly better fit for Co(II) and a much poorer fit for Co(III), and crudely it can be viewed as a trade-off in terms of steric strain for the two ions; that is, relative to [Co–(Me$_2$-sar)]$^{3+/2+}$ the total steric strain component for the reorganization energy of [Co(*fac*-Me$_5$-tricosaneN$_6$)]$^{3+/2+}$ is not very different, but is just distributed in a different manner. The structure of the Co(III) ion (Figure 4) shows three rather

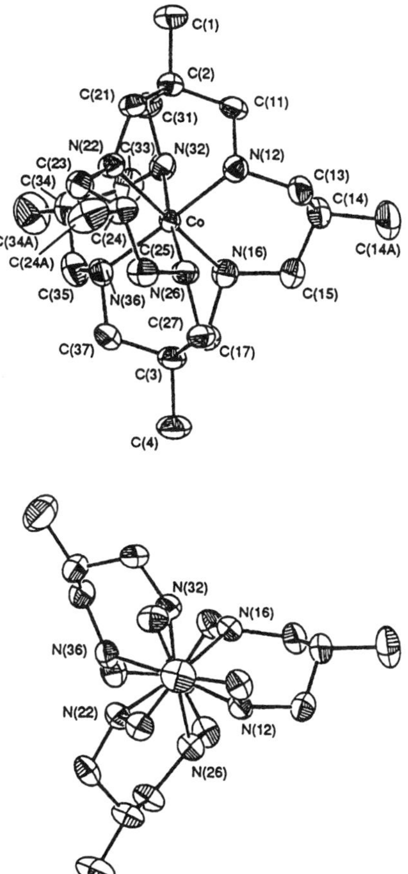

Figure 4. The structure of Λ-[Co(fac-Me$_5$-D$_{3h}$tricosaneN$_6$)]$^{3+}$ in its (PF$_6$)$_3$·H$_2$O salt. Selected bond lengths (Å) and angles (°) (averaged assuming C$_3$ symmetry over n = 1–3) are as follows. Bond lengths: Co–N(n2) 2.021 (±0.004), Co–N(n6) 2.024 (± 0.013); bond angles: N(n2)–Co–N(n6) 95.2 (±0.7), Co–N(n2)–C(n3) 120.1 (±0.5), Co–N(n6)–C(n5) 120.7 (±0.7); torsion angles: N(n6)–Co–N(n2)–C(n3) −19.9 (± 1.8), N(n2)–Co–N(n6)–C(n5) 15.9 (± 2.2).

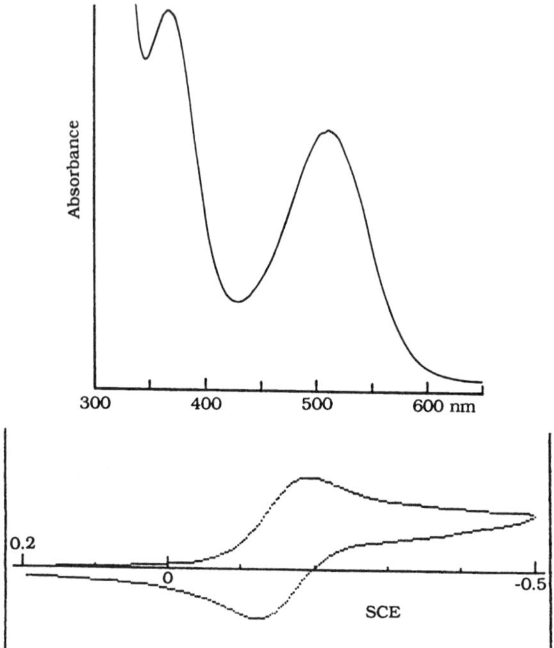

Figure 5. Visible spectra and cyclic voltammetry for [Co(fac-Me$_5$-D$_{3h}$tricos-aneN$_6$)]$^{3+}$ ion.

distorted chairlike six-membered rings with the methyl groups all equatorial and C_3 symmetry overall. The cyclic voltammetry also shows essentially a reversible couple (Figure 5), and the synthesis of this ion (Scheme I) is remarkably stereospecific with the overall yields at the moment being 25–30%. In summary then, we have this picture of a more positive redox potential for the larger cage but not much change in electron transfer rate.

What happens, then, if three more methyl groups are added to the annular six-membered rings as shown in Figure 3c? The added methyl group will clearly confuse the chelate in relation to which conformation it should adopt, and the steric forces overall should clearly increase. This synthesis can be readily accomplished by using 2-methyl propanal instead of propanal in the previous synthesis. The reaction works quite well, but in the final step we get a Co(II) complex whose structure is given in Figure 6, instead of the Co(III) ion that normally arises from the BH_4^- reduction of the imine groups in the initial triimine tricosane product. The structure has quite a different conformation to that of the [Co^{III}(Me$_5$-tricosaneN$_6$)]$^{3+}$ ion. Now the annular rings are skew–boat so the two methyl groups are equivalent, and a plane through the three skeletal C atoms of the ring is parallel to the C_3 axis of the ion. Moreover, the complex is now quite difficult to oxidize to the Co(III) state, which is now destabilized considerably.

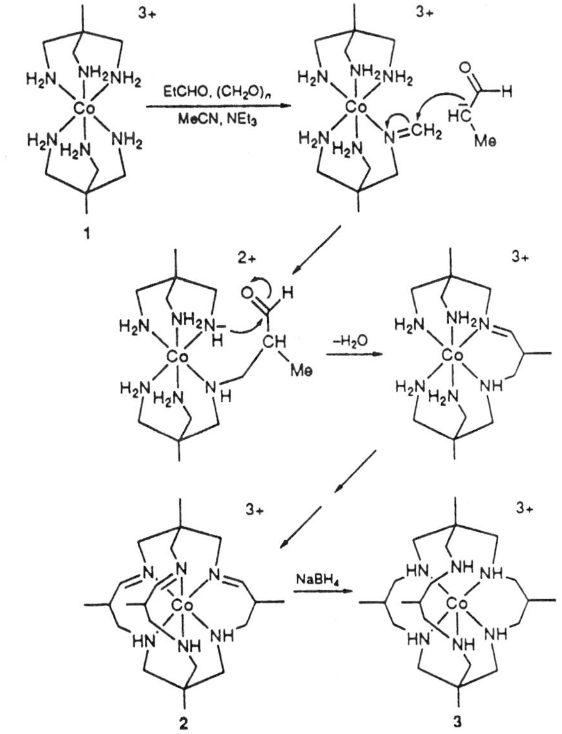

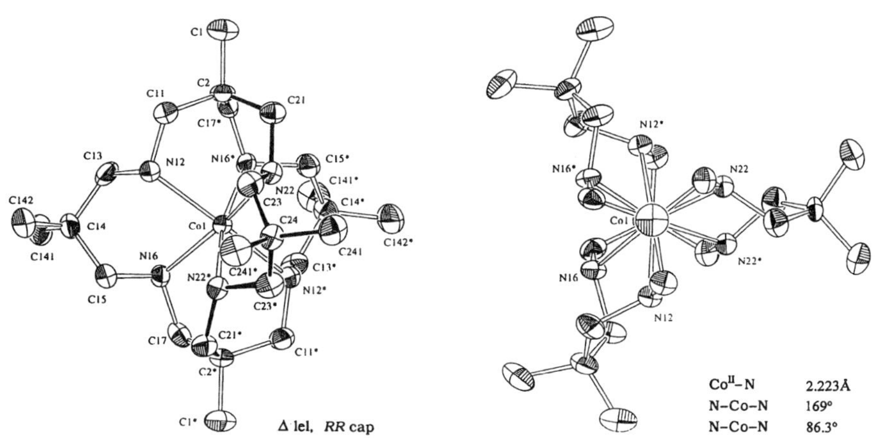

Scheme I. Synthetic scheme for [Co(fac-Me$_5$-D$_{3h}$tricosaneN$_6$)]$^{3+}$

Figure 6. Structure of the [Co^{II}(Me$_8$-D$_{3h}$tricosaneN$_6$)]$^{2+}$.

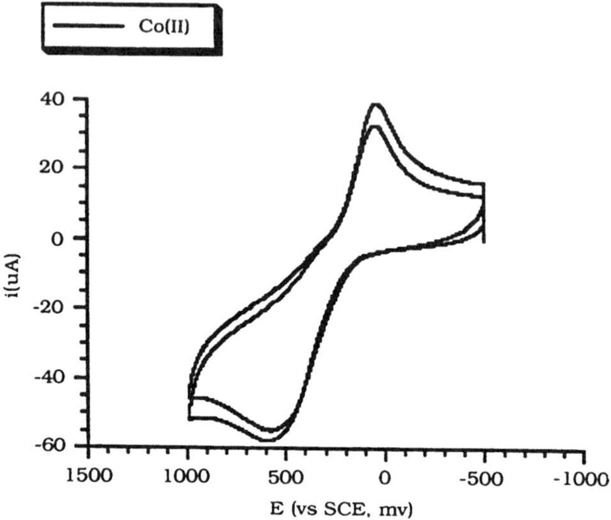

Figure 7. Cyclic voltammetry of [Co^{II}(Me_8-D_{3h}tricosaneN$_6$)]$^{2+}$ ion in water at 22 °C.

The cyclic voltammetry for the [Co^{II}(Me$_8$–tricosaneN$_6$)]$^{2+}$ ion in water (Figure 7) is no longer so reversible. It has an oxidation wave at 0.84 V (vs. standard hydrogen electrode (SHE)), and there is considerable asymmetry in the whole cyclic voltammogram. The system is chemically reversible, however, and it is possible to quantitatively oxidize the Co(II) ion to Co(III) coulometrically. Also, it can be seen that a distortion at ∼250 mV on the reduction wave [vs. saturated calomel electrode (SCE)] indicates the presence of another Co(III) species.

The oxidized solution was therefore examined in considerable detail, and two Co(III) salts have been isolated from the solution under different conditions, one orange and the other blue. These two salts have the same constitution except for water content. The orange form is isolated from water as the nitrate salt, whereas the blue form is isolated with ethanol from the concentrated mother liquor. If the solid orange form is dropped into ethanol it turns blue immediately, at least on the surface.

Are the two forms isomers, or is one a hexaamine and the other a partly dissociated cage complex? Reflectance spectra (Figure 8) show that they are clearly different. The orange form has an absorption maximum at 470 nm, whereas the blue form has a maximum at about 600 nm. The nitrate salt of the orange form gives crystals suitable for X-ray crystallographic analysis except that at ∼20 °C X-irradiation generates the blue form and crystallinity is lost. However, at low temperature (–60 °C) a good structure was obtained (Figure 9), quite different from that of the Co(II) complex. A skew–boat configuration of the annular six-membered rings is retained, but the complex now has an *ob$_3$*-

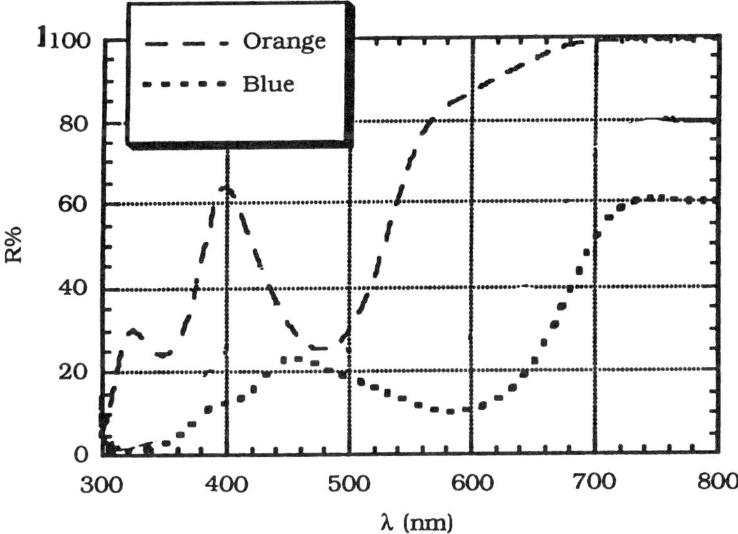

Figure 8. Reflectance spectra of the blue and orange isomers of $[Co(Me_8\text{-}D_{3h}tricosaneN_6)](NO_3)_3$.

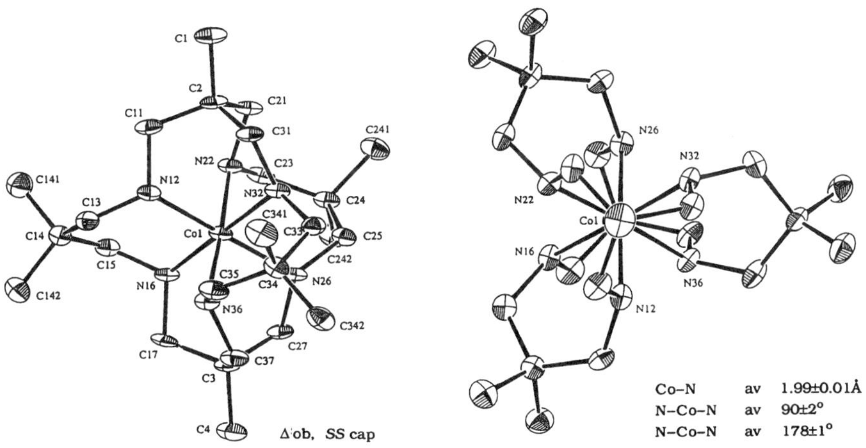

Figure 9. Structure of the orange isomer $[Co^{III}(Me_8\text{-}D_{3h}tricosaneN_6)]^{3+}(NO_3)_3\cdot5H_2O$.

like structure. The plane of the three C atoms in the annular six-ring chelates is oblique to the C_3 axis. Nice looking crystals of the blue form were also obtained, but an X-ray study yielded very weak diffraction data, which indicated considerable lattice disorder involving the cobalt atom.

The dissolution of each crystalline form in D_2O gives identical ^1H and ^{13}C solution NMR spectra (Figure 10), which also show that each form has the

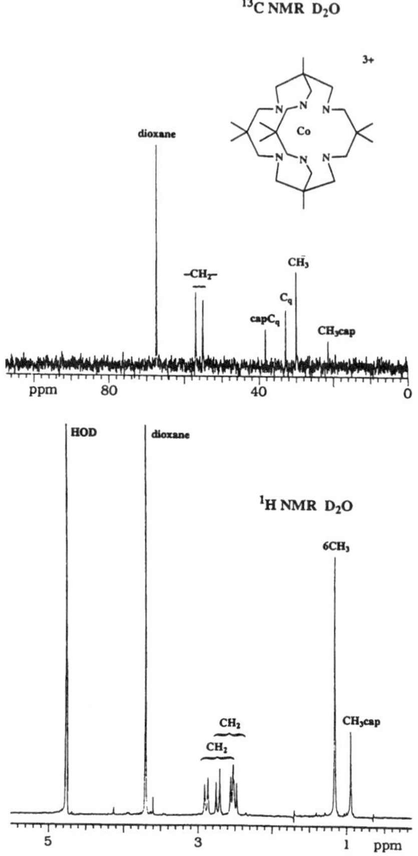

Figure 10. ^{13}C and ^{1}H NMR spectra of [Co^{III} (Me$_8$-D$_{3h}$tricosaneN$_6$)]$^{3+}$ ion in D$_2$O.

same average D_3 symmetry in solution. In contrast, the solid state ^{13}C (Figure 11) and ^{59}Co (Figure 12) magic angle spinning (MAS) NMR analyses of the orange and blue salts portray a different picture. The ^{13}C MAS NMR spectra of the orange form closely resemble its ^{13}C solution NMR spectrum, and are consistent with the presence of an essentially D_3 molecule with a similar structure. However, the ^{13}C MAS NMR spectra of the blue form show that the structure has been perturbed, with significant downfield shifts of the annular methylene resonances from ~54 ppm to ≥60 ppm, as well as some rearrangement of the 20–40-ppm resonances being observed. Although only six resonances are clearly observed for the blue salt, consistent with D_3 symmetry, the intrinsic broadness of the solid-state NMR resonances (in comparison to their solution NMR counterparts) may preclude the observation of further peak structure linked to a reduction from D_3 symmetry.

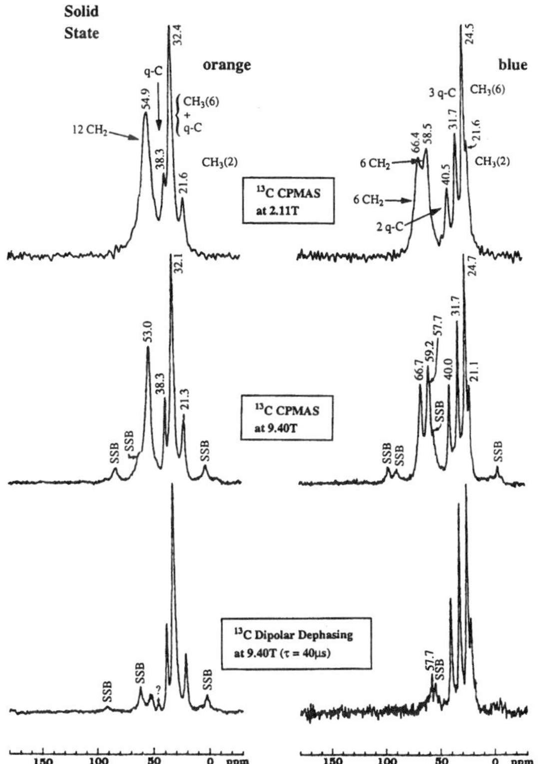

Figure 11. ^{13}C *solid-state NMR spectra of* $[Co(Me_8\text{-}D_{3h}tricosaneN_6)](NO_3)_3$: *blue (4H₂O) and orange (5H₂O) isomers.*

The ^{59}Co MAS NMR spectra (Figure 12) more clearly highlight large electronic and structural differences between the orange and blue forms. The very sharp ^{59}Co resonance obtained for the orange salt indicates a very symmetrical cobalt environment, which eliminates any substantial quadrupolar interaction at the metal site. The ^{59}Co spectrum for the blue form, however, reveals an enormous downfield shift of ~4000 ppm for the cobalt resonance relative to that of the orange form. This shift is coupled with a spectacular increase in linewidth attributed to an increased nuclear quadrupolar interaction experienced by the Co nucleus. These spectral changes are consistent with a reduced ligand field strength at the cobalt site(s) in the blue complex implied by the downfield shift to higher deshielding, and an increased electric field gradient reflecting a significant electronic distortion from D_3 symmetry about the Co nucleus.

In total, the spectroscopic data for the Co(III) complex imply that there are two conformational isomers that interconvert in solution and the blue form

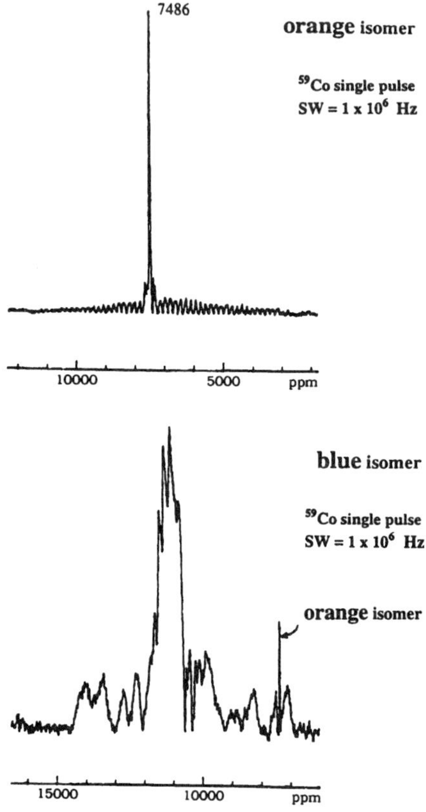

Figure 12. ^{59}Co *solid-state NMR spectra of* $[Co(Me_8\text{-}D_{3h}tricosaneN_6](NO_3)_3$: *blue (4H_2O) and orange (5H_2O) isomers.*

has a very weak ligand field, bordering on the high-spin state. Despite the lack of a structure for the blue form, it is clear that the average Co^{III}–N bond length should be ≥ 2.1 Å, much longer than has been observed hitherto. It is also likely that its structure is close to that of the Co(II) dinitrate salt. Simply, the two feasible structures are related to the *RR* or *SS* chirality of the two caps (Figure 13). (*RR* or *SS* describe the right- or left-handed helical relationship between the N–C bonds of the caps and the C_3 axis of the complex ion.) In the ΔSS (ΛRR) form, the ob_3 conformation is evident, whereas in the ΔRR (or ΛSS) form the lel_3 conformer obtains. The ΔRR (lel_3) conformer is essentially more extended along the C_3 axis and is thereby able to destabilize a smaller ion. The interconversion between the ΔRR and ΔSS forms therefore does not require a configuration change about Co or N and is simply a conformational change for the three six-membered annular chelates and the cap configurations. This

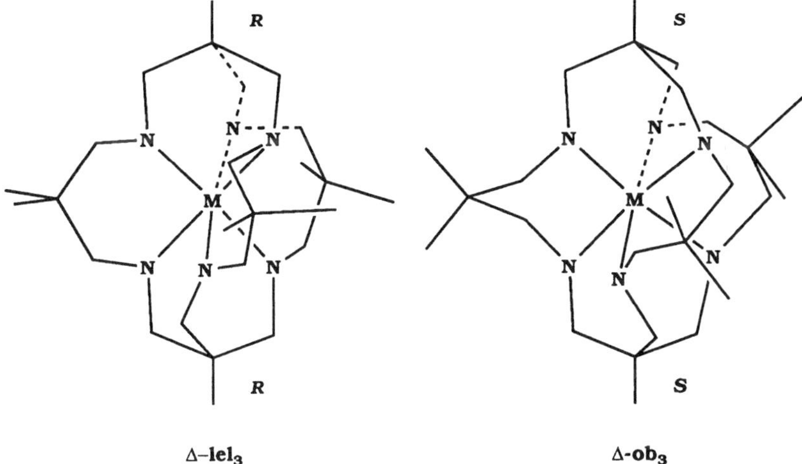

Figure 13. Projected structures of the blue (lel₃) and orange (ob₃) isomers of [Co(Me₈-D₃ₕtricosaneN₆)]³⁺.

change appears to occur on a time scale faster than a second but slower than a millisecond, and it may be possible to slow the rate of this process and observe both isomers independently. It is also evident that these two forms must be nearly equienergetic even in water although favoring the ob_3 form somewhat. It is also obvious that the blue form must be the good oxidant ($\sim$0.8 V vs. SHE).

Overall, the major factor in these studies is that the redox potential can be changed by at least 1.2 V with the same metal ion couple and the same set of ligating atoms merely by influencing the steric factors in a modest way in the cage. It is obvious also that the Co(III)–N bond lengths have been extended a considerable way toward the Co(II)–N bond length (2.22 Å), which itself is longer than normal for a high-spin Co(II) saturated amine complex. It is even possible that the Co(III)–N bond length in the blue isomer is close to the high-spin Co(III)–N value. It follows that the intramolecular reorganization energy component of the electron transfer barrier may be small and the rate of electron self-exchange fast between the blue Co(III) isomer and the Co(II) form. It remains to be seen what that self-exchange rate constant is for the cobalt(II)/(III) octamethyl tricosaneN$_6$ couple.

Acknowledgments

This chapter is based on a paper presented in honor of Henry Taube, a stimulating researcher, teacher, and friend, on the occasion of his 80th birthday Symposium.

References

1　Buckingham, D. A.; Sargeson, A. M. *Topics in Stereochemistry;* Wiley: New York, 1971; pp 219–277.

2.　(a) Creaser, I. I.; Geue, R. J.; Harrowfield, J. M.; Herlt, A. J.; Sargeson, A. M.; Snow, M. R.; Springborg, J. *J. Am. Chem. Soc.* **1982,** *104,* 6016, and references therein; (b) Geue, R. J.; Pizer, R.; Sargeson, A. M. *Abstracts of Papers,* National Meeting of the American Chemical Society, Las Vegas, NV; American Chemical Society: Washington, DC, 1982; INOR 62.

3.　Geselowitz, D. *Inorg. Chem.* **1981,** *20,* 4457.

4.　Vallee, B. L.; Williams, R. P. *Proc. Natl. Acad. Sci. U.S.A.* **1968,** *59,* 498.

5.　Geue, R. J.; Hendry, A. J.; Sargeson, A. M. *J. Chem. Soc. Chem. Commun.* **1989,** 1646.

6.　Geue, R. J.; Höhn, A.; Ralph, S. F.; Sargeson, A. M.; Willis, A. C. *J. Chem. Soc. Chem. Commun.* **1994,** 1513.

9

Intrinsic Barriers to Atom Transfer Between Transition-Metal Centers

Carol Creutz

Chemistry Department, Brookhaven National Laboratory, Upton, NY 11973–5000

The results of our mechanistic studies of one-equivalent, $C_5H_5(CO)_3M^I\bullet/$ $C_5H_5(CO)_3M^{II}$–X, and two-equivalent, $(C_5H_5(CO)_3M^0)^-/C_5H_5(CO)_3$-$M^{II}$–X, atom-transfer self-exchange reactions (M = Mo or W; X = Cl, Br, or I) are summarized. These studies are presented in the context of Henry Taube's contributions as a teacher and researcher, and the chapter provides an account of the interplay between ideas and experiments in the inorganic mechanistic community.

$\mathbf{I}$ CHOSE THE SUBJECT OF THIS CHAPTER—intrinsic barriers to atom transfer between transition metal centers—because Henry Taube was more excited about this work than any I have done since my student days and because I felt it owed so much to his way of thinking. In preparing this chapter, I have reconstructed the origin of "my" ideas for this work, and I have found that Taube's influence was even greater than that which I acknowledged at the symposium presentation on which this chapter is based. Thus, in this chapter, I make an effort to trace the origin and evolution of these ideas, so that this account is as much about my debt to Taube and the mechanistic community as it is about our research work itself.

All of us who have had the luxury of working with Henry Taube have to varying degrees come around to his ways of thinking. A near obsession with comprehensive descriptions of the thermodynamics and kinetics of the system is fairly common among us. I admit to being so obsessed. It has puzzled me how this happened to me because the focus of my doctoral work was spectroscopy rather than chemistry per se. Certainly these themes do pervade his papers, to which so many of us turn for help ranging from gripping accounts of descriptive chemistry to economical approaches to complex rate laws. After some reflection I recalled the undergraduate course Taube taught during the

1960s. In that course, which all of his graduate students attended, he provided a rich series of accounts of inorganic reactivity. I looked back to my notes from the course and on the first page I found

1. What is the state of equilibrium?

2. What governs the rate of reaction? Mechanism.

3. Understand observations.

Although I had completely forgotten the event, Taube had stated his philosophy of chemical research on the first day of that class. In subsequent lectures he taught us inorganic chemistry in these important terms. Following an initial section in which we learned to describe the principal net chemical change in systems of increasing complexity, he took on the oxidation state construct with illustrations first from the rich chemistry of sulfur.

In reviewing my notes I was startled to discover two issues that arise in the atom-transfer work. The first is the oxidation state convention for carbon. (Taube frequently uses the "metal center" carbon comparison as an effective teaching device.) In CO_2, because oxygen is so much more electronegative than carbon, the assignment of C as 4+ is natural. By contrast, in CH_4, "C^{4-}, H^{1+}" or "C^{4+}, H^{1-}" are equally valid, but retaining the convention H^+ (and C^{4-}) is convenient, at least in aqueous media. Later in that lecture he raised an issue that will emerge again when the theme of atom transfer is addressed:

$$BrF + Cl^- = BrCl + F^- \tag{1}$$

$$BrF + I^- = BrI + F^- \tag{2}$$

Despite the obvious similarity of the two halogen displacement reactions, eq 1 is classified as substitution (because bromine retains its $+1$ oxidation state, both chlorine and fluorine being more electronegative than bromine), whereas eq 2 is classified as a redox reaction (with bromine being reduced from $+1$ to -1).

On the other hand, I can trace the work presented in this chapter to very specific circumstances. In 1981, John Endicott and David Rorabacher organized an interdisciplinary mechanistic conference at Wayne State University. Although many of the talks focused on outer-sphere electron transfer, other mechanistic issues ranged from nucleophilic substitution on carbon centers in the gas phase [Pellerite and Brauman (1)] to proton transfer between transition-metal centers [Jordan and Norton (2)]. I was most influenced by those two contributions and that of Taube (3), entitled "Observations on Atom-Transfer Reactions". Indeed Taube provided a unifying intellectual theme for the mechanistic areas that lie outside that of outer-sphere electron transfer. He gave the following definition for "atom transfer":

> The term implies that an atom originating in the oxidizing agent is transferred to the reducing agent so that in the activated complex both centers are bound to the atom being transferred....The term "atom" is used only to distinguish the reactions being considered from the class in which large bridging ligands are transferred by remote attack.... Many of the latter involve only weak electronic interactions in the activated complexes and, for this reason, are set apart from the cases where oxidizing and reducing centers are separated by single atoms, so that electronic coupling in the activated complexes can be very large....Atom transfer reactions, as defined, are widely recognized as being extremely important.

We broadened (4) the definition to include the possibility that an atom originating in either oxidizing or reducing agent is transferred to the reaction partner. In the activated complex, both centers are attached to the atom being transferred. In this sense, "atom transfer" is intended as a broad reaction class and is not restricted to reactions in which a single neutral atom is transferred (e.g., a hydrogen atom or halogen atom abstraction). Reactions falling within this class include one-electron and two-electron inner-sphere electron-transfer reactions, halogen and hydrogen-atom abstractions, hydride-transfer reactions, and certain proton-transfer and nucleophilic substitution reactions.

This definition should be broadened further to include multiequivalent systems studied by Woo (5). The only substantive difference between these two definitions, and one which I assume Taube would support, is that the atom transferred may originally be attached to either the oxidant or the reductant. The broader definition includes (unusual) cases, normally classified as nucleophilic substitution, in which the group transferred is attached to the more electron-poor center. Taube (3) acknowledges this issue:

> Certain redox changes involving atom transfer can usefully be dealt with applying the ideas which have been developed for substitution. There is, in fact, no sharp distinction between a 2 e^- redox change involving atom transfer and an orthodox nucleophilic substitution. This point is illustrated by the two reactions

$$IBr + Cl^- = ICl + Br^-$$
$$BrCl + I^- = IBr + Cl^-$$

> where the second only is a redox process as ordinarily defined.

A broad definition for the atom-transfer process is helpful here because of the nature of the systems to be studied and compared. One comparison of interest is the reactivity of metal hydride complexes toward proton (H^+) transfer with halogen transfer systems. Furthermore, in the transition-metal systems, information and intuition about relative electronegativities is not so well-advanced as it is for the nonmetals.

Following the Wayne State Symposium, I was increasingly impressed and intrigued by the work of Norton and his colleagues (2, 6), which probed proton transfer between transition metals via measurements of self-exchange reactions, for example

$$Cp(CO)_3M^{0-} + Cp(CO)_3\mathbf{M}^{II}\text{–}H \rightleftharpoons Cp(CO)_3M^{II}\text{–}H + Cp(CO)_3\mathbf{M}^{0-} \qquad (3)$$

and the dependence of the rates of net reactions on the driving force for the reaction. Intellectually, this work paralleled the early studies of outer-sphere electron transfer made at Brookhaven National Laboratory. Dodson and colleagues evaluated self-exchange rate constants for the one- (7) and two-electron (8) aquo couples $Fe(H_2O)_6^{2+}/Fe(H_2O)_6^{3+}$ and Tl_{aq}^{+}/Tl_{aq}^{3+}. Sutin (9) systematically tested the driving force dependence of outer-sphere electron transfer rates and advanced our understanding of the factors that control the magnitude of the self-exchange barrier. I came to hope that we would see a comparably sophisticated understanding of the factors that control the rates of (at least some of) the atom-transfer reactions emerge in the next generation of mechanistic studies. When Smith et al. (*10*) reported their self-exchange studies (eq 4) for halogen (X) transfer between M=Os or Ru couples (Cp = cyclopentadienyl, $C_5H_5^-$)

$$Cp_2M^{II} + Cp_2\mathbf{M}^{IV}\text{–}X^+ \rightleftharpoons Cp_2M^{IV}\text{–}X^+ + Cp_2\mathbf{M}^{II} \qquad (4)$$

my colleague Morris Bullock and I saw an opportunity to make a new experimental contribution in the area. We chose to examine halogen transfer in $Cp(CO)_3M^{0-}/Cp(CO)_3\mathbf{M}^{II}\text{–}X$ systems (M = Mo or W) because of their relationship to systems examined by Norton (eq 3) and because their properties made them particularly suitable for a wide range of measurements.

Self-exchange reactions for outer-sphere electron transfer and for atom-transfer on the transition metal M and on carbon are depicted in Scheme I. Techniques ranging from "labeling" a metal with a radioactive isotope to EPR and NMR line broadening have been used for the determination of outer-sphere electron transfer self-exchange rates, which have been found to range over more than 14 orders of magnitude (*11*). The self-exchange reaction is, in principle, the simplest reaction because the driving force for the reaction (except for that of mixing) is nil. By studying the self-exchange process it is possible to focus on the properties of the reactants—their molecular structures and electronic properties—that determine their intrinsic reactivity with respect to the redox process.

It is instructive to compare the one- and two-equivalent atom-transfer processes for metal-centered and carbon-centered processes. The metal-centered two equivalent "X$^+$" transfer is analogous to the putative carbanion

Outer-Sphere Electron Transfer

$$*Fe(H_2O)_6{}^{2+} + Fe(H_2O)_6{}^{3+} \rightarrow *Fe(H_2O)_6{}^{3+} + Fe(H_2O)_6{}^{2+}$$

Atom Transfer

Metal-Centered

$$L_n*M^- + X\text{–}ML_n \rightarrow L_n*M\text{–}X + L_nM^- \qquad X^+ \text{ transfer}$$

$$L_n*M \; + X\text{–}ML_n \rightarrow L_n*M\text{–}X + L_nM \qquad X \text{ transfer}$$

Carbon-Centered

$$R_3*C^- + X\text{–}CR_3 \rightarrow R_3*C\text{–}X + R_3C^- \qquad X^+ \text{ transfer}$$

$$R_3*C \; + X\text{–}CR_3 \rightarrow R_3*C\text{–}X + R_3C \qquad X \text{ transfer}$$

Scheme I. Self-exchange reactions.

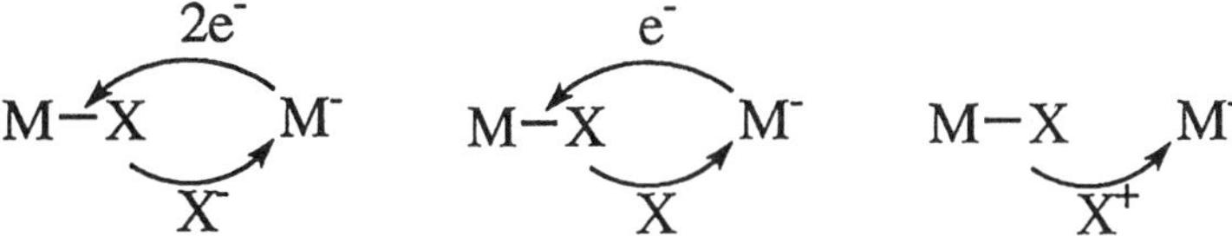

Scheme II.

process, while the one-equivalent "X•" transfer involves a metal radical with reactivity reminiscent of the organic radical analogue. The term "X⁺ transfer" is intended only to indicate stoichiometry. At least three mechanistic extremes (Scheme II) can be envisioned for the actual self-exchange.

We have studied self-exchange reactions for both one-equivalent atom-transfer reactions (*12, 13*) (eq 5) with M = W or Mo and X = Cl, Br, I, or H, and two-equivalent self-exchanges of metal anions (*4*) with their metal halide (eq 6) complexes.

$$Cp(CO)_3M^I + Cp(CO)_3M^{II}\text{–}X \rightleftharpoons Cp(CO)_3M^{II}\text{–}X + Cp(CO)_3M^I \qquad (5)$$

$$Cp(CO)_3M^{0-} + Cp(CO)_3M^{II}\text{–}X \rightleftharpoons Cp(CO)_3M^{II}\text{–}X + Cp(CO)_3M^{0-} \qquad (6)$$

These compounds have structures based on either the three- or four-legged piano stool structures depicted in structures **1–3**. Both the d^6 M(0) (**3**) and d^4 M(II) (**1**) complexes are 18-electron species, and both are extremely inert to substitution. By contrast, the d^5 M(I) (**2**) 17-electron species (*14*) are extremely substitution labile and, in the presence of appropriate ligands (L), rapidly equilibrate to yield 19-electron $Cp(CO)_3M^IL$ adducts (*15, 16*) of moderate stability.

$$M(II) \; d^4 \; 18e^- \qquad\qquad M(I) \; d^5 \; 17 \, e^- \qquad\qquad M(0) \; d^6 18e^-$$

$$\mathbf{1} \qquad\qquad\qquad\qquad \mathbf{2} \qquad\qquad\qquad\qquad \mathbf{3}$$

In the course of the two-equivalent, M–X/M⁻ self-exchange (eq 6), the X group is transferred from the "leg" site of one reactant to that in another, and only small changes in the intramolecular $Cp(CO)_3M$ distances and angles ensue. In the anion (*17*), the OC–M–CO angles and Cp–M–CO angles are typically 86 and 128°, respectively; in $Cp(CO)_3M–X$, both angles are somewhat smaller (for X = Cl, 78 and 110–125°, respectively) (*18*). The Mo–Cl and W–Cl distances are 2.498(1) and 2.490(2) Å (*18*). In the course of the one-equivalent, M–X/M• self-exchange (eq 5), the X group is also transferred between the "leg" sites of the reactants. For M = Mo and W, the structures of the M(I) radicals have not been characterized, but for M = Cr (*19–21*), the odd electron appears stereo-chemically active (as indicated in **2**). The metal radicals are not stable (K_D is extremely large for M = Mo and W)

$$3Cp(CO)_3M \;\; \rightleftharpoons [Cp(CO)_3M]_2 \qquad\qquad K_D \qquad (7)$$

but rather are produced via photolysis of the metal–metal-bonded M(I) dimer, which occurs with high quantum yield upon photolysis with even visible light.

To evaluate the rates of the self-exchange reactions eqs 5 and 6, we used a range of ^1H NMR methods that exploit chemical-shift differences between the protons on the cyclopentadienide ring in the different compounds. For the one-equivalent process, we simply measured the rate of incorporation of ^HCp (C_5H_5) in the deuterated ($^DCp = C_5D_5$) $^DCp(CO)_3M–X$ induced by photolysis of the dimer.

$$[Cp(CO)_3M]_2 + h\nu \rightarrow 2Cp(CO)_3M \qquad (8)$$

The dimer absorption spectra exhibit intense $\sigma_b \rightarrow \sigma^*$ transitions at 350–400 nm and moderately intense $d\pi \rightarrow \sigma^*$ transitions near 500 nm (*22*). Irradiation with ultraviolet or visible light induces scission of the metal–metal single bond (*see* Figure 1). The competition between exchange (eq 5), reaction with a "trap-

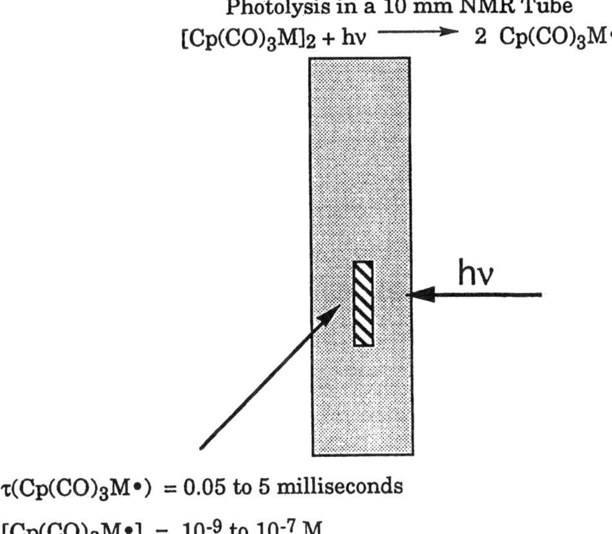

$\tau(Cp(CO)_3M\bullet) = 0.05$ to 5 milliseconds

$[Cp(CO)_3M\bullet] = 10^{-9}$ to 10^{-7} M

Figure 1. Photolysis in a 10-mm NMR tube.

ping" reagent T–Y (eq 9), and radical recombination (eqs 7 and 10) is then used to determine the exchange rate constant.

$$Cp(CO)_3M + T–Y \rightarrow Cp(CO)_3MY + T \qquad k_{TY} \qquad (9)$$

$$2Cp(CO)_3M \rightarrow [Cp(CO)_3M]_2 \qquad k_{di} \qquad (10)$$

These competition experiments were carried out in 10-mm NMR tubes, and product formation was monitored as a function of the extent of irradiation over a range of light intensities and concentrations.

By contrast, for the two-equivalent process ^1H NMR is applied to stable solutions of the reagents with the method being determined by the specific rate achievable when the reagents are 1–10 mM (*see* Scheme III). For the iodo systems, exchange is rapid over a range of temperatures, and the line broadening method was used. For the chloro systems, exchange was so slow that we resorted to mixing $Cp(CO)_3MCl$ with $^DCp(CO)_3M^-$. With X = Br, the specific rates and spin-lattice relaxation times $(T_1 \sim 60$ s) are such that the technique of magnetization transfer is applicable. The results of our studies are summarized in Table I.

The one-equivalent processes are very rapid, with even the slowest X = Cl self-exchanges being comparable to the outer-sphere $FeCp_2/FeCp_2^+$ ($k = 7.5 \times 10^6$ M^{-1} s^{-1}) or $CoCp_2/CoCp_2^+$ ($k = 3.8 \times 10^7$ M^{-1} s^{-1}) electron-transfer reactions (23). Indeed for X = I, the rate constants are within about a factor of 10 of

^{1}H NMR Studies

$$Cp(CO)^3M^{0-} + Cp(CO)^3M^{II}-X \rightarrow Cp(CO)^3M^{II}-X + Cp(CO)^3M^{0-}$$

Line Broadening	X = I	$k_{obs} = 10^2\ s^{-1}$
Magnetization Transfer	X = Br	$k_{obs} = 10^{-2}\ s^{-1}$
Labeling (C_5H_5/C_5D_5)	X = Cl	$k_{obs} = 10^{-4}\ s^{-1}$

Scheme III.

Table I. Rate Constants for Self-Exchange at 25 °C

Reactants		*M = Mo*	*M = W*
$Cp(CO)_3M\bullet +$	$^DCp(CO)_3M-Cl^a$	8.2×10^6	7.9×10^5
	$^DCp(CO)_3M-Br^a$	8.5×10^7	1.3×10^7
	$^DCp(CO)_3M-I^a$	2.6×10^9	8.2×10^8
	$^DCp(CO)_3M^{-\ a}$		$\geq 1 \times 10^6$
	$^DCp(CO)_3M-H^b$		$\geq 1 \times 10^6$
	$^DCp(CO)_3M-CH_3{}^a$		$\leq 5 \times 10^2$
$Cp(CO)_3M^- +$	$Cp(CO)_3M-Cl^c$	9×10^{-2}	2.1×10^{-3}
	$Cp(CO)_3M-Br^c$	16	2.8
	$Cp(CO)_3M-I^c$	1.5×10^4	4.5×10^3
	$Cp(CO)_3M-CH_3{}^{c,d}$		$\sim 1 \times 10^{-5}$
	$Cp(CO)_3M-H^e$	2.5×10^3	6.5×10^2

NOTE: Rate constant values are in $M^{-1}\ s^{-1}$. ^DCP is C_5D_5.
[a]In CD_3CN solvent.   [b]In C_6D_6 (*13*).   [c]In CD_3CN (*4*).   [d]At 62 °C.   [e]In CD_3CN (*6*).

the diffusion-controlled values, so that the reactions are subject to very little activation barrier. The reactivity with halogen follows the order I > Br > Cl, with the range being about 3 orders of magnitude. The reactivity ratio is reminiscent of that recorded in halogen abstraction by both carbon- and metal-centered radicals, as is illustrated in Figure 2.

For the two-equivalent processes the same reactivity order is observed, with, however, the rates much smaller than for the radical processes. The range encompassed for the M$^-$/M–X exchanges, about 6 orders of magnitude, is similar to that reported for $RuCp_2/RuCp_2X^+$ exchanges (*27–31*). It is reasonable that both one- and two-equivalent processes involve bridged transition states as indicated in structure 4. Although a linear M–X–M geometry is depicted, it is merely intended to be illustrative; similarly, M–X distances in the species are not expected to be the same. Interestingly, the transition-state geometry for interhalogen exchange in X + X_2 and X_2Y species (*32*) depends upon the halogen. Computations on M–X–M systems suggest that d^5–d^5 systems should be extremely bent because of the occupancy of the e_u^a orbitals (*33*). For the one-

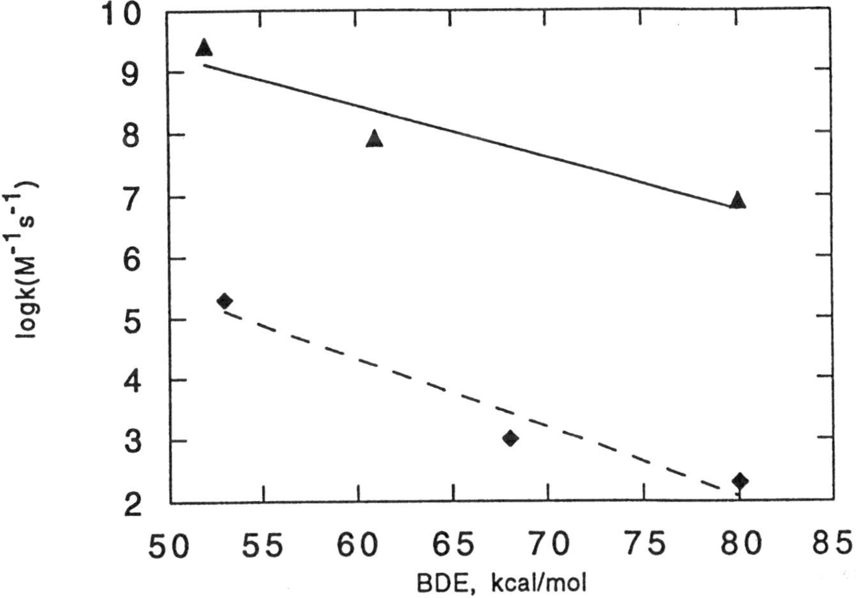

Figure 2. *Logarithm of the halogen abstraction rate constant for M = Mo complexes at 25 °C in CD$_3$CN (filled triangles) and for alkyl radicals (filled diamonds) in benzene (50 ± 2 °C) (adapted from references 24–26).*

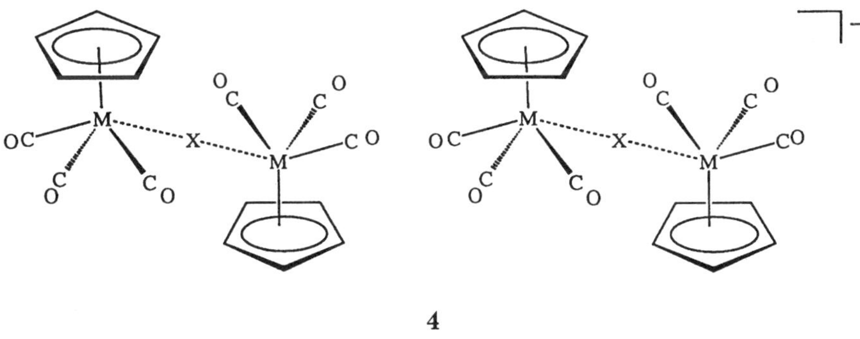

4

equivalent reaction, the stability of such 19-electron binuclear species is suggested by the affinity of the 17-electron radicals for a halide (seventh) ligand (*34*). For the two-equivalent process, 19-electron species formed through comproportionation

$$M^0 + M^{II}\text{–}X \rightarrow (M^I \text{ --}X\text{–}M^I)$$

could also be involved as indicated in Scheme IV.

$$M-X \quad \underset{X}{\overset{e^-}{\rightleftharpoons}} \quad M^-$$

Mechanism

Loose association	$*M^{0-} + X\text{–}M^{II} \rightarrow *M^{0-}/X\text{–}M^{II}$
One-electron transfer	$*M^{0-}/X\text{–}M^{II} \rightarrow (*M^{I}\text{--}X\text{--}M^{I})$
Second electron transfer	$(*M^{I}\text{--}X\text{--}M^{I}) \rightarrow *M^{II}\text{–}X/M^{0-}$
Dissociation	$*M^{II}\text{–}X/M^{0-} \rightarrow *M^{II}\text{–}X + M^{0-}$

Scheme IV.

By far the most stunning rate comparison for these data is that shown in Figure 3, a plot of the logarithm of the self-exchange rate for the two-equivalent self exchange vs. that for the corresponding one-equivalent process. Intriguingly, the slope of the plot is 2. Of course, although this case holds at 25 °C, it is unlikely to hold over a wide range of temperature; the two-equivalent self-exchanges have significant enthalpies of activation (6.4, 12.1, and 18.9 kcal mol^{-1} for Mo–I, Mo–Br, and Mo–Cl, respectively). The temperature dependences of the one-equivalent self-exchange rates have not yet been measured, but, in view of the rapid rates observed at 25 °C, the activation enthalpies must be much smaller than for the two-equivalent processes. [If activation entropy ($\Delta S^{\ddagger}$) is assumed to be –15 cal mol^{-1} K^{-1}, activation enthalpy ($\Delta H^{\ddagger}$) is 1–5 kcal mol^{-1}.] Furthermore, to the extent that comproportionation is an important feature of the activation in the M$^-$/M–X self-exchanges, the temperature dependence of the comproportionation thermodynamics may be the major determinant of $\Delta H^{\ddagger}$.

Since the original report from Taube's group the Cp$_2$M, two-equivalent systems (M = Ru or Os) have been more extensively studied (*27–30*), and related work on covalently coupled, mixed-valence systems has been reported (*35, 36*). Recent work (*31*) that addresses the halide-binding thermodynamics is particularly revealing. The high sensitivity of the self-exchange rates to the nature of the halogen is shown to derive from electronic effects in the transition states rather than to significant variations in the relative free energies of the reactants. An important role for the halogen valence p orbitals is invoked in discussing the electronic pathway.

Concluding Remarks

^{1}H NMR methods have proven immensely powerful for the study of atom-transfer self-exchange reactions, and transition-metal systems are now shown

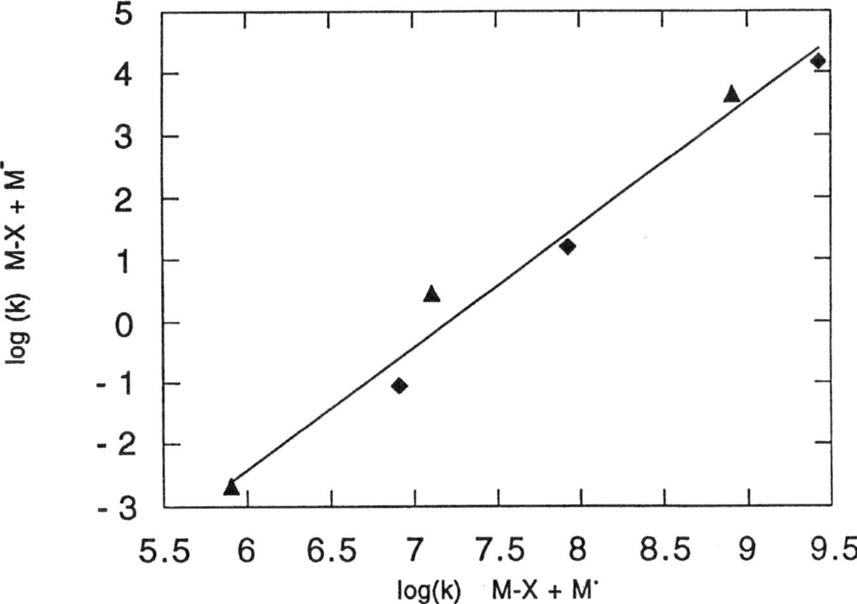

Figure 3. Logarithm of the rate constant for the M–X/M⁻ self-exchange vs. that for the M–X/M• self exchange. Filled triangles: M = W. Filled diamonds: M = Mo. Left to right: X = Cl, X = Br, and X = I. (Adapted from reference 13.)

to be excellent substrates for such studies. The one-equivalent, radical self-exchange reactions are exceptionally facile and do indeed proceed as rapidly as rapid outer-sphere electron transfer exchange reactions. Clearly very strong stabilization of the transition state compensates for the strong M–X bond broken when the exchange is culminated. In this context, the relative stability of 19-electron, seven-coordinate M(I) is relevant and contributes to stabilization of the association of the radical M• with M–X. The relative rapidity of the two-equivalent self-exchange reactions remains a very striking result. It is reasonable that here too the 19-electron radical species plays an important role, with comproportionation producing two such species. A clearer understanding of the nature of the activation process in both the one- and two-equivalent systems and the relationships between the two processes will require, at minimum, elucidation of the activation parameters for the one-equivalent systems and a detailed evaluation of the redox thermodynamics of the systems. Hopefully, these measurements will be possible in the near future.

In terms of the broader mechanistic issues defined by Taube, our work has contributed significantly to the descriptive chemistry of the systems we have studied and suggests that these atom-transfer reactions may proceed though transition states that lie in the very strong interaction limit. Our picture of the detailed mechanisms for the processes is very incomplete, however, and it remains to be seen if atom-transfer self-exchange data can lead to understanding at the level currently possible for outer-sphere electron-transfer reactions.

Acknowledgments

This chapter is based on a presentation I gave to the group assembled at Stanford University for the Henry Taube Symposium on March 30, 1995.

I thank my collaborators R. M. Bullock, C. L. Schwarz, and J.-S. Song for their irreplaceable contributions to this work and N. Sutin for helpful discussions on the preparation of this chapter.

This research was carried out at Brookhaven National Laboratory under contract DE–AC02–76 CH00016 with the U.S. Department of Energy and supported by its Division of Chemical Sciences, Office of Basic Energy Sciences.

References

1. Pellerite, M. J.; Brauman, J. I. In *Mechanistic Aspects of Inorganic Reactions*; Rorabacher , D. B.; Endicott , J. F., Eds.; ACS Symposium Series 198; Washington, DC: American Chemical Society, 1982; p 81.
2. Jordan, R. F.; Norton, J. R. In *Mechanistic Aspects of Inorganic Reactions*; Rorabacher, D. B ; Endicott, J. F., Eds; ACS Symposium Series 198; Washington, DC: American Chemical Society, 1982; p 403.
3. Taube, H. In *Mechanistic Aspects of Inorganic Reactions*, Rorabacher D. B. and Endicott, J. F. Eds.; ACS Symposium Series 198; Washington, DC: American Chemical Society, 1982; p 151.
4. Schwarz, C. L.; Bullock, R. M.; Creutz, C. *J. Am. Chem. Soc.* **1991**, *113*, 1225–1236.
5. Woo, L. K. *Chem. Rev.* **1993**, *93*, 1125–1136.
6. Edidin, R. T.; Sullivan, J. M.; Norton, J. R. *J. Am. Chem. Soc.* **1987**, *109*, 3945–3953.
7. Silverman, J.; Dodson, R. W. *J. Phys. Chem.* **1952**, *56*, 846–852.
8. Schwarz, H. A.; Comstock, D.; Yandell, J. K.; Dodson, R. W. *J. Phys. Chem.* **1974**, *78*, 488–493.
9. Sutin, N. *Acc. Chem. Res.* **1968**, *1*, 225–231.
10. Smith, T. P.; Iverson, D. J.; Droege, M. W.; Kwan, K. S.; Taube, H. *Inorg. Chem.* **1987**, *26*, 2882–2884.
11. Sutin, N. *Annu. Rev. Nucl. Sci.* **1962**, *12*, 285–328.
12. Song, J.-S.; Bullock, R. M.; Creutz, C. *J. Am. Chem. Soc.* **1991**, *113*, 9862–9864.
13. Creutz, C.; Song, J.-S.; Bullock, R. M. *Pure Appl. Chem.* **1995**, *65*, 47–54.
14. Trogler, W. C. In *Journal of Organometallic Chemistry Library 22; Organometallic Radical Processes*; Trogler, W. C., Ed.; Elsevier: Amsterdam, The Netherlands, 1990.
15. Steigman, A. E.; Tyler, D. R. *Coord. Chem. Rev.* **1985**, *63*, 217–240.
16. Tyler, D. R. *Acc. Chem. Res.* **1991**, *24*, 325–331.
17. Adams, M. A.; Folting, K.; Huffman, J. C.; Caulton, K. G. *Inorg. Chem.* **1979**, *18*, 3020–3023.
18. Bueno, C.; Churchill, M. R. *Inorg. Chem.* **1981**, *20*, 2197–2202.
19. Fortier, S.; Baird, M. C.; Preston, K. F.; Morton, J. R.; Ziegler, T.; Jaeger, T. J.; Watkins, W. C.; MacNeil, J. H.; Watson, K. A.; Hensel, K.; Le Page, Y.; Charland, J.- P.; Williams, A. J. *J. Am. Chem. Soc.* **1991**, *113*, 542–551.

20. Hoobler, R. J.; Hutton, M. A.; Dillard, M. M.; Castellani, M. P.; Rheingold, A. L.; Rieger, A. L.; Rieger, P. H.; Richards, T. C.; Geiger, W. E. *Organometallics* **1993**, *12*, 116–123.
21. MacNeil, J. H.; Roszak, A. W.; Baird, M. C.; Preston, K. F.; Rheingold, A. L. *Organometallics* **1993**, 4402–4412.
22. Wrighton, M. S.; Ginley, D. S. *J. Am. Chem. Soc.* **1975**, *97*, 4246–4251.
23. Nielson, R. M.; McManis, G. E.; Goloving, M. N.; Weaver, M. J. *J. Phys. Chem.* **1988**, *92*, 3441–3450.
24. Gordon, A. J.; Ford, R. A. *The Chemist's Companion*; Wiley: New York, 1972; p. 113.
25. Nolan, S. P.; de La Vega, R. L.; Hoff, C. D. *J. Organomet. Chem.* **1986**, *315*, 187–189.
26. Newcomb, M.; Sanchez, R. M.; Kaplan, J. *J. Am. Chem. Soc.* **1987**, *109*, 1195–1199.
27. Kirchner, K.; Dodgen, H. W.; Wherland, S.; Hunt, J. P. *Inorg. Chem.* **1989**, *28*, 604–605.
28. Kirchner, K.; Dodgen, H. W.; Hunt, J. P.; Wherland, S. *Inorg. Chem.* **1990**, *29*, 2381–2385.
29. Kirchner, K.; Han, L. F.; Dodgen, H. W.; Hunt, J. P.; Wherland, S. *Inorg. Chem.* **1990**, *29*, 4556–4559.
30. Anderson, K. A.; Kirchner, K.; Han, L. F.; Dodgen, H. W.; Hunt, J. P.; Wherland, S. *Inorg. Chem.* **1992**, *31*, 2605–2608.
31. Shea, T. M.; Deraniyagala, S. P.; Studebaker, D. B.; Westmoreland, T. D., submitted for publication, **1995**.
32. Proserpio, D. M.; Hoffmann, R.; Levine, R. D. *J. Am. Chem. Soc.* **1991**, *113*, 3217–3225.
33. Lin, Z.; Hall, M. B. *Inorg. Chem.* **1991**, *30*, 3817–3822.
34. Philbin, C. E.; Granatir, C. A.; Tyler, D. R. *Inorg. Chem.* **1986**, *25*, 4806–4807.
35. Watanabe, M.; Iwamoto, T.; Sano, H.; Motoyama, I. *Inorg. Chem.* **1993**, *32*, 5223–5227.
36. Watanabe, M.; Motoyama, I.; Shimoi, M.; Iwamoto, T. *Inorg. Chem.* **1994**, *33*, 2518–2521.

Nuclear Factors in Main-Group Electron Transfer Reactions

David M. Stanbury

Department of Chemistry, Auburn University, Auburn, AL 36849

The general field of aqueous electron transfer reactions involving main-group molecules is reviewed from the perspective of correlating self-exchange rate constants with calculated internal reorganization energies. Oxidation of SCN^- via SCN is used as an example of high reactivity, in which the rates are limited by diffusive product separation. The cross relationship of Marcus theory leads to a lower limit of about 5×10^4 M^{-1} s^{-1} for k_{11} for the SCN/SCN^- system. Oxidation of NH_2OH via NH_2OH^+ occurs quite differently, with a record-low self-exchange rate constant $(5 \times 10^{-13}$ M^{-1} $s^{-1})$. A comprehensive review of the self-exchange rates of main-group redox couples is presented, and the degree of correlation between $\log k_{11}$ and λ_i is examined.

REACTIVITY IS THE HEART OF CHEMISTRY, as Henry Taube has often said. Much of his early research was oriented toward the reactivity of aqueous main-group species such as O_3 and Br_2, with an important component being an interest in distinguishing between single- and multielectron redox reactions. This chapter provides a current overview of half of the picture (single-electron reactions of main-group species), with an emphasis on the role of nuclear factors in determining the rate constants. It is now evident that two important features have emerged. First, these nuclear factors can have an overwhelming effect. Second, simple methods now exist that can be used to estimate their magnitudes.

The present state of affairs is grounded in Taube's early distinction between inner-sphere and outer-sphere electron transfer reactions (*1*). One crucial aspect of this distinction is that reactions can be virtually assured of having single-electron steps if the mechanism is outer sphere. The only known exceptions to this rule are the reactions of Br_2 with a Ni(II) complex and of $S_2O_8^{2-}$ with $[Ru^{II}{}_2(NH_3)_{10}pz]^{4+}$ (2, 3). (In both cases the proposed mechanisms entail successive single-electron steps with the second step occurring before cage escape of the free radical.) As a consequence, by studying redox reactions between main-group species and "outer-sphere" metal complexes, one is investigating single-electron reactions of the main-group species, that is, free radical reactions.

Marcus theory provides the framework for analyzing the nuclear factors in such reactions, but making use of this theory generally requires accurate standard potentials for the species participating in the rate-limiting electron transfer step. Accurate standard potentials for the metal complexes are widely available, and formal potentials are readily determined by, for example, cyclic voltammetry. For inorganic free radicals, however, the situation is not as simple. The difficulties arise because most main-group one-electron redox processes are electrochemically irreversible. As an example, one-electron oxidation of SO_3^{2-} leads via SO_3^- to either SO_4^{2-} or $S_2O_6^{2-}$, and simple cyclic voltammetry shows an oxidation wave for SO_3^{2-} but no return wave. The ClO_2/ClO_2^- system is one of the very rare cases displaying reversible cyclic voltammetry.

Various methods have been developed to estimate the requisite reduction potentials, for example, by use of thermochemical cycles based on gas-phase heats of formation, but these methods are not generally of sufficient accuracy to assess the nuclear factors via Marcus theory. The concurrent developments of pulse radiolysis and flash photolysis have been essential here. These two complementary techniques have permitted the generation and characterization of most of the important free-radical intermediates, as well as the determination of the kinetics of their decay processes. These two methods have also made it possible to determine the standard potentials of the free radicals with considerable accuracy, as has been summarized in a review (4).

In its form as the widely used cross relationship, Marcus theory also makes extensive use of self-exchange rate constants. A major effort since the 1960s has been the development of a large library of "outer-sphere" coordination complex reagents and the measurement of their self-exchange rate constants. It is now possible to select a group of electron transfer reagents for a given main-group substrate that will have a high likelihood of generating useful kinetic data.

With this information in hand the stage is set for investigating the subject matter of this chapter. We begin with a description of two recent studies: (1) the oxidation of SCN^-, which is an example of a system with very little nuclear reorganization (5), and (2) the oxidation of NH_2OH, which suffers from such a large nuclear reorganization energy that major challenges beset its experimental study (6). We finish with a survey of the reported self-exchange rate constants for main-group redox couples, their estimated internal reorganization energies, and the degree of correlation between these two measures of reactivity.

Oxidation of SCN⁻

The general features of outer-sphere oxidation of SCN^- are well established and have been reviewed by Nord (7). The reactions have the stoichiometry

$$6M_{ox} + SCN^- + 4H_2O \rightarrow 6M_{red} + SO_4^{2-} + HCN + 7H^+ \qquad (1)$$

and they obey the general two-term rate law

$$-d[M_{ox}]/dt = (2k_1[SCN^-] + 2k_2[SCN^-]^2)[M_{ox}] \qquad (2)$$

The k_2 term is seen most commonly, but there are a few examples where the k_1 term appears. Most workers adopt the following mechanism to explain these results:

$$M_{ox} + SCN^- \rightleftarrows M_{red} + SCN \qquad k_1, k_{-1} \qquad (3)$$

$$M_{ox} + 2SCN^- \rightleftarrows M_{red} + (SCN)_2^- \qquad k_2, k_{-2} \qquad (4)$$

$$M_{ox} + (SCN)_2^- \rightarrow M_{red} + (SCN)_2 \qquad k_3 \qquad (5)$$

$$SCN + SCN^- \rightleftarrows (SCN)_2^- \qquad k_4, k_{-4}, K_{rad} \qquad (6)$$

$$3(SCN)_2 + 4H_2O \rightarrow 5SCN^- + HCN + SO_4^{2-} + 7H^+ \qquad fast \qquad (7)$$

The observed rate law is obtained when the k_3 step is fast and the reverse steps k_{-1} and k_{-2} can be ignored, as is usually the case.

Of greatest relevance to the present chapter is the k_1 rate constant, which corresponds to a simple outer-sphere electron transfer process. Prior to our recent study (5) there were only three reports of such rate constants for reactions that could reasonably be assigned outer-sphere mechanisms, and these were the reactions of $[IrCl_6]^{2-}$, $[Fe(bpy)_3]^{3+}$, and $[CoW_{12}O_{40}]^{5-}$ (8–10). We noticed as an interesting feature of these reactions that they all appeared to have values of k_{-1} that were approximately diffusion-controlled. These values of k_{-1} were estimated by using the published values of k_1, an estimate of the electron transfer equilibrium constant K_1 derived from the reduction potentials of the oxidants and the SCN radical, and the relationship $K_1 = k_1/k_{-1}$. This observation raised the question of whether, by appropriate selection of an oxidant, a reaction could be found in which the back electron transfer process (k_{-1}) would be slower than the diffusion limit.

This line of reasoning led to the choice of $[Ni(tacn)_2]^{3+}$ as an oxidant (tacn = 1,4,7-triazacyclononane). This complex is well-known as a potent outer-sphere oxidant with a standard reduction potential $(E°)$ of 0.94 V vs. normal hydrogen electrode (NHE) (*11*). More importantly, it has a low self-exchange rate constant ($k_{22} = 6 \times 10^3$ M^{-1} s^{-1}) (*11, 12*), which led to the hope that its oxidation of SCN$^-$ would proceed with activation-controlled kinetics (i.e., that k_{-1} would be less than diffusion-controlled).

We found (*5*) that the oxidation of SCN$^-$ by [Ni(tacn)$_2$]$^{3+}$ was conveniently rapid and that its products corresponded to the usual stoichiometry as in eq 1. However, the kinetics proved to be relatively complex. Under conditions of a large excess of SCN$^-$ the normal rate law, eq 2, leads to pseudo-first-order kinetics. But the reaction with [Ni(tacn)$_2$]$^{3+}$ was far from pseudo-first-order and gave a relatively good fit to a pseudo-*second*-order rate law. This effect was quickly traced to the powerful inhibitory effect of [Ni(tacn)$_2$]$^{2+}$, which is a product of the reaction. This mechanism, under conditions in which the k_{-1} and k_{-2} steps cannot be neglected, leads to the rate law

$$-\frac{d[M_{ox}]}{dt} = \frac{2(k_1[SCN^-] + k_2[SCN^-]^2)k_3 K_{rad}[M_{ox}]^2[SCN^-]}{k_{-1}[M_{red}] + k_{-2}K_{rad}[SCN^-][M_{red}] + k_3 K_{rad}[M_{ox}][SCN^-]} \tag{8}$$

When the denominator is dominated by the terms first order in [M$_{red}$], eq 8 predicts pseudo-second-order kinetics, as observed. We deduced that for some reason the k_{-1} and k_{-2} processes were quite important, leading to the product inhibition. A method that we have employed before in such situations is to use a spin trap as a free-radical scavenger (*13–15*); if the spin trap is sufficiently reactive it will intervene in the kinetics and prevent product inhibition. In the present case we found that PBN (*N–tert*-butyl-α-phenylnitrone) was not sufficiently reactive, but that excellent pseudo-first-order kinetics could be obtained with small concentrations (5 mM) of DMPO (5,5-dimethyl-1-pyrroline *N*-oxide).

In the presence of DMPO the reaction showed only mild inhibition by [Ni(tacn)$_2$]$^{2+}$, and the effect could be safely ignored under conditions in which no additional [Ni(tacn)$_2$]$^{2+}$ was present. Under these conditions the pseudo-first-order rate constants obeyed a simple two-term dependence on [SCN$^-$] as in eq 2, with $k_1 = 0.046 \pm 0.003$ M^{-1} s^{-1} and $k_2 = 2.04 \pm 0.12$ M^{-2} s^{-1} at 25 °C and 0.1 M ionic strength. At this stage we had met one important objective: to find another reaction in which the k_1 term of the rate law could be measured.

The next objective was to determine whether the k_{-1} process was at the diffusion limit. As for the other examples discussed previously, this rate constant was estimated by use of the known reduction potentials, along with the principle of detailed balancing and the measured value of k_1. The outcome was a calculated value of 9.7×10^9 M^{-1} s^{-1} for k_{-1}, which is essentially the diffusion-controlled result.

Although this result showed no evidence for an activation-controlled component to the value of k_1 it was still informative from the point of view of the cross relationship of Marcus theory. This cross relationship, as we use it, is given by (*16*)

$$k_{12} = (k_{11}k_{22}K_{12}f_{12})^{1/2}W_{12} \tag{9}$$

$$\ln f_{12} = \frac{[\ln K_{12} + (w_{12} - w_{21})/RT]^2}{4[\ln(k_{11}k_{22}/Z^2) + (w_{11} + w_{22})/RT]} \tag{10}$$

$$W_{12} = \exp[(-w_{12} - w_{21} + w_{11} + w_{22})/2RT] \tag{11}$$

$$w_{ij} = (4.23Z_iZ_j)/\left\{a[1 + 0.328a(\mu^{1/2})]\right\} \tag{12}$$

In eqs 9–12, k_{12} is the electron transfer rate constant for the cross reaction, while k_{11} and k_{22} are the component self-exchange rate constants, and K_{12} is the electron transfer equilibrium constant. This cross relationship yields rate constants that do not take into account the possibility of diffusion control. If one calculates a value of k_{11} by use of eqs 9–12, the result obtained will be a lower limit if the value of k_{12} (or k_{-12}) is limited by diffusion control. When applied to the oxidations of SCN^- by $[Ni(tacn)_2]^{3+}$, $[IrCl_6]^{2-}$, and $[Fe(bpy)_3]^{3+}$, the values of k_{11} obtained are 5×10^4, 3×10^3, and 2×10^2 M^{-1} s^{-1}, respectively. Because these are lower limits, the highest lower limit (5×10^4 M^{-1} s^{-1}) is the most meaningful. In a prior publication Nord et al. (17) reported that a k_{11} value of 1×10^7 M^{-1} s^{-1} gave a satisfactory account of the rates for $[IrCl_6]^{2-}$ and $[Fe(bpy)_3]^{3+}$, but this result did not take the work terms into account in applying the cross relationship.

Taken in the context of other main-group self-exchange rate constants, our inferred value of k_{11} ($> 5 \times 10^4$ M^{-1} s^{-1}) for the SCN/SCN^- system is among the highest, and is comparable to the value of 4×10^4 M^{-1} s^{-1} derived for the N_3/N_3^- system from similar reactions (14). In both cases there is virtually no structural difference between the two oxidation states of these linear species because the transferring electron resides in a nonbonding orbital (15, 18).

As noted already, a value for k_2 was also determined experimentally. This value corresponds to an electron transfer process that differs drastically from the k_1 process because it involves concurrent S–S bond formation in generating the $(SCN)_2^-$ intermediate. An interpretation of this process is outside the scope of the present chapter, but a full analysis in terms of Marcus theory has been presented (5). By analysis of the kinetic inhibition by $[Ni(tacn)_2]^{2+}$ it was also possible to extract the value of k_3, another electron transfer reaction, but this discussion too is outside the scope of the present chapter.

Oxidation of NH₂OH

Unlike for SCN^-, reports of outer-sphere oxidation of NH_2OH are rather rare, and prior to our study (6) the only examples were the oxidations by $[Fe(CN)_6]^{3-}$, $[IrCl_6]^{2-}$, and $[W(CN)_8]^{3-}$. The original report (19) of simple one-electron oxidation by $[Fe(CN)_6]^{3-}$ was subsequently shown to be in error because of unrecognized catalysis by trace amounts of copper ions (20). With this correction, the intrinsic rate of oxidation was immeasurably slow. Two studies had been published on the oxidation by $[IrCl_6]^{2-}$, but they differed in their rate laws (21, 22); neither of these papers reported the trace copper catalysis noted elsewhere (4). Thus only the oxidation by $[W(CN)_8]^{3-}$ appeared relevant, although even in this study there was no mention of whether a test for copper catalysis was performed (23). In our studies of the oxidation of $S_2O_3^{2-}$ by various outer-sphere reagents we found that copper catalysis could be suppressed by the addition of oxalate (24, 25), and so we decided to investigate whether this method could be applied to the direct oxidation of NH_2OH by $[IrCl_6]^{2-}$ (6).

Preliminary studies revealed that both Cu^{2+} and Fe^{2+} were highly effective catalysts in the oxidation of NH_2OH by $[IrCl_6]^{2-}$ but that their effects could be thoroughly inhibited by addition of $C_2O_4^{2-}$. With this precaution taken, a full study of the reaction showed that the stoichiometry and rate law between pH 4.2 and 6.8 are given by

$$[IrCl_6]^{2-} + NH_3OH^+ \rightarrow [IrCl_6]^{3-} + \tfrac{1}{2}N_2 + 2H^+ + H_2O \tag{13}$$

$$-\frac{d[IrCl_6^{2-}]}{dt} = \frac{k_1 K_a [N(-1)][IrCl_6^{2-}]}{[H^+] + K_a} \tag{14}$$

In this rate law, N(−1) refers to hydroxylamine, regardless of its state of protonation, and K_a refers to the acid dissociation of NH_3OH^+. A value of 24 ± 5 $M^{-1}\,s^{-1}$ was evaluated for k_1. The proposed mechanism was

$$NH_3OH^+ \rightleftarrows NH_2OH + H^+ \qquad\qquad K_a \tag{15}$$

$$[IrCl_6]^{2-} + NH_2OH \rightleftarrows [IrCl_6]^{3-} + NH_2OH^+ \qquad k_1, k_{-1}, K_1 \tag{16}$$

$$NH_2OH^+ \rightleftarrows NH_2O + H^+ \qquad\qquad K_{a,rad} \tag{17}$$

$$NH_2O \rightleftarrows NHOH \qquad\qquad K_{iso} \tag{18}$$

$$2NHOH \rightarrow N_2 + 2H_2O \qquad\qquad \text{fast} \tag{19}$$

According to this mechanism, the k_1 rate constant corresponds to an outer-sphere electron transfer process. An estimate of the equilibrium constant for this process was obtained by combining the known standard reduction potential for $[IrCl_6]^{2-}$ with an estimate of 0.42 V for $E°$ for the NH_2OH^+/NH_2OH system. This estimate was obtained by a thermochemical cycle that was based on an ab initio calculation of the standard enthalpy of formation ($\Delta_f H°$) for NH_2O (g), a semiempirical calculation of its entropy, a rough guess of its hydration free energy, a literature value for $K_{a,rad}$, and the National Bureau of Standards (NBS) value of the standard free energy of formation ($\Delta_f G°$) for NH_2OH (aq).

The outcome was a value of 9×10^7 for K_1, which explains why the reaction is not subject to product inhibition.

Application of the cross relationship of Marcus theory with eqs 9–12 to the measured value of k_1 led to a self-exchange rate constant of 5×10^{-13} M^{-1} s^{-1} for the NH$_2$OH$^+$/NH$_2$OH couple. This result stands in stark contrast to that found for the SCN/SCN$^-$ system, and in fact it is the lowest such rate constant ever reported (excluding reactions involving concerted bond cleavage).

We have just discussed self-exchange rate constants as a measure of the intrinsic reactivity of an outer-sphere redox couple. An alternative and commonly used method is to calculate the reorganizational barrier, λ, which is taken as $4\Delta G^{\ddagger}$. Moreover, it is generally considered that this barrier is given by the sum of the barriers arising from solvent reorganization, λ_o, and internal reorganization, λ_i, such that $\lambda = \lambda_o + \lambda_i$. It is to be expected that λ_o is principally determined by the size of the molecules involved; by analogy with other systems of comparable size (26) an estimate of 120 kJ mol^{-1} can be made for the NH$_2$OH$^+$/NH$_2$OH system. The self-exchange rate constant calculated from the Marcus cross relationship leads to a value of 530 kJ mol^{-1} for λ, which implies a value of 410 kJ mol^{-1} for λ_i.

We used ab initio calculations as an independent check of λ_i. The method taken was first to optimize the structures of NH$_2$OH and NH$_2$OH$^+$ and calculate their energies. Then, the energy of the transition state was determined by optimizing its structure under the double constraint that the two molecules have identical structures and that they be separated far enough from each other that there would be negligible electronic interaction between the two. The first set of calculations showed that there are major structural differences between the two molecules: NH$_2$OH is pyramidal at N and has a 1.45 Å N–O bond length, whereas NH$_2$OH$^+$ is planar and has a 1.30 Å N–O bond length. These qualitative features would lead one to expect a large value of λ_i. The quantitative results at the QCISD(T)/6-311+G(d,p) level of theory gave a value of 414 kJ mol^{-1} for λ_i, which is in remarkably close agreement with that deduced from the kinetics of the oxidation of NH$_2$OH by [IrCl$_6$]$^{2-}$.

Overview of Self-Exchange Rates and Nuclear Factors

The examples just reviewed describe approaches to evaluating the self-exchange rates for main-group electron transfer reactions. The two specific examples are notable in that one of them (SCN/SCN$^-$) appears to be near the upper limit of reactivity when expressed as k_{11}, whereas the other (NH$_2$OH$^+$/ NH$_2$OH) serves as an example of extremely low reactivity. These results, as well as all other such known data, are summarized in Table I, and alongside of these rate constants are presented the corresponding values of λ_i insofar as they are available. It should be recognized that the rate constants presented have all been determined by applying the Marcus cross relationship to reactions with coordination complexes. In the two cases for which the self-

Table I. Self-Exchange Rate Constants vs. Calculated λ_i

System	$\log k_{11}$	$\lambda_{i,calc}$ (kJ mol^{-1})
H^+/H	$<(-25.5)$	
NH_2OH^+/NH_2OH	-12.3	414
$ON(SO_3)_2^{2-/3-}$	-5.1	
CO_2/CO_2^-	-2.9	447
HO_2/HO_2^-	-1.5	70
NO_2/NO_2^-	-0.5	129
Cl_2/Cl_2^-	-0.2	305
CO_3^-/CO_3^{2-}	0.2	
SO_3^-/SO_3^{2-}	0.6	147
O_3/O_3^-	0.6	44
NO^+/NO	0.7	102
O_2/O_2^-	1.0	104
I_2/I_2^-	1.1	94
Br_2/Br_2^-	1.5	154
ClO_2/ClO_2^-	2.3	94
HCO_2/HCO_2^-	2.5	
OH/OH^-	2.5	
SO_2/SO_2^-	4.0	67
N_3/N_3^-	4.7	0
SCN/SCN^-	>4.7	0
$S_2O_3^-/S_2O_3^{2-}$	5.4	
I/I^-	8.3	0

NOTE: λ_i measures internal reorganizational barrier.

exchange rate constants have been measured directly (NO_2/NO_2^- and O_2/O_2^-), substantially greater k_{11} values have been recorded, the significance of which will be discussed. Another limitation is that the values of k_{11} tabulated have not generally been corrected for the effects of solvent-barrier nonadditivity. This effect arises when the cross-reaction takes place between two reagents of widely differing size, and we have shown that this effect can be a significant factor in reactions of the type considered here (15). Yet another limitation is that many of the self-exchange rate constants have not been corrected for work terms; as the O_2/O_2^- system shows, this omission can lead to errors as large as a factor of 100. However, the point of Table I is to discern rough trends, and for this purpose the existing data merit some consideration. In what follows we discuss the individual entries in Table I.

H^+/H. As recently reported by Kelly et al. (27), Schwarz has estimated the self-exchange rate constant for this redox couple ($<3 \times 10^{-26}$ M^{-1} s^{-1}) on the basis of the immeasurably slow reduction Fe^{3+} by H. This estimate should be recognized as tentative because it is based on a negative result for only one reaction. However, it raises a conceptual challenge: a proper treatment of λ_i

will have to model the transition between a proton that is specifically bound by the solvent and a hydrogen atom that is only weakly solvated.

NH_2OH^+/NH_2OH and SCN/SCN^-.   These entries have already been discussed in this chapter.

$ON(SO_3)_2^{2-}/ON(SO_3)_2^{3-}$.   The tabulated rate constant was estimated from the reaction of $[Fe(CN)_6]^{4-}$ with $ON(SO_3)_2^{2-}$ (*28*). It has the advantage of a directly measurable reduction potential for the main-group component, but it suffers from large work terms and because it derives from a single cross-reaction. No quantitative estimate of λ_i is available, but the authors noted that the barrier is expected to be large because of the large expected structural changes.

CO_2/CO_2^-.   An early estimate of 10^{-5} M^{-1} s^{-1} for this self-exchange rate constant was derived from a series of reductions of Co(II) complexes by CO_2^- (*29*). Venturi et al. (*30, 31*) derived a significantly greater self-exchange rate constant from reactions of CO_2^- with a series of $Ru(LL)_3^{2+}$ complexes. This latter result is given in Table I because it is more thoroughly documented and includes work terms in its calculation. The value of λ_i is one of the various estimates given by Bennett and Warlop (*32*) and derived from computed potential energy surfaces. Consistent with this large calculated value of λ_i, the bond angle expands from 135 to 180° and the C–O bond length decreases from 1.25 to 1.16 Å as CO_2^- is oxidized to CO_2 (*32*). In fact, the structural changes are so large that it is difficult to understand why the self-exchange rate constant is as large as it appears to be.

HO_2/HO_2^-.   Macartney (*33*) derived this rate constant from a series of reactions in which H_2O_2 is oxidized by various coordination complexes. These reactions display a two-term rate law, one of which is inverse in $[H^+]$. The latter term was inferred to indicate oxidation of HO_2^- to HO_2 and was used to calculate the self-exchange rate constant. Prior estimates that showed much greater scatter were imprecise because work terms were omitted (*34*). Macartney calculated a value of λ_i by use of a classical force field as applied to the O–O bond. His result appears to be somewhat low, perhaps because contributions involving the hydrogen atom were omitted.

NO_2/NO_2^-.   This system has been the subject of repeated scrutiny. In the first report (*9*), an estimate of 1×10^{-2} M^{-1} s^{-1} was given for k_{11} based on a series of four reactions. Because of a typographical error, the specific value derived from the reaction of NO_2^- with $[Fe(bpy)_3]^{3+}$ should be given as 5.5×10^{-3} M^{-1} s^{-1}, and the average value should be given as 5×10^{-3} M^{-1} s^{-1}. With the availability in 1984 of additional rate constants and a revised potential for the NO_2/NO_2^- couple, a small revision in k_{11} was reported (*35*). Yet another revision was reported in 1991, which was based on additional rate constants, further refinement of $E°$ for the NO_2/NO_2^- couple, and an adjustment to the collision rate used in the cross relationship (*36*); this final result ($k_{11} = 0.3$ M^{-1} s^{-1}) is the one presented in Table I. It is in good agreement with the value of 1 M^{-1} s^{-1} derived from the reaction of NO_2 with phenoxide (*37*). The NO_2/NO_2^- self-exchange rate constant has also been measured directly, by use of isotopic tracers (*38*). This directly measured rate constant, 580 M^{-1} s^{-1}, exceeds the Marcus result by a substantial margin. A large portion of the discrepancy can be removed by correcting for solvent-barrier nonadditivity, but the agreement is still not perfect (*15*).

Calculations of λ_i have been repeatedly published. The first such calculation was based on a classical force field, and gave a value of 181 kJ mol^{-1} (*26*). This finding led to the important realization that bond-angle deformation could be a major component of the nuclear reorganization energy. Concurrently, Eberson and Radner (*39*) calculated a value of 146 kJ mol^{-1} by the same method although with slightly different input data. A value of 74 kJ mol^{-1} (designated as R_b in the original paper) was calculated by similar methods and differs from the other estimates for unknown reasons (*40*). More recently a value of 129 kJ mol^{-1} was calculated by use of ab initio methods (*41*). This latter result (given in Table I) is probably the most accurate because it does not exclude contributions from off-diagonal elements of the force field, nor does it impose the harmonic approximation.

Cl_2/Cl_2^-.   The only estimate of k_{11} for this redox couple came from a Marcus-type analysis of the rates of reaction of Cl_2 with V^{2+}, Ti^{3+}, V^{3+}, Pu^{3+}, and $[Fe(phen)_3]^{2+}$ (*42*). The treatment omitted work terms and was based on a potential of 0.52 V for the Cl_2/Cl_2^- couple. More recent estimates place this $E°$ value at 0.70 V, although it is still shrouded in uncertainty (*4*). The value of k_{11} tabulated is that given in the original literature, but correction for the two effects noted here would reduce it by several orders of magnitude. Eberson et al. (*41*) have calculated λ_i by ab initio methods; their result is given in Table I, and it is quite incompatible with the cited value of k_{11}. We infer that when a reliable value of $E°$ becomes available the discrepancy will be minimized to some degree, although there may be other contributing factors.

CO_3^-/CO_3^{2-}, HCO_2/HCO_2^-, and OH/OH^-.   Self-exchange rate constants for these systems were derived from the rates of quenching of *$[Os(tmc)O_2]^{2+}$ by CO_3^{2-}, HCO_2^-, and OH^- (*43*). They differ from the other data discussed here in that one of the reactants is in an excited state. The result for the HCO_2/HCO_2^- system was based on an $E°$ value of 2.0 V for this couple, while the most recent estimate is 0.3 V larger (*44*). Correction for this would increase the tabulated value of k_{11} substantially. To our knowledge, no estimates of λ_i for these couples have been published. Meyerstein (*45*) argues that no reactions involving the OH radical can be described as "outer-sphere".

SO_3^-/SO_3^{2-}.   A self-exchange rate constant of 2×10^5 M^{-1} s^{-1} was derived for this couple from the reaction of ClO_2 with SO_3^{2-} (37). This parameter was the subject of a major revision, principally due to a revision in $E°$ for this couple (46). With this new $E°$ and a consideration of a series of reactions of SO_3^{2-} with outer-sphere oxidants a value of 4 M^{-1} s^{-1} was derived for k_{11} (46). A classical force field was used to estimate λ_i (47). Most of this barrier arises from the umbrella vibrational mode.

O_3/O_3^-.   The Marcus cross relationship was applied to the sole reaction of O_3 with $[IrCl_6]^{3-}$ to derive the tabulated self-exchange rate constant (32). Bennett and Warlop (32) estimated λ_i by use of a classical force field, and their result is in good agreement with the tabulated value that Eberson et al. (41) obtained more recently by ab initio methods. From the trends suggested by the data in Table I one might anticipate that further study of outer-sphere reactions of O_3 would lead to a significant revision of k_{11}.

NO^+/NO.   Substantial difficulties attend experimental studies of the NO^+/NO system in aqueous solution. If one attempts to measure the reduction of NO^+ by using nitrous acid the reactions often arise from other species that are also present in such solutions (35). If one is successful in this endeavor, as in the reaction with $[IrCl_6]^{3-}$, the reactions are quite likely to find inner-sphere pathways (48). When investigated via the oxidation of NO the problems with inner-sphere mechanisms remain challenging. For example, when $[Ni(tacn)_2]^{3+}$ is used as an oxidant the dominant term in the rate law is inverse in $[H^+]$ because of the high reactivity of the amido complex (36). It is only under quite acidic conditions that the direct oxidation can be detected, and it is from this sole reaction that the tabulated k_{11} value was estimated. An early estimate of λ_i by use of a classical force field has recently been confirmed and is in good agreement with ab initio results (39, 41, 49, 50).

O_2/O_2^-.   This particular main-group self-exchange reaction has been discussed more often than any other. To my knowledge, the first pertinent discussion was published by Marcus (51) in 1957, although a self-exchange rate constant per se was not calculated. A series of three papers by Taube and co-workers beginning in 1980 put forth the notion that a specific value of k_{11} could be assigned to the O_2/O_2^- couple. In the first paper (52) the cross relationship was applied to the rates of a series of reactions of O_2 with Ru(II) ammine complexes, and an estimate of 1×10^3 M^{-1} s^{-1} for k_{11} was derived. Work terms were omitted in this estimate. The second paper (34) confirmed that the reactions were indeed electron transfer reactions, but it pointed out that certain related reactions involving quinone/semiquinone systems led to anomalously large values of k_{11} while the reaction of O_2^- with $[Fe(CN)_6]^{3-}$ led to an anomalously small k_{11}. The third paper (53) described the autoxidations of several dinuclear Ru(II) ammines, which required inclusion of work terms in the calculation; this inclusion led to a revised estimate of about 5 M^{-1} s^{-1} for k_{11}.

Shortly after publication of these studies, McDowell et al. (54) published a report on the rates of reaction of O_2^- with several Co(III) ammines and ferrocenium. These workers rectified an error (53) in the prior calculation of k_{11} from the reaction of O_2 with $[Ru(NH_3)_6]^{2+}$; this rectification led to a more consistent range of k_{11} values calculated from the Ru(II) ammine reactions. However, the paper achieved notoriety principally by pointing out the gross deviations of k_{11} that derive from the rates of several non-ruthenium-based reactions. In particular, the reactions of O_2^- with $[Co(en)_3]^{3+}$ and $[Fe^{III}(edta)H_2O]^-$ were found to be much faster than would be consistent with the Ru(II)-based results, and the reactions of $[Fe(C_5H_5)_2]^+$ and $[Mo(CN)_8]^{3-}$ were much slower. It was suggested that the lack of agreement with Marcus theory came about because desolvation of $O_2(H_2O)_n^-$ would be a function of the specific reaction partners. A subsequent publication from the same group (55) described the kinetics of six reactions of O_2 with $[Cr^{II}(phen)_3]^{2+}$ and related Cr(II) complexes. They found that these reactions, along with those of the Ru(II) ammines and $[Co(sep)]^{2+}$, were consistent with a k_{11} value in the range of 1–10 M^{-1} s^{-1}. They also pointed out that all of the deviating reactions noted previously used O_2^- as a reductant rather than O_2 as an oxidant. This outcome is not universal, since the reaction of O_2 with $[Ru(NH_3)_5isn]^{2+}$ and its reverse are both consistent with Marcus theory (34). Specific reasons have been put forth as to why certain of the reactions should be so deviant (55), but in most of the cases definitive explanations are not yet at hand.

A significantly different point of view was offered by Lind et al. (56), who directly measured a value of 450 M^{-1} s^{-1} for k_{11} by isotopic labeling. These workers mentioned that solvent-barrier nonadditivity could explain qualitatively why their result was substantially greater than that derived from reactions with large molecules. Recently, Merényi et al. (57) examined the rates of autoxidation of a wide range of closed-shell organics and found that they were consistent with a k_{11} value of 2 M^{-1} s^{-1}, that is, that they were consistent with the results derived from autoxidation of coordination complexes. Moreover, they presented quantitative evidence that solvent-barrier nonadditivity could account for the difference between the directly measured value of k_{11} and that derived from the cross relationship.

Estimates of λ_i have been reported by several groups. An early value of 16 kJ mol^{-1} based on a harmonic force field suffered from a mathematical error in its calculation (52), and it was corrected to 62 kJ mol^{-1} (53). Similar results were reported by Eberson (58), Bennett and Warlop (32), and by Lind et al. (56). A significantly greater value of 89 kJ mol^{-1} was recently reported by Eberson et al. (41); the difference is due principally to an improved bond length for O_2^-. An alternative approach, based on ab initio methods, yielded 104 kJ mol^{-1} for λ_i (41). Because of anharmonicities in the potential energy surfaces, this last result is probably the most accurate.

I_2/I_2^-.   An estimate of 8.5×10^4 M^{-1} s^{-1} for k_{11} was first reported in 1974 (59). McDowell et al. (54), however, have rejected this result on the grounds that the reaction from which it was derived ($I_2 + [Fe^{II}(edta)]^{2-}$) is inner sphere. Another estimate ($\sim 10^2$ M^{-1} s^{-1}) was reported by Nord et al. (60), on the basis of the diffusion controlled reaction of I_2^- with $[Os(bpy)_3]^{3+}$. Rudgewick-Brown and Cannon (61) found that the earlier value of k_{11} gave a satisfactory fit for the reaction of I_2 with $[Co(sep)]^{2+}$, but their calculation was based on a superseded value of E° for the I_2/I_2^- couple (4). Use of a more correct potential (0.21 V) gives 13 M^{-1} s^{-1} for k_{11}. As a side note, we mention that Ishikawa et al. (62) found that $k_{11} = 8.5 \times 10^4$ M^{-1} s^{-1} gave a good account of the rates of reaction of I_2 with a series of ferrocenes in acetonitrile. An early estimate of λ_i based on the harmonic force-field approximation was incorrect due to a mathematical error (52); a corrected estimate is 94 kJ mol^{-1}, which seems qualitatively in line with the lower self-exchange rate constant (13 M^{-1} s^{-1}).

Br_2/Br_2^-.   As with the I_2/I_2^- system, the first estimate of k_{11} should be rejected because it was based on inner-sphere reactions (54, 59). A more recent estimate by Ige et al. (42) was based on the reactions of Br_2 with $[Fe(phen)_3]^{2+}$ and with four metal aquo ions. As McDowell et al. (54) point out, the data on which this latter estimate were based require a reverse rate constant that exceeds the diffusion limit, and so this estimate of k_{11} must also be rejected. Certainly part of the difficulties resides in the value of E° selected for the calculation (0.51 V). We have recommended (albeit tentatively) a value of 0.58 V for this parameter (4), and this recommendation would remove the objection regarding diffusion limits. Correction for E° would lower the calculated value of k_{11}, and inclusion of work terms would lead to a further reduction of k_{11}. Correction of the original estimate (52) of λ_i for a mathematical error leads to the tabulated value; its magnitude supports our opinion that an optimal value of k_{11} will prove to be lower than current estimates.

ClO_2/ClO_2^-.   The first estimate of k_{11} ($= 78$ M^{-1} s^{-1}) was based on the reaction of ClO_2 with $[Fe(phen)_3]^{2+}$ (63). The data base was then expanded to include three other outer-sphere reactants, which adjusted the estimate to 160 M^{-1} s^{-1} (26). Huie and Neta (37) estimated 440 M^{-1} s^{-1} from the reaction of ClO_2 with phenoxide (37). These results were confirmed in a study of the reactions of ClO_2 with a series of phenolic compounds (64). A correction for the effects of solvent-barrier nonadditivity was subsequently reported (15), but, for the purpose of comparisons, Table I gives the uncorrected value. An early estimate of 70 kJ mol^{-1} for λ_i was based on a harmonic force field (26). Loeff et al. (40) obtained 53 kJ mol^{-1} with the use of slightly different structural parameters. A somewhat larger value (94 kJ mol^{-1}) was obtained by ab initio methods (41).

SO_2/SO_2^-.   Outer-sphere reactions of this couple are generally accessed through reactions in which $S_2O_4^{2-}$ acts as a reducing agent, although occasionally SO_2 in strong acid is used as an oxidant. When reactions of $S_2O_4^{2-}$ show terms in the rate laws that are proportional to $[S_2O_4^{2-}]^{1/2}$ they are taken to imply that SO_2^-, which is in rapid equilibrium with $S_2O_4^{2-}$, is the species undergoing oxidation. The first estimate of k_{11} was based on this concept (65), but it was flawed by the use of $E°$ for HSO_3^-/SO_2^- rather than $E°$ for SO_2/SO_2^-. Correcting for this led to a value of 340 M^{-1} s^{-1} for k_{11} (26). Bradic and Wilkins (66) observed that reductions by SO_2^- are typically 10^3-fold greater than for the corresponding reactions of O_2^-; they argued that this is to be expected from Marcus theory if the differences in reduction potentials are taken into account and if k_{11} for SO_2/SO_2^- exceeds that for O_2/O_2^- by a factor of 10^4, which would lead to a value of 10^7 for k_{11} for SO_2/SO_2^- if a value of 10^3 M^{-1} s^{-1} were taken for k_{11} for O_2/O_2^-.

Part of the reason for this excessive result is the high self-exchange rate constant used for the O_2/O_2^- system. An even more excessive value (10^8–10^9 M^{-1} s^{-1}) was obtained from the rates of reaction of SO_2^- with a series of viologens (67). Atherton et al. (68) investigated the reactions of a series of bis-viologen compounds and confirmed the previously noted trend that SO_2/SO_2^- reactions are faster by a factor of 10^3 than those of O_2/O_2^-. Balahura and Johnson (69) argued that the viologen reactions are fast because they have an inner-sphere mechanism, and we may infer that this would also apply to the more recently studied bis-viologen systems. By limiting their scope to reactions of outer-sphere complex oxidants and substantially increasing the number of such reactions, Balahura and Johnson (69) derived an estimate of $\sim 10^3$ M^{-1} s^{-1} for k_{11}.

Concurrently, Neta et al. (70) reported a pulse radiolysis study of the reactions of SO_2^- with several metalloporphyrins, organics, and $IrCl_6^{2-}$; they drew attention to the wide range of derived k_{11} values, most of which were quite large. In a rather different type of experiment, rate constants for reduction of SO_2 by photochemically generated $[Cr(phen)_3]^{2+}$ and related complexes were used to derive a k_{11} value of 5×10^4 M^{-1} s^{-1} (71). These workers noted that k_{11} values derived from reactions with viologens and metalloporphyrins should be discounted because they quite likely have inner-sphere mechanisms. However, some uncertainty attends their derived k_{11} value because they used an $E°$ value for the SO_2/SO_2^- that was determined (70) from irreversible cyclic voltammograms of 20% 2-propanol solutions that showed a significant pH dependence.

By use of a harmonic force field an estimate of 60 kJ mol^{-1} was derived for λ_i (26), although it had a large uncertainty because of uncertainties in the structure of SO_2^-. The smaller value of 34 kJ mol^{-1} reported by Bennett and Warlop (32) differed principally because of the use of a more recent ab initio geometry for SO_2^-. When Eberson et al. (41) used this method with the recent experimental geometry for SO_2^- they obtained a value of 81 kJ mol^{-1}, and when they used a relatively sophisticated ab initio calculation they got 67 kJ mol^{-1} for λ_i.

N_3/N_3^-. Electron transfer rate constants for the oxidation of N_3^- by a series of four outer-sphere oxidants were used to derive a value of 4×10^4 M^{-1}

s^{-1} for k_{11} (*14*). A correction to this result was subsequently applied in order to account for solvent-barrier nonadditivity (*15*), but the uncorrected value is used here in order to facilitate comparisons. This couple has generally been taken as having a value of zero for λ_i because the two oxidation states differ negligibly in their structures (*14, 15, 32, 41*). The low value of k_{11}, relative to the highest values found for transition metal complexes, is ascribed to the solvent barrier, λ_o, which should be large for ions having small ionic radii.

$S_2O_3^-/S_2O_3^{2-}$.   The initial attempt to apply the Marcus cross relationship to this redox couple focused on a series of oxidations of $S_2O_3^{2-}$ by $[Os(phen)_3]^{3+}$ and related complexes (*24*). Unfortunately, it was not possible to resolve the question of whether electron transfer or diffusive separation was the rate-limiting step, and so only a lower limit to k_{11} could be derived. This ambiguity was resolved in a followup study, in which $[Ni(tacn)_2]^{3+}$ was used as the oxidant (*25*). Because of its slow self-exchange reaction this oxidant gave good evidence for rate-limiting electron transfer. The derived value of k_{11} was 2×10^5 M^{-1} s^{-1}. Accurate structural data for $S_2O_3^-$ are lacking, although electron-spin resonance spectra and molecular orbital calculations indicate that there are minimal structural differences between $S_2O_3^-$ and $S_2O_3^{2-}$. It has been inferred from such considerations that λ_i is close to zero (*9, 25*).

I/I⁻.   Most of the rate constants pertaining to the I/I⁻ system come from studies of the oxidation of I⁻. The first such studies to analyze these data in terms of the Marcus cross relation come from the overlapping work of three different research groups. The first of these (*17*) reported on the kinetics of oxidation by $[Os(phen)_3]^{3+}$ and $[Os(bpy)_3]^{3+}$, and evaluated the rate constants together with those for reactions of five other oxidants. The result obtained (k_{11} = 10^7 M^{-1} s^{-1}) should be viewed with caution because its derivation did not include work terms and did not recognize the limitations of diffusive separation. The second of these papers (*72*) reported on the reactions of $[Fe(phen)_3]^{3+}$, five related complexes, and $[IrCl_6]^{2-}$; from these data (and ignoring work terms and diffusive separation) a k_{11} value of 7×10^7 M^{-1} s^{-1} was derived. The third paper (*8*) showed that diffusive separation was the rate-limiting step for the reactions considered, that is, that the reverse step was diffusion controlled. Such a mechanism leads to a strictly linear relationship between $\log k$ and $\log K$ when K is taken as the equilibrium constant for formation of the iodine atom. A lower limit of 10^9 M^{-1} s^{-1} for k_{11} was proposed, although work terms were again ignored (*8*).

Several years later Fairbank and McAuley (*73*) measured the rates of four reactions of Ni^{III} macrocycles and included several other literature rate constants to compile a list of 24 rate constants for one-electron oxidation of iodide. Unfortunately, these workers overlooked the paper (*8*) where the limitations of diffusive separation were discussed, they analyzed the data without taking these effects into account, and they ignored work terms. This analysis led to a value of 2×10^8 M^{-1} s^{-1} for k_{11}. Wells's recent mammoth review (74) of metal–ligand redox reactions discusses the oxidation of I^- and supports the viewpoint of Fairbank and McAuley. In contrast, Nord (7) has reviewed the same data plus some additional reactions; she has reaffirmed that *all* of the reactions are limited by diffusive separation of the products (7). The only exceptions to this rule are some fast inner-sphere reactions of Cu^{III} peptides and the reactions of $[Ni^{III}(\text{cyclam})]^{3+}$ and $[Ni^{III}(\text{dimethylcyclam})]^{3+}$, which have rate constants a factor of 2–3 less than expected.

In summary, there has been no Marcus analysis of the rates of oxidation of I^- that has taken into account the effects of work terms and diffusive separation. We can only conclude that k_{11} for the I/I^- system must be very large. As this system is monatomic and not specifically solvated, a value of zero is assigned to λ_i.

Conclusions

Efforts to apply Marcus theory to main-group electron transfer reactions now cover a considerable number of molecular systems. In some cases, notably for the O_2/O_2^- and SO_2/SO_2^- systems, indiscriminate application of the cross relationship has led to highly divergent results. The theory is much more consistent when care is taken to limit the range of reactions to those that have outer-sphere mechanisms. Further concern about the applicability of the classical theory relates to the possibility of nuclear tunneling and the inadequacy of the dielectric continuum solvation model for such molecules (43).

The perspective of the present chapter is to view the broad picture and not dwell on details of the model. From this vantage we see that the theory works reasonably well. The range of small-molecule self-exchange rate constants is now as wide as that for coordination complexes, with SCN/SCN^- and NH_2OH^+/NH_2OH at the extremes. A moderate correlation of log k_{11} vs. λ_i serves to rationalize many of the self-exchange rate constants and highlight those systems that need further work. Essential to the whole endeavor is Taube's insight that redox reactions can have either inner- or outer-sphere mechanisms (1).

Acknowledgments

This research was supported by a grant from the National Science Foundation.

D. M. Stanbury is a Sloan Research Fellow. Cliff Kubiak (at Purdue University) is thanked for his suggestions regarding the chemistry of the CO_2/CO_2^- and SO_2/SO_2^- systems. A. Bakac (Iowa State), R. E. Huie (National Institute of Standards and Technology), and C. F. Wells (Birmingham) are thanked for their helpful comments on the chapter.

References

1. Taube, H.; Myers, H.; Rich, R. L. *J. Am. Chem. Soc.* **1953**, *75*, 4118–4119.
2. Fürholz, U.; Haim, A. *Inorg. Chem.* **1987**, *26*, 3243–3248.
3. Lappin, A. G.; Osvath, P.; Baral, S. *Inorg. Chem.* **1987**, *26*, 3089–3094.
4. Stanbury, D. M. *Adv. Inorg. Chem.* **1989**, *33*, 69–138.
5. Hung, M.-L.; Stanbury, D. M. *Inorg. Chem.* **1994**, *33*, 4062–4069.
6. Hung, M.-L.; McKee, M. L.; Stanbury, D. M. *Inorg. Chem.* **1994**, *33*, 5108–5112.
7. Nord, G. *Comments Inorg. Chem.* **1992**, *13*, 221–239.
8. Stanbury, D. M.; Wilmarth, W. K.; Khalaf, S.; Po, H. N.; Byrd, J. E. *Inorg. Chem.* **1980**, *19*, 2715–2722.
9. Wilmarth, W. K.; Stanbury, D. M.; Byrd, J. E.; Po, H. N.; Chua, C.- P. *Coord. Chem. Rev.* **1983**, *51*, 155–179.
10. Olatunji, M. A.; Ayoko, G. A. *Polyhedron* **1984**, *3*, 191–197.
11. McAuley, A.; Norman, P. R.; Olubuyide, O. *Inorg. Chem.* **1984**, *23*, 1938–1943.
12. McAuley, A.; Norman, P. R.; Olubuyide, O. *J. Chem. Soc. Dalton Trans.* **1984**, 1501–1505.
13. Ram, M. S.; Stanbury, D. M. *Inorg. Chem.* **1985**, *24*, 4233–4234.
14. Ram, M. S.; Stanbury, D. M. *J. Phys. Chem.* **1986**, *90*, 3691–3696.
15. Awad, H. H.; Stanbury, D. M. *J. Am. Chem. Soc.* **1993**, *115*, 3636–3642.
16. *Inorganic Reactions and Methods;* Zuckerman, J. J., Ed.; VCH: Deerfield Beach, FL, 1986; Vol. 15, pp 13–47.
17. Nord, G.; Pedersen, B.; Farver, O. *Inorg. Chem.* **1978**, *17*, 2233–2238.
18. Bradforth, S. E.; Kim, E. H.; Arnold, D. W.; Neumark, D. M. *J. Chem. Phys.* **1993**, *98*, 800–810.
19. Jindal, V. K.; Agrawal, M. C.; Mushran, S. P. *J. Chem. Soc. A* **1970**, 2060–2062.
20. Bridgart, G. J.; Waters, W. A.; Wilson, I. R. *J. Chem. Soc. Dalton Trans.* **1973**, 1582–1584.
21. Sen, P. K.; Maiti, S.; Sen Gupta, K. K. *Indian J. Chem.* **1980**, *19A*, 865–868.
22. Rekha, M.; Prakash, A.; Mehrotra, R. N. *Can. J. Chem.* **1993**, *71*, 2164–2170.
23. Van Wyk, A. J.; Dennis, C. R.; Leipoldt, J. G.; Basson, S. S. *Polyhedron* **1987**, *6*, 641–643.
24. Sarala, R.; Rabin, S. B.; Stanbury, D. M. *Inorg. Chem.* **1991**, *30*, 3999–4007.
25. Sarala, R.; Stanbury, D. M. *Inorg. Chem.* **1992**, *31*, 2771–2777.
26. Stanbury, D. M.; Lednicky, L. A. *J. Am. Chem. Soc.* **1984**, *106*, 2847–2853.
27. Kelly, C. A.; Mulazzani, Q. G.; Venturi, M.; Blinn, E. L.; Rodgers, M. A. J. *J. Am. Chem. Soc.* **1995**, *117*, 4911–4919.
28. Balasubramanian, P. N.; Gould, E. S. *Inorg. Chem.* **1983**, *22*, 1100–1102.
29. Schwarz, H. A.; Creutz, C.; Sutin, N. *Inorg. Chem.* **1985**, *24*, 433–439.
30. Venturi, M.; Mulazzani, Q. G.; D'Angelantonio, M.; Ciano, M.; Hoffman, M. Z. *Radiat. Phys. Chem.* **1991**, *37*, 449–456.
31. D'Angelantonio, M.; Mulazzani, Q. G.; Venturi, M.; Ciano, M.; Hoffman, M. Z. *J. Phys. Chem.* **1991**, *95*, 5121–5129.
32. Bennett, L. E.; Warlop, P. *Inorg. Chem.* **1990**, *29*, 1975–1981.
33. Macartney, D. H. *Can. J. Chem.* **1986**, *64*, 1936–1942.
34. Stanbury, D. M.; Mulac, W. A.; Sullivan, J. C.; Taube, H. *Inorg. Chem.* **1980**, *19*, 3735–3740.
35. Ram, M. S.; Stanbury, D. M. *J. Am. Chem. Soc.* **1984**, *106*, 8136–8142.
36. deMaine, M. M.; Stanbury, D. M. *Inorg. Chem.* **1991**, *30*, 2104–2109.
37. Huie, R. E.; Neta, P. *J. Phys. Chem.* **1986**, *90*, 1193–1198.
38. Stanbury, D. M.; deMaine, M. M.; Goodloe, G. *J. Am. Chem. Soc.* **1989**, *111*, 5496–5498.

39. Eberson, L.; Radner, F. *Acta Chem Scand. Ser. B* **1984**, *38*, 861–870.
40. Loeff, I.; Treinin, A.; Linschitz, H. *J. Phys. Chem.* **1992**, *96*, 5264–5272.
41. Eberson, L.; González-Luque, R.; Lorentzon, J.; Merchán, M.; Roos, B. O. *J. Am. Chem. Soc.* **1993**, *115*, 2898–2902.
42. Ige, J.; Ojo, J. F.; Olubuyide, O. *Can. J. Chem.* **1979**, *57*, 2065–2070.
43. Schindler, S.; Castner, E. W., Jr.; Creutz, C.; Sutin, N. *Inorg. Chem.* **1993**, *32*, 4200–4208.
44. Yu, D.; Rauk, A.; Armstrong, D. A. *J. Chem. Soc., Perkins Trans. 2* **1994**, 2207–2215.
45. Meyerstein, D. *Acc. Chem. Res.* **1978**, *11*, 43–48.
46. Sarala, R.; Stanbury, D. M. *Inorg. Chem.* **1990**, *29*, 3456–3460.
47. Sarala, R.; Islam, M. S.; Rabin, S. B.; Stanbury, D. M. *Inorg. Chem.* **1990**, *29*, 1133–1142.
48. Ram, M. S.; Stanbury, D. M. *Inorg. Chem.* **1985**, *24*, 2954–2962.
49. Boughriet, A.; Wartel, M. *Int. J. Chem. Kinet.* **1993**, *25*, 383–397.
50. Bu, Y.; Song, X. *J. Phys. Chem.* **1994**, *98*, 5049–5051.
51. Marcus, R. A. *J. Chem. Phys.* **1957**, *26*, 872–877.
52. Stanbury, D. M.; Haas, O.; Taube, H. *Inorg. Chem.* **1980**, *19*, 518–524.
53. Stanbury, D. M.; Gaswick, D.; Brown, G. M.; Taube, H. *Inorg. Chem.* **1983**, *22*, 1975–1982.
54. McDowell, M. S.; Espenson, J. H.; Bakac, A. *Inorg. Chem.* **1984**, *23*, 2232–2236.
55. Zahir, K.; Espenson, J. H.; Bakac, A. *J. Am. Chem. Soc.* **1988**, *110*, 5059–5063.
56. Lind, J.; Shen, X.; Merényi, G.; Jonsson, B. Ö. *J. Am. Chem. Soc.* **1989**, *111*, 7654–7655.
57. Merényi, G.; Lind, J.; Jonsson, M. *J. Am. Chem. Soc.* **1993**, *115*, 4945–4946.
58. Eberson, L. *Adv. Free Radical Biol. Med.* **1985**, *1*, 19–90.
59. Woodruff, W. H.; Margerum, D. W. *Inorg. Chem.* **1974**, *13*, 2578–2585.
60. Nord, G.; Pedersen, B.; Floryan-Løvborg, E.; Pagsberg, P. *Inorg. Chem.* **1982**, *21*, 2327–2330.
61. Rudgewick-Brown, N.; Cannon, R. D. *J. Chem. Soc. Dalton Trans.* **1984**, 479–481.
62. Ishikawa, K.; Fukuzumi, S.; Tanaka, T. *Inorg. Chem.* **1989**, *28*, 1661–1665.
63. Lednicky, L. A.; Stanbury, D. M. *J. Am. Chem. Soc.* **1983**, *105*, 3098–3101.
64. Tratnyek, P. G.; Hoigné, J. *Water Res.* **1994**, *28*, 57–66.
65. Mehrotra, R. N.; Wilkins, R. G. *Inorg. Chem.* **1980**, *19*, 2177–2178.
66. Bradic, Z.; Wilkins, R. G. *J. Am. Chem. Soc.* **1984**, *106*, 2236–2239.
67. Tsukahara, K.; Wilkins, R. G. *J. Am. Chem. Soc.* **1985**, *107*, 2632–2635.
68. Atherton, S. J.; Tsukahara, K.; Wilkins, R. G. *J. Am. Chem. Soc.* **1986**, *108*, 3380–3385.
69. Balahura, R. J.; Johnson, W. D. *Inorg. Chem.* **1987**, *26*, 3860–3863.
70. Neta, P.; Huie, R. E.; Harriman, A. *J. Phys. Chem.* **1987**, *91*, 1606–1611.
71. Simmons, C. A.; Bakac, A.; Espenson, J. H. *Inorg. Chem.* **1989**, *28*, 581–584.
72. Adedinsewo, C. O.; Adegite, A. *Inorg. Chem.* **1979**, *18*, 3597–3601.
73. Fairbank, M. G.; McAuley, A. *Inorg. Chem.* **1987**, *26*, 2844–2848.
74. Wells, C. F. *Prog. React. Kinet.* **1995**, *20*, 1–184.

11

Calculation of Rate Constants from Spectra

Nonradiative Decay and Electron Transfer

Darla Graff, Juan Pablo Claude, and Thomas J. Meyer

Department of Chemistry, The University of North Carolina at Chapel Hill, Chapel Hill, NC 27599–3290

Relative nonradiative rate constants for the metal-to-ligand charge transfer (MLCT) excited states of $[Os(bpy)(py)_4]^{2+}$, $[(Os(bpy)_2(py)_2]^{2+}$, and $[(Os(bpy)_3]^{2+}$ have been calculated by analysis of resonance Raman spectral profiles and for electron transfer in a series of Re^I chromophore-quencher complexes by emission spectral analysis.

IN A NOW-FAMOUS REVIEW ON MIXED-VALENCE CHEMISTRY, Hush (*1*) pointed out in 1967 the connection between rate constants, electron transfer barriers, and charge transfer spectra. He derived relationships for the reorganizational energy from band energies and widths and for the electronic delocalization energy from the integrated band intensity. For the experimentalist these derivations opened the possibility of using simple absorption band measurements to calculate rate constants for electron transfer (*1*).

The Hush theory was classical in its treatment of the electron transfer barrier and incorporated Kubo and Toyozawa line-shape analysis for radiationless transitions. Later work by Jortner and co-workers (*2, 3*) was based on quantum mechanics. They were able to show that, within certain limits, other useful relationships could be derived between spectral properties and rate constants and barriers for electron transfer, energy transfer, or nonradiative decay.

These theories pointed to the possibility of using spectrally derived parameters to calculate dynamic quantities. There are good reasons why this possibility is of interest to the experimentalist. It reduces the problem of determining rate constants from what are often difficult time-dependent measurements to simple spectroscopic measurements. It allows the usually complex rate expressions derived by time-dependent perturbation theory to be parameterized in terms of independently measured spectroscopic quantities.

The first inorganic molecule deliberately designed to test Hush's theoretical results was the Creutz–Taube ion $[(NH_3)_5Ru(pz)Ru(NH_3)_5]^{5+}$ (pz = pyrazine) (4). One-electron oxidation of the precursor Ru(II)–Ru(II) ion led to the mixed-valence form. Spectral measurements in the near infrared (NIR) revealed an intense, relatively narrow absorption band at ~1300 nm that was assigned to an intervalence transfer (IT) transition (eq 1).

$$(NH_3)_5Ru^{III}(pz)Ru^{II}(NH_3)_5^{5+} \xrightarrow{h\nu} (NH_3)_5Ru^{II}(pz)Ru^{III}(NH_3)_5^{5+} \quad (1)$$

This interpretation assumes localized oxidation states—in retrospect, a debatable assumption since there is extensive electronic coupling across pyrazine. A fully adequate description of the Creutz–Taube ion remains elusive. It appears that electronic coupling is sufficiently strong so that it lies near the localized–delocalized crossover. Other examples are known whose properties are well-described by the Hush model, including once-oxidized biferrocene and the 4,4'-bipyridine-bridged analog of the Creutz–Taube ion.

Much progress has been made in using spectra to calculate rate constants. This area is one of many inspired by the genius of Henry Taube, and it is an appropriate one for inclusion in this volume. The goal of this account is to present two examples. They reveal how much has been learned and illustrate how the close interplay between theory and experiment has led to some remarkable advances in our understanding of dynamic processes in complex molecules.

Nonradiative Decay in MLCT Excited States

Complexes such as $[Ru(bpy)_3]^{2+}$ (bpy = 2,2'-bipyridine) are strong visible light absorbers because they have intense metal-to-ligand charge transfer (MLCT) absorption bands (5). The resulting excited states decay to the ground state by a combination of radiative (k_r) and nonradiative (k_{nr}) processes (eq 2). The electronic configuration of the excited state is $d\pi^5\pi^{*1}$. It is largely triplet in character with some singlet character mixed in by spin-orbit coupling (6). There are many examples of this kind based on polypyridyl complexes of Ru(II), Os(II), and Re(I).

$$Ru^{III}(bpy)_3^{2+} \xrightarrow{h\nu} Ru^{III}(bpy^{\bullet-})(bpy)_2^{2+*} \xrightarrow{k_r+k_{nr}} Ru(bpy)_3^{2+} \quad (2)$$

Absorption and emission spectra for a representative example, $[Os(bpy)(py)_4]^{2+}$, are shown in Figure 1. The low-energy absorption spectrum consists of a series of overlapping Os(II) → bpy bands. The bands at lowest energy arise from direct transitions to the largely triplet emitting excited states. The

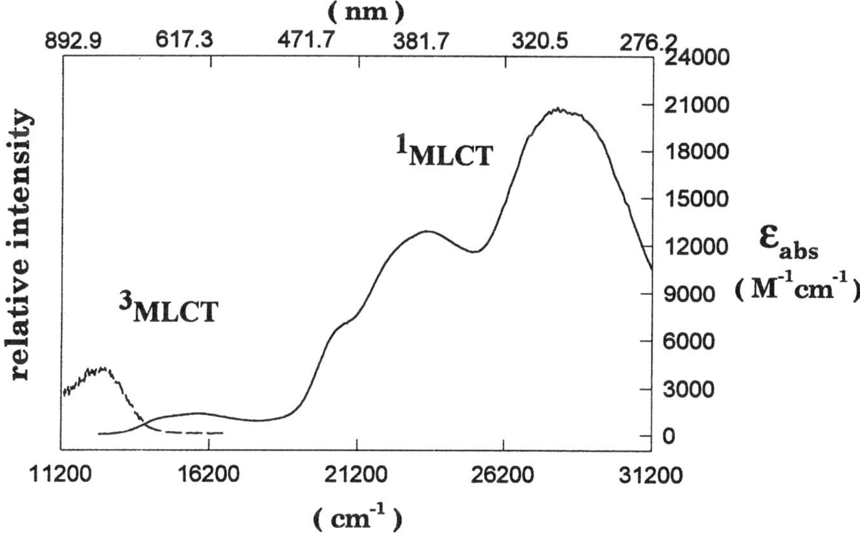

Figure 1. Absorption and emission spectra of [Os(bpy)(py)$_4$]$^{2+}$ in water at 295 K.

more intense bands at higher energy arise from transitions to the corresponding singlets. Os(II) → py bands appear at even higher energy. For the emitting excited state, k_r and k_{nr} can be calculated from excited state lifetime (τ) and emission quantum yield measurements (ϕ_{em}) by $\tau^{-1} = k_r + k_{nr}$, and $\phi_{em} = k_r\tau^{-1}$.

It is known from resonance Raman measurements that nonradiative decay is dominated by energy dispersal into a series of ν(bpy) ring stretching vibrations with quantum spacings between 1000 and 1600 cm^{-1}. The resonance Raman experiment gives direct insight because only those vibrations that are coupled to the transition between states (the acceptor modes) are resonantly enhanced. Coupling occurs if there is a change in equilibrium displacement between states ($\Delta Q_e \neq 0$). The ring-stretching vibrations are obvious candidates because the addition of the excited electron to a π^* level of the bpy ligand increases the average C–C and C–N bond lengths.

Within the harmonic approximation, radiative and nonradiative decay interconnect the same states, and analysis of emission is relevant to both. The nonradiative transition occurs between excited and ground states whose electronic wave functions are solutions of the same Born–Oppenheimer hamiltonian. To zero order, they cannot mix. They can be mixed, and the transition between states induced, by vibrations of appropriate symmetry, the "promoting modes". When activated, they perturb the electron clouds and mix the states. Nonradiative decay must occur with energy conservation. The initial electronic energy of the excited state appears in the coupled vibrations and the solvent.

Application of time-dependent perturbation theory to nonradiative decay results in expressions for the rate constant such as eq 3 (2, 3). It applies when a single high- or medium-frequency acceptor vibration and the solvent are coupled to the transition. In this equation it is assumed that the vibrational quantum spacing ($\hbar\omega = h\nu$) is the same in the initial and final states ($\hbar\omega = \hbar\omega'$), that $\hbar\omega >> k_BT$, and that the vibration is a harmonic oscillator. The vibrational quantum spacing for the promoting mode is $\hbar\omega_k$. E_0 is the energy gap between states, and $\lambda_{0,L}$ is the solvent reorganizational energy, including low-frequency modes treated classically. E_0 is related to the free energy of the excited state above the ground state, ΔG^0_{ES}, by $E_0 = \Delta G^0_{ES} - \lambda_{0,L}$, and χ_0 and χ_v are the vibrational wave functions for the initial ($v = 0$) and final (v') vibrational levels.

$$k_{nr} = \frac{2\pi V_k^{\,2}}{\hbar}\left(\frac{2}{\hbar\omega_k}\right)\sum_{v'}\langle\chi_0|\chi_{v'}\rangle\exp\left\{-\left[\frac{-\left(E_0 - v'\hbar\omega - \lambda_{0,L}\right)^2}{4\lambda_0 k_BT}\right]\right\} \tag{3}$$

The first term in eq 3 contains the vibrationally induced electronic coupling integral, V_k. It introduces the dynamic role of the promoting mode(s) in mixing the states. The second is the sum of the squares of the vibrational overlap integrals between the initial ($v = 0$) level in the excited state and the v' acceptor levels in the ground state. The vibrational overlap integrals quantify the extent to which the two states are coincident along the normal coordinate. The sum is over all of the acceptor vibrational levels. In practice, only a few are important, those for which there is near energy conservation, $\Delta G^0_{ES} \sim v'\hbar\omega$. Each integral in the sum represents a separate reaction channel to a different v' acceptor level. The third term includes the solvent as an energy acceptor and the requirement that energy be conserved. Equation 3 can be generalized and extended to include any number of acceptor vibrations.

When evaluated, the vibrational integrals in eq 3 are parameterized in terms of $\hbar\omega$ and the electron-vibrational coupling constant, S. S is related to the reduced mass of the normal mode (M or μ), the change in equilibrium displacement (ΔQ_e), and ω by

$$S = \frac{1}{2}\left(\frac{M\omega}{\hbar}\right)(\Delta Q_e)^2 \tag{4}$$

The same vibrations that are coupled to the transition between states are resonantly enhanced in the Raman experiment. The requirement for resonance enhancement is that $\Delta Q_e \neq 0$. Because of this connection, resonance Raman measurements can be used to assess the role of the acceptor vibrations on a mode-by-mode basis (7).

Figure 2 shows a series of resonance Raman spectra of $[\text{Os(bpy)(py)}_4]^{2+}$ acquired in water at four different excitation wavelengths. Reference to the absorption spectrum in Figure 1 reveals that these excitation wavelengths cor-

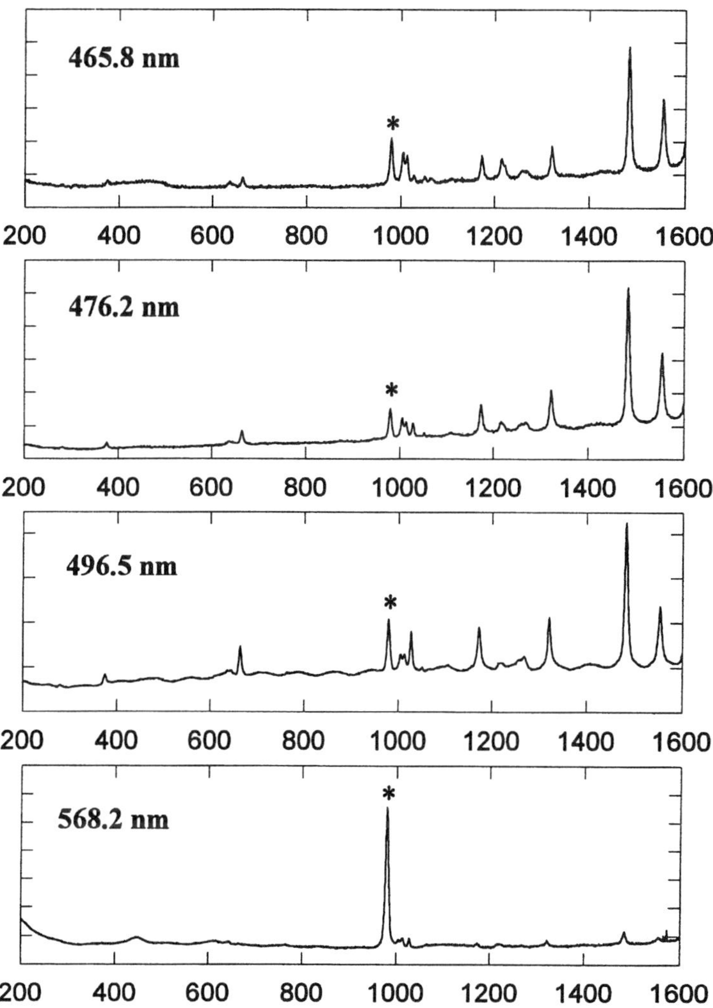

Figure 2. Resonance Raman spectra at 298 K in water 0.5 M in Na_2SO_4 at different excitation wavelengths. The SO_4^{2-} band used as an internal reference is labeled with an asterisk.

respond to the onset and near maximum of the MLCT absorptions that give rise to the lowest, largely singlet MLCT excited states.

The spectra reveal immediately the quantum spacings of the coupled vibrations from the Raman band energies. Thirteen bands are resonantly enhanced significantly above background with energies ranging from 1606 to 375 cm^{-1}. From Figure 2 it is evident that relative Raman intensities are dependent on the excitation energy. This effect is understandable qualitatively, and can be treated quantitatively by using time correlation methods and the

Heller theory (8). Qualitatively, the contributions of the individual vibrations to the absorption band shape vary with the excitation energy. This variation occurs because the extent of vibrational overlap between the ground and excited state vibrational wave functions depends on the energy difference between states and the broadening of individual vibronic lines by the solvent.

The vibrational wave functions are parameterized in terms of S and $\hbar\omega$. Heller has derived a general procedure for analyzing Raman excitation profiles (plots of absolute Raman intensity as the excitation energy is varied) in terms of these parameters and two solvent parameters, Γ and σ, for homogeneous and inhomogeneous broadening. To apply the theory and account for absolute intensities, it is necessary to apply an appropriate electronic structural model to the polarizability tensor for Raman scattering. With this model available, it is possible to fit the Raman profiles for all thirteen resonantly enhanced bands in Figure 2 (9). Figure 3 shows calculated profiles for the four bands in Figure 2. Experimental Raman intensities at the excitation wavelengths in Figure 2 are shown for comparison. There are 8–10 Raman excitation lines available in the MLCT region in Figure 1 by using a combination of Ar^+ and Kr^+ ion lasers.

The results of the Heller analysis and the S and $\hbar\omega$ values that result can be applied to the dynamics of radiative and nonradiative decay by making a series of approximations. They include the neglect of anharmonicity and the

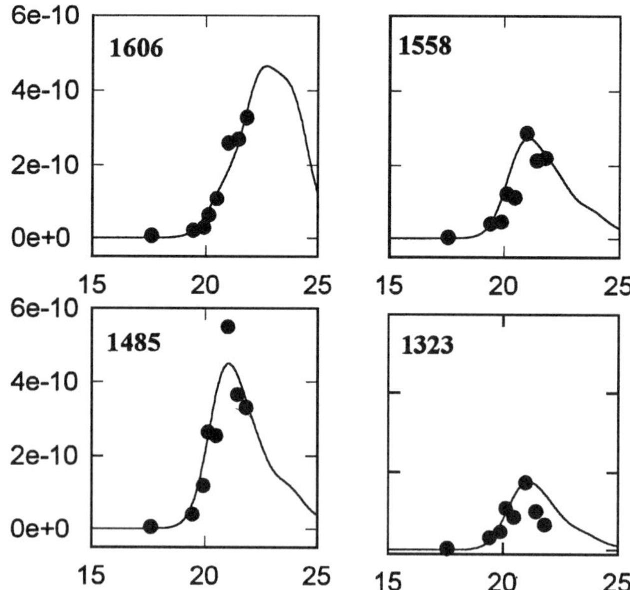

Figure 3. Calculated resonance Raman profiles for the bands at 1606, 1555, 1485, and 1322 cm⁻¹ in Figure 2. Experimental intensities for the excitation wavelengths in Figure 2 are shown for comparison.

assumption that $\hbar\omega = \hbar\omega'$. It is also assumed that the S_j and $\hbar\omega_j$ values for the individual modes are the same for the largely triplet emitting MLCT state as for the higher energy MLCT singlets that dominate the resonance effect. Figure 4 shows calculated and experimental emission spectra and a listing of S and $\hbar\omega$ values for the vibrations coupled to emission from $[Os(bpy)(py)_4]^{2+*}$. The spectrum was calculated by using the S_j and $\hbar\omega_j$ values listed in Figure 4, $E_{00} = 12{,}560$ cm^{-1}, and a full width at half maximum (fwhm) for each of the separate vibronic lines of $\Delta\bar{v}_{1/2} = 1780$ cm^{-1}. E_{00} is the energy of the excited state above the ground state with both states in $v = 0$ vibrational levels plus the solvent reorganizational energy, λ_0. The bandwidth is related to λ_0 by

$$(\Delta\bar{v}_{1/2})^2 = 16k_\mathrm{B}T\lambda_0 \ln 2 \tag{5}$$

The emission spectral fitting procedure provides $\Delta\bar{v}_{0,1/2}$ and E_{00}. They, along with the values of S_j and $\hbar\omega_j$ from the Raman profiles, provide all of the parameters required to evaluate the vibrational overlap integrals and exponential terms in the multimode version of eq 3. The terms that remain to be evaluated in the expression for k_nr are V_k and $\hbar\omega_k$. V_k was evaluated in an earlier study of nonradiative decay in more than 30 Os(II) polypyridyl complexes that gave $V_k = 1300$ cm^{-1} (2d). Assuming that Os–N stretches dominate the role of promoting mode(s), a reasonable value for $\hbar\omega_k$ is 300 cm^{-1}.

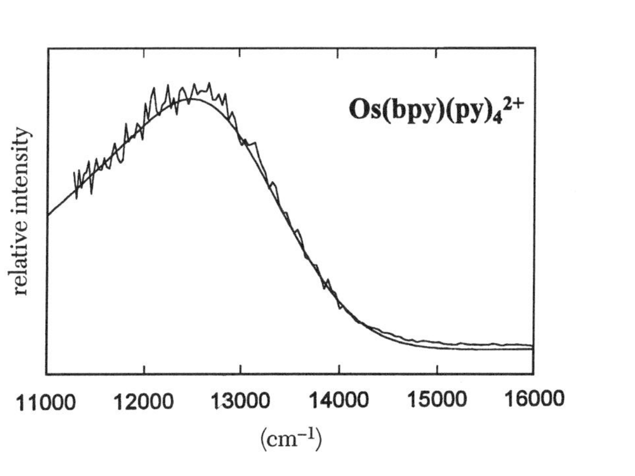

$E_{00} = 12{,}560$ cm^{-1}	
fwhm $= 1780$ cm^{-1}	
$\hbar\omega_j$ (cm^{-1})	S_j
1606	.071
1558	.142
1485	.275
1323	.082
1265	.038
1174	.051
1108	.005
1066	.007
1053	.007
1030	.021
1015	.019
1006	.025
666	.019
375	.015
280	.009

Figure 4. Experimental and calculated emission spectra for $[Os(bpy)(py)_4]^{2+*}$ in H_2O at 298 K. The spectrum was calculated by using the parameters listed. E_{00} is the $v_j = 0$ excited-state energy, and fwhm is the full-width at half-maximum for the individual vibronic lines.

**Table I. Comparison of Calculated and Experimental Values
of k_{nr} in H_2O at 290 K**

Excited State	$k_{nr} \times 10^{-8}$ (s^{-1})	
	Exp	Calc
$[Os(bpy)(py)_4]^{2+*}$	1.0	4.0
$[Os(bpy)_2(py)_2]^{2+*}$	0.74	1.4
$[Os(bpy)_3]^{2+*}$	0.50	1.4

NOTE: With $V_k = 1300$ cm^{-1} from reference 2d and $\hbar\omega_k = 300$ cm^{-1}.

The analysis described here for $[Os(bpy)(py)_4]^{2+*}$ has been extended to $[(Os(bpy)_2(py)_2)]^{2+*}$ and $[Os(bpy)_3]^{2+*}$ as well (9). Experimental values of k_{nr} are compared with calculated values in Table I.

The comparison shows that there is close agreement between the experimental calculated values. These results, and those obtained earlier, for example, on electron transfer in donor–acceptor complexes and metal complexes (10), show that it is possible to include each separate, coupled vibration in an explicit manner in these inherently multimode, dynamic processes.

The initial results need to be viewed with some skepticism. A number of approximations were used in the analysis. Resonance Raman excitation profiles were analyzed from data obtained by resonance enhancement from the MLCT singlet absorption bands. The excited state is largely triplet. The use of resonance Raman data to fit the emission spectra assumes that anharmonicities are negligible and that $\hbar\omega = \hbar\omega'$. Nonetheless, the initial results are encouraging, and these methods promise to open new insights into excited state structure and reactivity by application of resonance Raman.

Calculation of Electron Transfer Rate Constants from Emission Spectra

In the original Hush formulation, classical relationships were derived between absorption band energies, widths, and integrated intensities and the rate constant and activation energy for electron transfer. Despite the straightforward connections between spectral properties and rate constants, there have been surprisingly few reports in which both rate and spectral measurements are available to test the theory. The absence of such measurements is a consequence of the experimental difficulties involved. For example, in mixed-valence molecules where intervalence transfer (IT) bands can be observed, intramolecular electron transfer is usually rapid and difficult to measure. In molecules for which electron transfer is measurable, the corresponding absorption bands often are not.

The successes have come largely with electron transfer reactions that occur in the "inverted region" (11). In this region the driving force $(-\Delta G^0)$ is greater than the reorganizational energy: $-\Delta G^0 > \lambda$. The energy-coordinate curve for

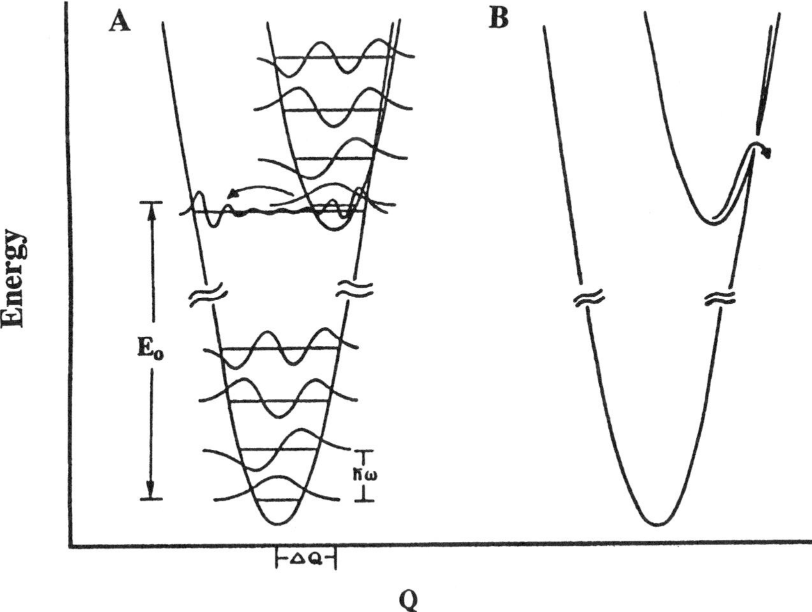

Figure 5. Energy-coordinate curves for electron transfer in the inverted region. Curve A illustrates the energy-coordinate curve and quantum spacings for a high- or medium-frequency vibration coupled to the transition. Curve B illustrates the classical surface crossing.

the reactants is "embedded" in the curve for the products, as illustrated in Figure 5, and electron transfer is closely related to nonradiative decay.

In the classical limit, thermal activation occurs in order to reach the intersection between energy curves where electron transfer can occur with energy conservation. This condition holds for the solvent and coupled low-frequency vibrations treated classically. If there are coupled medium- or high-frequency vibrations, the reaction is dominated by channels from $v = 0$ vibrational levels in the reactants to v' in the products. Vibrational overlap is high because the energy curves are embedded. Most of the initial energy in the reactant ends up in v' levels in the product with $-\Delta G^0 \sim v'\hbar\omega$.

The quantum spacings for these vibrations are large relative to the background thermal energy, $\hbar\omega >> k_\mathrm{B}T$ ($\sim$200 cm^{-1} at room temperature). Populations above $v = 0$ in the reactant are negligible, and there is no requirement for thermal activation.

Inverted electron transfer has similarities with nonradiative decay of excited states. In either, high- and/or medium-frequency vibrations coupled to the transition play the dominant role as energy acceptors. There are also differences. In nonradiative decay the surrounding vibrations and solvent are coupled to the electron donor–acceptor pair in the excited state. In electron transfer they are coupled separately to the electron donor and acceptor, which are weakly coupled electronically.

A slightly modified form of eq 3 provides a quantitative basis for analyzing electron transfer in the inverted region, but with V_k in the preexponential term replaced by H_{DA}. H_{DA} is the electron transfer matrix element. It is the resonance energy arising from overlap and mixing of the electronic wave functions of the electron transfer donor and acceptor. The electrostatic perturbation between the two causes the mixing.

Equation 3 can be applied in the closed form shown in eq 6, which is valid if $E_0 >> S\hbar\omega$ and $\hbar\omega >> k_BT$. Equation 6 is the electron transfer version of the famous "energy gap law" for nonradiative decay with one vibration and the solvent coupled to the transition. In eq 6 ν_{ET} is the frequency factor for electron transfer, E_0, S, and $\hbar\omega$ were defined earlier, and $E_0 = |\Delta G^0| - \lambda_{0,L}$.

$$\ln(k_{ET}) = \ln(\nu_{ET}) + \ln(F) \tag{6a}$$

$$\nu_{ET} = \frac{2\pi H_{DA}^2}{\hbar}\left(\frac{1}{2\pi\hbar\omega E_0}\right)^{1/2} \tag{6b}$$

$$\ln(F) = -S_M - \frac{\gamma E_0}{\hbar\omega} + \left[\frac{\gamma+1}{\hbar\omega}\right]^2 (k_BT\lambda_{0,L}) \tag{6c}$$

$$\gamma = \ln\left[\frac{E_0}{S\hbar\omega}\right] - 1 \tag{6d}$$

In general, there are many vibrations coupled to electron transfer or nonradiative decay. The thirteen modes coupled to nonradiative decay in $[Ru(bpy)_3]^{2+*}$ is a case in point. The single mode result in eq 6 can be extended to include multiple modes by mode averaging. In this procedure S for the coupled medium-frequency vibrations is the sum of S values for the individual contributors, S_j,

$$S = \sum_j S_j$$

and $\hbar\omega$ is the weighted sum of the associated vibrational quantum spacings,

$$\hbar\omega = \sum_j S_j \hbar w_j \Big/ \sum_j S_j$$

$\lambda_{i,L}$ is the sum of the reorganizational energies for the n coupled low-frequency vibrations treated classically, $\lambda_{i,L} = \Sigma_n \lambda_{n,L}$. $\lambda_{0,L}$ is the sum of λ_{iL} and the solvent reorganizational energy,

$$\lambda_{0,L} = \lambda_0 + \lambda_{i,L} \tag{7}$$

One example where this theory has been applied successfully is to intramolecular electron transfer in eq 8 (*12, 13*).

$$[(4,4'\text{-}(X)_2\text{bpy}^{\bullet-})(CO)_3\,Re^I(py\text{-}PTZ^+)]^+$$

$$\underset{h\nu}{\overset{k_{ET}}{\rightleftarrows}}[(4,4'\text{-}(X)_2\text{bpy})(CO)_3\,Re^I(py\text{-}PTZ)]^+ \tag{8}$$

$$(X = CO_2Et, (CO)NEt_2, H, CH_3, OCH_3)$$

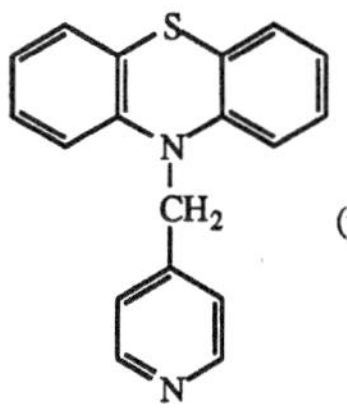

(py-PTZ)

These reactions were studied by nanosecond transient absorption measurements following $Re^I \rightarrow 4,4'(X)_2$bpy laser flash photolysis. $Re^{II}(4,4'(X)_2\text{bpy}^{\bullet-})$ MLCT excited states are formed by initial excitation, and the final, redox-separated states by subsequent $-PTZ \rightarrow Re^{II}$ electron transfer.

By varying the substituent $-X$ from electron withdrawing (CO_2Et) to electron donating, (OCH_3), ΔG^0 was varied from -1.37 to -1.93 eV in 1,2-dichloroethane. Over this range in driving force, k_{ET} decreased by ~30. Absorption bands were observed for the optically induced reverse of k_{ET}, $h\nu$ in eq 8. These bands appear at low energies and are of low absorptivity. H_{DA} was calculated from integrated band intensities by using an equation derived by Hush. The remaining terms in eq 6 were evaluated by a combination of electrochemical, spectral, and kinetic measurements. Calculated values of k_{ET} were within a factor of ~10 of the experimental values, for example, $k_{obs} = 3.1 \times 10^7$ s^{-1} and $k_{calc} = 3.0 \times 10^8$ s^{-1} for $X = CH_3$ in propylene carbonate at room temperature.

Similar agreement has been obtained for back electron transfer in a series of organic donor–acceptor complexes such as hexamethylbenzene–1,2,4,5-tetracyanobenzene (HMB–TCB) (*14*). In these cases, laser flash excitation is followed by electron transfer and formation of contact radical ion pairs (CRIP). Back electron transfer occurs to give the ground state both radiatively and by electron transfer.

$$HMB, TCB \xrightarrow{h\nu} HMB^+, TCB^- \xrightarrow{k_r + k_{ET}} HMB, TCB \tag{9}$$

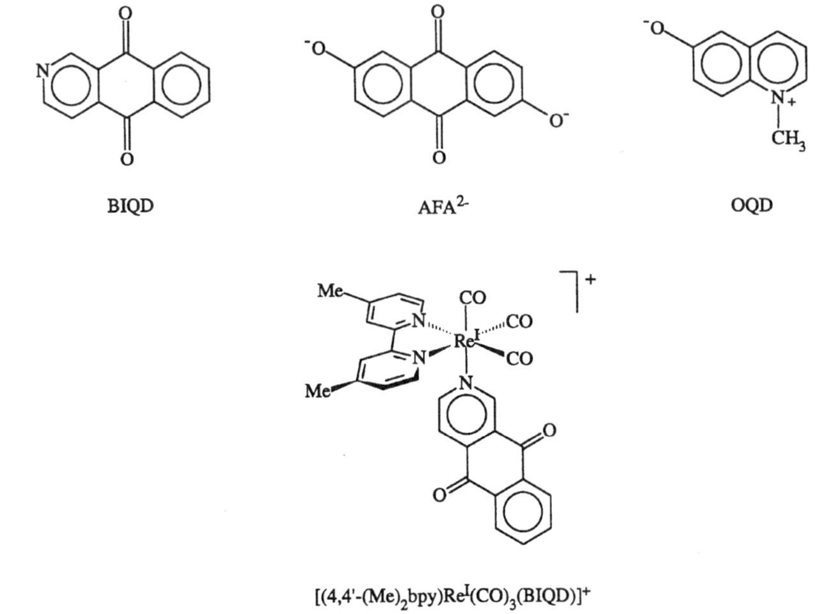

Figure 6. Structures of the acceptor ligands and of $[(4,4'-Me_2(bpy)Re(CO)_3-(BIQD)]^+$.

The various quantities in eq 6c were evaluated by Franck–Condon analysis of the emission spectrum.

We have used emission measurements to study back electron transfer in a series of Re(I) complexes containing bound acceptor ligands (15). Their structures and the structure of one of the complexes are illustrated in Figure 6. In these complexes the usual (bpy$^{\bullet-}$) $\rightarrow$ Re^{II} emissions are quenched, but other, weak emissions appear at lower energies. The quenching mechanism was investigated by transient absorption measurements following laser flash excitation. Following 420-nm excitation of the BIQD complex in 1,2-dichloroethane, a bleach appears at ~470 nm in the transient absorption difference spectrum due to loss of Re^I $\rightarrow$ BIQD absorption. New features appear at ~350 nm and at ~570 nm. They are consistent with formation of BIQD$^-$. Similar features appear in the spectrum of $[(4,4'-Me_2bpy)Re(CO)_3(BIQD^{\bullet-})]^0$, generated electrochemically.

The spectral changes are consistent with the reactions in Scheme I, which uses $[(4,4'-Me_2bpy)Re(CO)_3(BIQD)]^+$ as the example. The final state is reached by a combination of direct Re $\rightarrow$ BIQD excitation and Re $\rightarrow$ 4,4'-Me$_2$bpy excitation followed by 4,4'-Me$_2$bpy$^{\bullet-}$ $\rightarrow$ BIQD electron transfer. Subsequent return to the ground state by back electron transfer, k_{ET} in Scheme I, was monitored by transient absorption and emission measurements.

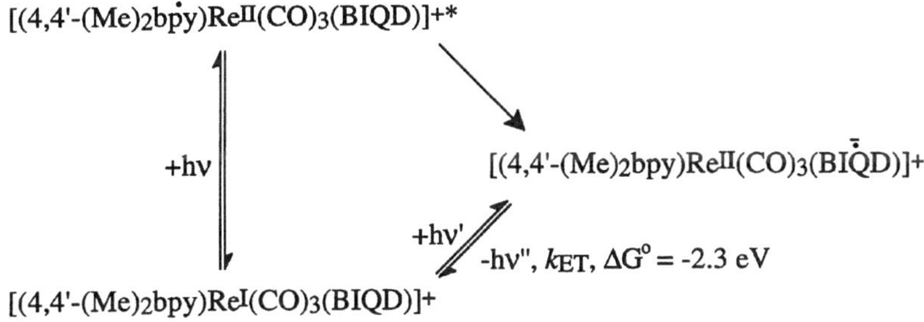

Scheme I.

The appearance of emission allows the emission band shape to be analyzed by Franck–Condon analysis. In this analysis the spectrum is constructed from a series of vibronic lines spaced by $\hbar\omega$. The lines appear at E_0, $E_0 - \hbar\omega$, $E_0 - 2\hbar\omega$, ...The relative intensities of the lines depend on S, and each line is broadened by a solvent distribution function that includes $\lambda_{0,L}$ and energy conservation. This analysis provides the parameters—E_0, S, $\hbar\omega$, and $\lambda_{0,L}$— required to calculate the electron transfer barrier in eq 6c. Emission from the BIQD complex in 1,2-dichloroethane at room temperature was successfully fit with the parameters $E_0 = 14{,}920$ cm^{-1}, $S = 0.81$, $\Delta\bar{v}_{0,1/2} = 2900$ cm^{-1} ($\lambda_{0,L} = 3660$ cm^{-1}), and $\hbar\omega = 1620$ cm^{-1}.

In order to complete the analysis and calculate k_{ET}, it is necessary to evaluate H_{DA}; k_r and H_{DA} are related as in eq 10, which follows from the Einstein equation for spontaneous emission, and an equation derived by Hush that relates H_{DA} to the transition moment for the optical transition between states. Combining the various relationships and assuming a Gaussian band shape for the optical transition gives,

$$H_{DA}^2 = 1.39 \times 10^5 \left[\frac{(E_0 + 2\lambda_{0,L})}{n \cdot d} \right]^2 \left\langle \bar{v}^{-3} \right\rangle k_r \tag{10}$$

where n is the index of refraction, d is the electron transfer distance, and $\bar{v}$ is the average emission energy. This provides the last quantity required to calculate k_{ET} by using eq 6. Calculated and experimental k_{ET} values and other relevant kinetic parameters for the three Re^I complexes in Figure 6 are compared in Table II.

There are several comparisons of interest in these data. First, there is good agreement between calculated and experimental values of k_{ET}. The calculated values are systematically too low but within a factor of 10 of the experimental values. All things considered, the agreement is extraordinary given the approximations involved. This agreement verifies the theory and shows that spectral

Table II. Spectral and Kinetic Parameters for Electron Transfer (Scheme I) in 1,2-Dichloroethane at Room Temperature

Parameters	$[(4,4'-Me_2bpy)Re-(CO)_3(BIQD)]^+$	$[(4,4'-Me_2bpy)Re-(CO)_3]_2(AFA)$	$[(4,4'-t-Bu_2bpy)Re(CO)_3(OQD)]^+$
E_{em} (nm)	688	664	748
τ (ns)	25	5.6×10^3	1.5×10^4
ϕ_{em}	1.8×10^{-3}	1×10^{-3}	5×10^{-4}
E_0 (cm^{-1})^a	14,920	15,490	13,610
$S_M{}^a$	0.81	0.95	1.11
$\Delta\bar{v}_{1/2}$ (cm^{-1})^a	2900	2830	1770
$\hbar\omega$ (cm^{-1})^a	1625	1500	1376
$\lambda_{0,L}$ (cm^{-1})	3660	3500	1360
E_{abs} (cm^{-1})	23,560	23,910	17,870
ε_{calc} (M^{-1} cm^{-1})	2.9×10^1	1.1×10^{-1}	2.8×10^{-2}
H_{DA} (cm^{-1})	153	9.4	3.9
$\ln(F)$	-13.82	-15.05	-13.61
$k_{ET, obs}$ (s^{-1})	4.0×10^7	1.78×10^5	6.58×10^4
$k_{ET, calc}$ (s^{-1})^b	1.1×10^7	1.2×10^4	8.8×10^3

NOTE: E_{em} is the emission maximum; τ is the lifetime; ϕ_{em} is the emission quantum yield; E_0, S, $\hbar\omega$, and $\Delta\bar{v}_{1/2}$ are emission spectral fitting parameters definied in the text; $\lambda_{0,L}$ is the solvent reorganization energy including low-frequency vibrations treated classically (*see* eq 7); E_{abs} and ε are the energy and molar extinction for absorption hv' in Scheme I; $\ln(F)$ is defined in eq 6c; and $k_{ET, obs}$ and $k_{ET, calc}$ are the experimental and calculated electron transfer rate constants.
^{a}From emission spectral fitting.
^{b}Calculated from eq 6.

parameters can be used to calculate electron transfer rate constants to a high degree of accuracy.

LA$^-$ → Re^{II} back electron transfer in the aryloxy-linked complexes is remarkably slow, τ = 5.6 and 15 µs, given their small energy gaps. This is a consequence of weak electronic coupling (note the small magnitudes of H_{DA} in Table II). Initial transient infrared results in the v(CO) region suggest that the orbital origin of these states may be σ–π* $(d\sigma^1(Re-O)\pi^{*1})$ rather than $d\pi^5\pi^{*1}$ with the electron donor and acceptor orbitals orthogonal to each other. For all three, electronic coupling is weak to moderate. There are separate electron donor and acceptor sites separately coupled to surrounding vibrations and the solvent. These are redox-separated (RS) states in contrast to MLCT excited states such as $[Ru(bpy)_3]^{2+*}$.

The spectral fitting parameters are also revealing; $\hbar\omega$ is larger for the quinone acceptors than for the pyridinium. The value for the pyridinium falls in line with cases for which bpy or phen are the acceptor ligands ($\hbar\omega$ = 1200–1400 cm^{-1}). $\hbar\omega$ is larger for the quinones because there are two coupled, quinone-based vibrations in the region 1600–1700 cm^{-1} in addition to lower frequency ring-stretching modes. The fact that these vibrations are coupled was shown by resonance Raman measurements. There is also a significant increase in $\lambda_{0,L}$ for the quinone acceptors. This comes from strong, specific solvent interactions with the O atoms of the semiquinone radical anion in the redox-separated state.

Acknowledgments

The work in this chapter on nonradiative decay was supported by Department of Energy Grant DE–FGO5–86ER 13633, and the work on electron transfer by National Science Foundation Grant CHE–9321413.

References

1. (a) Hush, N. S. *Prog. Inorg. Chem.* **1967**, *8*, 391; (b) Hush, N. S. *Electrochim. Acta* **1968**, *3*, 1005; (c) Creutz, C. *Prog. Inorg. Chem* **1983**, *30*, 1; (d) Hush, N. S. *Coord Chem. Rev.* **1985**, *64*, 135.

2. (a) Englman, R.; Jortner, J. *Mol. Phys.* **1970**, *18*, 145; (b) Freed, K. F.; Jortner, J. *J. Chem. Phys.* **1970**, *52*, 6272; (c) Heller, E. J.; Brown, R. C. *J. Chem. Phys.* **1983**, *79*, 3336; (d) Kober, E. M.; Caspar, J. V.; Lumpkin, R. S.; Meyer, T. *J. Phys. Chem.* **1986**, *90*, 3722; (e) Kubo, R.; Toyozawa, Y. *Prog. Theor. Phys.* **1955**, *13*, 160.

3. (a) Freed, K. F. *Top. Curr. Chem.* **1972**, *31*, 65; (b) Englman, R.; Jortner, J. *Mol. Phys.* **1970**, *18*, 145; (c) Freed, K. F.; Jortner, J. *J. Chem. Phys.* **1970**, *52*, 6272; (d) Bixon, M.; Jortner, J. *J. Chem. Phys.* **1968**, *48*, 715.

4. (a) Creutz, C.; Taube, H. *J. Am. Chem. Soc.* **1969**, *91*, 3988; (b) Creutz, C.; Taube, H. *J. Am. Chem. Soc.* **1973**, *95*, 1086; (c) Creutz, C. *Prog. Inorg. Chem* **1983**, *30*, 1.

5. (a) Juris, A.; Balzani, V.; Baragelletti, F.; Campagna, S.; Belser, P.; Von Zewelsky, A. *Coord. Chem. Rev.* **1988**, *84*, 85; (b) Krause, R. A. *Struct. Bonding (Berlin)* **1987**, *67*, 1; (c) Meyer, T. J. *Pure Appl. Chem.* **1986**, *58*, 1576; (d) Kalyanasundaram, K. *Coord. Chem. Rev.* **1982**, *46*, 159; (e) Seddon, K. R. *Coord. Chem. Rev.* **1982**, *42*, 79; (f) Ferguson, J.; Herren, F.; Krausz, E. R.; Maeder, M.; Vrbancich, J. *Coord. Chem. Rev.* **1985**, *64*, 21; (g) Sutin, N.; Creutz, C. *Pure Appl. Chem.* **1980**, *52*, 2717.

6. Kober, E. M.; Meyer, T. J. *Inorg. Chem.* **1986**, *56*, 1193.

7. (a) Mallick, P. K.; Danzer,G. D.; Strommen, D. P.; Kincaid, J. R. *J. Phys. Chem.* **1988**, *92*, 5628; (b) Danzer,G. D.; Golus, J. A.; Strommen, D. P.; Kincaid, J. R. *J. Raman Spectrosc.* **1990**, *21*, 3; (c) Caspar, J. V.; Westmoreland, T. D.; Allen, G. H.; Bradley, P.G.; Meyer, T. J; Woodruff, W. H. *J. Am. Chem. Soc.* **1984**, *106*, 3492; (d) Morris, D. E.; Woodruff, W. H. In *Spectroscopy of Inorganic -Based Materials;* Clark, R. H.; Hester, R. E., Eds.; John Wiley and Sons: New York, 1987, p 285.

8. (a) Heller, E. J.; Sundberg, R. L.; Tannor, D. *J. Phys. Chem.* **1982** , *86*, 1822; (b) Heller, E. J. *Acc. Chem. Res.* **1981**, *14*, 368; (c) Lee, S.; Heller, E. J. *J. Chem. Phys.* **1979**, *71*, 4777; (d) Myers, A. B.; Mathies, R. A. In *Biological Applications of Raman Spectroscopy;* Spiro, T. G., Ed.; Wiley and Sons: New York, 1988; p. 1.

9. Graff, D. Ph.D. Dissertation, University of North Carolina at Chapel Hill, 1994.

10. (a) Markel, F.; Ferris, N. S.; Gould, I. R.; Myers, A. B. *J. Am Chem. Soc.* **1992**, *114*, 6208; (b) Wynne, K.; Galli, C.; Hochstrasser, R. M. *J. Chem. Phys.* **1994**, *100*, 4797; (c) Kliner, D. V.; Tominga, K.; Walker, G. C.; Barbara, P. F. *J. Am. Chem. Soc.* **1992**, *114*, 8323; (d) Tominga, K.; Kliner, D. V.; Johnson, A. E.; Levinger, N. E.; Barbara, P. F. *J. Chem. Phys.* **1993**, *98*, 1228; (e) Fisher, S. F.; Van Duyne, R. P. *Chem. Phys.* **1977**, *26*, 9; (f) Lu, H.; Petrov, V.; Hupp, J. T. *Chem. Phys Lett.* **1995**, *235*, 521.

11. (a) Miller, J. R.; Beitz, J. V.; Huddleston, R. K. *J. Am. Chem. Soc.* **1984**, *106*, 5057; (b) Gloss, G.L.; Calcaterra, L.T.; Green, N. J.; Penfield, K.W.; Miller, J. R. *Phys. Chem.* **1986**, *90*, 3673; (c) Marcus, R. A. *Rev. Mod. Phys.* **1993**, *65*, 599.

12. Chen, P.; Duesing, R; Graff, D. K.; Meyer, T. J. *J. Phys. Chem.* **1991**, *95*, 5850.

13. Katz, N. E.; Mecklenburg, S. L.; Graff, D. K.; Chen, P. Y.; Meyer, T. J. *J. Phys. Chem.* **1994**, *98*, 8959.

14. (a) Gould, I. R.; Young, R. H.; Mueller, L. J.; Albracht, A. C.; Farid, S. *J. Am. Chem. Soc.* **1994**, *116*, 3147; (b) Gould, I. R.; Noukakis, D.; Gomez-Jahn, L.; Young, R. H.; Goodman, J. L.; Farid, S. *Chem. Phys.* **1993**, *176*, 439.

15. (a) Claude, J. P. Ph.D. Dissertation,. University of North Carolina at Chapel Hill, 1995; (b) Claude, J. P.; Williams, D.; Meyer, T. J. *J. Am. Chem. Soc.* **1996**, *118*, 9782.

12

Ligand-Induced, Stereochemical Relaxation of Electronic Constraints in a Simple Chemical Process

Examples from Hexaam(m)ine Cr(III) Photophysics

John F. Endicott, Marc W. Perkovic, Mary Jane Heeg, Chong Kul Ryu, and David Thompson

Department of Chemistry, Wayne State University, Detroit, MI 48202

The photophysical behavior of several hexaam(m)ine chromium(III) complexes has demonstrated that ligand stereochemistry can perturb the $(^2E)Cr(III)$ excited state sufficiently to greatly accelerate or inhibit its rate of nonradiative decay in ambient solutions. The stereochemical perturbations that are most effective, and some features of the emission spectroscopy, indicate that one of the most effective excited-state relaxation channels for this class of complex involves a trigonal deformation of the electronic excited state. The effectiveness of this relaxation channel is attributed to a combination of (1) the similar energies of the $^2E_g(O_h)$ and $^2E'(D_{2h})$ states in simple complexes, (2) the smaller energy differences between the low-energy electronic states and their greater relative nuclear displacements in D_{2h} than in O_h symmetry, and (3) the relaxation of some electronic selection rules in the D_{2h} geometry. The nuclear coordinate associated with this relaxation channel, a trigonal twist, does not correlate with the differences between the initial and final state equilibrium geometries; rather, it is required because the $^2E_g \rightarrow {}^4A_{2g}$ relaxation process is electronically forbidden. It is proposed that the reorganizational energies found for electronically forbidden processes in transition metal systems will often contain contributions from nuclear coordinates that are required only because the overall process is electronically forbidden.

THIS CHAPTER DESCRIBES HOW THE BEHAVIOR OF ELECTRONIC EXCITED STATES of transition metal complexes can be altered using logic and strategies common to most areas of ambient chemical kinetics. That is, we will treat transition metal excited states as ordinary chemical reagents that are subject to common chemical perturbations. More specifically, we describe how one can use molecular structure–reactivity arguments to design approaches for the regulation of excited-state lifetimes. Electronic excited states are generated by the absorption of electromagnetic radiation, visible light for most transition metal complexes. Thus, although one can view light absorption as a convenient technique for the preparation of reactant species with unusual electronic configurations, one must also recognize that only those excited states that are in vibrational equilibrium with their environment can be treated with the chemical approaches that have evolved for the usual, ground-state reaction systems (*1, 2*). This consideration will necessarily restrict straightforward chemical approaches largely to the lowest energy electronic excited states with lifetimes greater than a few hundred picoseconds. For these and other reasons we have selected the ^2E excited state of chromium(III) complexes for discussion in this chapter.

The displacement of the nuclei of reactant molecules away from their equilibrium positions is a feature common to most descriptions of chemical reaction pathways (*3–5*). In terms of a semiclassical formalism (*6*) the reaction rate constant, k, is given by eq 1

$$k = \kappa_{el}\kappa_{nu}\upsilon_{nu} \tag{1}$$

where the κ_i are transmission coefficients or retardation factors; subscripts $i =$ el and nu mean electronic and nuclear, respectively; and υ_{nu} is the frequency of the reaction process (or the nuclear motion in the transition state) when there is no electronic or nuclear retardation. The nuclear displacements are discussed in terms of their effect on κ_{nu} assuming that κ_{el} is constant. (This assumption is equivalent to the Born–Oppenheimer approximation.) In most cases the important nuclear displacements are those implicated by the structural differences between reactants and products. We will call such distortions "concerted". This need not be the case when the chemical process in question is electronically forbidden ($\kappa_{el} << 1$). For example, it has been proposed that nonconcerted distortions could contribute to the rates of reduction of some low-spin cobalt(III) complexes by high-spin cobalt(II) complexes (*7*). In principle, nonconcerted distortions could be a feature of other multielectron processes (*8*) or in reactions that involve a change in spin multiplicity. Such effects are difficult to establish in most chemical reaction systems, but they can be modeled in the nonradiative relaxation behavior of certain electronic excited states. Such an issue arises in the relaxation, $^2E_g \rightarrow {}^4A_{2g}$, of the lowest energy electronic excited states of octahedral chromium(III) complexes.

This chapter will examine information bearing on the questions of how thermally activated distortions of the ground-state molecular geometry can facilitate $^2E \rightarrow {}^4A_2$ relaxation and how these distortions may be enhanced or blocked by making use of ligand stereochemistry. (Skeletal structures of the ligands employed in our studies are shown in Figure 1.) A related approach has been employed by Hendrickson and co-workers (*9*), who have used ligand stereochemical propensities to promote the high-spin–low-spin crossover in Fe(II) complexes. Before we address these issues we will review some features of Cr(III) electronic structure and some general properties of electronically excited Cr(III) complexes.

The Nature and Dynamic Behavior of the Lowest Energy Excited States of Cr(III)

The Cr(NH$_3$)$_6^{3+}$ Paradigm.   The Cr(NH$_3$)$_6^{3+}$ complex is a well-characterized example of the issues considered here (*1, 2, 10*). Owing to the O$_h$ microsymmetry this is a relatively simple system to consider, and since the d-orbital splitting energy, 10 *Dq*, is relatively large, the electronic states are reasonably well-separated (*2, 10–12*). The lowest energy excited state, ^2E_g, has the same orbital configuration (t_{2g}^3) as the $^4A_{2g}$ ground state, while the lowest

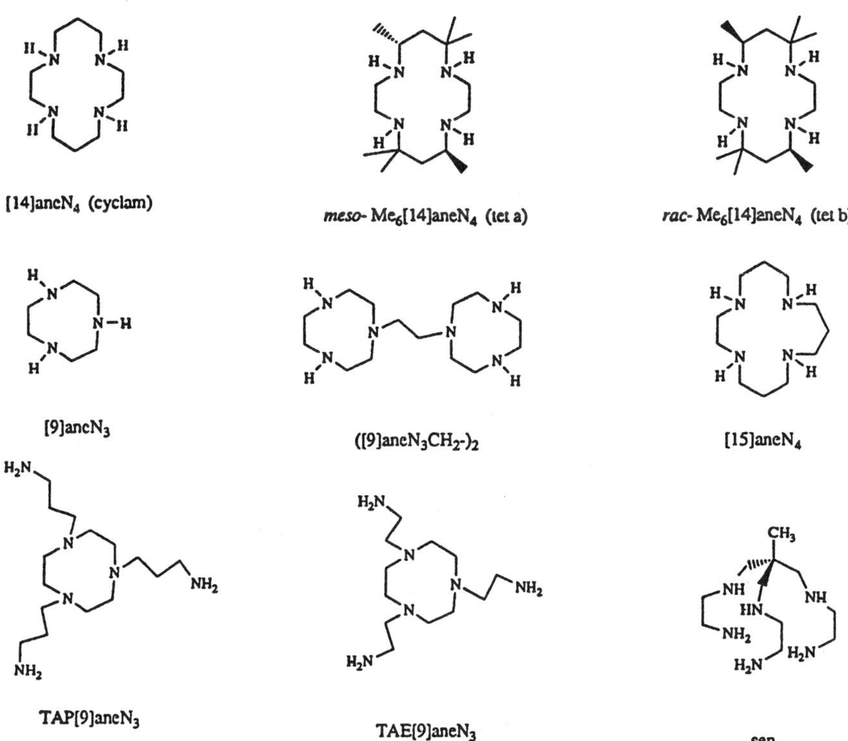

Figure 1. Skeletal structures of macrocyclic ligands.

energy quartet excited state, $^4T_{2g}$, must involve the promotion of an electron to an antibonding e_g orbital. As a consequence of this difference in orbital population, the $^4T_{2g}$ state at vibrational equilibrium would be tetragonally distorted (D_{4h} symmetry), while the ^2E_g excited state would be expected to have a molecular geometry nearly identical to that of the ground state (2, *10–12*) (*see* Figure 2). This contrast in the excited state distortions is evidenced in the broad (full width at half-height, $\Delta v_{1/2} \cong 5 \times 10^3$ cm^{-1}) $^4A_{2g} \rightarrow {}^4T_{2g}$ absorption band and narrow band ($\Delta v_{1/2}$ typically 20 cm^{-1}), structured $^4A_{2g} \rightarrow {}^2E_g$ absorptions (and $^2E_g \rightarrow {}^4A_{2g}$ emission) (2, *10–12*). From the point of view developed in this chapter, one can infer that the reorganizational energy required for the nonradiative relaxation of the ^2E_g excited state to the ground state would be expected to be $\lambda_{reorg} \cong 0$, and κ_{nu} should be approximately unity. The energy difference between the potential energy (PE) minima of the ^2E_g and $^4T_{2g}$ excited states of $Cr(NH_3)_6^{3+}$ can be estimated to be in the range of $(4–5) \times 10^3$ cm^{-1} (2, *10, 12*). In addition to these states there is a $^2T_{1g}$ excited state about 1×10^3 cm^{-1} higher energy than the ^2E_g excited state and a $^2T_{2g}$ excited state somewhat higher in energy than the $^4T_{2g}$ state. These higher energy doublet states, also t_{2g}^3 electronic configurations, are not usually considered in mechanistic discussions of Cr(III) photophysics. However, they have some relevance to the following discussion because some of their contributing microstates contain doubly occupied $d\pi$ orbitals (*13*).

Limiting Low-Temperature ^2E Excited-State Relaxation Behavior of Hexaam(m)ine Cr(III) Complexes. The mean (^2E) Cr(III) excited-state lifetime, τ_d, is inversely related to the sum of the first-order rate constants, k_i, for all the possible different pathways by means of which the ^2E excited state can dispose of its excess electronic energy (eq 2).

$$\tau_d^{-1} = \sum_i k_i \tag{2}$$

It is generally found that the ^2E lifetime is strongly temperature-dependent in the ambient range and approximately temperature-independent at low temperatures (Figure 3) (2, *10*). Thus the sum in eq 2 can be separated into the collection of those rate constants that are temperature-dependent, $k_d(T)$, and those that are temperature-independent, $k_d°$, as in eq 3.

$$k_d = \tau_d^{-1} = k_d° + k_d(T) \tag{3a}$$

$$k_d° = k_r° + k_{nr}° \tag{3b}$$

Because radiative decay is usually not strongly temperature-dependent, the rate constant for radiative decay, $k_r°$, is usually included in $k_d°$, and it can be

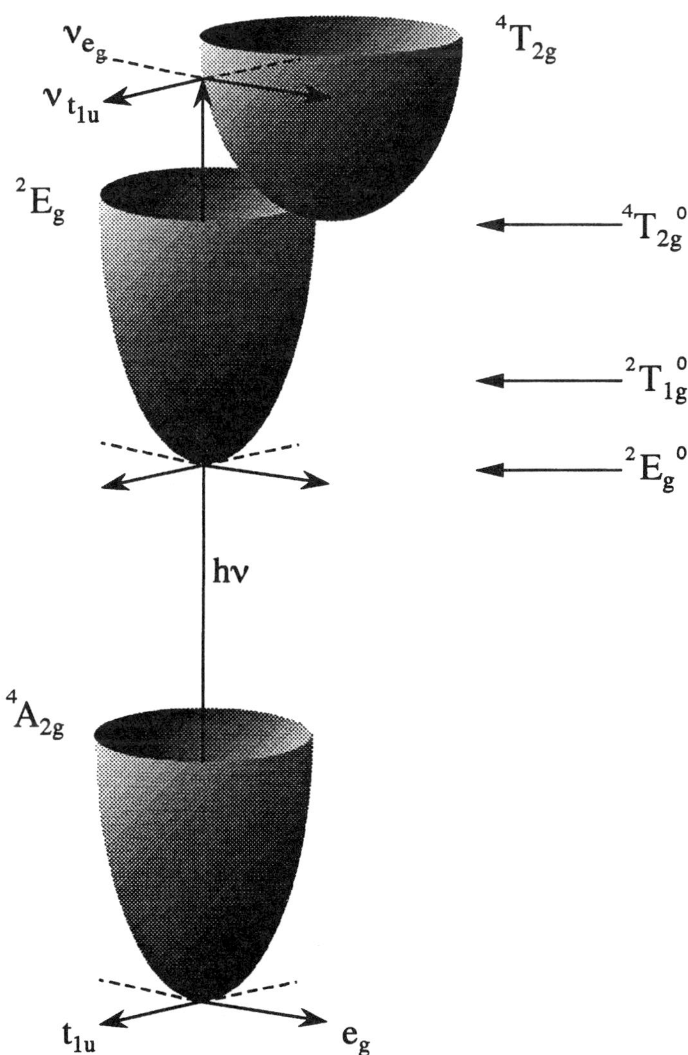

Figure 2. Qualitative representation of lowest energy potential energy surfaces for (^2E) $Cr(NH_3)_6^{3+}$ (based on Figure 9). The $^2T_{1g}$ state has been omitted and the surfaces have been assumed to be parabolic for simplicity. Only the two normal stretching coordinates, t_{1u} and e_g, are illustrated.

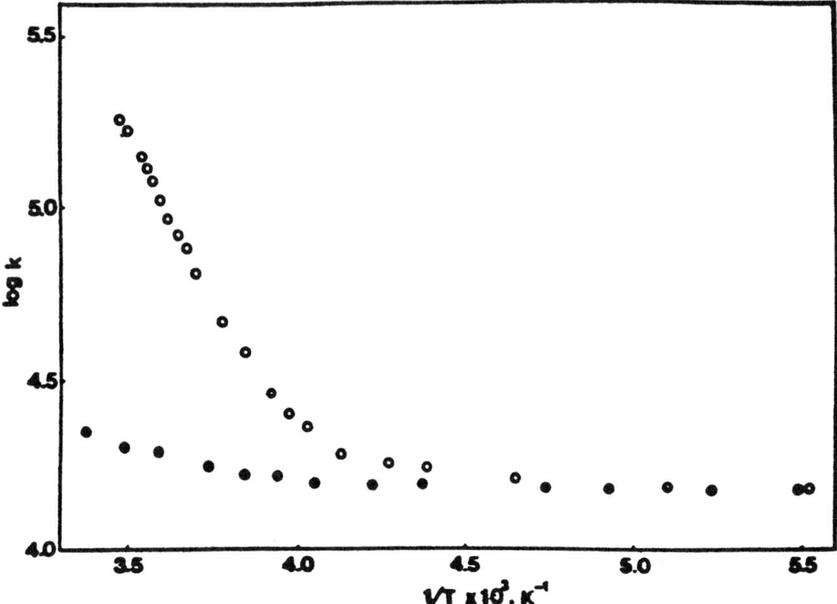

Figure 3. Temperature dependence of $\tau(^2E)$ for $Cr(NH_3)_6^{3+}$. Open circles for DMF/CHCl$_3$ solutions; closed circles for $Cr(NH_3)_6^{3+}$ doped into [Ru(NH_3)_6]-(ClO_4)_3. (Reproduced with permission from reference 33. Copyright 1984 Elsevier Science.)

experimentally distinguished from any nearly temperature-independent non-radiative decay pathways whose rate constants are included in $k_{nr}°$ (eq 3b).

For $(^2E_g)Cr(NH_3)_6^{3+}$, $k_d° \cong 2 \times 10^4$ s^{-1} (*3, 10*). Because ν_{nu} must be on the order of 10^{12}–10^{13} s^{-1}, and $\lambda_{reorg} \sim 0$, one might infer from eq 1 that $\kappa_{el} \sim 2 \times 10^{-9}$ for the nonradiative $^2E_g \rightarrow {}^4A_{2g}$ relaxation process. Certainly this transition is very strongly forbidden by conventional electronic selection rules (i.e., the change in spin multiplicity does not equal zero and the g $\rightarrow$ g transition is dipole-forbidden, so κ_{el} should be very small). Furthermore the $^2E \rightarrow {}^4A_2$ non-radiative relaxation in Cr(III) complexes may be considered a limiting example of "inverted region" behavior (*6, 14*), and there is no classical pathway (i.e., there is no formal PE surface crossing) for the nonradiative relaxation process.

As a consequence, eq 1 is not really a good basis for discussing $k_{nr}°$; that is, a quantum mechanical, rather than a classical, model should be used. Because the ^2E_g and $^4A_{2g}$ PE surfaces of $Cr(NH_3)_6^{3+}$ are nested, there can be no formal intersection (unless the ^2E_g surface is distorted; *see* discussion that follows) and the formal PE barrier to the surface crossing is infinite (so, from this perspective $\kappa_{nu} \rightarrow 0$). The intrinsic deficiencies of a classical estimate of κ_{nu} for this process can be corrected by replacing κ_{nu} in eq 1 by a nuclear tunneling coefficient, Γ (*15, 16*). Nuclear tunneling is clearly most important for vibrational modes whose quanta are very large compared to $k_B T$ (i.e., for $\hbar\omega_i >> k_B T$, where k_B is Boltzman's constant). In rigorous treatments, the magnitude of Γ has also been shown to be largest (for vibrational modes with comparable $\hbar\omega_i$) for the modes with the largest difference in ground state and excited state nuclear coordinates (or for the largest λ_{reorg}). For hexaam(m)ine chromium(III) complexes, the vibrational modes that make the largest contribution to k_{nr}° involve N–H stretching vibrations. This is readily demonstrated by the 10–70-fold decreases in k_d° when all the NH functions in these complexes are replaced by ND (2, 10, 17, 18).

For many of the hexaam(m)ine complexes, the dependence of k_d° on the number of N–H oscillators in the complex is nearly linear (Figure 4). However, Figure 4 also illustrates that factors other than the number of high-frequency oscillators in the molecule can also contribute significantly to k_d°. The most likely other factors that might contribute to deviations of k_d° from a simple dependence on the number of N–H oscillators are (1) differing electronic factors (since κ_{el} is expected to depend on symmetry-based selection rules) (19); and (2) contributions from lower frequency vibrational modes that have unusually large displacements (i.e., modes along which the ^2E PE surface is distorted) (2, 20). Although the origin of deviations of k_d° from a simple dependence on the number of N–H oscillators has not been clearly established, any significant contribution to these deviations from low-frequency modes would be very important to the present discussion because some of these modes are likely to contribute to $k_d(T)$.

Thermally Activated Decay of $(^2E)Cr(Am)_6^{3+}$ Complexes

The ^2E excited states of most Cr(III) complexes relax to the ground state more quickly under ambient conditions than at very low temperatures. The thermally activated ^2E excited-state relaxation rate can usually, but not always, be described by an exponential temperature dependence (8, 10). The apparent activation energy in many cases (e.g., for $Cr(NH_3)_6^{3+}$) is comparable to estimates of the energy difference between the ^2E and $^4T_2$ excited states (10, 11), and back intersystem crossing, $^2E \rightarrow {}^4T_2$, followed by very rapid and barrierless $^4T_2 \rightarrow {}^4A_2$ relaxation, has been the most commonly proposed mechanism for ^2E relaxation (2, 10, 21). This pathway is certainly important in many systems (2, 10, 20–25), but there are many other complexes for which it is not appropriate (2, 20, 22, 26, 27). In this chapter we focus on these other complexes.

Because electronic excited states are transients, their structures must be inferred from evidence that is more often circumstantial than direct. Consequently, one needs to consider a variety of evidence. We will first review some spectroscopic clues about the nature of potential $(^2E)Cr(III)$ excited state distortions, we will then review the use of straightforward stereochemical

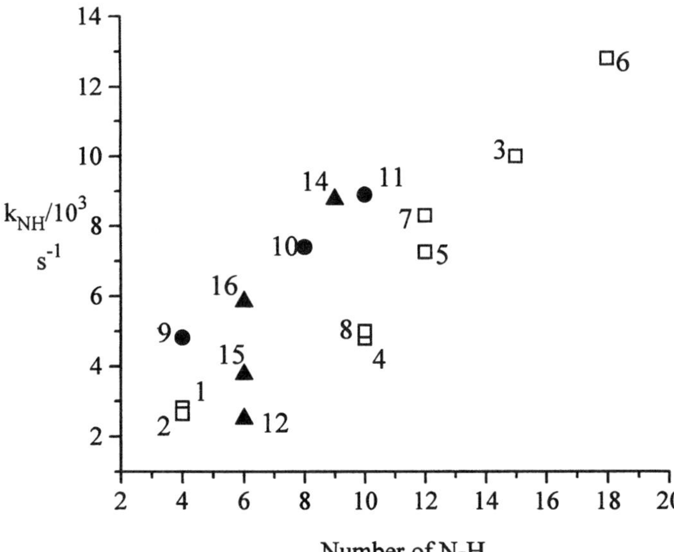

Figure 4. Dependence of τ_{77} (^2E) on the number of high-frequency oscillators for some hexaam(m)ine and cyanoam(m)ine Cr(III) complexes. Ligands: 1 is trans-([14]aneN$_4$)(CN)$_2$ (20), 2 is trans-(ms-Me$_6$[14]aneN$_4$)(CN)$_2$ (20); 3 is (NH$_3$)$_5$CN$_2$; 4 is 1,5,9-triazanonane (ditn) (15); 5 is (tn)$_3$ where tn is 1,3-diaminopropane (15); 6 is (NH$_3$)$_6$ (2, 8, 15); 7 is (en)$_3$ where en is 1,2-diaminoethane (2, 15); 8 is trans-([14]aneN$_3$)(NH$_3$)$_2$ (20, 22); 9 is cis-(rac-Me$_6$[14]aneN$_4$)(CN)$_2$ (20); 10 is cis-([14]aneN$_4$)(en) (15);11 is cis-([14]aneN$_4$)(NH$_3$)$_2$ (15);12 is ([9]aneN$_3$)$_2$ (15, 23); 13 is ([9]aneN$_3$)(CH$_2$−)$_2$ (24); 14 is 4,4',4''-ethylidyne-tris(3-azabutan-1-amine) (sen) (23); 15 is (TAP)[9]aneN$_3$) (23); and 16 is (TAE[9]aneN$_3$) (23). In this figure, $[\tau(^2E)]^{-1} = k_{NH}$, and most measurements were made in glassy solutions.

approaches for promoting and inhibiting these distortions, and then we will examine some of the consequences.

Intensities of Vibronic Origins in Cr(III) Complexes. There are some hints of possible excited-state distortions in the intensities of the vibronic origins of the emission spectra of the Cr(III) complexes. The assignment of these vibronic origins is most definitive for the high-symmetry complexes, and the details have been especially well-documented for Cr(NH$_3$)$_6^{3+}$ (28–30). Because the $^2E_g \rightarrow {}^4A_{2g}$ transition is electronically forbidden, both spin- and symmetry-forbidden, selection rule issues enter into the interpretation of intensities of the vibronic origins, and there have been proposals that one should use octapole rather than dipole selection rules (*1*). We will take the simpler view that dipole selection rules are adequate and that, all else being equal, the relative intensities of vibronic origins is an approximate measure of the relative amplitude of the distortion along the respective symmetry coordinates (*12, 31*). On this basis some interesting qualitative features of the spectra can be identified.

Qualitative Spectral Features. *$(^2E)Cr(NH_3)_6^{3+}$ Emission Spectra.* A portion of the emission spectrum of $Cr(NH_3)_6^{3+}$ in fluid solution is shown in Figure 5. Because this spectrum was obtained at relatively high temperatures (233 K) there are vibronic origins centered around the 0–0′ electronic origin: (1) intense bands at lower energy corresponding to population of coupled ground-state vibrational modes, and (2) weaker bands at higher energy arising from emission from thermally populated vibrational excited states of $(^2E)Cr(NH_3)_6^{3+}$. The lowest energy vibronic origins are CrN_6 skeletal modes, and those for which coupling is predicted by the dipole selection rules are the t_{1u} (asymmetric stretch) and the $t_{1u'}$ and t_{2u} (deformation modes) (*28–30*). The much greater intensity of the $t_{1u'}$ (deformation) than the t_{1u} (stretch) is strongly suggestive of a small displacement of the ^2E_g PE surface along the $t_{1u'}$ coordinate. The similarities of intensities of their vibronic origins also suggest some small distortion along the t_{2u} deformation coordinate. As noted already, this interpretation of the intensities of the vibronic origins is not unique, but some relative enhancement of deformation intensities is certainly consistent with the arguments developed in this chapter. The very sharp vibronic bands in the $Cr(NH_3)_6^{3+}$ emission spectrum do indicate that any distortion of the ^2E PE surface must be very small near its minimum.

Stokes Shifts and Band Widths. The energy difference between the emission maximum and the emission band threshold (0–0′ line) and the emission band widths can be interpreted as measures of distortions in the excited state PE surface. The small, if any, stokes shifts and narrow lines (usually ≤ 20 cm^{-1}) characteristic of ^2E Cr(III) emissions indicate that distortions near the PE minimum are small in amplitude. For perspective, it is useful to make note of the broad (usually $\Delta v_{1/2} = (2–4) \times 10^3$ cm^{-1}) $^4A_2 \rightarrow {}^4T_1$ absorption bands, for which the excited state distortions would characteristically lead to stokes shifts (if emission could be observed) of $(4–5) \times 10^3$ cm^{-1} (*11*). The broadest (^2E) Cr(III) (hexaam(m)ine) emission that we have observed is for $Cr([9]$ane-$N_3CH_2-)_2^{3+}$ for which $\Delta v_{1/2} \sim (1–4) \times 10^2$ cm^{-1}, depending on the matrix, and which is broad even when doped into an inert, isostructural matrix (*see* Figure 6 and the discussion that follows). Real, but much more subtle, effects are found in the emission band shapes of many Cr(III) complexes. Thus, the emission band widths usually increase and the relative intensities of vibronic components

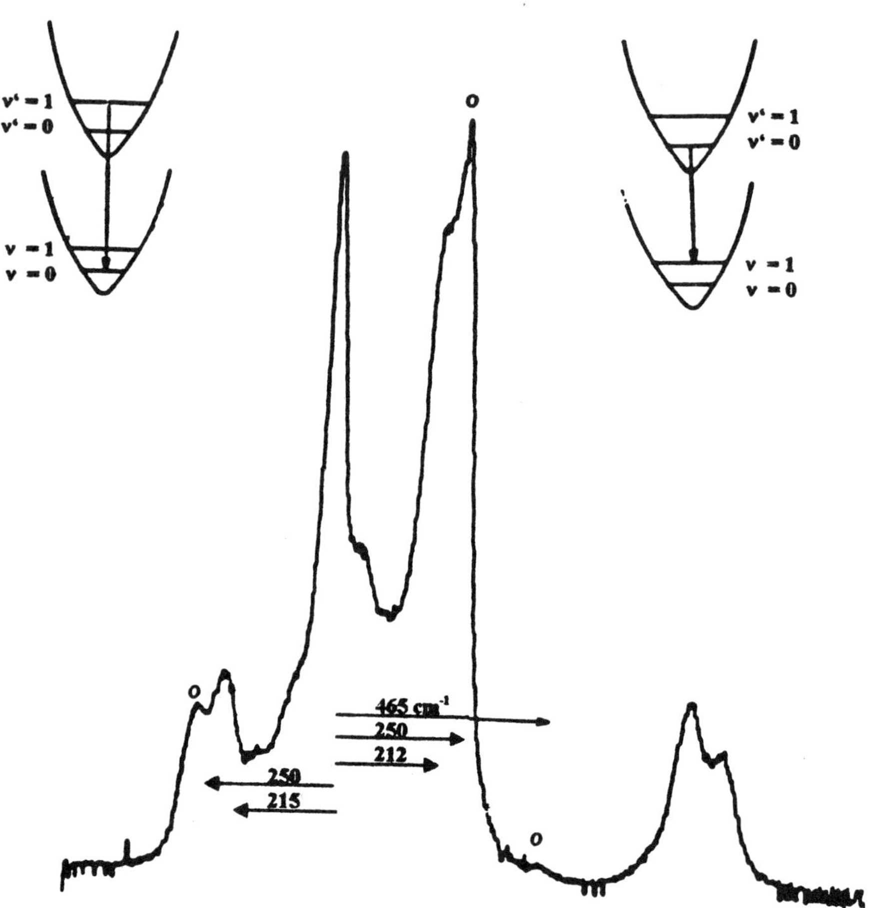

Figure 5. The (^{2}E) Cr(NH$_3$)$_6^{3+}$ emission lines near to the electronic origin (at 15.00 × 10^3 cm^{-1}) in fluid DMF at –40 °C. The skeletal vibrations of t$_{1u}$ symmetry are indicated (°) at 259 and 465 cm^{-1} lower energy than the origin, and at 258 cm^{-1} higher energy.

often change as the temperature is increased. An example of this behavior, Cr([9]aneN$_3$)$_2^{3+}$ doped into [Rh([9]aneN$_3$)$_2$](PF$_6$)$_3$, is shown in Figure 7. The effects observed are probably a result of phonon (i.e., lattice vibrations) coupling with the low-frequency (deformation) molecular vibrational modes (*31*), but this coupling implies some distortion in the PE surface (*31, 32*). Thus, in a relatively rigid complex one would expect much smaller effects on band widths and relative vibronic component intensities, and this is illustrated in the spectrum of the related Cr(TAP[9]aneN$_3$)$^{3+}$ complex in Figure 7. We presented a solution spectrum of the latter complex because solution spectra are almost always broader than spectra obtained in doped crystalline solids. Obviously, the

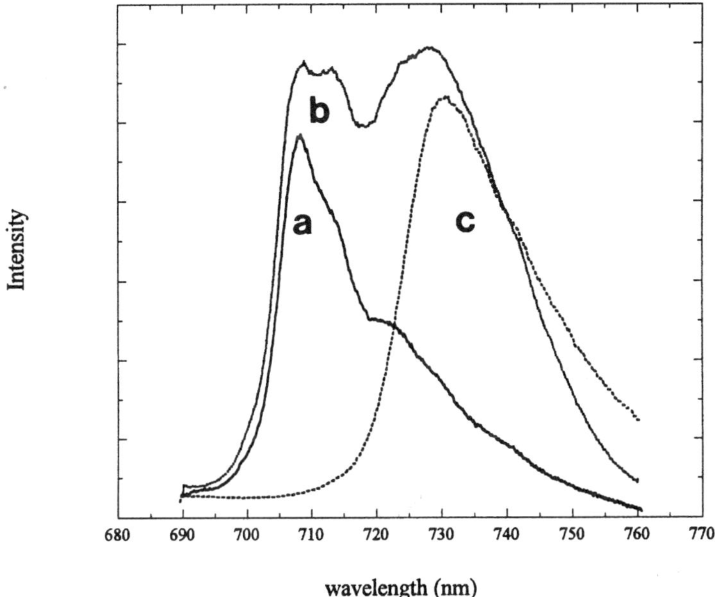

Figure 6. A comparison of the 77 K, (^2E), $Cr([9]aneN_3CH_2-)_2^{3+}$ emission spectra: a, doped (3%) into $[Rh([9]aneN_3CH_2)_2(ClO_4)_3^{3+}$; b, doped (3%) into $[Rh([9]aneN_3)_2-(PF_6)_3$;and c, in $DMSO/H_2O$ solution.

emission band widths and relative peak intensities are not very temperature-dependent for $Cr(TAP[9]aneN_3)^{3+}$, consistent with the argument, summarized subsequently, that distortions are stereochemically restricted in this complex.

Implications of the Spectroscopic Observations

Overall, the (^2E)Cr(III) emission spectra indicate that the ^2E excited state, and the $^4A_2$ ground state of hexam(m)ine complexes have very similar nuclear coordinates at their PE minima. Temperature-dependent line broadening and a very small (≤ 400 cm^{-1}) temperature dependence of apparent "Stokes" shifts do suggest a tendency for the ^2E excited state in at least some complexes to distort at the higher temperatures. Overall, the spectroscopic observations, very briefly summarized here, suggest that the (^2E)Cr(III) PE surface is often distorted along a low-frequency deformation coordinate and that the distortions may become important only away from the PE minimum. This inference would suggest certain excited-state vibrational frequencies are less harmonic than their ground-state analogs, but current information does not allow us to evaluate this point. The information summarized previously does provide some clues about where one should look for an excited-state quenching channel: a distortion along a low-frequency skeletal distortion mode, such that the coordi-

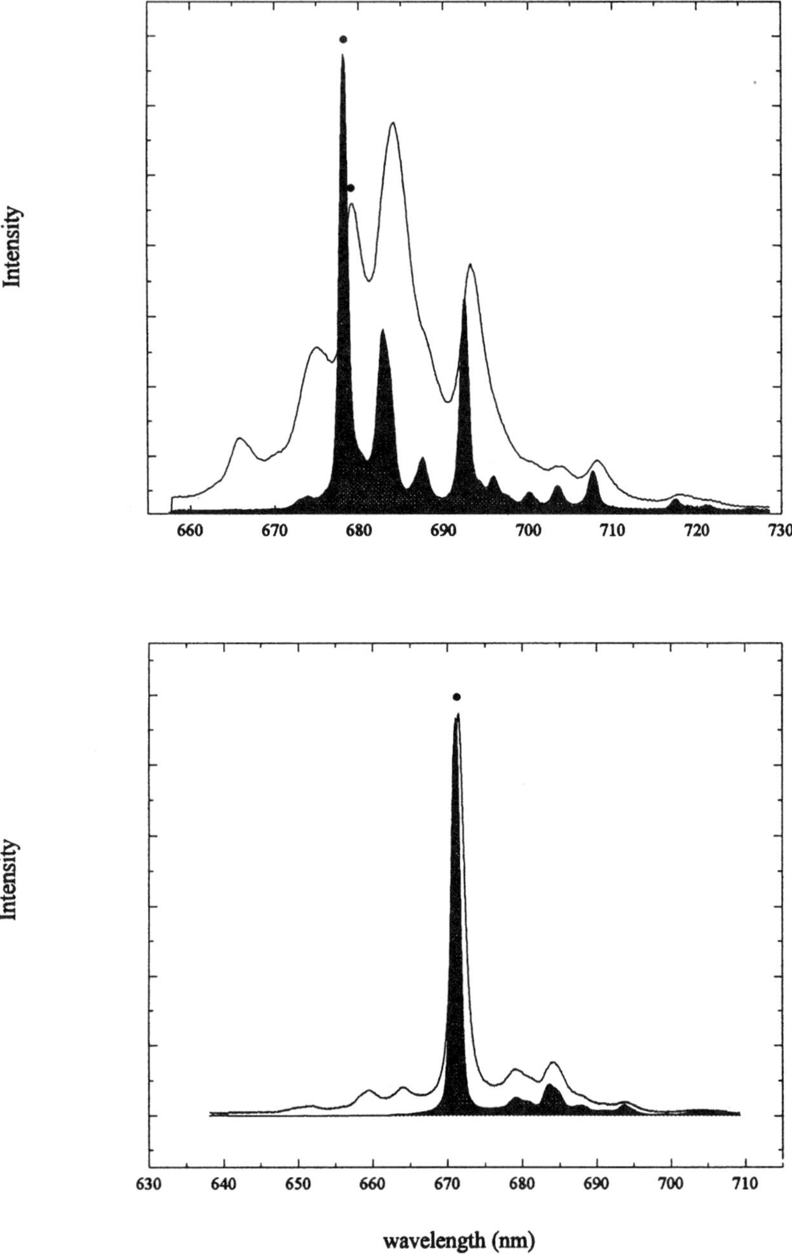

Figure 7. A comparison of the ambient and 77 K (shaded) emission spectra of $(Cr[9]aneN_3)^{3+}$ (2.9%) doped into $[Rh([9]aneN_3)_2(PF_6)_3$ (top); and $Cr(TAP[9]ane-N_3)^{3+}$ in $DMSO/H_2O$ solution (bottom). The electronic origins are marked with a filled circle.

nation isomer generated when the distortion is carried to completion can relax very rapidly to the ground electronic state. A very interesting distortion in the present context is a trigonal twist because (1) it would be composed of low-frequency deformations, (2) the Laporte selection rule would be relaxed in D_{3h} symmetry due to p/d mixing, and (3) none of the ligand-field excited states in D_{3h} is expected to be nested with the ground state.

Stereochemical Perturbations of the Thermally Activated ^2E Decay.   The spectroscopy already summarized and some of the variations in $k(T)$ (2) suggest that stereochemical constraints imposed by the coordinated ligands might be useful in altering $k(T)$. This inference is reinforced by the observation (33) that doping $Cr(NH_3)_6^{3+}$ into the crystal lattice of the inert, but isostructural, $Rh(NH_3)_6^{3+}$ complex effectively quenches $k(T)$ (Figure 3). One interpretation of this observation is that intermolecular, lattice forces restrict the amplitude of a molecular distortion, which promotes the nonradiative, excited-state relaxation. Similar effects ought to be achievable using only molecular stereochemistry. We have built on this hypothesis by examining the ^2E excited-state decay and luminescence behavior of several complexes containing ligands that would be trigonally strained if coordinated octahedrally to Cr(III). Details of the synthesis and characterization of these complexes and their ligands can be found elsewhere (26, 34, 35). It is useful to consider pairs of complexes that are closely related in their coordinates of the Cr(III) $^4A_2$ ground state, but only one of which contains a ligand whose internal stereochemical repulsions would be minimized by a large amplitude twist around a C_3 axis.

Cr(en)$_3^{3+}$ and Cr(sen)$^{3+}$ (23b).   The sen ligand is generated from (en)$_3$ by placing a neopentyl cap on one trigonal face of M(en)$_3$ (Figure 8). The bond angles of the capping atoms (CCC and CCN) are somewhat flattened in Cr(sen)$^{3+}$ indicating that there is appreciable trigonal strain in the neopentyl cap (27). The CrN_6 microsymmetry is very nearly the same for both of these complexes, with the NCrN bond angles in Cr(sen)$^{3+}$ slightly closer to octahedral angles than those of Cr(en)$_3^{3+}$. Thus the neopentyl cap can be thought of as a compressed spring whose tension can be released when resistance to a trigonal distortion is reduced. The similarity of the CrN_6 coordination environments is reflected in the very similar absorption spectra of these complexes (Table I). Despite the similarities in their CrN_6 microstructure and in their spectroscopy the ^2E excited states of these complexes differ by at least 4 orders of magnitude in their ambient lifetimes (Table I), with the ambient lifetime of (^2E)Cr(sen)$^{3+}$ less than 0.01% that of the Cr(en)$^{3+}$ complex.

Cr(TAP[9]aneN$_3$)$^{3+}$ and Cr(TAE[9]aneN$_3$)$^{3+}$ (26, 34, 35).   Both ground-state molecular structures show the effects of stereochemical distortion. In both complexes the Cr–N–C and N–C–C bond angle of the pendant amines

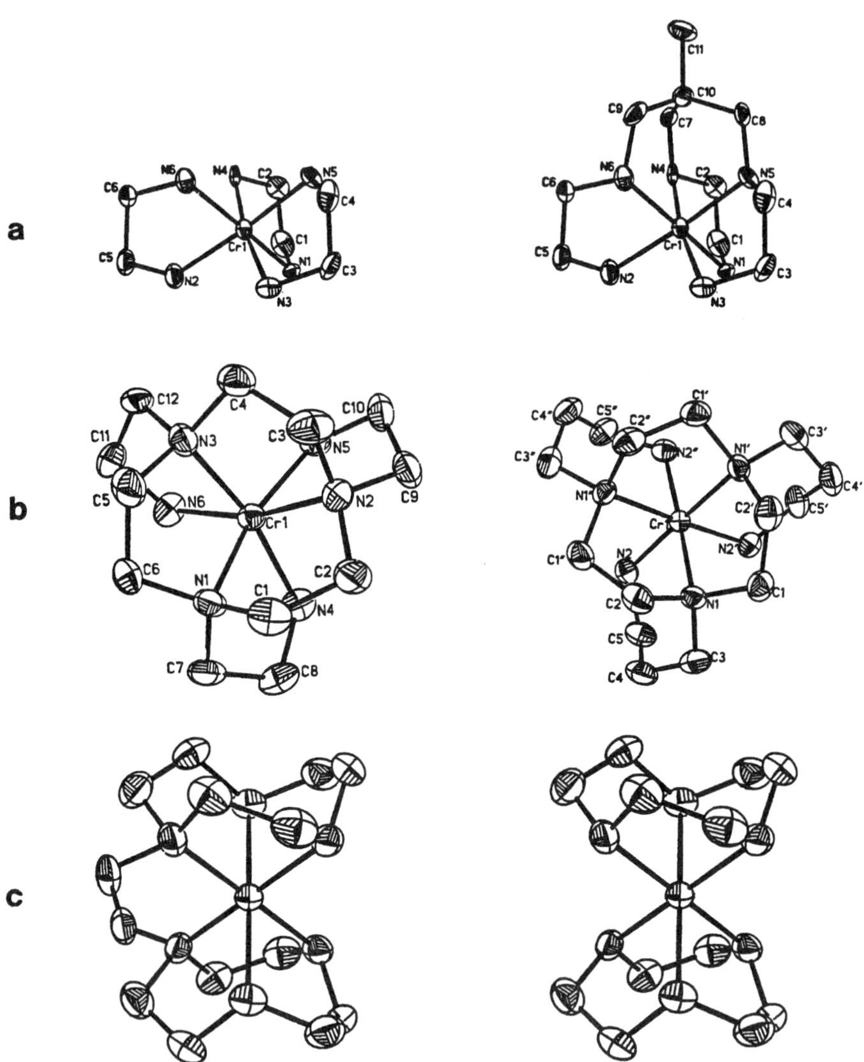

Figure 8. A comparison of structures of pairs of complexes discussed in the text: a, X-ray structures of $Cr(en)_3^{3+}$ (left) and $Cr(sen)^{3+}$ (right); b, X-ray structures of $Cr(TAE[9]aneN_3)^{3+}$ (left) and $Cr(TAP[9]aneN_3)^{3+}$ (right); c, structures of $Cr([9]aneN_3CH_2-)_2^{3+}$ (left) and $Cr([9]aneN_3)_2^{3+}$ (right) (based on the X-ray structures of $Rh([9]aneN_3)_2^{3+}$); an ethylene linkage was added to this structure for the $([9]aneN_3CH_2-)_2$ complex.

Table I. Photophysical Properties of Some Hexaam(m)inechromium(III) Complexes

	$E({}^4T_2)^a_{max}$ $(cm^{-1}/10^3)^a$	$E({}^2E)_o$ $(cm^{-1}/10^3)^b$	$[E({}^4T_2)_o]$ $(cm^{-1}/10^3)^c$	$\tau(298)$ $(\mu s)^d$	$\tau(77)$ $(\mu s)^e$	$k(T)$ $(s^{-1}/10^3)^e$	$k(T)[1-\phi]$ $(s^{-1}/10^3)^f$
$Cr(NH_3)_6^{3+}$	21.64	15.20	4.4	2.2	78	453	243
$Cr(en)_3^{3+}$	21.88	14.96	4.8	1.2	120	833	606
$Cr(sen)_3^{3+}$	22.2	14.81	5.4	$\leq10^{-4}$	171	$\geq10^7$	$\geq0.7\times10^7$
$Cr(TAE[9]aneN_3)^{3+}$	21.41	14.49	3.9	1×10^{-2}	114	1×10^5	0.9×10^5
$Cr(TAP[9]aneN_3)^{3+}$	21.65	14.90	3.7	179 $(850)^g$	265 $(4300)^g$	5.6	5.5
$Cr([9]aneN_3)_2^{3+}$	22.78	14.71	6.2	30.2	400	33	33
$Cr([9]aneN_3CH_2-)_2^{3+}$	20.79	14.0	4.6	$[6.5\times10^{-5}]$	50	$\geq10^7$	$\geq10^7$

[a]Data are from refs. 24, 25, and 33 and references cited therein.

[b]Energy of electronic origin (0–0').

[c]Energy of quartet absorption origin (0–0') estimated as energy where absorption ~5% that of the absorption maxima. This probably underestimates the ${}^4T_2/{}^2E$ energy difference in many cases (2).

[d]In DMSO/H_2O solutions (238 K) or glasses (77 K). Data are from refs. 24, 25, and 33.

[e]$k(T) = [\tau({}^2E)]^{-1}$. Data are from refs. 33–35.

[f]Quantum yields, 0, for photosubstitutions from refs. 33–35.

[g]Lifetimes of perdeuterated complexes in D_2O solutions.

are larger than normal; this effect is appreciably larger for $Cr(TAP[9]aneN_3)^{3+}$ than for $Cr(TAE[9]aneN_3)^{3+}$. In both of these complexes, there is some distortion of the CrN_6 microsymmetry. The unusually intense $^4A_2 \rightarrow {}^4T_2$ transitions of these complexes may reflect these distortions, but it seems likely that misorientation (*36, 37*) of the nitrogen lone pair also contributes to this intensity. However, the two complexes have $^4A_2 \rightarrow {}^4T_2$ absorptions that are very similar in energy and intensity, and the estimated differences in energy between $E(^4T_2)_0$ and $E(^2E)$ are also reasonably similar. Yet there is a dramatic contrast in the ambient lifetimes of their ^2E excited states, with the $Cr(TAE[9]aneN_3)^{3+}$ complex being shorter lived by a factor of about 10^4. The $Cr(TAP[9]aneN_3)^{3+}$ complex is particularly remarkable in that $k(T)$ is so small that there is still a very large isotope effect ($k_H/k_D = 4.7$) on perdeuteration of its amines even in solution (D_2O) at 300 K. This is unique among the hexaam(m)ine complexes. These complexes differ structurally only in the length of the pendant amine arms. Modeling of the $TAE[9]aneN_3$ complex using an MM2 program (*38*) has shown (*34, 35*) that the pendant ethylamine linkage is appreciably strained in octahedral coordination because the chain length is not quite adequate to span the opposing trigonal (antiprismatic) faces and that much of this stereochemical stress is reduced in a large-amplitude trigonal twist. In contrast, most of the coordination sphere strain in $Cr(TAP[ane]N_3)^{3+}$ arises from packing so many atoms into the coordination sphere, and the stereochemical repulsions are increased by a large-amplitude trigonal twist. We conclude that the unusually long lifetime of $Cr(TAP[ane]N_3)^{3+}$ results because stereochemical repulsions make an important relaxation pathway nearly inaccessible.

$Cr[9]aneN_3)_2^{3+}$ and $Cr([9]aneN_3CH_2-)_2^{3+}$.   In contrast to the two pairs of complexes just considered, the complexes of this pair are electronically dissimilar: (1) the $^4A_2 \rightarrow {}^4T_2$ absorption maximum is 2×10^3 cm^{-1} lower in energy and three times as intense in $Cr([9]aneN_3CH_2-)_2^{3+}$ as in $Cr([9]aneN_3)_2^{3+}$, and (2) the zero–zero energy of the ^2E excited state, $E_{00}(^2E)$, is about 600 cm^{-1} lower in $Cr([9]aneN_3CH_2-)_2^{3+}$. Although the absorption and emission energies of the $Cr([9]aneN_3CH_2-)_2^{3+}$ complex are at the lower end of the ranges commonly observed (*2, 10, 11, 17, 21*) for CrN_6 chromophores, the $^4T_2 - {}^2E$ energy difference appears to be reasonably typical of CrN_6 complexes (Table I). However, there is a 10^6-fold difference in the ambient lifetimes of these complexes. The emission spectra of both complexes are matrix-sensitive (*see* Figures 6 and 7 and the preceding discussion); however, the emission behavior of $Cr([9]aneN_3CH_2-)_2^{3+}$ is unique for CrN_6 species (*20*): (1) the vibrational fine structure is difficult to detect, (2) there are unusually large shifts in the emission maximum with the nature of the surrounding matrix, (Figure 6), and (3) when doped into the matrix of an isostructural Rh(III) complex, the emission spectra are excitation energy-dependent (*2*).

The emission spectra resulting from excitation of $Cr([9]aneN_3CH_2-)_2^{3+}$ doped into various salts of $Rh([9]aneN_3CH_2-)_2^{3+}$ were only slightly more structured in these salts than in DMSO/H$_2$O glasses. However, the emission spectra in the doped solids always included a broad band component of higher energy ($\lambda_{max} \cong 708$ nm) than the maximum (~720 nm) or the apparent emission threshold in DMSO/H$_2$O glasses (Figure 6). Qualitatively, the peculiar emission behavior of Cr([9]aneN$_3$CH$_2$–)$_2^{3+}$ suggests that the ^{2}E excited state is fluxional; that is, that the PE surface for this excited state has only very shallow local minima, some of which are determined by the crystal or solvent environment and others by stereochemical requirements of the ligand and metal. It is appealing to regard the emission spectrum that was obtained in DMSO/H$_2$O as characteristic of the relaxed state because the local heating of the solvent, which accompanies intersystem crossing and vibrational relaxation from the initial Franck–Condon excited state, should result in some adjustment of the solvent environment to accommodate stereochemical changes in the first coordination sphere. Then the higher energy emission spectrum of Cr([9]ane-N$_3$CH$_2$–)$_2^{3+}$ doped into [Rh([9]aneN$_3$CH$_2$–)$_2$](ClO$_4$)$_3$ could be interpreted as characteristic of the (^{2}E)Cr[9]aneN$_3$CH$_2$–)$_2^{3+}$ excited state trapped in a geometry dictated by the crystal matrix; the apparent emission origin observed in this matrix is typical of the CrN$_6$ chromophores. This interpretation would imply that the Rh([9}aneN$_3$CH$_2$–)$_2$](PF$_6$)$_3$ matrix is a little less restrictive and that these two geometries can be somewhat preferentially accessed by rapid intersystem crossing from different vibrational or electronic components of the ^{4}T$_2$ state (weakly split into three electronic components as a consequence of the approximately C$_2$ microsymmetry).

The emission behavior of (^{2}E)Cr([9]aneN$_3$CH$_2$–)$_2^{3+}$ is certainly what one would expect of a stereochemically flexible excited state, and it contrasts dramatically with the spectra of the stereochemically rigid (^{2}E)Cr(TAP[9]ane-N$_3$)$^{3+}$.

Possibility of a Thermally Activated (^{2}E)Cr(III) Quenching Channel Accessed by a Large-Amplitude Trigonal Twist.   *Correlations with Stereochemical Constraints.*

The information already presented strongly implicates a thermally activated (^{2}E) Cr(III) quenching channel for which the critical nuclear coordinate is a trigonal twist. Thus, we have demonstrated that coordination of Cr(III) to ligands that are compressed along a trigonal twisting coordinate can greatly increase $k(T)$ relative to a complex with nearly the same electronic structure. Conversely, the coordination of Cr(III) to ligands in which stereochemical repulsions inhibit a large-amplitude trigonal twist can effectively quench $k(T)$. This argument can be made more systematically by using MM2 calculations (38) to model the contribution of changes in the ligand's stereochemical repulsions for a large-amplitude twist. To do this we have compared the steric energies (E_{steric}) of the ligands when twisted 15° around the C$_3$ axis (i.e., half-way between the O$_h$ and D$_{3h}$ stereochemistries and holding the Cr–N distance constant) to that of the complexes in their ground states. On the basis of these calculations the ligands can be grouped into three categories: (1) ligands for which E_{steric} is smaller in the twisted complex than in the ground state ($\Delta E_{\text{steric}} < 0$ for TAE[9]aneN$_3$ ([9]aneN$_3$CH$_2$–)$_2$ and sen); (2) ligands for which $\Delta E_{\text{steric}} \sim 0$ [(NH$_3$)$_6$, (en)$_3$, ([9]aneN$_3$)$_2$]; and (3) one ligand, TAP[9]aneN$_3$, for which ΔE_{steric} is significantly greater than zero. The results are summarized in Figure 9.

Factors That Might Make a Trigonal Twist an Effective ^2E Excited State Quenching Channel.   The relaxation of the Laporte selection rule in D$_{3h}$ or D$_3$ symmetry could be a factor, as mentioned already. Simple angular overlap arguments (*11*) can be used to estimate approximately the state energies in the D$_{3h}$ limit (Figure 10). None of the lower energy electronic states in D$_{3h}$ symmetry will be nested, the electronic states are more closely spaced, and normal

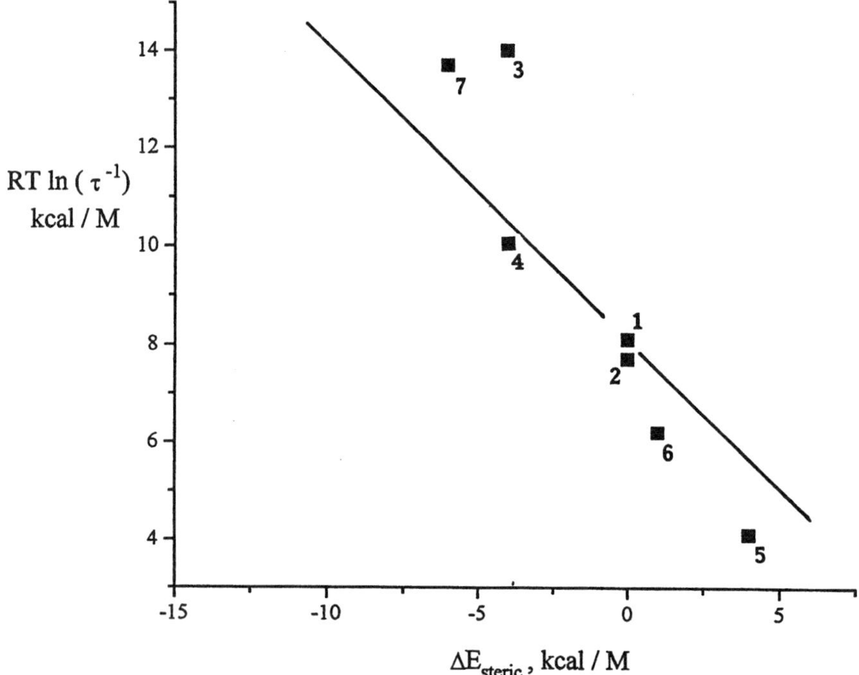

Figure 9. *Correlation of* k(T) *with differences in ligand steric energy between the ground-state geometry and the geometry with a 15° trigonal twist and mean ground-state bond lengths. All* k(T) *values are from Table I. Points are numbered 1–7 from the top of Table I. The line is drawn with a unit slope. Ligand steric energies were based on MM2 calculations (38) of the energy changes within the ligand as the CrN$_6$ geometry was changed incrementally from that of the ground state (based on X-ray crystal structures where available) toward a trigonal prismatic geometry; M–N bond lengths were kept constant (2.06 Å) during changes of geometry (26, 27, 34, 35).*

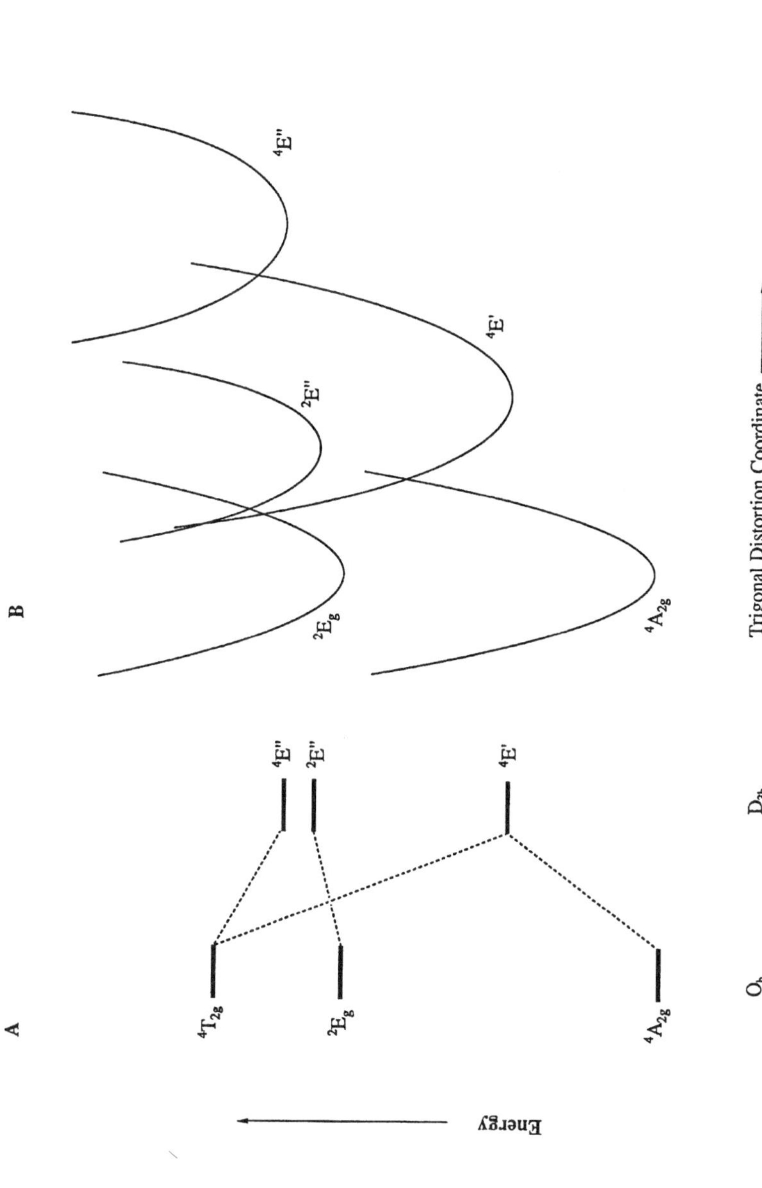

Figure 10. *Approximate energy level diagrams (A) and potential energy surfaces (B) for Cr^{III}(N$_6$) in O$_h$ and D$_{3h}$ coordination geometries. (Reproduced from reference 27. Copyright 1991 American Chemical Society.)*

$(|\Delta G_i^\circ| < \lambda_{\text{reorg},i})$ surface crossing pathways will predominate. In the sense of this idealized, limiting model, a major advantage of this quenching pathway is to move the system away from the tunneling–relaxation pathways (or inverted region behavior) characteristic of $^2E(O_h)$, into a manifold of PE surfaces among which normal, and more rapid, surface crossings can occur.

The D_{3h} intermediate geometry discussed in the previous paragraph is useful for gaining some insight into the twist-induced relaxation process, but there may be no need for a trigonal prismatic geometry to be formed during the relaxation process: a large-amplitude twist (e.g., of 10–15°) may induce sufficient mixing of the electronic wave functions of the excited and ground states to result in relaxation.

Summary and Prospects

We have shown that simple stereochemical features of the coordinated ligands can be used to regulate the ambient excited-state lifetimes of $(^2E)Cr(III)$ complexes. One of the most effective stereochemical features that can be used in this way is the tendency of ligands to twist toward a trigonal prismatic geometry when they are constrained to a sphere the size of Cr^{3+}. The implied excited-state relaxation channel would involve a large-amplitude, thermally activated trigonal twist away from the initially generated antiprismatic geometry of the $(^2E)Cr(III)$ excited state. This relaxation pathway is very different from the back intersystem crossing pathway, which must involve tetragonal stretching modes rather than deformation modes. We have discussed the effects of ligand stereochemistry (especially in Figure 9) on the twisting channel in terms of the effect on λ_{reorg} for this channel. Alternatives, which present observations do not distinguish, are (1) the distortion of the ^2E PE surface along an angular trigonal coordinate (this distortion should be manifested in anharmonicities in the overtones of the twisting coordinate), and (2) stereochemical stabilization of a trigonal prismatic intermediate. The concept of using ligand stereochemistry in an attempt to stabilize an actual antiprismatic intermediate deserves some more systematic exploration. In our one related attempt to do this, we were unable to detect any intermediates following the 20-ps excitation of $Cr([9]aneN_3CH_2-)_2^{3+}$. Detecting such an intermediate would probably require a different kind of strategy [although few Cr systems have actually been examined (*24, 25*) for the existence of chemical intermediates].

This work, using stereochemical constraints to alter the nonradiative relaxation dynamics of $(^2E)Cr(III)$ excited states, supports the hypothesis that the most effective reaction channels for electronically forbidden processes can involve distortion (or nuclear reorganization) along a nuclear coordinate (or coordinates) that is different from the nuclear coordinates that are required for the concerted transformation of the reactant species structures into the product species structures. In some sense, related intermediates have been proposed in photochemical mechanisms of d^6 complexes, in which a trigonal bipyramidal intermediate has been proposed that would tend to be most stable with a triplet spin configuration and lead to stereospecific net chemical reaction (*39, 40*).

In most of the systems discussed here, photochemistry plays a minor role [$Cr(NH_3)_6^{3+}$ is the exception], and corrections of $k(T)$ for the photodecomposition pathway have little effect on the issues involved (*see* Table I). The issue examined here pertains to the dominant relaxation channel for nonreactive, nonradiative, thermally activated relaxation of (^{2}E)Cr(III). Much more closely related to our work are studies of d^6 Fe(II) complexes in which a low-energy deformation mode has been implicated in the quintet-excited-state–singlet-ground-state relaxation process (*9*). Our approach, treating $k(T)$ as the rate constant for a normal chemical reaction process, ought to be applicable to other systems, but it does require careful attention to the interrelation between complex stereochemistry and electronic structure. The application to bimolecular reactions is even more complicated, but the observations discussed here tend to support the previous hypothesis (*5*) that some contributions to λ_{reorg} for electronically retarded bimolecular reactions might originate from distortions required to reduce the electronic constraints.

Acknowledgments

We gratefully acknowledge partial support of this work by the Division of Chemical Sciences, Office of Basic Energy Sciences, Office of Energy Research, U.S. Department of Energy. We also thank Professor Karl Weighardt for providing us, many years ago, with the original samples of $Cr([9]aneN_3)^{3+}$ and $Cr([9]aneN_3CH_2-)_2^{3+}$ that helped stimulate the work discussed in this chapter.

References

1.   Hollebone, B. R.; Langford, C. H.; Serpone, N. *Coord. Chem. Rev.* **1981**, *39*, 181.
2.   Endicott, J. F.; Ramasami, T.; Tamilarasan, R.; Lessard, R. B.; Ryu, C. K.; Brubaker, G. R. *Coord. Chem. Rev.* **1987**, *77*, 1.
3.   Steinfeld, J. I.; Francisco, J. S.; Hase, W. L. *Chemical Kinetics and Dynamics*; Prentice Hall: Englewood Cliffs, NJ, 1989.
4.   Wilkins, R. G. *Kinetics and Mechanisms of Reactions of Transition Metal Complexes*, 2nd Ed.; VCH: New York, 1991.
5.   Taube, H. *Electron Transfer Reactions of Complex Ions in Solution*; Academic: New York, 1970.
6.   Newton, M. D.; Sutin, N. *Annu. Rev. Phys. Chem.* **1984**, *35*, 437.
7.   Endicott, J. F.; Ramasami, T. *J. Phys. Chem.* **1986**, *90*, 3740.
8.   Taube, H. in *Mechanistic Aspects of Inorganic Reactions*; Rorabacher, D. B.; Endicott, J. F., Eds.; ACS Symposium Series 198; American Chemical Society: Washington, DC, 1982; p 151.
9.   McCusker, J. K.; Toftland, H.; Rheingold, A. L.; Hendrickson, D. N. *J. Am. Chem. Soc.* **1993**, *115*, 1797.
10.   Forster, L. *Chem. Rev.* **1990**, *90*, 331.

11. Lever, A. P .B. *Inorganic Electronic Spectroscopy*, 2nd Ed.; Elsevier: New York, 1984.
12. Wilson, R. B.; Solomon, E. I. *Inorg. Chem.* **1978**, *17*, 1729.
13. Cuelemans, A.; Bonagaerts, N.; Vanquickenborne, L. G. *Inorg. Chem.* **1987**, *26*, 1566.
14. Marcus, R. A. *Annu. Rev. Phys. Chem.* **1964**, *15*, 155.
15. Englman, R.; Jortner, J. *Mol. Phys.* **1970**, *18*, 145.
16. Marcus, R. A.; Sutin, N. *Biochim. Biophys. Acta.* **1985**, *811*, 265.
17. (a) Kuhn, K.; Wasgestian, F.; Kupka, H. *J. Phys. Chem.* **1981**, *85*, 655; (b) Muele, M.; Wasgestian, F. *Inorg. Chim. Acta.* **1986**, *119*, 25.
18. Robbins, D. J.; Thomson, A. J. *Mol. Phys.* **1973**, *25*, 1103.
19. Newton, M. D. *Chem. Rev.* **1992**, *92*, 767.
20. Endicott, J. F.; Lessard, R. B.; Lynch, D.; Perkovic, M. W.; Ryu, C. K. *Coord. Chem. Rev.* **1990**, *97*, 65.
21. Kirk. A. D. *Coord. Chem. Rev.* **1981**, *39*, 225.
22. Lessard, R. B.; Heeg, M. J.; Perkovic, M. W.; Schwarz, C. L.; Rudong, Y.; Endicott, J. F. *Inorg. Chem.* **1992**, *31*, 3091.
23. Kane-Maguire, N. A. P.; Wallace, K. C.; Miller, D. B. *Inorg. Chem.* **1988**, *24*, 597.
24. Vincze, L.; Friesen, D. A.; Mezyk, S. P.; Waltz, W. L. *Inorg. Chem.* **1992**, *31*, 4950.
25. Friesen, D. A.; Lee, S. H.; Lilie, J.; Waltz, W. L.; Vincze, L. *Inorg. Chem.* **1991**, *30*, 1975.
26. Perkovic, M. W.; Endicott, J. F. *J. Phys. Chem.* **1990**, *94*, 1217.
27. Perkovic, M. W.; Heeg, M. J.; Endicott, J. F. *Inorg. Chem.* **1991**, *30*, 3140.
28. Flint, C. D.; Greenough, P.; Matthews, A. P. *J. Chem. Soc. Faraday Trans 2* **1983**, *69*, 23.
29. Flint, C. D.; Greenough, P. *J. Chem. Soc. Faraday Trans 2* **1972**, *68*, 897.
30. Urushiyama, A.; Schonherr, T.; Schmidke, H. H. *Ber. Bunsenges. Phsy. Chem.* **1986.** *90*, 1188.
31. Denning, R. G. In *Vibronic Processes in Inorganic Chemistry;* Flint, C. D., Ed.; Kluwer: Dordrecht, Netherlands, 1989; p 111.
32. Solomon, E. I. *Comments Inorg. Chem.* **1984**, *3*, 300–318.
33. Endicott, J. F.; Tamilarasan, R.; Lessard, R. B. *Chem. Phys. Lett.* **1984**, *112*, 381.
34. Perkovic, M. W. Ph.D. Dissertation, Wayne State University, 1989.
35. Perkovic, M. W.; Heeg, M. J.; Endicott, J. F.; Ryu, C. K.; Thompson, D., unpublished.
36. Fenton, N. D.; Gerloch, M. *Inorg. Chem.* **1989**, *28*, 2975.
37. Brown, C. A.; Gerloch, M.; McMeeking, R. F. *Mol. Phys.* **1988**, *64*, 771.
38. Allinger, N. L.; Yuh, Y. H. *Molecular Mechanics: Operating Instructions for MM2 and MMP2 Programs 1980 Force Field;* Indiana University Chemistry Department: Bloomington, IN; QCPE Program No. 395.
39. Vanquickenborne, L. G.; Ceulemans, A. *Coord. Chem. Rev.* **1983**, *48*, 157.
40. Ford, P. C. *Rev. Chem. Intermed.* **1979**, *2*, 267.

13

Time-Resolved Infrared Studies of Migratory Insertion Mechanisms in Manganese Carbonyls

Peter C. Ford and William T. Boese

Department of Chemistry, University of California, Santa Barbara, CA 93106

This chapter presents an overview of time-resolved spectroscopic studies of intermediates in the migratory insertion of carbon monoxide into the metal–alkyl bond of the prototype compound $CH_3Mn(CO)_5$. Transients were generated by flash photolysis of the acyl complex, $CH_3C(O)Mn(CO)_5$, and it was found that their natures are strongly medium-dependent. Comparison to thermal reaction kinetics led to the conclusion that the transients generated photochemically were the same as key intermediates along the thermal reaction coordinate. In THF, the lowest energy intermediate along the reaction coordinate was shown to be the solvento acyl species, $CH_3C(O)Mn(CO)_4(S)$, whereas in cyclohexane, the chelated acyl complex, $(\eta^2\text{-}CH_3CO)Mn(CO)_4$, dominated. Regardless of medium, the rate of methyl migration to form $CH_3Mn(CO)_5$ from the intermediate generated by photolysis proved to be solvent-dependent, consistent with previous views that methyl migration is solvent-assisted. Ligand reaction kinetics for these intermediates are compared to the dynamics of "unsaturated" species generated by flash photolytic CO dissociation from $CH_3Mn(CO)_5$.

ONE OF THE MOST IMPORTANT REACTIONS of organometallic chemistry is the migratory insertion of carbon monoxide into metal–carbon bonds (eq 1).

$$L'_nM-CO \xrightarrow{\ +L\ } L'_nM-C\underset{O}{\overset{R}{\diagdown}} \tag{1}$$

It is by migratory insertion that C–C bonds can be formed from the C_1 starting material CO; thus this reaction is often invoked in proposed catalysis mechanisms for methanol carbonylation, alkene hydroformylation, and other processes of industrial interest (*1, 2*). Despite considerable mechanistic investigation (*3–18*), the reactive intermediates and roles of the solvent and of the acyl group in defining the properties of such species remain to be fully elucidated. The low steady-state concentrations of reactive intermediates achieved during thermal migratory insertion preclude direct spectroscopic observations, so the properties of such species are inferred from less direct criteria such as the response of overall reaction kinetics to system variables, and stereochemistry.

Photochemical techniques provide a possible entry into the formation of non-steady-state concentrations of reactive intermediates. The strategy uses flash photolysis to generate highly reactive transients with stoichiometries consistent with intermediates proposed for analogous thermal reactions and studies the structures and dynamics of these reactive species with time-resolved spectroscopy (*19–24*). The present chapter will be an overview of time-resolved IR (TRIR) and time-resolved optical (TRO) investigations of the oft-studied carbonylation of the methyl manganese compound **M** (eq 2). For such systems, a major advantage of the TRIR method is that IR bands in the ν_{co} region can be quite specific to different carbonyl complexes. Thus, formation and/or decay of one species might be studied without interference from absorbances characteristic of another undergoing similar processes at the same or different rate.

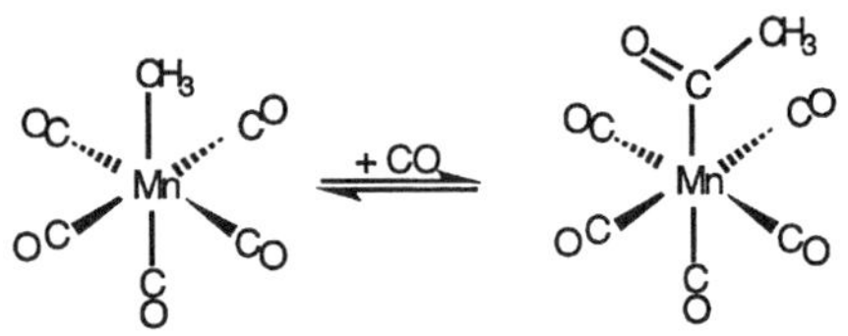

$$\mathbf{M} \; + \text{CO} \; \rightleftharpoons \; \mathbf{A} \tag{2}$$

To provide systematic guidelines for the accumulation and interpretation of time-resolved spectra, we generally use a rubric such as:

1. What do transient spectra indicate about structures of intermediates?

2. What reactions do these intermediates undergo?

3. How do medium and substrate influence the reaction dynamics?

4. Are the intermediates generated photochemically relevant to the thermal mechanism of interest?

Experimental Procedures

Materials. Solvents and gases were purchased and purified as described previously (*21*). The complexes $CH_3Mn(CO)_5$ and $CH_3C(O)Mn(CO)_5$, were synthesized by an adaptation of the method of Gladysz (*25*).

TRIR and TRO Studies.   Sample solutions were prepared under deaerated conditions in a manner analogous to that reported (*21*). The TRIR apparatus has a XeCl excimer laser (308 nm) as the pulsed excitation source, lead salt diode infrared lasers as probe sources, and a Hg–Cd–Te photovoltaic detector (*26*). TRIR experiments were carried out on a flowing samples with the flow rate sufficient to ensure that successive pulses (1 Hz) irradiated fresh solution. Typically, data from 25–50 flashes were averaged.

TRO experiments were also carried out using the XeCl excimer laser as the excitation source (*26*). The optical probe source was a tungsten–halogen filament lamp plus monochromator. The monochromatic probe beam was detected with an IP28 PMT whose signal was amplified and sent to a LeCroy 9400 digital oscilloscope. The detection range was 350–600 nm. The greater sensitivity of the TRO detection system required only 5–10 acquisitions. Data workup was carried out using software custom designed by UGI Scientific.

Results and Discussion

TRIR and FTIR Spectra from Flash Photolysis of A.   In probing this mechanism the operational method was to run the reaction backwards by photoexcitation of the acetyl complex **A**. This method led first to photodissociation of CO to generate an "unsaturated" and vibronically excited species that decayed promptly to a much longer-lived transient, **I**. This transient underwent reaction with CO to regenerate **A**, rearrangement to **M**, or trapping by a ligand L to give a substituted acyl derivative, $\mathbf{A_L}$ (eq 3).

$$\mathbf{A} \underset{k_{co}[CO]}{\overset{h\nu\ (-CO)}{\rightleftharpoons}} \mathbf{I} \quad \overset{k_m}{\nearrow}\ \mathbf{M} \qquad \underset{k_L[L]}{\searrow}\ \mathbf{A_L} \tag{3}$$

Simultaneous with the appearance of **I**, small yields of **M** were also seen to be formed promptly, presumably as one pathway for decay of the vibronically excited species initially formed upon CO photodissociation (Figure 1). As noted already, the TRIR method allowed following these processes individually, although prompt formation of **M** was more clearly observed by using the ^{13}C-labeled starting material $CH_3{}^{13}C(O)Mn(CO)_5$.

Figure 2 is the difference spectrum of a 1.0 mM solution of **A** in cyclohexane under argon observed 100 µs after being subjected to 308-nm flash photolysis from an XeCl excimer laser. In the TRIR experiment IR bands corresponding to **A** [2113 (w), 2051 (m), 2012 (s), 1661(w) cm^{-1}] were bleached within the shortest possible observation time (~150 ns), and a new species (**I**) with IR absorbances at 1991 (s), 1952 (s), and 1607 (w) cm^{-1} was formed promptly. The transient generated displayed little reactivity over a period of several hundred microseconds, although it eventually decayed to a mixture of **M** and **A**. Even under CO (1 atm), **I** proved to be quite long-lived and decayed little over the time window available to the TRIR experiment. Similar transient

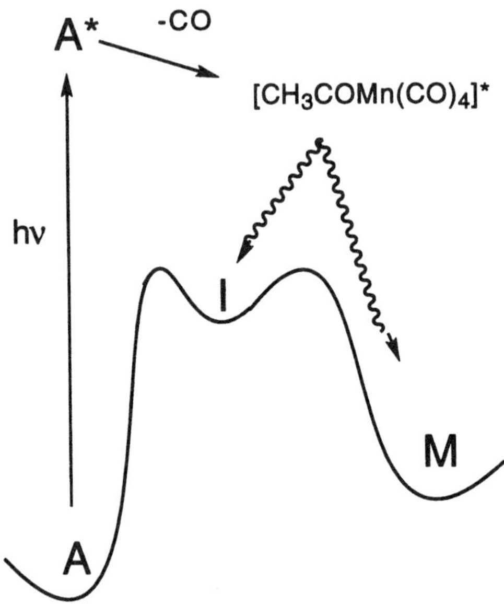

*Figure 1. Illustration of prompt pathways following photoexcitation of **A**.*

spectra were seen in other solvents, and in all cases the intermediates formed proved to be relatively long-lived (hundreds of microseconds). Analogous spectra were observed for experiments carried out in 195 K solutions, where the intermediates generated had indefinitely long lifetimes (Table I).

The positions of the ν_{co} bands observed for the terminal carbonyls of **I** are modestly dependent on the solvent medium (Table I). For example, these appeared at lower frequencies in THF than in cyclohexane. [In weakly coordinating solvents, only *cis*-tetracarbonyl intermediates were seen under the experimental conditions. In THF some (<20%) trans isomer was observed, but this decayed rapidly to the cis analog (*20, 21*). The ensuing discussion will focus on the reactivity of the cis isomer.] The band seen at ~1,600 cm^{-1} suggests that these species continue to have an acyl-type functionality. Four logical structures for an intermediate generated by labilization of a *cis*-carbonyl are shown in Chart I. Of these, the unsaturated species **U** is considered the most unlikely. Earlier flash photolysis studies of other d^6 carbonyls, such as $Cr(CO)_6$, showed the initial pentacarbonyl formed by flash photolytic dissociation of CO has a lifetime of a few picoseconds before relaxing to give much longer-lived solvento complexes (*19*).

The modest sensitivity of the TRIR spectra to the nature of the solvent might suggest the solvento species **S** is indeed the one formed when CO is photodissociated from **A**. The shift of terminal ν_{co} bands to lower frequencies is consistent with the expected pattern for solvento complexes. However, the

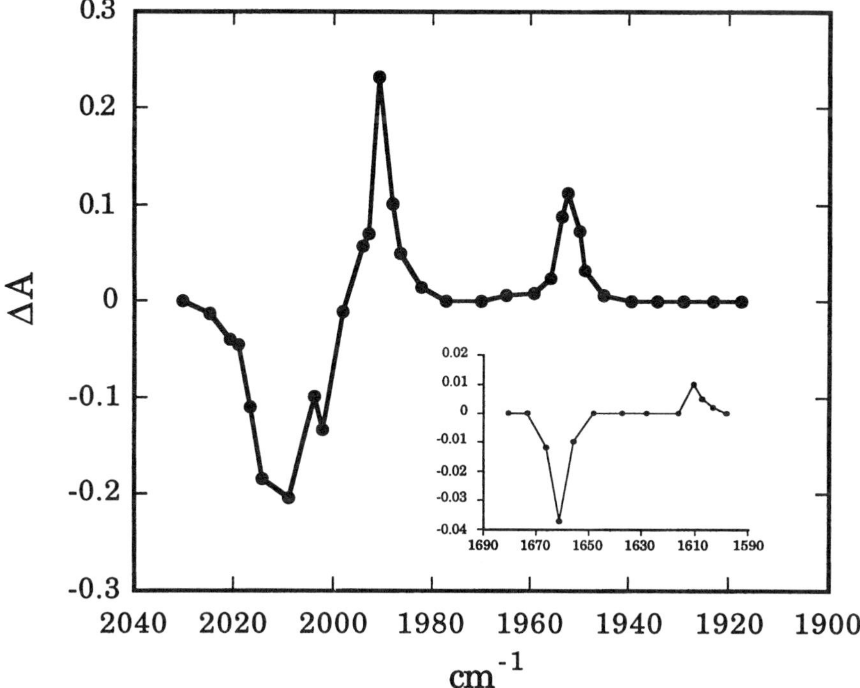

Figure 2. TRIR spectrum 100 μs after 308 nm excitation of 1.0 mM $CH_3C(O)Mn$-$(CO)_5$ in cyclohexane under 1 atm argon. Inset: Transient IR spectrum in the acyl-stretching region acquired using a more concentrated (3.0 mM) solution.

Table I. Carbonyl Bands for Intermediate I Formed by 308-nm Excitation of A in Various Solvents at Ambient T and at 195 K as Measured by TRIR and FTIR

Solvent[a]	TRIR ν_{co} (296 K)[b]	FTIR ν_{co} (195 K)[c]
Perfluoromethylcyclohexane	1997, 1959	2083(w), 1998, 1958
Cyclohexane	1990, 1952, 1607(w)	
Methylcyclohexane	1990, 1952	2080(w), 1988, 1941, 1607(w)
Dichloromethane	1987(br), 1940(br)	
Toluene	1984(br), 1941(br)	
Tetrahydrofuran	1981(br), 1931(br)	2077(w), 1977, 1928, 1602(w)
2,2,5,5-Me₄THF	1984(br), 1945(br)	

NOTE: Carbonyl band (ν_{co}) values are in cm^{-1}; w means weak; br means broad.

[a]All solvents were dried and redistilled before use.

[b]Room T data taken from time-resolved infrared (TRIR) spectra 100 μs after 308-nm flash excitation.

[c]Low T data recorded on a Bio-Rad FTS-60 Fourier transform infrared (FTIR) spectrometer immediately after excitation.

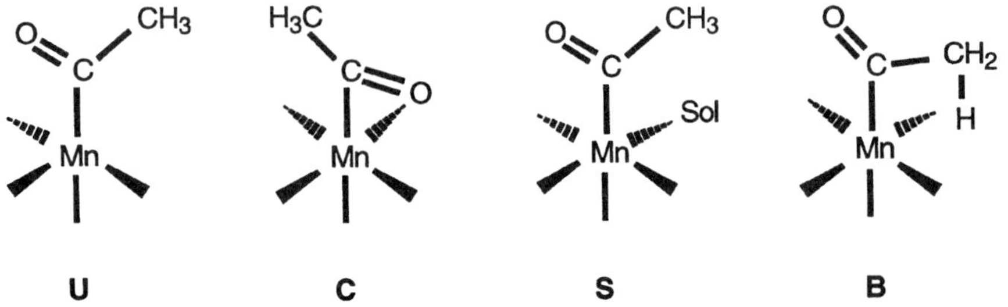

U C S B

Chart I.

sluggish behavior of **I** toward CO in weakly coordinating solvents such as cyclohexane was unexpected given the dramatically higher reactivities observed by us and others for "unsaturated" transients formed by photodissociation of CO from other metal carbonyls (*19, 27–33*). These are largely solvento species, so the relative unreactivity of **I** implies either that there is some unusual electronic character of a Mn–carbon bond (other than Mn–CO) in an **S**-type intermediate or that some other characteristic of the acetyl group imparts added stability. Two logical examples in which the acyl group might give additional stabilization are as the η^2 chelate CO illustrated by structure **C** and the agostically bound methyl illustrated by structure **B**.

However, the possibility of unusual electronic character resulting from the Mn–C bond of **S** could not be ignored; accordingly, it was desirable to probe the reactivity of a suitable model under similar conditions. In this context, the time-resolved spectra and dynamics of transient species formed by the flash photolysis of **M** were examined as the closest analog for **A** not having the acyl function (*21, 24*). That this is reasonable derives from the similarity of the ν_{co} regions of the IR spectra of **M** and **A**, suggesting there are no exceptional electronic differences between the methyl and acyl substituents.

Photoreactions of the Methyl Complex $CH_3Mn(CO)_5$.   Figure 3 shows the TRIR spectra resulting from the flash photolysis of **M** in cyclohexane solution under CO. The notable features are the prompt formation of a transient species **X** that decays exponentially within a few microseconds, that is, much faster than the intermediate **I** under analogous conditions. A plot of k_{obs} vs. [CO] proved to be linear with slope k_{co} of 4.8×10^8 M^{-1} s^{-1} and a zero intercept consistent with the rate law

$$-\frac{d[\mathbf{X}]}{dt} = k_{obs}[\mathbf{X}] = k_{co}\,[\mathrm{CO}][\mathbf{X}] \tag{4}$$

Reformation of **M** occurred at the same rate (although some permanent bleaching was observed owing to competing photochemical cleavage of the Me–Mn

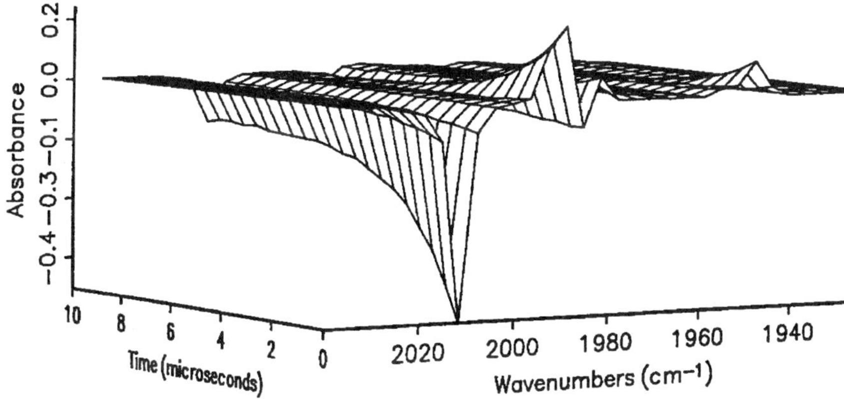

Figure 3. Transient IR spectral changes following 308-nm laser flash photolysis of Mn(CO)₅CH₃ in 295 K cyclohexane under 10% CO (0.001 M). TRIR spectra are shown at 500-ns intervals (adapted from reference 21).

bond). The TRIR spectra of **X** proved to be strongly dependent on the nature of the solvent medium (Table II). The ν_{co} bands shifted to lower frequencies as the solvent became more electron-donating. This pattern is consistent with that shown already for **I** formed by the photolysis of **A** in the various solvents, but the magnitude of the spectral changes were larger for **X**.

The reaction of **X** with CO in cyclohexane (Figure 3) is over in a few microseconds; thus it is orders of magnitude faster than the corresponding reaction of CO with **I**. For **X**, k_{co} ranged from a nearly diffusion-limited value in perfluoromethylcyclohexane (PFMC) solution to a value 8 orders of magnitude smaller in THF (Table II). The kinetics data and TRIR spectra can be interpreted in terms of the photodissociation of CO leading (eventually) to formation of a reactive solvento species, *cis*-CH₃Mn(CO)₄S (**X**), which reacts with CO to regenerate the starting complex (eq 5, L = CO). As noted previously, the kinetics observations parallel those made previously for the "isoelectronic" chromium carbonyl, that is, Cr(CO)₅S (*28–33*).

$$\text{(5)}$$

Kinetics Studies of I: Implications Regarding Structure.   Because of the relatively sluggish reactivity of **I** with CO, the TRO technique, which has a longer time window (0.5 s), was used to study this reaction. These experiments demonstrated the prompt formation of **I** followed by exponential decay

**Table II. TRIR Spectra of the Intermediate X in Selected Solvent
Media, and Rate Constants k_{co} for Reaction with CO**

Solvent	$v_o \ (cm^{-1})$	$k_{co} \ (M^{-1} \ s^{-1})$
C_7F_{14}	2008, 1964	1.0×10^{10}
c-C_6H_{12}	1992, 1986, 1952	4.5×10^8
THF	1974, 1964, 1921	1.4×10^2

NOTE: Intermediate X is $CH_3Mn(CO)_4(S)$.
SOURCE: Data from ref. 21.

over several milliseconds. The k_{obs} values determined proved to be both [CO]-
and medium-dependent. According to eq 3, the disappearance rate of **I** at a
fixed [CO] and [L] = 0 would be the sum of the substitution and methyl migration terms:

$$-\frac{d[\mathbf{I}]}{dt} = k_{obs}[\mathbf{I}] = (k_{co}[CO] + k_m)[\mathbf{I}] \qquad (6)$$

Thus, a plot of k_{obs} vs. [CO] should be linear with slope k_{co} and intercept k_m;
k_{co} is the second-order rate constant for CO substitution, and k_m is the first-
order rate constant for methyl migration. Such relationships were indeed
observed in weakly coordinating solvents (Figure 4), and k_m and k_{co} values
obtained in several hydrocarbon and halocarbon solvents are summarized in
Table III. These determined k_{co} values proved to be relatively insensitive to
solvent, ranging from 3.3×10^3 M^{-1} s^{-1} in benzene to 1.5×10^4 M^{-1} s^{-1} in
PFMC. In contrast, k_m varied by more than a factor of 50, and the weakest
coordinating solvent (PFMC) displayed the smallest value.

In THF a k_m of 8.8 ± 1 s^{-1} was determined in the absence of added CO,
but reaction of **I** with CO was too slow to observe by TRO or TRIR techniques. Under $P_{co} = 1.0$ atm ([CO] ~0.008 M), k_{obs} was marginally higher, but
the increase was not sufficiently outside experimental uncertainty to attribute
confidently to a $k_{co}[CO]$ contribution. From these observations, the upper limit
for k_{co} in THF was calculated to be ~5×10^2 M^{-1} s^{-1}. Other data suggest a k_{co} of
~4×10^2 M^{-1} s^{-1}, in reasonable agreement with this upper limit.

Reactions of **I** with more nucleophilic ligands such as PPh_3, $P(OMe)_3$, and
4-phenylpyridine were much faster and gave cis-$CH_3C(O)Mn(CO)_4(L)$ complexes $\mathbf{A_L}$ in each case. TRIR spectra were constructed by carrying out the
flash experiment at incrementally varied detection frequencies and assembling
a composite of the v_{co} region (Figure 5). Product formation occurred concomi-
tantly with loss of **I** via pseudo-first-order kinetics in excess L (Figure 6). In
accord with eq 7, plots of k_{obs} versus [L] for disappearance of **I** and for appearance of cis-$\mathbf{A_L}$ were linear with intercepts near the origin because k_m was but a
small component under the experimental conditions (excess L). Slopes of these
plots gave the k_L values for substitution by various ligands (Table IV).

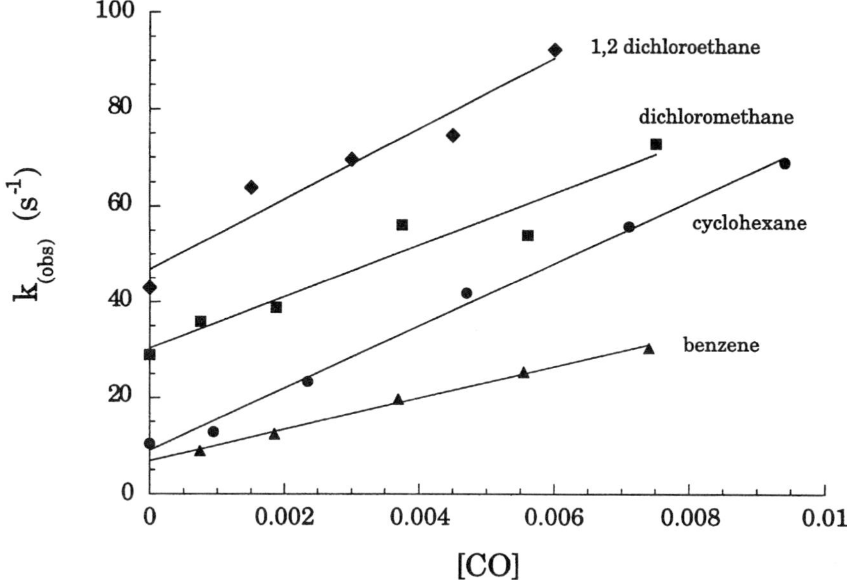

Figure 4. Plots of the observed rate constants for decay of I monitored at 400 nm in various solvents as a function of [CO].

Table III. Rate Constants for CO Addition (k_{co}) and Methyl Migration (k_m) Reactions of I in Various Solvents Determined from Optical Flash Photolysis Experiments

Solvent	k_{co} ($M^{-1}s^{-1}$)	k_m (s^{-1})
Perfluoromethylcylohexane	$(1.5 \pm 0.2) \times 10^4$	<1.0
Benzene	$(3.3 \pm 0.3) \times 10^3$	6.7 ± 0.7
Cyclohexane	$(6.5 \pm 0.7) \times 10^3$	9.0 ± 0.9
Dichloromethane	$(5.3 \pm 0.5) \times 10^3$	30.4 ± 3.0
1,2-Dichloroethane	$(7.3 \pm 0.7) \times 10^3$	46.8 ± 4.7
THF	$<5 \times 10^2$	8.8 ± 1

NOTE: k_{co} and k_m are rate constants in various solvents for CO addition and methyl migration reactions of I, respectively. Conditions were [A] = 5×10^{-4} M, [CO] = $0-1 \times 10^{-2}$ M, 296 K. For k_{co} values, concentrations of CO in various solvents were corrected for differences in solubility.

$$-\frac{d[\mathbf{I}]}{dt} = k_{obs}[\mathbf{I}] = (k_L[L] + k_m)[\mathbf{I}] \qquad (7)$$

Kinetics data for reaction of **I** with $P(OMe)_3$ in THF to give *cis*-$\mathbf{A_L}$ plus **M** were collected using TRO detection. In accord with eq 7, a linear relationship between k_{obs} and [L] was obtained with slope k_L and intercept k_m values (Table IV). In THF, k_L is considerably slower than in cyclohexane. However, the

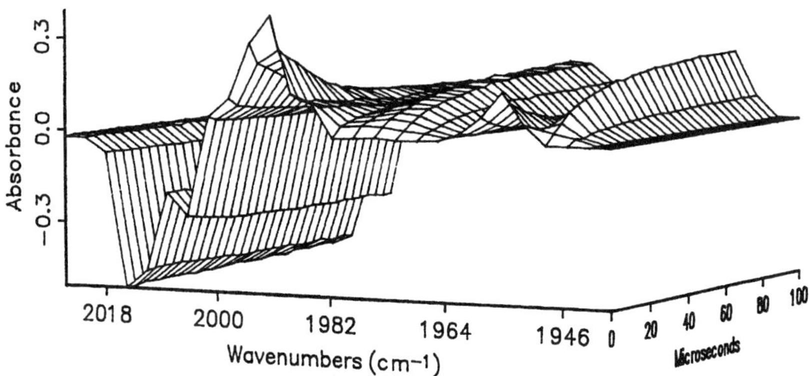

Figure 5. TRIR spectral changes following 308-nm photolysis of **A** *in cyclohexane in the presence of 6.0 mM 4-phenylpyridine. Spectra are spaced at 4-μs intervals.*

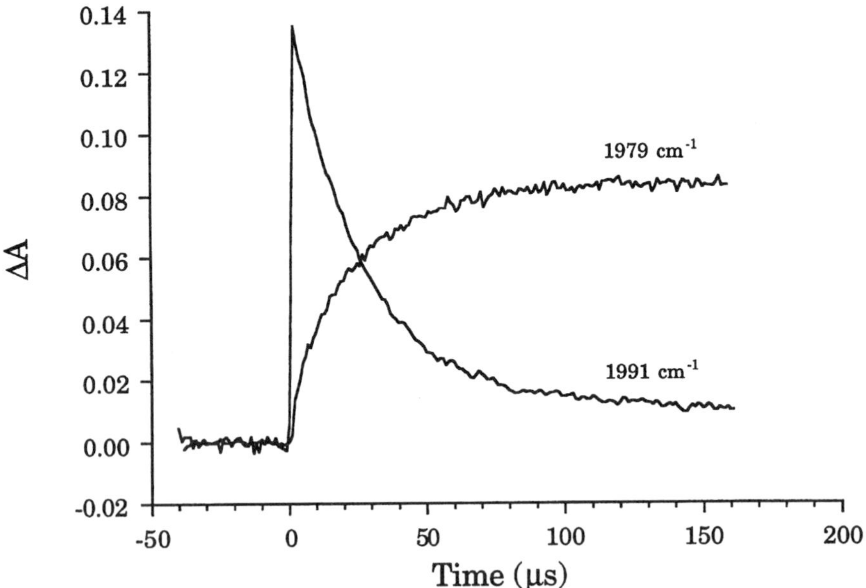

Figure 6. Temporal IR absorbance changes corresponding to the loss of **I** *(1991 cm⁻¹) and formation of* cis-CH₃C(O)Mn(CO)₄(4-Phpy) *(1979 cm⁻¹).*

Table IV. Reactivity of I with Various L in Cyclohexane, PFMC, and THF, and Comparison of Reactivity of I versus that of X under Similar Conditions

		k_L $(M^{-1}\,s^{-1})$	
Solvent	*Ligand*	*I*	*X^b*
PFMC	CO	$(1.5 \pm 0.3) \times 10^4$	$(1.0 \pm 0.5) \times 10^{10}$
Cyclohexane	CO	$(6.5 \pm 1.3) \times 10^3$	$(4.5 \pm 0.5) \times 10^8$
Cyclohexane	Phpy	$(7.5 \pm 1.5) \times 10^6$	$(2.5 \pm 0.3) \times 10^9$
Cyclohexane	P(Ph)$_3$	$(2.3 \pm 0.5) \times 10^6$	$(1.0 \pm 0.2) \times 10^9$
Cyclohexane	P(OMe)$_3$	$(1.4 \pm 0.3) \times 10^6$	$(1.1 \pm 0.2) \times 10^9$
THF	CO	$<5 \times 10^2$	$(1.4 \pm 0.3) \times 10^2$
THF	P(OMe)$_3$	$(1.7 \pm 0.3) \times 10^3$	
2,2,5,5-Me$_4$THF	P(OMe)$_3$	$\sim 6 \times 10^5$	

NOTE: Conditions are as follows: $[A] = 1 \times 10^{-3}$ M; $[Phpy] = 4.0 \times 10^{-3}$ to 1.8×10^{-2} M; $[P(Ph)_3] = 4.0 \times 10^{-3}$ to 1.5×10^{-2} M; $[P(OMe)_3] = 6.0 \times 10^{-3}$ to 1.5×10^{-1} M; $[CO] = 1.0 \times 10^{-3}$ to 1.0×10^{-2} M; room temperature.

SOURCE: k_L values for **X** (CH$_3$Mn(CO)$_4$Sol) are from ref. 21.

behavior was somewhat different in 2,2,5,5-Me$_4$THF solution; k_L for reaction with P(OMe)$_3$ ($\sim 6 \times 10^5$ M^{-1} s^{-1}) proved to be much closer to that seen in cyclohexane than in THF.

We have interpreted the lower reactivity of **I** toward various ligands in THF in terms of the solvento species **S**. **I** reactivity is quite similar to that of **X** in this medium. In contrast, it is orders of magnitude less reactive than **X** toward various L when the solvent is a poor donor such as PFMC or cyclohexane (Table IV). This finding argues for a structure such as the η^2-acyl chelate **C** under these circumstances, although the agostic chelate **B** cannot be excluded off-hand by the current experiments. Nonetheless, the former possibility gains support from ab initio calculations, which demonstrate much greater stability for **C** than for **B** (or the truly unsaturated species **U**) (34–37).

Thermal Reaction of CH$_3$Mn(CO)$_5$ with P(OMe)$_3$ in THF: Comparison with Photochemical Results.   The thermal reaction of **M** with P(OMe)$_3$ was investigated to provide comparative thermal and photochemical data under closely analogous conditions. Earlier kinetics studies of reactions with other L to give the respective acetyl complexes *cis*-A$_L$ in various solvents were interpreted in terms of the model illustrated in Scheme I (6), which gives the relationship described by eq 8. Because the $k_3[L]$ term for direct reaction of L with **M** is relatively small in THF, values of k_1 as well as the ratio k_{-1}/k_2 can be obtained from the slopes and intercepts of double reciprocal k_{obs}^{-1} vs. $[L]^{-1}$ plots.

$$k_{obs} = \frac{k_1 k_2 [L]}{k_{-1} + k_2 [L]} + k_3 [L] \qquad (8)$$

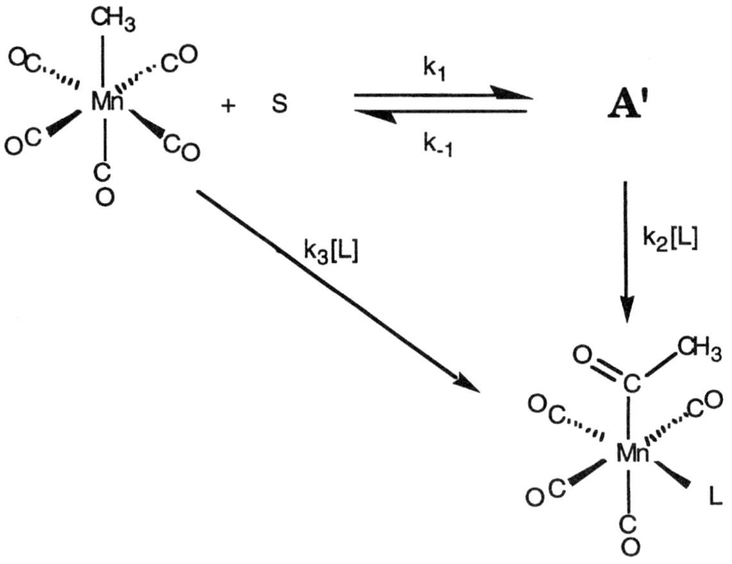

*Scheme I. Possible thermal reactions of **M** with L*

In the present studies, reaction of **M** (1.1×10^{-3} M) with $P(OMe)_3$ (1.2×10^{-2} to 2.3×10^{-1} M) in 25 °C THF to give *cis*-$CH_3C(O)Mn(CO)_4(P(OMe)_3)$ followed pseudo-first-order kinetics. From the linear k_{obs}^{-1} vs. $[L]^{-1}$ plot were determined $k_1 = (8.5 \pm 0.9) \times 10^{-4}$ s^{-1} and $k_{-1}/k_2 = (6.6 \pm 1.3) \times 10^{-3}$ M. This k_1 value is in good agreement with the average k_1 value (9.6×10^{-4} s^{-1}) previously obtained for reactions with other L (6).

If the transient **I** identified photochemically is indeed the same as the intermediate **A'** argued to be involved in the thermal carbonylation of **M** (Scheme I), these two species must have identical reactivities regardless of how they were individually generated. If **I** and **A'** are the same, then k_m is k_{-1} and k_L is k_2; therefore, k_m/k_L and k_{-1}/k_2 should be equal. As noted already, the ratio, $k_{-1}/k_2 = (6.6 \pm 1.3) \times 10^{-3}$ M, was determined from the slope/intercept ratio of the linear k_{obs}^{-1} vs. $[L]^{-1}$ plot for the thermal reaction of **M** plus $P(OMe)_3$ in THF. The photochemical experiment allows one to determine k_m and k_L independently. From these data, the k_m/k_L ratio is $(5.5 \pm 1.5) \times 10^{-3}$ M, which within experimental uncertainty is the same as k_{-1}/k_2. Thus, the assertion that the photochemically generated intermediates are relevant to thermally induced migratory insertion is supported.

Comments on the Role of Solvent in the Methyl Migration Pathway.　When one examines the kinetics of the pathways leading to the decay of **I,** it is notable that methyl migration is much more sensitive to solvent nature than is trapping by CO. This condition holds even when the solvent is a

Free energy profile for k_m pathway from I

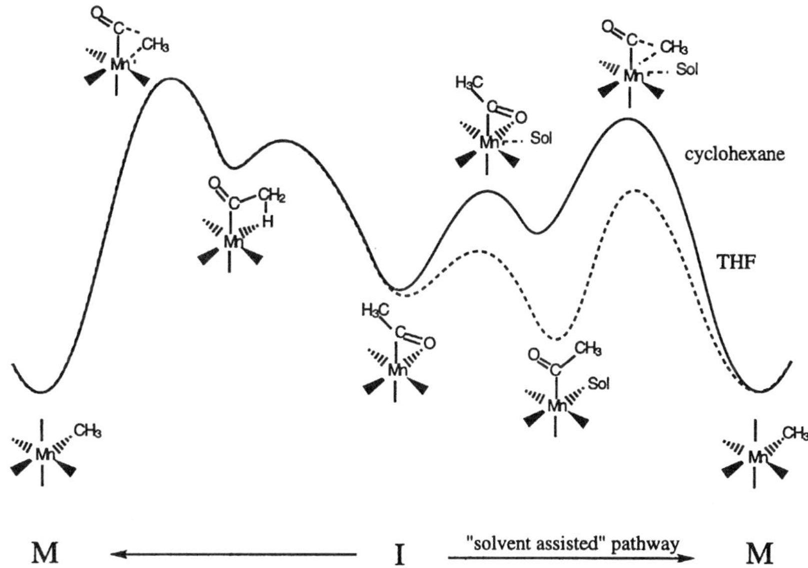

Figure 7. Reaction coordinate diagram for methyl migration via solvent-assisted and unassisted pathways.

relatively weak donor, that is, when **I** is the η^2-acyl chelate **C**. What are the possible explanations? One argument might be that the transition state of the k_m pathway is more polar than **I** and thus is stabilized in a higher-dielectric-constant medium. However, this view would not explain the marked difference between k_m values in cyclohexane and in PFMC, so more direct solvent involvement appears likely.

Figure 7 illustrates methyl migration in the context of a free-energy reaction coordinate diagram. The solid curve describes the case for a weakly coordinating solvent in which the η^2-acyl chelate **C** is the most stable form of **I**. Unassisted rearrangement of **C** (perhaps via **U** and **B**) must be a less favorable pathway to **M** than is reaction of **C** with solvent to give first the solvento species **S** followed by methyl migration to give **M** via a transition state having significant metal–solvent interactions. Solvent assistance of k_m would result from such interactions being stronger in the transition state than in **C**. Thus, increasing solvent coordinating ability lowers the free energy of activation ($\Delta G^{\ddagger}$) and enhances k_m.

The situation changes in a medium such as THF, in which **S** is the lowest energy form of **I**. Certainly the reaction of **M** via a solvent-assisted pathway to give **I** (i.e., the k_1 path of Scheme I) is even more favorable owing to stabilization of the methyl migration transition state by the stronger donor THF. However, the k_m step (i.e., the microscopic reverse) is not similarly accelerated; indeed, under otherwise comparable conditions, k_m is essentially identical in THF and cyclohexane. This observation is easily explained in terms of the solvent-induced change in the nature of **I**, from the chelated structure **C** in cyclohexane (and other relatively weak donors) to the solvento complex **S** in THF. In the latter case, the free energies of both **S** and the transition state are strongly affected by the solvent, with **S** likely being the more stabilized (Figure 7). Because the $\Delta G^{\ddagger}$ is the difference between the two, the similar k_m values in these two media would appear to be fortuitous.

A Free-Energy Diagram for Migratory Insertion with Added P(OMe₃)₃. The thermal and photochemical kinetics data described previously can be combined to generate the free-energy surfaces depicted in Figure 8 for reaction of **M** with 0.1 M P(OMe)₃ in THF or cyclohexane. For comparison, the free energy of **M** was arbitrarily set at zero as the reference point, and solvent-dependent solvation energy differences between **M** and **A** were ignored. The reactions were proposed to proceed from **M** to **S** to **C** to the substituted acyl complex $\mathbf{A_L}$ in cyclohexane and from **M** to **S** to $\mathbf{A_L}$ in THF. The overall reaction to form $\mathbf{A_L}$ was estimated to have $\Delta G_{rxn} \sim -0.5$ kcal mol⁻¹; however, exact position of the equilibrium is not crucial to discussion of intermediates.

For the reaction coordinate in THF, $\Delta G^{\ddagger}_{1,THF} = 21.6$ kcal mol⁻¹ was derived directly from the thermal kinetics ($k_{1,THF} = 8.5 \times 10^{-4}$ s⁻¹). If **I** (i.e., $\mathbf{S_{THF}}$ observed photochemically) and **A'** from the thermal experiment are indeed the same, the k_1 and k_m steps must have the same transition state. Thus, $\Delta G(\mathbf{S_{THF}}) = \Delta G^{\ddagger}_{1,THF} - \Delta G^{\ddagger}_{m,THF}$ and is +5.5 kcal mol⁻¹ relative to **M**. The $\Delta G^{\ddagger}_{1,THF}$ (14.4 kcal mol⁻¹) for reaction of $\mathbf{S_{THF}}$ with L was derived from the k_{obs} in the presence of 0.1 M P(OMe)₃ ($k_L[L] = 1.7 \times 10^2$ s⁻¹).

For cyclohexane, the *lower limit* for $\Delta G^{\ddagger}_{1,CH}$ was derived from the estimated free energy difference between **C** and **M** and the $\Delta G^{\ddagger}_{m,CH}$ measured by flash photolysis. The equilibrium constant between **M** and **C** was estimated from the k_1 (2.2×10^{-6} s⁻¹) determined earlier in mesitylene (*13*) and k_m (6.7 s⁻¹) in benzene solution (Table III), $K_{est} = k_1/k_m = 3.3 \times 10^{-7}$, because **M** and **C** should be solvated similarly in aromatic and aliphatic hydrocarbons. The estimate for $\Delta G(\mathbf{C_{CH}})$ is therefore +8.9 kcal mol⁻¹. In cyclohexane, $\Delta G^{\ddagger}_{m,CH} = 16.1$ kcal mol⁻¹; accordingly, the barrier for the reverse reaction would be $\Delta G^{\ddagger}_{1,CH} = \Delta G(\mathbf{C_{CH}}) + \Delta G^{\ddagger}_{m,CH} = 25.0$ kcal mol⁻¹. This value corresponds to a hypothetical $k_{1,CH}$ value in cyclohexane of 1.2×10^{-6} s⁻¹, about half that determined for mesitylene. The ΔG of the $\mathbf{S_{CH}}$ (not observed) should be above that of $\mathbf{S_{THF}}$ by the differences in Mn–solvent bond strengths (assuming entropic differences are small). The estimated difference is ~12 kcal mol⁻¹ (38), and thus $\mathbf{S_{CH}}$ would be ~7.6 kcal mol⁻¹ above $\mathbf{C_{CH}}$.

The estimate of $\Delta G^{\ddagger}_{1,CH}$ is about 4 kcal mol⁻¹ higher than $\Delta G^{\ddagger}_{1,THF}$. Thus the effect of solvent on the barrier for methyl migration (**M** to **I**) is substantial but is significantly attenuated from energy differences estimated for the full

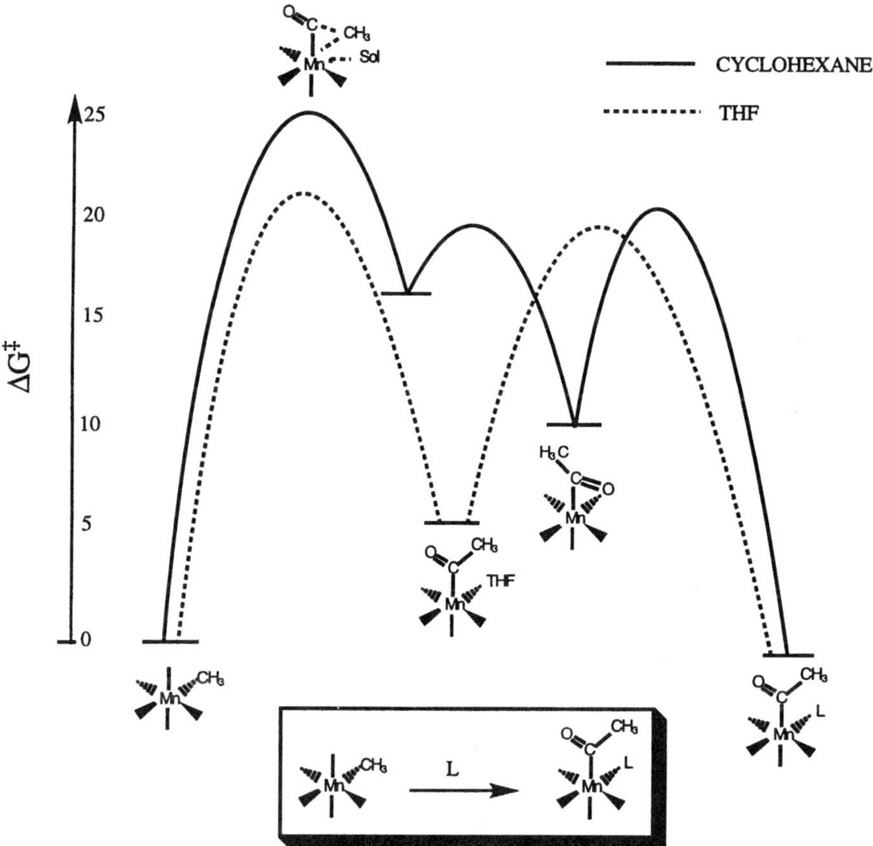

Figure 8. Free energy diagrams for reaction of $Mn(CO)_5CH_3$ with $P(OMe)_3$ (0.1 M) to give cis-$CH_3C(O)Mn(CO)_4(P(OMe)_3)$ in cyclohexane and in THF. $\Delta G^{\ddagger}$ values are calculated from the appropriate rate constants at 25 °C. ΔG_{rxn} is approximated as –0.5 kcal mol⁻¹ on the basis of an equilibrium constant (23.8 M⁻¹) for reaction of M with N-methylcyclohexylamine in methanol. Symbols: L = $P(OMe)_3$, S_{THF} = $CH_3(O)Mn(CO)_4(THF)$, S_{CH} = $CH_3(O)Mn(CO)_4(c$-$C_6H_{12})$ plus those defined in the text. $\Delta G^{\ddagger}_{1,THF}$ is from k_1 for thermal carbonylation of M in THF with $P(OMe)_3$, $\Delta G^{\ddagger}_{M,THF}$ is from k_m for methyl migration of I in THF, $\Delta G^{\ddagger}_{L,THF}$ is from k_{obs} for trapping of I with 0.1 M $P(OMe)_3$ in THF, $\Delta G^{\ddagger}_{M,CH}$ is from k_m for methyl migration of I in cyclohexane, and $\Delta G^{\ddagger}_{L,CH}$ is from k_{obs} for trapping of I with 0.1 M $P(OMe)_3$ in cyclohexane.

Mn–solvent bonds. In contrast, the reverse reactions (k_m steps) show little difference between rates in cyclohexane and in THF, because **C** is the dominant form of the intermediate in cyclohexane, and **S** is the dominant form of the intermediate in THF.

Summary

The spectroscopic and kinetic results obtained from TRIR and TRO experiments demonstrate the η^2-acyl chelate **C** to be the dominant form of the "unsaturated" intermediate **I** formed by photodissociation of CO from **A** in weakly coordinating solvents, but the solvento complex **S** to be the predominant form in THF.

Quantitative comparison of the photochemical and thermal reactivity patterns substantiates the view that the photochemically generated intermediates are indeed relevant to the thermal reaction mechanisms.

The methyl migration kinetics indicates solvent assistance in the rearrangement of **I** to **M**, even for those media where the lower energy form of **I** is the η^2-acyl chelate **C**. This finding validates earlier arguments that solvent assistance plays a key role in the thermal carbonylation of $CH_3Mn(CO)_5$.

Acknowledgments

This research was sponsored by a grant (DE–FG03–85 ER13317) to P. C. Ford from the Division of Chemical Sciences, Office of Basic Energy Sciences, U.S. Department of Energy. Some laser flash photolysis experiments were carried out on a time-resolved optical system constructed with support from U.S. Department of Energy University Research Instrumentation Grant DE–FG05–91 ER79039.

Brian Lee of the Department of Chemistry, University of California, Santa Barbara, contributed significantly to the success of these studies.

This chapter is dedicated to Henry Taube: mentor, role model, and friend

References

1. Collman, J. P.; Hegedus, L. S.; Norton, J. R.; Finke, R. G. *Principles and Applications of Organotransition Metal Chemistry*; University Science Books: Mill Valley, CA, 1987.
2. Parshall, G. W.; Ittle, S. D. *Homogeneous Catalysis*, 2nd Ed.; Wiley Interscience: New York, 1992.
3. Calderazzo, F. *Angew. Chem. Int. Ed. Engl.* **1977**, *16*, 299–311.
4. Wojcicki, A. *Adv. Organomet. Chem.* **1973**, *11*, 87–145.
5. Flood, T. C. *Top. Stereochem.* **1981**, *12*, 37–118.
6. Mawby, R. J.; Basolo, F.; Pearson, R. G. *J. Am. Chem. Soc.* **1964**, *86*, 3994–3999.
7. Calderazzo, F.; Cotton, F. A. *Inorg. Chem.* **1962**, *1*, 30–36.
8. Noack, K.; Calderazzo, F. *J. Organomet. Chem.* **1967**, *10*, 101–104.
9. Noack, K.; Ruch, M.; Calderazzo, F. *Inorg. Chem.* **1968**, *7*, 345–349.

10. Butler, I. S.; Basolo, F.; Pearson, R. G. *Inorg. Chem.* **1967**, *11*, 2074–2079.
11. Green, M.; Westlake, D. J. *J. Chem. Soc. A* **1971**, 367–371.
12. Nicholas, K.; Raghu, S.; Rosenblum, M. *J. Organomet. Chem.* **1974**, *78*, 133–137.
13. Flood, T. C.; Jensen, J. E.; Statler, J. A. *J. Am. Chem. Soc.* **1981**, *103*, 4410–4414.
14. Flood, T. C.; Campbell. K. C. *J. Am. Chem. Soc.* **1984**, *106*, 2853–2860.
15. Cotton, J. D.; Bent, T. L. *Organometallics* **1991**, *10*, 3156–3160.
16. Wax, M. J.; Bergman, R. G. *J. Am. Chem. Soc.* **1981**, *103*, 7028–7030.
17. Cawse, J. N.; Fiato, R. A.; Pruett, R. L. *J. Organomet. Chem.* **1979**, *172*, 405–413.
18. Webb, S.; Giandomenico, C.; Halpern, J. *J. Am. Chem. Soc.* **1986**, *108*, 345–347.
19. Ford, P. C.; Boese, W.; Lee, B.; MacFarlane, K. In *Photosensitization and Photocatalysis by Inorganic and Organometallic Compounds;* Graetzel, M.; Kalyanasundaram, K., Eds.; Kluwer Academy: Netherlands, 1993; pp 359–390.
20. Boese, W. T.; Ford, P. C. *J. Am. Chem. Soc.* **1995**, *117*, 8381–8391.
21. Boese, W. T.; Ford, P. C. *Organometallics* **1994**, *13*, 3525–3531.
22. Boese, W. T.; Lee, B. L.; Ryba, D. W.; Belt, S. T.; Ford, P. C. *Organometallics* **1993**, *12*, 4739–4741.
23. Belt, S. T.; Ryba, D. W.; Ford, P. C. *J. Am. Chem. Soc.* **1991**, *113*, 9524–9528.
24. Belt, S. T.; Ryba, D. W.; Ford, P. C. *Inorg. Chem.* **1990**, *29*, 3633–3634.
25. Gladysz, J. A.; Williams, G. M.; Tam, W.; Johnson, D. L.; Parker, D. W.; Selover, J. C. *Inorg. Chem.* **1979**, *18*, 553–557.
26. Ford, P. C.; DiBenedetto, J. A.; Ryba, D. W.; Belt, S. T. *SPIE Proc.* **1992**, *1636*, 9–16.
27. DiBenedetto, J. A.; Ryba, D. W.; Ford, P. C. *Inorg. Chem.* **1989**, *28*, 3503–3507.
28. Kelly, J. M.; Bent, D. V.; Hermann, H.; Schulte-Frohlinde, D.; Koerner von Gustorf, E. *J. Organomet. Chem.* **1974**, *69*, 259–269.
29. Bonneau, R.; Kelly, J. M. *J. Am. Chem. Soc.* **1980**, *102*, 1220–1221.
30. Kelly, J. M.; Long, C.; Bonneau, R. *J. Phys. Chem.* **1983**, *87*, 3344–3349.
31. Church, S. P.; Grevels, F. W.; Hermann, H; Schaffner, K. *Inorg. Chem.* **1984**, *23*, 3830–3833.
32. Church, S. P.; Grevels, F. W.; Hermann, H; Schaffner, K. *Inorg. Chem.* **1985**, *24*, 418–422.
33. Horton-Mastin, A.; Poliakoff, M.; Turner, J. *Organometallics* **1986**, *5*, 405–408.
34. Marynick, D. S., personal communication to P. C. Ford, 1994.
35. Axe, F. U.; Marynick, D. S. *Organometallics* **1987**, *6*, 572–580.
36. Axe, F. U.; Marynick, D. S. *J. Am. Chem. Soc.* **1988**, *110*, 3728–3734.
37. Ziegler, T.; Versluis, L.; Tschinke, V. *J. Am. Chem. Soc.* **1986**, *108*, 612–617.
38. Yang G. K.; Vaida, V.; Peters, K. S. *Polyhedron* **1988**, *7*, 1619–1622.

14

Redox Reactions of Binuclear Complexes of Ruthenium and Iron

Kinetic vs. Thermodynamic Control and Activation Effects

Albert Haim

Department of Chemistry, State University of New York, Stony Brook, NY 11794–3400

Kinetic measurements of peroxydisulfate oxidations of several ligand-bridged binuclear complexes of ruthenium and iron are reviewed. The kinetically controlled products of the one-electron oxidations of $(NC)_5$-$Fe^{II}pzRu^{II}(NH_3)_5^-$ (pz = pyrazine) and $(NC)_5Fe^{II}bpaRu^{II}(NH_3)_5^-$ [bpa = μ-1,2-bis(4-pyridyl)ethane] are also the thermodynamically stable electronic isomers, $(NC)_5Fe^{II}pzRu^{III}(NH_3)_5$ and $(NC)_5Fe^{II}bpaRu^{III}(NH_3)_5$. In contrast, the primary product of oxidation of $(EDTA)Ru^{II}pzRu^{II}(NH_3)_5$ is the unstable electronic isomer $(EDTA)Ru^{II}pzRu^{III}(NH_3)_5^+$, which subsequently undergoes rapid isomerization to the stable product $(EDTA)$-$Ru^{III}pzRu^{II}(NH_3)_5^+$. Peroxydisulfate oxidation of the equilibrium mixture of the mixed-valence isomers $(EDTA)Ru^{III}pzRu^{II}(NH_3)_5^+$ and $(EDTA)$-$Ru^{II}pzRu^{III}(NH_3)_5^+$ proceeds via the former, stable isomer. In the analogous reaction for the equilibrium mixture of electronic isomers $(NC)_5$-$Fe^{II}bpaRu^{III}(NH_3)_5$ and $(NC)_5Fe^{III}bpaRu^{II}(NH_3)_5$, the latter, unstable isomer is the reactive species. The reactions of $(EDTA)Ru^{II}pzRu^{II}(NH_3)_5$ and $(NC)_5Fe^{II}bpaRu^{III}(NH_3)_5$ represent examples of activation of the oxidation of a metal center by intramolecular electron transfer.

$\mathbf{A}$FTER THE DISCOVERY OF THE DISCRETE, MIXED-VALENCE COMPOUND μ-pyrazinedecaamminediruthenium(II,III) by Creutz and Taube (*1, 2*) (structure I) there was a publication explosion dealing with compounds of this type (*3*). The bulk of the published work has placed emphasis on synthetic (*4*), thermodynamic (*5*), structural (*6*), and, particularly, spectroscopic aspects (*7, 8*). There

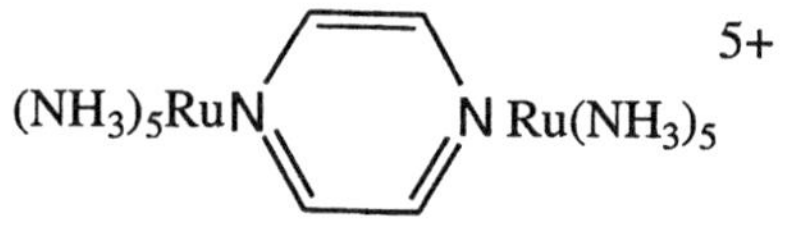

$$(NH_3)_5RuN \overbrace{} N\,Ru(NH_3)_5 \quad 5+$$

I

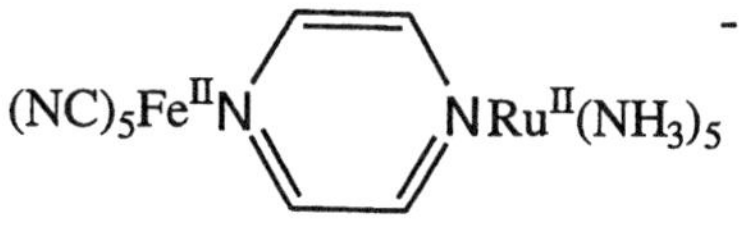

$$(NC)_5Fe^{II}N \overbrace{} N\,Ru^{II}(NH_3)_5 \quad -$$

II

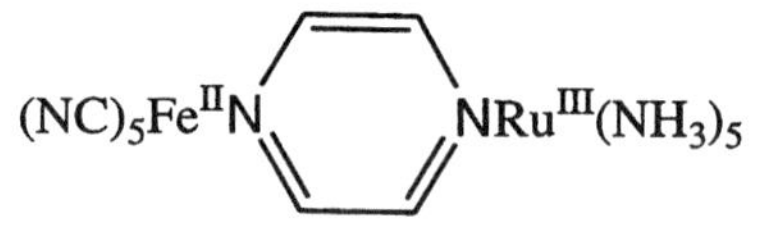

$$(NC)_5Fe^{II}N \overbrace{} N\,Ru^{III}(NH_3)_5$$

III

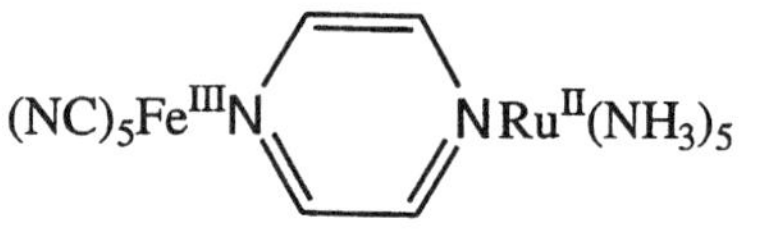

$$(NC)_5Fe^{III}N \overbrace{} N\,Ru^{II}(NH_3)_5$$

IV

have been some studies of intramolecular electron transfer (*9, 10*), but the dynamic solution behavior of such compounds with external redox reagents has been neglected.

We have a program of mechanistic studies in which mixed-valence compounds are either reactants or products of reactions that involve external redox reagents. In most of our studies, the mixed-valence compounds are unsymmetrical and valence-localized (*11–14*), although in some of our work we also deal with symmetric complexes (*15–17*). We have explored several questions. Consider first the complex ion in structure **II**, which contains two reducing sites, namely the ruthenium(II) and the iron(II) centers. Two related questions are pertinent. First, which of the two sites is thermodynamically the stronger one-electron reductant? Second, which of the electronic isomers (structures **III** or **IV**) is the kinetically controlled product of the one-electron oxidation of **II**? In addition, we have investigated the question of the effect of one metal center on the reactivity of the other metal center. Specifically, consider the unsymmetrical binuclear complex, structure **V**, with the indicated localized valences. Does the presence of the ruthenium(III) center affect the rate of oxidation of the ruthenium(II) center? Similarly, does the ruthenium(III) center in structure **VI** affect the reactivity toward oxidation of the iron(II) center?

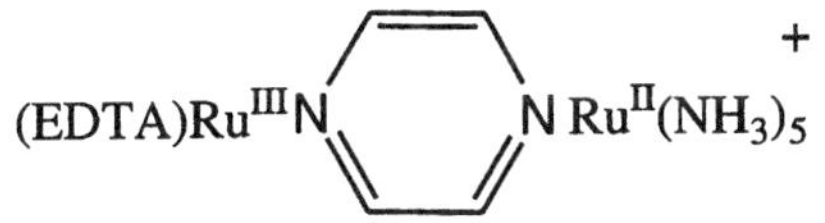

$$(\text{EDTA})\text{Ru}^{\text{III}}\text{N} \diagdown \text{pyrazine} \diagup \text{N Ru}^{\text{II}}(\text{NH}_3)_5 \quad +$$

V

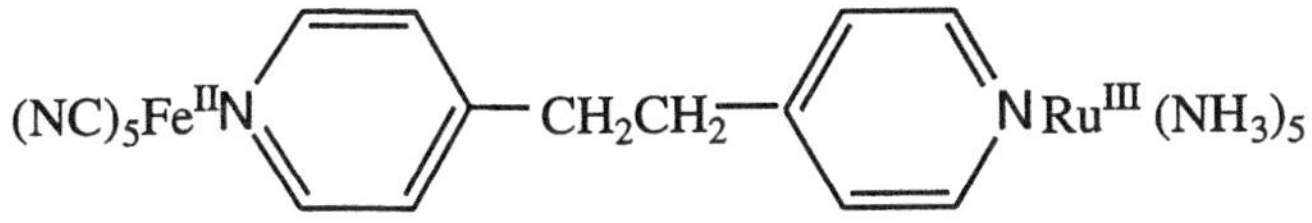

$$(\text{NC})_5\text{Fe}^{\text{II}}\text{N} \diagdown \text{pyridine} \diagup -\text{CH}_2\text{CH}_2- \diagdown \text{pyridine} \diagup \text{N Ru}^{\text{III}} (\text{NH}_3)_5$$

VI

$$(\text{NC})_5\text{Fe}^{\text{II}}\text{pzRu}^{\text{III}}(\text{NH}_3)_5 \xrightarrow{\ E_I^o\ } (\text{NC})_5\text{Fe}^{\text{III}}\text{pzRu}^{\text{II}}(\text{NH}_3)_5$$

III   **IV**

$$E_1^o \qquad e \qquad\qquad e \qquad E_2^o$$

$$(\text{NC})_5\text{Fe}^{\text{III}}\text{pzRu}^{\text{III}}(\text{NH}_3)_5{}^+$$

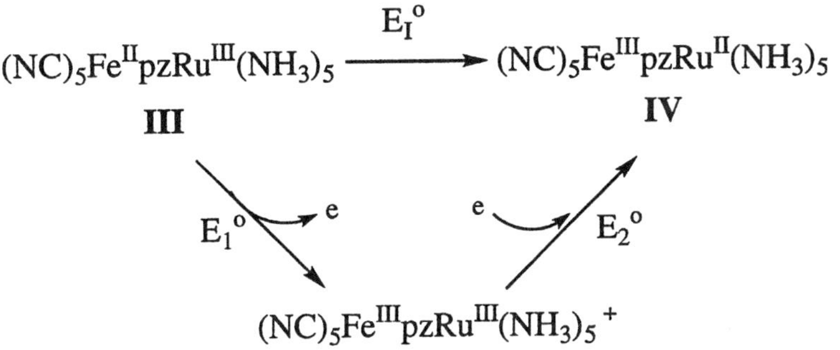

Scheme I.

In the present chapter we review the results of our inquiries into these questions. The methodology used, the results obtained, and the proposed interpretations are summarized herein.

Relative Stabilities of Electronic Isomers

Consider the equilibrium between isomers **III** and **IV** in Scheme I (pz is pyrazine). First, we identify the more stable isomer by constructing models and by taking advantage of the sensitivity to substituent effects of the metal-to-ligand charge transfer bands characteristic of pentaammineruthenium(II) and pentacyanoferrate(II) complexes with nitrogen heterocycles (*18, 19*). In Table I we list the maximum wavelength of the MLCT bands of a series of pyrazine complexes of ruthenium and iron. It will be seen that addition of an electropositive substituent (a proton or a metal center in the 3+ oxidation state) results in a shift of the MLCT band to lower energy, the effect being more substantial for the iron(II) than for the ruthenium(II) complexes. Next we model the two electronic isomers in Scheme I by replacing the 3+ metal center by another 3+

**Table I. MLCT Bands of Iron(II) and Ruthenium(II) Complexes of Pyrazine and
μ-1,2-Bis(4-pyridyl)ethane in Aqueous Solution**

Complex	λ (nm)	log ε	Ref.
$Fe^{II}(CN)_5pz^{3-}$	452	3.70	19
$Fe^{II}(CN)_5pzH^{2-}$	625	—	19
$Ru^{II}(NH_3)_5pz^{2+}$	474	4.3	2
$Ru^{II}(NH_3)_5pzH^{3+}$	528	4.1	2
$(NC)_5Fe^{II}pzRu^{II}(NH_3)_5$	522	4.36	11
$(NC)_5Fe^{II}pzRu^{III}(NH_3)_5$	590	~4	11
$(NC)_5Fe^{II}pzRh^{III}(NH_3)_5$	576	4.0	11
$(NC)_5Co^{III}pzRu^{II}(NH_3)_5$	520	4.2	11
$Fe^{II}(CN)_5bpa^{3-}$	365	3.7	13
$Ru^{II}(NH_3)_5bpa^{2+}$	410	3.9	13
$(NC)_5Fe^{II}bpaRu^{III}(NH_3)_5$	368	3.6	13
$(NC)_5Fe^{II}bpaRu^{II}(NH_3)_5^-$	408, 368	3.9, 3.8	13
$(NC)_5Fe^{II}bpaRh^{III}(NH_3)_5$	367	3.5	13
$(NC)_5Co^{III}bpaRu^{II}(NH_3)_5$	403	—	13
$Ru^{II}(EDTA)pz^{2-}$	463	4.1	9
$(EDTA)Ru^{III}pzRu^{II}(NH_3)_5^-$	520	4.2	9
$(EDTA)Ru^{II}pzRu^{II}(NH_3)_5^-$	544	4.4	9
$(EDTA)Ru^{II}pzRh^{III}(NH_3)_5^-$	533	4.3	9
$(EDTA)Rh^{III}pzRu^{II}(NH_3)_5^-$	528	4.3	9

NOTE: MLCT is metal-to-ligand charge transfer; pz is pyrazine; bpa is μ-1,2-bis(4-pyridyl)ethane;
λ is absorption wavelength; ε is molar absorptivity.

metal center, namely, Rh(III) for Ru(III) and Co(III) for Fe(III). The Rh(III)
and Co(III) centers were chosen because the corresponding 2+ oxidation
states are thermodynamically precluded for the indicated coordination
spheres. It will be seen that $(NC)_5Fe^{II}pzRh^{III}(NH_3)_5$ and $(NC)_5Co^{III}pzRu^{II}$-
$(NH_3)_5$ have their MLCT bands at 576 and 520 nm, respectively. The bands are
shifted by 124 and 46 nm, respectively, from the corresponding mononuclear
complexes. The stable electronic isomer in Scheme I has its MLCT band at
590 nm, for a 138-nm shift with respect to the Fe(II) mononuclear complex or a
116-nm shift with respect to the Ru(II) mononuclear complex. Evidently, the
stable isomer must have Fe(II) and Ru(III) localized valences. The somewhat
larger shift for Ru(III) as compared to Rh(III) is consistent with the π acceptor
properties of the former.

An entirely analogous argument is invoked to establish that the stable
electronic isomer in Scheme II is $(EDTA)Ru^{III}pzRu^{II}(NH_3)_5^+$. In the case of
the isomers in Scheme III, since the bridging ligand transmits electronic
effects to a minor extent, if at all, the MLCT bands of the constituent mononuclear units [MLCT bands at 410 nm and 365 nm for Ru(II) and Fe(II), respectively; column 2 of Table I] clearly indicate that $(NC)_5Fe^{II}bpaRu^{III}(NH_3)_5$ [bpa

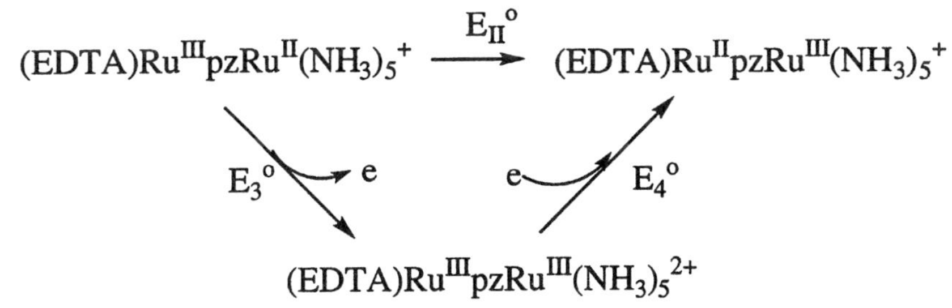

$$(EDTA)Ru^{III}pzRu^{II}(NH_3)_5{}^+ \xrightarrow{\quad E_{II}{}^o \quad} (EDTA)Ru^{II}pzRu^{III}(NH_3)_5{}^+$$

$$E_3{}^o \searrow \rightarrow e \qquad e \nwarrow \nearrow E_4{}^o$$

$$(EDTA)Ru^{III}pzRu^{III}(NH_3)_5{}^{2+}$$

Scheme II.

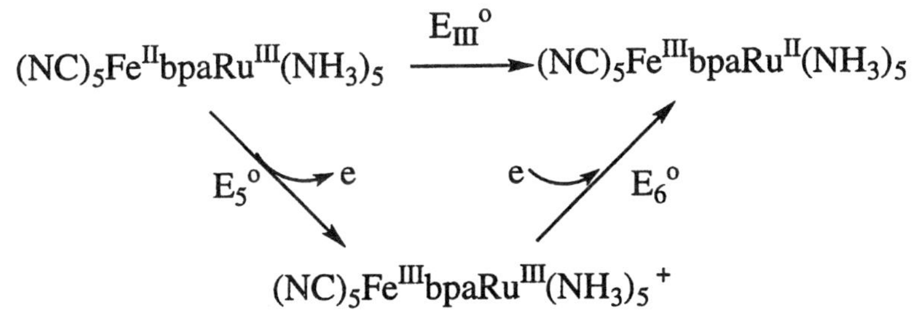

$$(NC)_5Fe^{II}bpaRu^{III}(NH_3)_5 \xrightarrow{\quad E_{III}{}^o \quad} (NC)_5Fe^{III}bpaRu^{II}(NH_3)_5$$

$$E_5{}^o \searrow \rightarrow e \qquad e \nwarrow \nearrow E_6{}^o$$

$$(NC)_5Fe^{III}bpaRu^{III}(NH_3)_5{}^+$$

Scheme III.

= μ-1,2-bis(4-pyridyl)ethane, MLCT band at 368 nm] is the stable electronic isomer. This conclusion is strongly reinforced by comparing the MLCT bands of the models $(NC)_5Fe^{II}bpaRh^{III}(NH_3)_5$ (367 nm) and $(NC)_5Co^{III}bpaRu^{II}$-$(NH_3)_5$ (403 nm).

Next, we estimate the difference in stability between the isomers by constructing simple thermodynamic cycles. Consider first Scheme I. $E_1{}^o$, measured by cyclic voltammetry (*11*), has a value of -0.72 V. $E_2{}^o$ is estimated at 0.64 V from the measured (*11*) reduction potential (Ru^{III} to Ru^{II}) of $(NC)_5Co^{III}pz$-$Ru^{III}(NH_3)_5^+$. Therefore, $E_I{}^o$ is -0.08 V. Confidence in the adequacy of the estimate for $E_2{}^o$ comes from a comparison of the measured (*11*) reduction potentials, 0.72 V and 0.71 V for the couples $(NC)_5Fe^{III/II}pzRu^{III}(NH_3)_5^{+/0}$ and $(NC)_5Fe^{III/II}pzRh^{III}(NH_3)_5^{+/0}$. The excellent agreement indicates that the reduction potential of the electroactive center is rather insensitive to the identity of the metal center (provided it is not a π acceptor) bound at the remote nitrogen.

Estimates of the relative stabilities of the electronic isomers in Schemes II and III have also been carried out. Values of $E_3{}^o$ and $E_4{}^o$ are -0.57 V (measured) and 0.37 V (estimated from the measured value for the model $(EDTA)Ru^{III/II}pzRh^{III}(NH_3)_5^{2+/+}$). Therefore, $E_{II}{}^o = -0.20$ V. For Scheme III, $E_5{}^o$ is measured (*13*) as -0.45 V. In the present case, $E_6{}^o$ could not be estimated from model studies because the very low solubility of $(NC)_5Co^{III}bpa$-

$Ru^{II}(NH_3)_5$ (*13*) precludes measurements in solution. However, in the case of bpa as a bridging ligand, coupling of the two metal centers is negligible and thus the reduction potentials of $(NC)_5Fe^{II}bpaRu^{III}(NH_3)_5$, 0.290 V, or of $Ru^{III}(NH_3)_5bpa^{3+}$, 0.293 V (*13*), represent good estimates for $E_6°$. Support for the approximation comes from a comparison of the reduction potentials (Fe^{III} to Fe^{II}) for $(NC)_5Fe^{III}bpaRu^{III}(NH_3)_5^+$, 0.45 V, $(NC)_5Fe^{III}bpaRh^{III}(NH_3)_5^+$, 0.44 V, and $Fe^{III}(CN)_5bpa^{2-}$, 0.44 V (*13*). Thus, we estimate $E_{III}°$ as −0.16 V.

In every case, the stable electronic isomer is the one predicted on the basis of the reduction potentials of the pertinent mononuclear complexes. This observation is not surprising for the binuclear complexes bridged by bpa, a ligand that does not couple the metal centers. For such bridging ligands the reduction potentials of mononuclear and binuclear complexes are virtually identical, and therefore the trends for mononuclear and binuclear complexes must be the same. However, for cases in which the bridging ligand (pyrazine) provides strong coupling between the metal centers, the reduction potentials of binuclear and mononuclear complexes differ considerably. Nevertheless, the relative stabilities of electronic isomers in the binuclear systems follow the order predicted from the potentials of the mononuclear complexes. This result occurs because the effects of the two metal centers on the reduction potentials of each other are rather similar.

One-Electron Oxidation of Binuclear Complexes with Both Metals in the 2+ Oxidation State: Kinetic or Thermodynamic Control?

The oxidation of **II** by peroxydisulfate proceeds according to the stoichiometry given in eq 1 and obeys monophasic, second-order kinetics with a rate constant of 3.8×10^3 M^{-1} s^{-1} at 25 ° (*11*).

$$2(NC)_5Fe^{II}pzRu^{II}(NH_3)_5^- + S_2O_8^{2-} = 2(NC)_5Fe^{II}pzRu^{III}(NH_3)_5 \\ + 2SO_4^{2-} \tag{1}$$

To be sure, because **III** is the stable electronic isomer, it is the thermodynamic product. The question arises whether it is also the kinetically controlled product or, alternatively, isomer **IV** is first produced and then rapidly undergoes isomerization to the stable isomer. We attempt to answer this question by measuring and comparing the rate constants for oxidation of iron(II) and ruthenium(II) complexes in systems for which there is no ambiguity as to the site of oxidation. Results are collected in Table II. It will be seen that rate constants for the oxidation of pentacyanoferrate(II)–pyrazine complexes fall in the vicinity of ~2 M^{-1} s^{-1} whereas for pentaammineruthenium(II)–pyrazine complexes rate constants are of the order of ~10^3 M^{-1} s^{-1}.

Table II. Rate Constants for Oxidation of Fe(II) and Ru(II) Complexes by Peroxydisulfate at 25 °C and Ionic Strength 0.10 M

Complex	k $(M^{-1}\,s^{-1})$	E^o (V)[a]	$k/K^{1/2}$	Ref.
$Fe^{II}(CN)_5pz^{3-}$	2.5	0.55	6.2×10^{-8}	11
$(NC)_5Fe^{II}pzRh^{III}(NH_3)_5$	1.5	0.71	8.4×10^{-7}	11
$Ru^{II}(NH_3)_5pz^{2+}$	3.6×10^3	0.52	5.0×10^{-5}	16
$(NC)_5Co^{III}pzRu^{II}(NH_3)_5$	1.4×10^2	0.64	2.0×10^{-5}	11
$(NC)_5Fe^{II}pzRu^{II}(NH_3)_5^-$	$3.8\times10^{3\,b}$	$0.49^{\,c}$, $0.72^{\,d}$	$2.9\times10^{-5\,b}$	11
$Fe^{II}(CN)_5bpa^{3-}$	0.10	0.440	2.9×10^{-10}	13
$Ru^{II}(NH_3)_5bpa^{2+}$	1.0×10^5	0.293	1.7×10^{-5}	13
$(NC)_5Fe^{II}bpaRh^{III}(NH_3)_5$	0.50	0.440	1.5×10^{-9}	13
$(NC)_5Fe^{II}bpaRu^{II}(NH_3)_5^-$	$5.0\times10^{4\,b}$	$0.290^{\,c}$	7.9×10^{-6}	13
$(NC)_5Fe^{II}bpaRu^{III}(NH_3)_5$	$4.3\times10^{2\,d}$	$0.447^{\,e}$	1.4×10^{-6}	13
$(NC)_5Fe^{III}bpaRu^{II}(NH_3)_5$	1.6×10^5	$0.29^{\,f}$	2.5×10^{-5}	13
$Ru^{II}(EDTA)pz^{2-}$	<8	0.24	$<5\times10^{-10}$	14
$(EDTA)Ru^{II}pzRu^{II}(EDTA)^{4-}$	6.1×10^2	0.18	1.1×10^{-8}	14
$(EDTA)Ru^{II}pzRu^{III}(EDTA)^{3-}$	26	0.32	7.3×10^{-9}	14
$(EDTA)Ru^{III}pzRu^{II}(NH_3)_5^+$	5.5×10^2	$0.57^{\,g}$	2.0×10^{-5}	14
$(EDTA)Ru^{II}pzRu^{III}(NH_3)_5^+$	1.4×10^6	$0.37^{\,h}$	1.0×10^{-3}	14
$(EDTA)Ru^{II}pzRu^{II}(NH_3)_5$	$2.5\times10^{4\,i}$	$0.21^{\,j}$	8.3×10^{-7}	14
$(EDTA)Ru^{II}pzRu^{II}(NH_3)_5$	$2.5\times10^{4\,k}$	$0.41^{\,l}$	4.0×10^{-5}	14

[a]Reduction potential for oxidized form of species indicated.

[b]Oxidation of Ru(II).

[c]For reduction of $(NC)_5Fe^{II}bpaRu^{III}(NH_3)_5$ to $(NC)_5Fe^{II}bpaRu^{II}(NH_3)_5^-$.

[d]Oxidation of Fe(II).

[e]For reduction of $(NC)_5Fe^{III}bpaRu^{III}(NH_3)_5^+$ to $(NC)_5Fe^{II}bpaRu^{III}(NH_3)_5$.

[f]For reduction of $(NC)_5Fe^{III}bpaRu^{III}(NH_3)_5^+$ to $(NC)_5Fe^{III}bpaRu^{II}(NH_3)_5$.

[g]For reduction of $(EDTA)Ru^{III}pzRu^{III}(NH_3)_5^{2+}$ to $(EDTA)Ru^{III}pzRu^{II}(NH_3)_5^+$.

[h]For reduction of $(EDTA)Ru^{III}pzRu^{III}(NH_3)_5^{2+}$ to $(EDTA)Ru^{II}pzRu^{III}(NH_3)_5^+$.

[i]Oxidation of $Ru^{II}(EDTA)$.

[j]For reduction of $(EDTA)Ru^{III}pzRu^{II}(NH_3)_5^+$ to $(EDTA)Ru^{II}bzRu^{II}(NH_3)_5$.

[k]Oxidation of $Ru^{II}(NH_3)_5$.

[l]For reduction of $(EDTA)Ru^{II}pzRu^{III}(NH_3)_5^+$ to $(EDTA)Ru^{II}pzRu^{II}(NH_3)_5$.

Intrinsic (e.g., self-exchange) factors are fairly similar for the classes of ruthenium and iron pyrazine complexes under consideration (*11*, *15*), but thermodynamic (e.g., reduction potentials) factors vary. Therefore, before the rate constants can be taken as diagnostic of oxidation of a Ru(II) or Fe(II) center, they must be corrected for the differences in the thermodynamic driving forces. This correction is accomplished by dividing the measured rate constant by $K^{1/2}$, where K is the equilibrium constant for the one-electron oxidation of the complex under consideration by peroxydisulfate. This correction is based on the Marcus cross-relationship, which has been shown to be obeyed for the peroxydisulfate oxidation of a variety of ruthenium(II) complexes (*16*). It will be seen that corrected rate constants (column 4 of Table II) for oxidation of Fe(II)–pz and Ru(II)–pz fall in the vicinity of 10^{-7} and 10^{-5} $M^{1/2}$ s^{-1}, respectively.

One of the factors that governs this difference in reactivity is undoubtedly the charge difference between ruthenium and iron complexes. This point will be discussed in more detail later. The corrected rate constant for **II** is 2.9×10^{-5} $M^{1/2}$ s^{-1}. Comparison of this value with the values diagnostic of oxidation of Fe(II) or Ru(II) provides a strong indication that oxidation of **II** takes place at the Ru(II) reducing site. Therefore, the kinetically controlled and thermodynamically controlled products of the oxidation of **II** by peroxydisulfate are one and the same, namely, the electronic isomer **III**. Moreover, the similarity in the rate constants for oxidation of sites that are localized Ru(II) centers and the rate constant for oxidation of **II** leads us to suggest that the electron that is transferred from **II** is in an orbital of predominantly ruthenium(II) character, and therefore that there is little mixing of iron and ruthenium orbitals.

The oxidation of $(NC)_5Fe^{II}bpaRu^{II}(NH_3)_5^-$ by peroxydisulfate exhibits biphasic kinetics that were assigned (*13*) to eqs 2 and 3, respectively.

$$2(NC)_5Fe^{II}bpaRu^{II}(NH_3)_5^- + S_2O_8^{2-} = 2(NC)_5Fe^{II}bpaRu^{III}(NH_3)_5 \atop + 2SO_4^{2-} \tag{2}$$

$$2(NC)_5Fe^{II}bpaRu^{III}(NH_3)_5 + S_2O_8^{2-} = 2(NC)_5Fe^{III}bpaRu^{III}(NH_3)_5^+ \atop + 2SO_4^{2-} \tag{3}$$

At this point we consider eq 2, in which the thermodynamically controlled product is the electronic isomer **VI**. Equation 3 will be discussed later. Rate comparisons (*see* Table II) similar to the ones invoked for the $(NC)_5Fe^{II}$-$pzRu^{II}(NH_3)_5$–$S_2O_8^{2-}$ system (Ru^{II}–bpa is ca. 10^4 more reactive than Fe^{II}–bpa, after correction for thermodynamics) lead us to infer that the Ru(II) site in $(NC)_5Fe^{II}bpaRu^{II}(NH_3)_5^-$ is the initial site of oxidation. As for the analogous pyrazine system, the primary one-electron oxidation product is also the stable electronic isomer. Here again absence of mixing between Fe and Ru orbitals is indicated by the kinetic results, but this interpretation is expected from the spectroscopic results that show (*see* Table I) that $(NC)_5Fe^{II}bpaRu^{II}(NH_3)_5^-$ exhibits two MLCT bands at wavelengths unshifted with respect to those for the MLCT bands of the parent Ru(II) and Fe(II) mononuclear complexes.

Biphasic kinetics are also observed for the peroxydisulfate oxidation of $(EDTA)Ru^{II}pzRu^{II}(NH_3)_5$. The two phases are ascribed to eqs 4 and 5, respectively (*14*).

$$2(EDTA)Ru^{II}pzRu^{II}(NH_3)_5 + S_2O_8^{2-} = 2(EDTA)Ru^{III}pzRu^{II}(NH_3)_5^+ \atop + 2SO_4^{2-} \tag{4}$$

$$2(EDTA)Ru^{III}pzRu^{II}(NH_3)_5^+ + S_2O_8^{2-} = 2(EDTA)Ru^{III}pzRu^{III}(NH_3)_5^{2+}$$
$$+ 2SO_4^{2-} \tag{5}$$

At this point we discuss eq 4 in the context of identifying the primary site of oxidation. We adopt the approach described already for the Fe–Ru systems to provide a calibration of rate constants for oxidation of $Ru^{II}(EDTA)$–pz sites and $Ru^{II}(NH_3)_5$–pz sites. As was the case for the Fe–Ru systems, intrinsic factors are similar for the two metal centers in $(EDTA)Ru^{II}pzRu^{II}(NH_3)_5$–$S_2O_8^{2-}$ (rate constants for self-exchange in the precursor complexes of the $Ru(EDTA)pz^{-/2-}$ and $Ru(NH_3)_5pz^{3+/2+}$ systems are 13×10^5 and 8.3×10^5 M^{-1} s^{-1}, respectively), but the reduction potentials vary widely. Therefore, the measured rate constants are corrected for thermodynamic factors (*14*) by dividing the measured rate constants by $K^{1/2}$, where K is the equilibrium constant for the one-electron oxidation of the complex by peroxydisulfate.

According to this treatment (*see* column 4 of Table II), corrected rate constants for loss of an electron from a $Ru^{II}(EDTA)$–pz site or from a Ru^{II}-$(NH_3)_5$–pz site are approximately 1×10^{-9} and 1×10^{-5} $M^{1/2}$ s^{-1}, respectively. The correction for thermodynamics to be applied to the measured rate constant for oxidation of $(EDTA)Ru^{II}pzRu^{II}(NH_3)_5$ depends on whether the electron is removed from the Ru(II) bound to EDTA or the Ru(II) bound to ammonia. In the former case, the appropriate reduction potential measured (*14*) for the couple $(EDTA)Ru^{III/II}pzRu^{II}(NH_3)_5^{+/0}$ is 0.21 V. In the latter case, the appropriate reduction potential estimated for the couple $(EDTA)Ru^{II}pz$-$Ru^{III/II}(NH_3)_5^{+/0}$ is 0.41 V. The estimate is obtained from Scheme IV, where $E_8°$ (measured) is 0.21 V and $E_7°$ (estimated in Scheme II) is 0.20 V.

The corrected rate constants for eq 4 on the assumption that the electron is removed from the $Ru^{II}(EDTA)$ or $Ru^{II}(NH_3)_5$ moieties are 8.3×10^{-7} or 4.0×10^{-5} $M^{1/2}$ s^{-1}, respectively. The former value is in poor agreement with the value 1×10^{-9} $M^{1/2}$ s^{-1} characteristic of oxidation of $Ru^{II}(EDTA)$. The 4.0×10^{-5} $M^{1/2}$ s^{-1} value is in excellent agreement with the values diagnostic of oxidation of

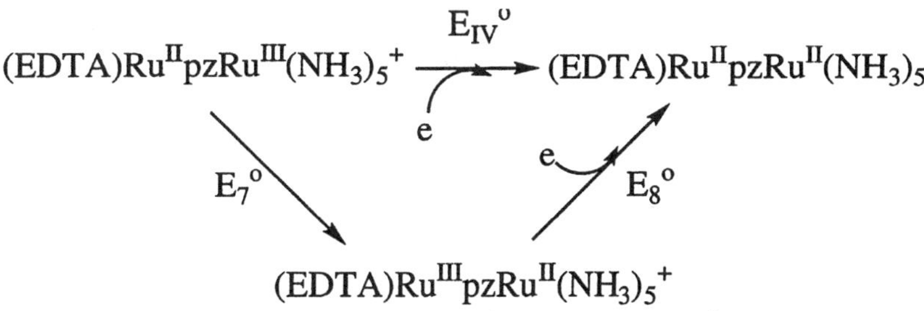

Scheme IV.

$Ru^{II}(NH_3)_5$ and we infer that direct oxidation of the $Ru^{II}(EDTA)$ moiety in eq 4 does not obtain. Instead, the mechanism depicted in eqs 6 and 7 is operative.

$$(EDTA)Ru^{II}pzRu^{II}(NH_3)_5 + S_2O_8^{2-} = (EDTA)Ru^{II}pzRu^{III}(NH_3)_5^+ \\ + SO_4^{2-} + SO_4^- \tag{6}$$

$$(EDTA)Ru^{II}pzRu^{III}(NH_3)_5^+ = (EDTA)Ru^{III}pzRu^{II}(NH_3)_5^+ \tag{7}$$

In the proposed sequence, eq 6 is rate-determining and is followed by eq 7, the very rapid (9) isomerization via intramolecular electron transfer (rate constant 8×10^9 s^{-1}). Here, in contrast with the previous systems, the kinetically controlled product is not the thermodynamic product.

Activation Effects: Intramolecular Electron Transfer Assistance of Bimolecular Redox Reactions?

The mechanism embodied in eqs 6 and 7 represents a form of intramolecular "catalysis" of the oxidation of the $Ru^{II}(EDTA)$ center by the $Ru^{II}(NH_3)_5$ moiety. The $Ru^{II}(EDTA)$ reacts sluggishly. On the other hand, the electron in $Ru^{II}(NH_3)_5$ is readily accessible to the external oxidant and is given up. Subsequently, the electron of $Ru^{II}(EDTA)$ is transferred to the $Ru^{III}(NH_3)_5$ moiety and the overall reaction in eq 4 is consummated. The process is not truly catalytic because the "catalyst" is the reactant itself, which, of course, is consumed in the reaction. Perhaps a better description is that the net oxidation of the $Ru^{II}(EDTA)$ site is activated by the facile intramolecular electron transfer between the metal centers. The proposed mechanism, Scheme V, bears some resemblance to the classical chemical mechanism (20) for inner-sphere electron transfer.

Kinetic studies of eq 5 yield the rate law given in eq 8.

$$-\frac{d\ln[(EDTA)Ru^{III}pzRu^{II}(NH_3)_5^+]}{dt} = a + b[S_2O_8^{2-}] \tag{8}$$

The mechanism proposed (14) to account for the two-term rate law is depicted in Scheme VI. On the basis of Scheme VI, the derived rate law is given by eq 9.

$$-\frac{1}{2}\frac{d\ln[(EDTA)Ru^{III}pzRu^{II}(NH_3)_5^+]}{dt} = k_{-1} + \left(k_3 + \frac{k_{et}}{k_{-et}}k_3'\right)[S_2O_8^{2-}] \tag{9}$$

According to the proposed mechanism, complex **V** disappears via parallel paths, one, independent of $[S_2O_8^{2-}]$, by dissociation into its mononuclear moi-

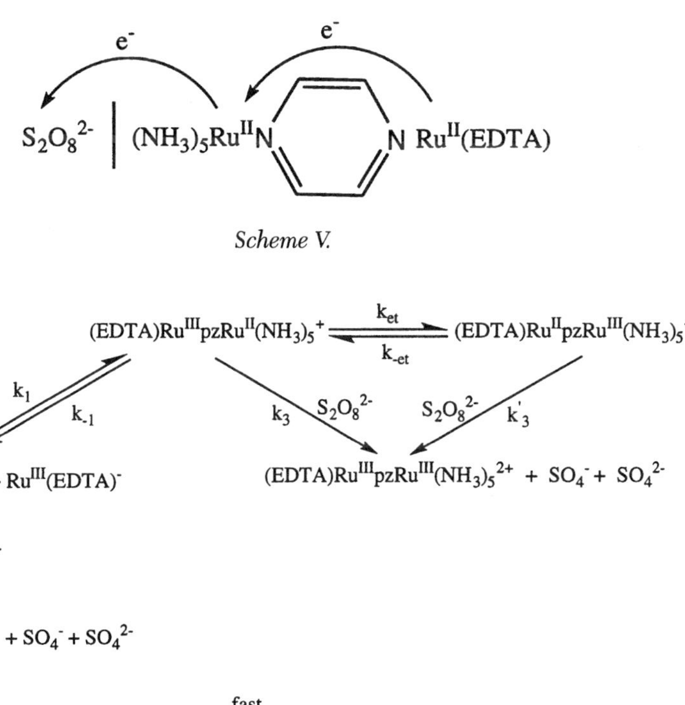

Scheme V.

$$(EDTA)Ru^{III}pzRu^{II}(NH_3)_5{}^+ \underset{k_{-et}}{\overset{k_{et}}{\rightleftharpoons}} (EDTA)Ru^{II}pzRu^{III}(NH_3)_5{}^+$$

$$Ru^{II}(NH_3)_5pz^{2+} + Ru^{III}(EDTA)^-$$

$$(EDTA)Ru^{III}pzRu^{III}(NH_3)_5{}^{2+} + SO_4^- + SO_4^{2-}$$

with k_1, k_{-1}, k_3 ($S_2O_8^{2-}$), k'_3 ($S_2O_8^{2-}$)

$$Ru^{III}(NH_3)_5pz^{3+} + SO_4^- + SO_4^{2-}$$

with k_2 ($S_2O_8^{2-}$)

$$(EDTA)Ru^{III}pzRu^{II}(NH_3)_5{}^+ + SO_4^- \xrightarrow{\text{fast}} (EDTA)Ru^{III}pzRu^{III}(NH_3)_5{}^{2+} + SO_4^{2-}$$

$$Ru^{II}(NH_3)_5pz^{2+} + SO_4^- \xrightarrow{\text{fast}} Ru^{III}(NH_3)_5pz^{3+} + SO_4^{2-}$$

Scheme VI.

eties, and one, dependent on $[S_2O_8^{2-}]$. In the latter path, initially we assume that both electronic isomers are reactive and therefore, $b = k_3 + k_{et}k'_3/k_{-et} = 5.5 \times 10^2$ M^{-1} s^{-1}. The proposed interpretation is supported by the excellent agreement between the value of k_{-1} (0.95 s^{-1}) measured in the redox studies (*14*) with the value (0.90 s^{-1}) measured directly (*3*) in substitution studies.

We turn to an analysis of the $[S_2O_8^{2-}]$-dependent path. The approach is the same as the one we utilized to identify the site of oxidation in (EDTA)Ru^{II}pzRu^{II}(NH$_3$)$_5$. If EDTA)Ru^{II}pzRu^{III}(NH$_3$)$_5^+$, the unstable electronic isomer, is unreactive, then $k_3 = 5.5 \times 10^2$ M^{-1} s^{-1}. Alternatively, if the stable isomer is unreactive, then $k_{et}k'_3/k_{-et} = 5.5 \times 10^2$ M^{-1} s^{-1}. Using $E_{II}°$ estimated in Scheme II to calculate k_{et}/k_{-et}, we compute $k'_3 = 1.4 \times 10^6$ M^{-1} s^{-1}. To compare k_3 and k'_3 with the values diagnostic of oxidation of a Ru^{II}(EDTA)–pz site (1 × 10^{-9} M$^{1/2}$ s^{-1}) or a Ru^{II}(NH$_3$)$_5$–pz site (1 × 10^{-5} M$^{1/2}$ s^{-1}), we correct the measured values for free-energy differences and obtain $k_3/K^{1/2} = 2.0 \times 10^{-5}$ M$^{1/2}$ s^{-1} and $k'_3/K^{1/2} = 1.0 \times 10^{-3}$ M$^{1/2}$ s^{-1}. The former value falls nicely in the range characteristic of oxidation of $Ru^{II}(NH_3)_5$–pz, whereas the latter value is several orders of magnitude off the range of values diagnostic of oxidation of $Ru^{II}(EDTA)$–pz. On this basis, we infer that oxidation of $(EDTA)Ru^{III}pzRu^{II}(NH_3)_5^+$ is operative, and thus that the more stable isomer is the reactive species as well. For both $(EDTA)Ru^{II}pzRu^{II}(NH_3)_5$ and for the equilibrium mixture of $(EDTA)Ru^{III}pzRu^{II}(NH_3)_5^+$ and $(EDTA)Ru^{II}pzRu^{III}(NH_3)_5^+$, the site of oxidation is the $Ru^{II}(NH_3)_5$–pz center. In both cases, the thermodynamically weaker reducing site is being oxidized. Some of the implications of these reactivity patterns will be discussed later in this chapter.

Equation 3 represents another example of activation toward oxidation of one metal center by another linked to the first by facile electron transfer. The oxidations of $Fe^{II}(CN)_5$–bpa complexes by peroxydisulfate proceed with corrected rate constants in the vicinity of 1.0×10^{-9} $M^{1/2}$ s^{-1} (cf. Table II). For $Ru^{II}(NH_3)_5$–bpa complexes, rate constants are about 1×10^{-5} $M^{1/2}$ s^{-1}. The corrected rate constant for complex **VI** is 1.4×10^{-6} $M^{1/2}$ s^{-1}, which is anomalously high for removal of an electron from Fe(II). The ca. 10^3 difference between **VI** and $(NC)_5Fe^{II}bpaRh^{III}(NH_3)_5$ is particularly noteworthy because the two complexes have equal charges and reduction potentials. To account for the high reactivity of **VI**, we proposed (*13*) that the oxidation of **VI** proceeds via rapid and reversible isomerization (eq 10)

$$(NC)_5Fe^{II}bpaRu^{III}(NH_3)_5 \rightleftarrows (NC)_5Fe^{III}bpaRu^{II}(NH_3)_5 \quad Q_{10} \quad (10)$$

and is followed by the rate-determining reaction of the unstable isomer with $S_2O_8^{2-}$ in eq 11.

$$(NH)_5Fe^{III}bpaRu^{II}(NH_3)_5 + S_2O_8^{2-} \xrightarrow{k_{11}} (NH)_5Fe^{III}bpaRu^{III}(NH_3)_5^+$$
$$+ SO_4^{2-} + SO_4^- \quad (11)$$

In this interpretation, the measured rate constant is $Q_{10}k_{11} = 4.6 \times 10^2$ M^{-1} s^{-1}. $E_{III}°$, the potential associated with eq 10, was estimated as -0.16 V by constructing the cycle depicted in Scheme III. Therefore, Q_{10} is 2.9×10^{-3} and $k_{11} = 1.6 \times 10^5$ M^{-1} s^{-1}, which after correction for thermodynamics becomes 2.5×10^{-5} $M^{1/2}$ s^{-1}, a perfectly reasonable value for oxidation of a ruthenium(II) center. The electron to be lost from a given site (Fe) is first transferred to a site (Ru) that is oxidized more readily. Again, this mechanism can be viewed as a form of intramolecular catalysis, although the process cannot be truly catalytic because the catalyst (the reactant itself) is consumed. As was the case with the oxidation of $(EDTA)Ru^{II}pzRu^{II}(NH_3)_5$, the mechanism embodied in eqs 10 and 11, depicted in Scheme VII, is the outer-sphere equivalent of the chemical mechanism for inner-sphere electron transfer.

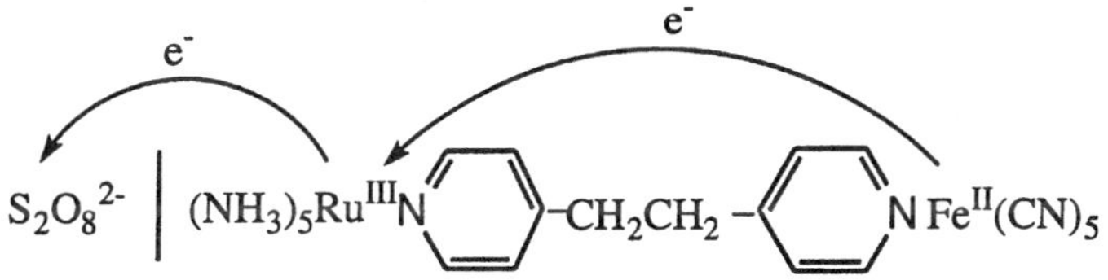

Scheme VII.

Conclusion

Oxidation potentials for the two reducing sites in the binuclear complexes $(NC)_5Fe^{II}bpaRu^{II}(NH_3)_5^-$, $(NC)_5Fe^{II}pzRu^{II}(NH_3)_5^-$, and $(EDTA)Ru^{II}pzRu^{II}$-$(NH_3)_5$ are given in Schemes VIII–X. Regardless of the thermodynamic driving forces for oxidation of the two sites, the reactions of these binuclear complexes with peroxydisulfate proceed via oxidation at the $Ru^{II}(NH_3)_5$ site. Similarly, the reactions of the mixed-valence complexes $(NC)_5Fe^{II}bpaRu^{III}$-$(NH_3)_5$ and $(EDTA)Ru^{III}pzRu^{II}(NH_3)_5^{\pm}$ also feature oxidation of a $Ru^{II}(NH_3)_5$ site. For the latter complex, the dominant electronic isomer already contains the center that undergoes reaction. For the former complex, however, reaction occurs by oxidation of the Ru(II) center of the unstable electronic isomer $(NC)_5Fe^{III}bpaRu^{II}(NH_3)_5$ assumed to be present in small concentrations in equilibrium with the dominant isomer.

From an operational point of view, these results are a consequence of the fact that, regardless of thermodynamic driving forces, the $Ru^{II}(NH_3)_5^{2+}$ center is considerably more reactive toward peroxydisulfate than the other two reducing centers, namely, $Fe^{II}(CN)_5^{3-}$ and $Ru^{II}(EDTA)^{2-}$. The higher reactivity of $Ru^{II}(NH_3)_5^{2+}$ as compared to $Fe^{II}(CN)_5^{3-}$ and $Ru^{II}(EDTA)^{2-}$ is not a consequence of a more favorable intrinsic factor. Self-exchange rate constants, after correction for electrostatics, in the three systems are comparable (*11, 14*). Most likely, the higher reactivity of $Ru^{II}(NH_3)_5^{2+}$ for oxidation by the negatively charged peroxydisulfate is related to the positive charge of the ruthenium complex, and the decreased reactivity of $Fe^{II}(CN)_5^{3-}$ and $Ru^{II}(EDTA)^{2-}$ are related to their negative charges. However, charge effects are insufficient to account quantitatively for the differences in reactivity (*21*). After correction for thermodynamics and charges, the $Ru^{II}(NH_3)_5^{2+}$ complexes still react about ca. 10^2 times faster than corresponding $Fe^{II}(CN)_5^{3-}$ or $Ru^{II}(EDTA)^{2-}$ complexes.

Hydrogen bonding between peroxydisulfate and the ammonia ligands of $Ru^{II}(NH_3)_5^{2+}$ and nonadiabaticity in the reactions of $Fe^{II}(CN)_5^{3-}$ or Ru^{II}-$(EDTA)^{2-}$ complexes have been invoked to account for the relative reactivities (*21*). There are three supporting studies for the postulated specific interaction of the negatively charged peroxydisulfate with the positively charged ruthenium–ammine end of the binuclear complexes via hydrogen bonding. (a) Outer-sphere charge transfer studies in pentaammineruthenium(III)–hexa-

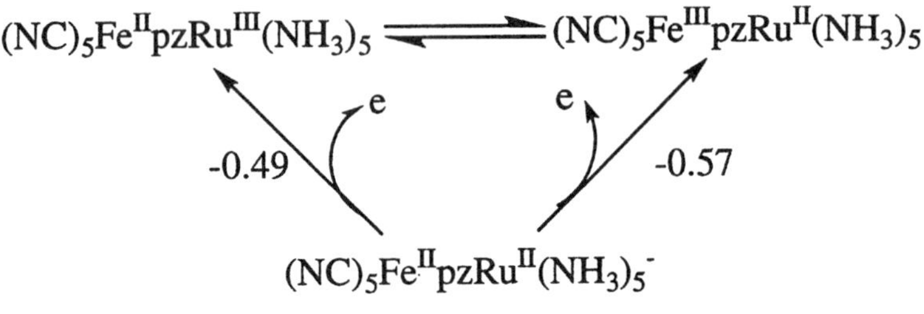

Scheme VIII.

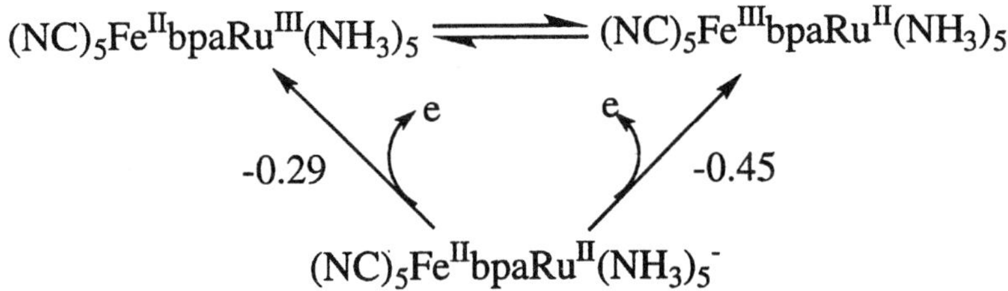

Scheme IX.

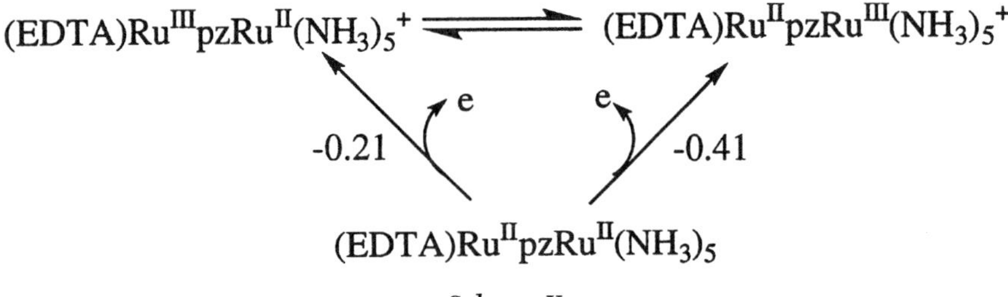

Scheme X.

cyanometallates(II) (Fe, Ru, Os) (*22*) have been interpreted on the basis of van der Waals contacts between the hexacyanide moiety and the ammine face of the ruthenium complex. (b) Solvatochromism studies of the charge transfer transitions of pentaammineruthenium complexes (*23*) have led to the suggestion that the solvent effects are associated with a hydrogen-bonding interaction via electron-pair donation from the solvent to the ammine hydrogens. (c) From circular dichroism studies of ion pairs between tris(diamine)cobalt(III) complexes and various tetrahedral oxyanions (*24, 25*), it has been suggested that hydrogen bonding from the amine groups to the oxyanion obtains, and stereospecific outer-sphere complexation along the C_3 axis was postulated.

On the basis of all this information, an appropriate model for the specific interaction between the $Ru^{II}(NH_3)_5^{2+}$ end of the binuclear complexes and $S_2O_8^{2-}$ is depicted in structure **VII**. The hydrogen bonding interaction would

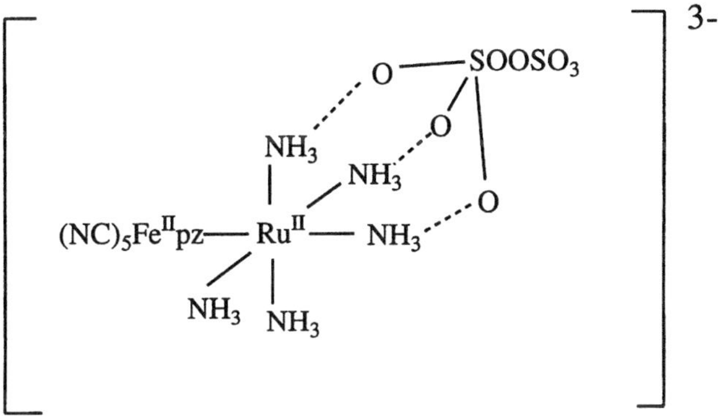

$$[(NC)_5Fe^{II}pz\!-\!Ru^{II}(NH_3)_5 \cdots O\!-\!SOOSO_3]^{3-}$$

VII

facilitate electron transfer from Ru(II) to the antibonding orbital of the peroxide bond. Such interactions, electrostatic or specific, would be impeded or absent for the $Fe^{II}(CN)_5^{3-}$ or $Ru^{II}(EDTA)^{2-}$ ends, and coupling between the reaction centers would be diminished. Thus, the postulate of nonadiabatic electron transfer appears reasonable.

The reactions discussed display four reactivity patterns. For $(NC)_5Fe^{II}pzRu^{II}(NH_3)_5^-$ and $(NC)_5Fe^{II}bpaRu^{II}(NH_3)_5^-$, thermodynamic and kinetic products are the same. In contrast, for $(EDTA)Ru^{II}pzRu^{II}(NH_3)_5$, the kinetically controlled product is not the thermodynamically favored product. In the oxidation of $(EDTA)Ru^{III}pzRu^{II}(NH_3)_5^+$ the thermodynamically favored electronic isomer is the reactive species. In contrast, the oxidation of $(NC)_5Fe^{II}bpaRu^{III}(NH_3)_5$ proceeds by reaction via the unstable electronic isomer $(NC)_5Fe^{III}bpaRu^{II}(NH_3)_5$. These four patterns, which involve kinetic vs. thermodynamic control and relative reactivities of stable vs. unstable electronic isomers, fall into two fundamental categories once microscopic reversibility is taken into consideration. When the kinetic product of oxidation in the forward reaction is also the thermodynamically favored product, the reverse reaction must proceed by reduction of the stable electronic isomer. Alternatively, when the kinetically controlled product of the forward reaction is the unstable electronic isomer, microscopic reversibility requires that the reactive species in the reverse direction be the unstable electronic isomer. Therefore, the first category includes reactions in which the thermodynamic and kinetic product are the same and/or reactions of electronic isomers in which the dominant form is reactive. The second category includes reactions in which the primary product is kinetically controlled and/or reactions of electronic isomers in which the minor isomer is reactive. Viewed in this manner, the reactions of $(NC)_5Fe^{II}pzRu^{II}(NH_3)_5^-$, $(NC)_5Fe^{II}bpaRu^{II}(NH_3)_5^-$, and $(EDTA)Ru^{III}pzRu^{II}(NH_3)_5^+$ fall into the first category, whereas reactions of $(EDTA)Ru^{II}pzRu^{II}(NH_3)_5$ and $(NC)_5Fe^{II}bpaRu^{III}(NH_3)_5$ fall into the second category.

Acknowledgment

This chapter is dedicated to Henry Taube, model scientist, master teacher, and friend.

References

1. Creutz, C.; Taube, H. *J. Am. Chem. Soc.* **1969**, *91*, 3988.
2. Creutz, C.; Taube, H. *J. Am. Chem. Soc.* **1973**, *95*, 1086.
3. Creutz, C. *Prog. Inorg. Chem.* **1983**, 30, 1.
4. Siddiqui S.; Henderson, W. W.; Shepherd, R. F. *Inorg. Chem.* **1987**, *26*, 3191.
5. Richardson, D. E.; Taube, H. *Coord. Chem. Rev.* **1984**, *60*, 107.
6. Furholz, U.; Joss. S.; Burgi, H. B.; Ludi, A. *Inorg. Chem.* **1985**, *24*, 943.
7. Reimers, J. R.; Hush, N. S. *Inorg. Chem.* **1990**, *29*, 4510.
8. Ribon, A.-C.; Launay, J.-P.; Takahashi, K.; Nihira, T.; Tarutani, S.; Spangler, C. W. *Inorg. Chem.* **1994**, *33*, 1325.
9. Creutz. C.; Kroger, P.; Matsubara, T.; Netzel, T. L.; Sutin, N. *J. Am. Chem. Soc.* **1979**, *101*, 5442.
10. Ohno, T.; Nozaki, K.; Ikeda, N.; Haga, M. *Adv. Chem. Ser.* **1991**, *228*, 215.
11. Yeh, A.; Haim, A. *J. Am. Chem. Soc.* **1985**, *107*, 369.
12. Burewicz, A.; Haim, A. *Inorg. Chem.* **1988**, *27*, 1611.
13. Olabe, J.; Haim, A. *Inorg. Chem.* **1989**, *28*, 3277.
14. Ram, M. S.; Haim, A. *Inorg. Chem.* **1991**, *30*, 1319.
15. Furholz, U.; Haim, A. *J. Phys. Chem.* **1986**, *90*, 3686.
16. Furholz, U.; Haim, A. *Inorg. Chem.* **1987**, *26*, 3243.
17. Akhtar, M.; Haim, A. *Inorg. Chem.* **1988**, *27*, 1608.
18. Ford, P. C.; Rudd, D. F. P.; Gaunder, R. G.; Taube, H. *J. Am. Chem. Soc.* **1968**, *90*, 1187.
19. Toma, H.; Malin, J. M. *Inorg. Chem.* **1973**, *12*, 1039.
20. Nordmeyer, F.; Taube, H. *J. Am. Chem. Soc.* **1968**, *90*, 1162.
21. Yeh, A., Tunghai University, Taichung, Taiwan, ROC, personal communication, 1995.
22. Curtis, J. C.; Meyer, T. J. *Inorg. Chem.* **1982**, *21*, 1562.
22. Curtis, J. C.; Sullivan, B. P.; Meyer, T. J. *Inorg. Chem.* **1983**, *23*, 1562.
23. Zhang, X. L.; Kankel, C. R.; Hupp, J. T. *Inorg. Chem.* **1994**, *33*, 4738.
24. Sarneski, J. E.; Urbach, F. L. *J. Am. Chem. Soc.* **1971**, *93*, 884.
25. Hermer, R. E.; Douglas, B. E. *J. Coord. Chem.* **1977**, *7*, 43.

15

Electron Delocalization Through the Disulfide Bridge

Icaro de Sousa Moreira[1] and Douglas Wagner Franco[2]

[1]Departamento de Química Orgânica e Inorgânica, Universidade Federal do Ceará, Fortaleza-CE, Brasil

[2]Instituto de Química de São Carlos-USP, Caixa Postal 780, 13560–970, São Carlos–SP, Brasil

The –S–S– bridge capability for conducting electrons was investigated using 4,4′-dithiodipyridine (DTDP)-type molecules as bridging ligands and the metal centers of ruthenium, iron, and osmium. Acid–base properties of the coordinated DTDP in the monomers and near-infrared and electrochemical data for the binuclear complexes [{Ru(NH₃)₅}₂DTDP]-(PF₆)₄ and Na₆[{Fe(CN)₅}₂DTDP] and their respective mixed-valence complexes indicate intense electron delocalization between the two metal centers. The Mössbauer spectral data for the mononuclear and binuclear ironpentacyano DTDP complexes also are consistent with the assignment of a valence-delocalized system.

$\mathbf{T}$HE BASIC CHEMISTRY OF SULFUR LIGANDS has received less attention than the chemistry of the corresponding oxo and nitrogen species, despite the well-recognized importance of sulfur compounds in biological processes and for defining properties in a wide range of different materials. This sulfur chemistry began to be systematically investigated only in the 1960s (*1, 2*).

Henry Taube recognized the relevance of the sulfur-containing ligands and devoted efforts toward the development of their chemistry. The first well-characterized H_2S transition metal complex was characterized at Taube's laboratory by C. G. Kuehn (*3*), and a series of related sulfur ligands were then investigated (*3–6*). The reduction of the *trans*-[Ru(NH₃)₄ClSO₂]Cl to [Ru-(NH₃)₄ClS]₂Cl₂ has been cited (*7*) as one of the few examples of oxygen removal from coordinated SO_2 and demonstrates the Ru–S bond strength.

$$[(NH_3)_5Ru^{(III)} - N\!\!\bigcirc\!\!\rangle - S - \langle\!\!\bigcirc\!\!N - Ru^{(II)}(NH_3)_5]^{5+}$$

I

II   **III**

The existence of an intervalence transition for the species in structure **I** led Henry Taube to suggest the availability (8) of $3d_\pi$ sulfur orbitals for electron delocalization between the two centers. No experimental evidence for electron delocalization has been obtained in studies with the $[isn(NH_3)_4RuS]_2^{4+}$ ion (4) and related species (9–11). The fact that disulfides react with $[Ru(NH_3)_5\text{-}(H_2O)]^{2+}$ fairly rapidly to produce Ru(III) thiolate (12) almost quantitatively and the advances in the field of superconductors using dithiols (13, 14) as synthetic starting material led to the interest in investigating ligands where the sulfur atoms are spaced apart by one or more carbon atoms (12, 15). As a consequence, quite remarkable examples of sulfur-to-sulfur through-space interaction have been described (12, 15–20).

The reduction of the –S–S– bridge occurs (12, 21) when the 2,2′-dithiodipyridine molecule is allowed to react with $[Ru(NH_3)_5(H_2O)]^{2+}$. However, this reduction does not occur when the ligand is 4,4′-dithiodipyridine [*see* structure **II** (4,4′-dithiodipyridine) and structure **III** (2,2′-dithiodipyridine)]. With 4,4′-dithiodipyridine, quite stable mono and binuclear species have been isolated (22, 23). Furthermore the disulfide bridge, when coordinated to the metal centers (15, 16), becomes resistant to reductive attack (zinc amalgam, ascorbic acid).

Therefore, dealing with such simple and easy implementable models, where the spacers are 4,4′-dithiodipyridine (DTDP)-type ligands and the metal centers are ruthenium (22, 23), iron (23), and osmium (24) complexes, we investigate the –S–S– bridge capability for conducting electrons.

In this chapter we describe the Mössbauer parameter isomer shift and quadrupole splitting for the $[Fe^{II}(CN)_5\text{–}DTDP\text{–}Fe^{II}(CN)_5]^{6-}$ and $[Fe^{II}(CN)_5\text{–}DTDP\text{–}Fe^{III}(CN)_5]^{5-}$ species, the changes in the basicity of the uncoordinated pyridinic nitrogen (pK_a), the determination of the comproportionation constants (K_c), and the near-infrared spectrum characteristics (λ, ε) for the intervalence absorption exhibited by the mixed-valence species.

Isomer Shift and Quadrupole Splitting

The Mössbauer spectra of mononuclear ironpentacyano complexes are doublets due to the presence of an electric field gradient with noncubic symmetry in the ^{57}Fe nucleus (25). The isomer shift δ reflects the changes in the electronic density at the nucleus, which are caused by modifications in the electron populations of valence orbitals of the Mössbauer atom. In the iron pentacyano complexes the noncubic electronic configuration of the central ions comes from the asymmetric σ- and π-bonding involving only the axial ligand L. The σ-donation mechanism increases the s-electron population causing consequently an increase of the electronic density at the metal. The σ-acceptance mechanism also increases the s-electron density at the metal due to the decrease in the shielding effects by the donation from metal dπ orbital to ligand pπ* orbital.

For iron pentacyano complexes the σ- and π-bonding involving the axial ligand L are responsible for the appearance of the electric field gradient (EFG) at the iron nucleus and consequently for the quadrupole splitting. In this case the quadrupole splitting is caused by the interaction of the nuclear quadrupole moment of the iron atom with the Z component of the EFG (25, 26).

The isomer shift and quadrupole splitting values for the mononuclear and binuclear species are shown in Table I.

Fe(III)–DTDP has a d^5 low-spin configuration decreasing the shielding effect of the s-electrons compared with Fe(II)–DTDP d^6 low-spin configuration. On the other hand, d^5 low-spin configuration leads to a bigger EFG due to the asymmetric configuration. Thus the isomer shift is much smaller, and the quadrupole splitting of the Fe(III)–DTDP monomer is much bigger, than those observed for the Fe(II)–DTDP mononuclear complex.

The Mössbauer spectra of Fe(II)–DTDP–Fe(II) and Fe(II)–DTDP–Fe(III) complexes, illustrated in Figure 1, showed two absorption lines indicating identical coordination sites of DTDP to the ^{57}Fe. Since there is no evidence of either Fe(III) or Fe(II) quadrupole splitting (at least at room temperature), it is an indication of an electronic delocalized system. The bigger value of quadrupole splitting (Δ_{QS}) observed for the mixed-valence complex compared with that of the fully reduced species is consistent with the existence, for the metal

Table I. Mössbauer Parameter Data

Complex	$\delta \pm 10^{-4}$ *(mm/s)*	$\Delta_{QS} \pm 10^{-3}$ *(mm/s)*
$Na_3[(CN)_5Fe(DTDP)]\cdot 4H_2O$	0.291	0.864
$Na_5[(CN)_5Fe(DTDP)Fe(CN)_5]\cdot 5H_2O$	0.278	0.740
$Na_6[(CN)_5Fe(DTDP)Fe(CN)_5]\cdot 6H_2O$	0.300	0.705
$Na_2[(CN)_5Fe(DTDP)]\cdot 3H_2O$	0.220	1.709

NOTE: δ means isomer shift; Δ_{QS} means quadrupole splitting. $T = 300$ K; sodium nitroprusside was used as standard. DTDP is 4,4'-Dithiodipyridine.

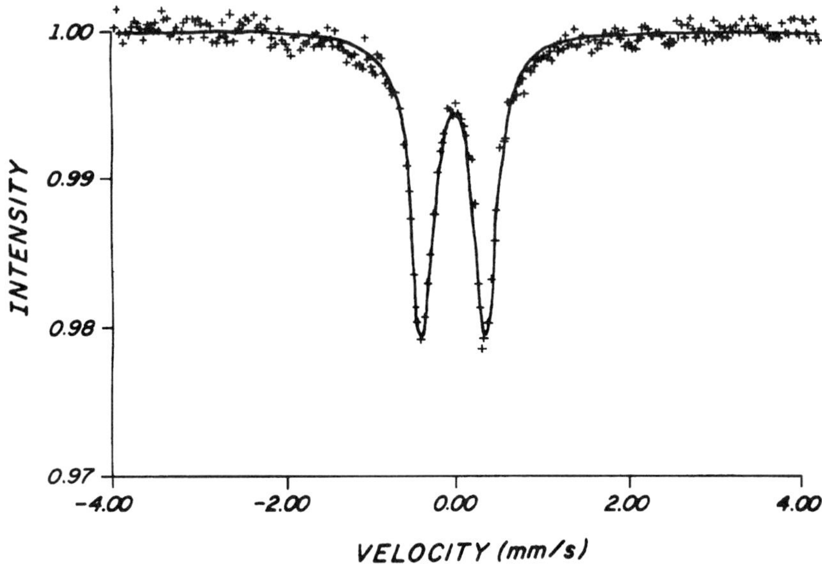

Figure 1. Mössbauer spectrum of $Na_5[(CN)_5FeDTDPFe(CN)_5]\cdot6\,H_2O$.

centers in the Fe(II)–DTDP–Fe(III) species, of an intermediate formal valence state between Fe(II)–DTDP–Fe(II) and Fe(III)–DTDP–Fe(III).

The 3d–3d electronic population in the mixed-valence species is smaller than that of the Fe(II)–Fe(II) complex, decreasing the shielding effect. Thus, the smaller isomer shift observed in the spectrum of the mixed-valence species not only is in agreement with this observation but is also consistent with the assignment of a valence-delocalized system.

Acid–Base Properties and pK_a

Acid–base properties of the coordinated ligands, when compared with those of the uncoordinated ligand (*27–30*), can provide information on the interplay between the σ and π components of the M–L bond. Thus the pK_a for the reaction types shown in Scheme I has been extensively (*25–28*) used to estimate how the σ and π components of the $Ru^{II,III}$–L bond change with the change in the metal-center oxidation state.

As can be observed in Table II, for L = pz (pyrazine), as a consequence of the coordination to the Ru(II) center, the pzH^+ fragment exhibits an increase in its pK_a value of 1.9. units. This is interpreted (*27–30*) as due to the Ru(II) → pz, $4d_\pi \to \pi^*$ backbonding interaction, which will be responsible for transferring electron density from the metal center to the heterocyclic ring, and then

$$\text{N} \bigcirc \text{NH}^+ + H_2O \rightleftharpoons H_3O^+ + \text{N} \bigcirc \text{N}$$

$$(H_3N)_5 M^{II,III} \, \text{N} \bigcirc \text{NH}^+ + H_2O \rightleftharpoons (H_3N)_5 M^{II,III} \, \text{N} \bigcirc \text{N} + H_3O^+$$

Scheme I.

Table II. pK_a Data for Coordinated and Uncoordinated DTDPH$^+$ and pzH$^+$ Acids at 25 °C

Acid	pK_a	Ref.
$[Ru(NH_3)_5DTDPH]^{3+}$	5.25	23
$[Ru(NH_3)_5DTDPH]^{4+}$	3.20	23
$[Os(NH_3)_5DTDPH]^{3+}$	5.50	24
$[Os(NH_3)_5DTDPH]^{4+}$	3.20	24
DTDPH$^+$	4.80	23
pzH$^+$	0.6	27
$[Ru(NH_3)_5pzH]^{3+}$	2.5	27
$[Ru(NH_3)_5pzH]^{4+}$	−0.8	27

NOTE: DTDP is 4,4′-dithiodipyridine; pz is pyrazine.

to the uncoordinated nitrogen atom. The polarization effects in this case are overcome by the backbonding (*27–30*).

However, for the Ru(III) complex, in the absence of the $4d_\pi$ electrons available for backbonding to the pzH$^+$ ligand, the inductive polarization effects of the good Lewis acid Ru(III), $4d^5$, dominates. Thus, electron density is withdrawn from the ligand, and the electron pair of the uncoordinated nitrogen becomes less available for the proton. Accordingly, a decrease of 1.4 pK_a units is exhibited for the acid pzH$^+$ when coordinated to Ru(III). A comparison between the DTDPH$^+$ and the pzH$^+$ system pK_a values showed that the same tendency is observed for both cases (*22, 31*). However, it should be emphasized that on the DTDPH$^+$ the protonated nitrogen is located on another pyridinic ring separated from the (NH$_3$)$_5$Rupz moiety by the −S−S− bridge. Thus, the electronic effects are transmitted to the protonated nitrogen, through the −S−S− bridge, distant from the coordinated nitrogen at least by 10 Å.

Comproportionation Constants and Near-Infrared Data

Comproportionation constants (K_c) and near-infrared band characteristics (λ, ε) are well-accepted as good parameters (*32–35*) to analyze the mediator ability of a ligand (L) that bridges two metal centers on different oxidation states M^{II}–L–M^{III}.

Based on the voltammetric spectra (23) of the $[\{Ru(NH_3)_5\}_2DTDP]^{4+}$ ions $[(E_{1/2})_1 = -130$ V and $(E_{1/2})_2 = 0.160$ V$]$ and of the $[\{Fe(CN)_5\}_2DTDP]^{6-}$ species $[(E_{1/2})_1 = 0.155$ V and $(E_{1/2})_2 = 0.275$ V$]$, the comproportionation constants $K_c = 8 \times 10^4$ and 1×10^2 have been calculated for the reaction in eq 1:

$$(II, II) + (III, III) \underset{}{\overset{K_c}{\rightleftharpoons}} 2(III, II) \tag{1}$$

The intervalence band (23) exhibited by the $[\{Ru(NH_3)_5\}_2DTDP]^{5+}$ (see Figure 2) has the characteristics $\lambda_{max} = 1.500$ cm^{-1}, $\varepsilon = 4.3 \times 10^3$ M^{-1} cm^{-1}, bandwidth at half-intensity $(\Delta v_{1/2}) = 1.24 \times 10^3$ cm^{-1} (calc. = 3,924) (33–34), and electronic coupling value $(H_{AB}) = 855$ cm^{-1}. For the $[\{Fe(CN)_5\}_2 DTDP]^{5-}$ species, the intervalence band is centered at $\lambda_{max} = 1.195$ nm, $\varepsilon = 9.0 \times 10^2$ M^{-1} cm^{-1}, $\Delta v_{1/2} = 4.02 \times 10^3$ cm^{-1} (calc. = 4.396) (32–33), and $H_{AB} = 466$ cm^{-1} (see Figure 3). The energetic barrier (ΔG^*_{th}) estimated (33–34) on the basis of the energy (E_{op}) of the intervalence band is 4.7 and 5.9 kcal mol^{-1} for the ruthenium and iron DTDP complex data, respectively, which are comparable with the value of 5.6 kcal mol^{-1} calculated (34) for the $[\{Ru(NH_3)_5\}pz_2]^{5+}$ system.

These data collected from electrochemical and spectroscopic measurements clearly indicate a strong coupling between the metal centers bridged through the DTDP ligand.

As observed in Table III, the –S–S– bridge provides more electron delocalization than NH, methylene, ethylene, and acetylene groups (23). Among the listed examples, the DTDP efficiency on conducting electrons is only sur-

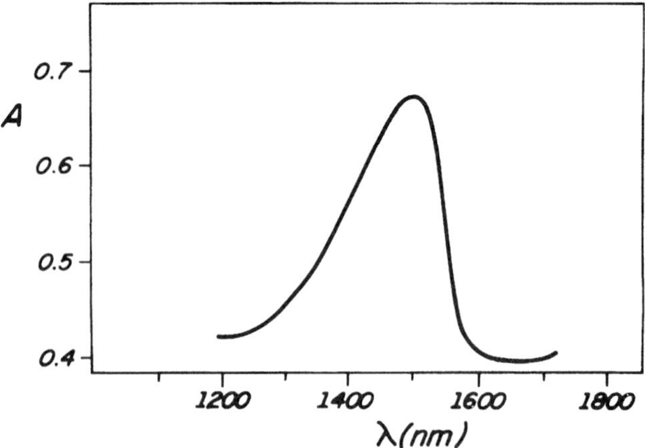

Figure 2. Near-infrared spectra for the mixed-valence species [Ru^{II}, Ru^{III}] in D$_2$O. (Reproduced from reference 23. Copyright 1994 American Chemical Society.)

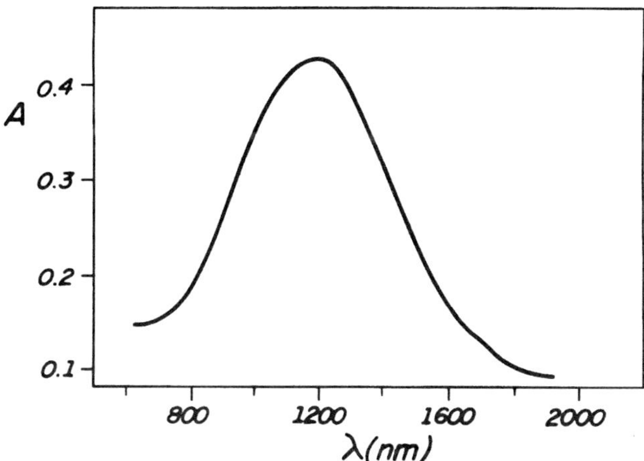

Figure 3. Near-infrared spectra for the mixed-valence species [Fe^{II}, Fe^{III}] in D_2O. (Reproduced from reference 23. Copyright 1994 American Chemical Society.)

passed by pyrazine. Again, it is worthwhile to keep in mind that in the pz ligand, both nitrogen atoms are in the same ring and provide a bridge of about 6 Å between the metal centers.

Table IV summarizes K_c, λ_{max}, and ε for binuclear ruthenium mixed-valence complexes, where the bridging ligand is a bidentate sulfur ligand in which the S atoms are apart one from the other. For all these systems, some electronic coupling between the metal centers, through L, has been observed. On these ligands, the sulfur atoms are not directly linked one to the other, and since only a σ-bonded framework is involved, the bridge mediation effects have been explained either on the basis of sulfur–sulfur through-space interaction (*12, 15–17*) for the small ligands, or assuming some electron tunneling for the spiro-ring-type ligands (*19, 20*).

The bridging efficiency of DTDP is quite remarkable compared to the other sulfur ligands (Table III–V). Although no X-ray structure is available at the moment for the binuclear species MA^{III}_5–DTDP – MA^{II}_5, (M = Ru and Fe; A = NH_3 and CN^-), it is reasonable to assume that the two pyridine rings, coordinated each one to a charged metal center, will be apart, one from the other, to minimize the electrostatic repulsion. Therefore, through-space interactions are unlikely to occur in such systems.

The introduction of one CH_2 group on each side of the –S–S– bridge acts as an insulator, since no coupling between the two ruthenium centers could be observed (*36*) in the corresponding binuclear complex.

The bis(4-pyridyl)sulfide (DPS) ligand resembles the DTDP species, but it has only one sulfur atom linking the two pyridine rings. On the DPS ruthe-

Table III. Near-Infrared Data and Comproportionation Constants

Metal System	L	NIR Bands		K_c
		λ_{max} (nm)	ε (M^{-1} cm^{-1})	
$[(NH_3)_5Ru]_2$	pyrazine (N⎓N)	1570	5000	4×10^6
	4,4'-dipyridyl disulfide (py–S–S–py)	1500	4260	8×10^4
	di(4-pyridyl)amine (py–N–py)	920	1010	5×10^2
	1,2-bis(4-pyridyl)ethylene (py–CH=CH–py)	960	760	2×10
	bis(4-pyridyl)acetylene (py–C≡C–py)	920	640	14
	4,4'-bipyridine (py–py)	1030	920	2×10
	bis(4-pyridyl)methane (py–CH$_2$–py)	810	30	9.8
$[(CN)_5Fe]_2$	pyrazine (N⎓N)	1200	2200	5×10^2
	4,4'-dipyridyl disulfide (py–S–S–py)	1195	900	1×10^2
	1,2-bis(4-pyridyl)ethylene (py–CH=CH–py)	1300	600	<10
	4,4'-bipyridine (py–py)	1200	1100	<10

NOTE: L means ligand; NIR means near infrared; λ_{max} is wavelength of maximum absorption; ε is molar absorptivity. Comproportionation constant (K_c) for (III,III) + (II,II) $\xrightleftharpoons{K_c}$ 2(III, II) is equal to exp ($\Delta E_{1/2}/25.68$), T = 298 K, $n_1=n_2=1$.
SOURCE: Data are adapted from refs. 23 and 24.

Table IV. Near-Infrared Bands and Comproportionation Constants for Binuclear Ruthenium (II, III) Mixed-Valence Compounds, $[(NH_3)_5Ru^{II}–L–Ru^{III}(NH_3)_5]^{5+}$

L	*NIR Bands*			
	λ_{max} *(nm)*	ε *($M^{-1}\ cm^{-1}$)*	K_c	*Ref.*
	1220	86	890	12
	1000	64	~185	12
	996	55	~100	12
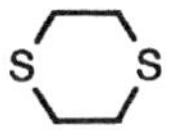	972	6	~40	12
	1210	80	<10	12
	928	45	125	12
	840	21	≤10	12
	775	5	<10	12
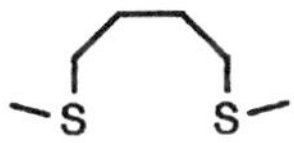	615	0.75	≤10	12
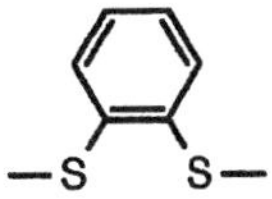	1132	35	~48	12
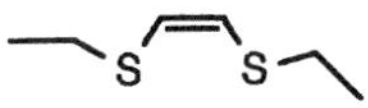	964	22	~30	12
	910	43	≤10	12, 18
	808	9		19
	690	2.3±0.7		19
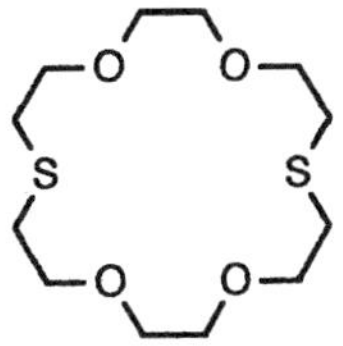	860	3	≤10	12

SOURCE: Data are adapted from ref. 10.

Table V. Near-Infrared Bands and Comproportionation Constants for Binuclear Ruthenium (II,III) Mixed-Valence Compounds, $[(NH_3)_5Ru^{II}-L-Ru^{III}(NH_3)_5]^{5+}$

L	NIR Bands		K_c	Ref.
	λ_{max} (nm)	ε (M⁻¹ cm⁻¹)		
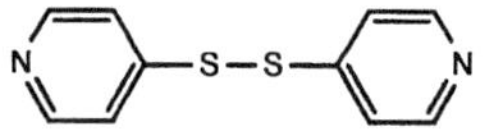 (4-pyridyl)–S–S–(4-pyridyl)	1.500	4.260	8×10^4	22
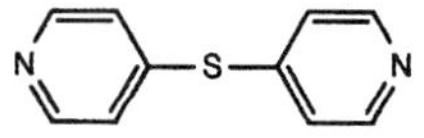 (4-pyridyl)–S–(4-pyridyl)	855	70	<10	8
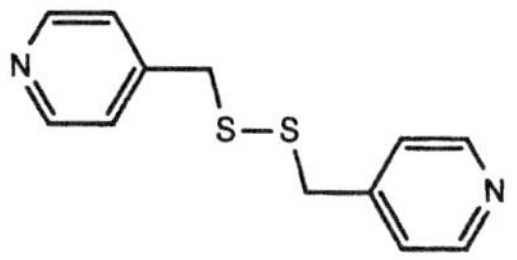 (4-pyridyl-CH₂)–S–S–(CH₂-4-pyridyl)			<10	31
—S—S—			<10[a]	4

[a]For the complex $[ins(NH_3)RuSSRu(NH_3)isn]^{5+/4+}$.

nium binuclear complexes, the coupling between the two metal centers is so weak that only one redox process can be observed in the $[(NH_3)_5Ru$ DPS $Ru(NH_3)_5]^{4+}$ voltammetric spectra (8).

These results will suggest that the same degree of coupling will be expected for the $[A_5RuS–SRuA_5]^{5+/4+}$ system. As far as the electrochemistry is concerned, no coupling has been observed (4) for the isonicotinamide (isn) derivative $[isn(NH_3)_4Ru–S]_2^{5+/4+}$ system, and no mixed-valence transition has yet been reported for the $[(NH_3)_5Ru–S–S–Ru(NH_3)_5]^{4+}$ species (9, 11).

It is tempting to explain the difference in behavior of the S–S bridge in the $[A_5RuS–S–RuA_5]^{5+/4+}$ and $[A_5RuDTDPRuA_5]^{5+/4+}$ species on the basis of the torsion or dihedral angle on the X–S–S–X fragment. For $[(NH_5)Ru–S]_5^{4+}$, this angle has been measured (9) as 180°, therefore precluding any π interaction along the Ru–S–S–Ru core.

A great deal of the metal ion–sulfur ligands chemistry has been explained using the $nd_\pi \rightarrow 3d_\pi$ backbonding interaction model. There are, however, examples (10) in which σ^* orbitals, derived from atomic 2p on C and 3p on S, have been claimed to substantially contribute to the metal–sulfur bonds. Unfortunately, no quantum mechanic calculations or X-ray crystallographic date are available at the moment for the DTDP complexes. Therefore the question of deciding which orbitals are predominantly involved in the –S–S– linkage mediating the coupling still remains.

Thus, more data on these systems are required before these suggestions can be put forward.

Acknowledgments

The authors thank Fundação de Amparo a Pesquisa do Estado de São Paulo (FAPESP), Conselho Nacional de Denselvolvimento Cientifico e Tecnológico (CNPq), and Financiadora de Estudos e Projetos (FINEP) for financial support.

References

1. Cotton, F. A.; Wilkson, G. *Advanced Inorganic Chemistry*, 5th ed.; Wiley: New York, 1988.
2. Lockyer, T. N.; Martin L. R. *Prog. Inorg. Chem.* **1979,** *27,* 223.
3. Kuehn, C. G.; Taube, H. *J. Am. Chem. Soc.* **1976,** *98,* 689.
4. Brulet, C. R.; Isied, S. S.; Taube, H. *J. Am. Chem. Soc.* **1973,** *95,* 4758.
5. Isied, S. S.; Taube, H. *Inorg. Chem.* **1974,** *13,* 1545.
6. Isied, S. S.; Taube, H. *J. Am. Chem. Soc.* **1973,** *95,* 8198.
7. Ryan, R. R.; Kubas, J. R.; Moody, D. C.; Eller, P. G. *Struct. Bonding (Berlin)* **1981,** *46,* 47.
8. Fischer, H.; Tom. G. M.; Taube, H. *J. Am. Chem. Soc.* **1976,** *98,* 5512.
9. Elder, R. C.; Trkula, M. *Inorg. Chem.* **1977,** *16,* 1048.
10. Kuehn, C. G.; Isied, S. S. *Prog. Inorg. Chem.* **1980,** *27,* 153.
11. Kim, S.; Otterbein, E. S.; Rava, R. P.; Isied, S. S.; Fillipp Jr., J. J.; Waszcyak, J. V. *J. Am. Chem. Soc.* **1983,** *105,* 336.
12. Stein, C. A. PhD Dissertation, Stanford University, 1978.
13. Willians, J. M. *Prog. Inorg. Chem.* **1988,** *33,* 183.
14. Williams, J. M.; Wang, H. H.; Emge, T. J.; Geiser, V.; Beno, M. A.; Levng, P. C. W.; Carlson, D .K.; Thom, R. J.; Schultz, A. J.; Whangbo, M. H. *Prog. Inorg. Chem.* **1987,** *35,* 51.
15. Stein, C. A.; Taube, H. *J. Am. Chem. Soc.* **1978,** *100,* 1635.
16. Stein, C. A.; Taube, H. *Inorg. Chem.* **1979,** *18,* 1168.
17. Stein, C. A.; Taube, H. *Inorg. Chem.* **1979,** *18,* 2212.
18. Stein, C. A.; Taube, H. *J. Am. Chem. Soc.* **1981,** *103,* 693.
19. Stein, C. A.; Lewis, N. A.; Seitz, G. *J. Am. Chem. Soc.* **1982,** *104,* 2596.
20. Lewis, N. A.; Obeng, S. Y.; Taveras, D. V.; van Eldik, R. *J. Am. Chem. Soc.* **1989,** *111,* 924.
21. Marinho, L. A.; Franco, D. W., unpublished observation.
22. Moreira, I. S.; Franco, D. W. *J. Chem. Soc. Commun.* **1992,** *5,* 450.
23. Moreira, I. S.; Franco, D. W. *Inorg. Chem.* **1994,** *33,* 1607.
24. Lima, J. B.; Neto, A.M.; Gandra, F. C. G.; Franco, D. W. *Proceedings of the 30th International Conference on Coordination Chemistry;* Kyoto, Japan, 1994; pp 2–152.
25. Coelho, A. L.; Moreira, I. S.; Araujo, J. H.; Araujo, M. A. B. *Radional. Chem. Lett.* **1989,** *136,* 299.
26. Dickson, D. P. E.; Berry, F. J. *Mössbauer Spectroscopy;* Cambridge University Press: London, 1986.
27. Ford, P. C.; Rudd, F. P.; Gaunder, R.; Taube, H. *J. Am. Chem. Soc.* **1968,** *90,* 1187.
28. Taube, H. *Surv. Prog. Chem.* **1973,** *6,* 1.
29. Taube, H. *Comments Inorg. Chem.* **1981,** *1,* 17.
30. Seddon, A. E.; Seddon, K. In *Topics in Inorganic and General Chemistry;* Clark, R. J. H., Ed.; Elsevier: London, 1984.
31. Moreira, I. S.; Araujo, M. A. B.; Franco, D. W., submitted to *Inorg. Chim. Acta.*
32. Taube, H. *Ann. N. Y. Acad. Sci.* **1978,** *313,* 481.

33. Hush, N. S. *Prog. Inorg. Chem.* **1967**, *8*, 391.
34. Creutz, C. *Progr. Inorg. Chem.* **1983**, *30*, 1.
35. Isied, S. S. *Metal Ions in Biological Systems;* Sigel H., Ed.; Dekker: New York, 1991; pp 1–56.
36. Moreira, I. S.; Lima, E. C. unpublished observations.

16

Kinetic and Equilibrium Studies of the Reactions of Cysteine and Penicillamine with Aqueous Iron(III)

M. J. Sisley and R. B. Jordan*

Department of Chemistry, University of Alberta, Edmonton, Alberta, Canada T6G2G2

The reactions of cysteine and penicillamine (H_3L^+) with aqueous iron(III) have been studied in dilute acid solutions (0.01–0.2 M H^+) with iron(III) concentrations (0.01–0.1 M) in excess over the amino acid (~1.5–8.5 × 10^{-4} M) at 25 °C in 1.0 M $LiClO_4/HClO_4$. Stopped-flow and standard spectrophotometry were used to measure the kinetics of the complexation and oxidation–reduction reactions, respectively. The absorbance immediately after the complexation reaction has been used to determine complexation equilibrium constants, defined as $\{[(H_2O)_n\text{-}FeLH^{2+}][H^+]^2\}/\{[Fe(OH_2)_6^{3+}][LH_3^+]\}$ and $\{[(H_2O)_m(\mu\text{-}OH)_2Fe_2LH^{3+}][H^+]^2\}/\{[Fe_2(OH_2)_8(\mu\text{-}OH)_2^{4+}][LH_3^+]\}$, with values of 0.063 and 0.38 M for penicillamine, and 0.025 and 0.07 M for cysteine, respectively. The dependence of the complexation rate on iron(III) and H^+ concentrations indicates that the dominant reaction pathways and their rate constants (M^{-1} s^{-1}) are (1) for $(H_2O)_5Fe(OH)^{2+}$ + H_3L^+: k = 1.4 × 10^3 (cys); 1.2 × 10^3 (pen); (2) for $(H_2O)_5Fe(OH)^{2+}$ + H_2L: k = 7.4 × 10^3 (cys); 7.4 × 10^3 (pen); and (3) for $Fe_2(OH_2)_8(\mu\text{-}OH)_2^{4+}$ + H_2L: k = 4.25 × 10^3 (cys); 8.1 × 10^3 (pen). The oxidation of cysteine is second-order in cysteine, with terms in the rate law first, second, and third order in $Fe(OH_2)_6^{3+}$. From the $[H^+]$ dependence, these are assigned to reactions of $(H_2O)_4FeLH^{2+}$ + H_2L, $(H_2O)_4FeLH^{2+}$ + $(H_2O)_4FeLH^{2+}$ or $Fe_2(OH_2)_6(\mu\text{-}OH)_2(LH)^{3+}$ + H_2L, and $Fe_2(OH_2)_6(\mu\text{-}OH)_2(LH)^{3+}$ + $(H_2O)_4FeLH^{2+}$. Penicillamine reacts similarly but more slowly, and the first pathway is not observed, but one fourth-order in iron(III) is found. With excess penicillamine the reaction is second-order in penicillamine and is inhibited by H^+ and iron(II), and the reactive species is suggested to be $Fe(LH)_2^+$.

*Corresponding author.

$\mathbf{M}$ERCAPTOCARBOXYLIC ACIDS OF THE GENERAL FORMULA ($HSRCO_2H$) react with aqueous iron(III) to form blue complexes that subsequently undergo oxidation–reduction to the corresponding disulfide and iron(II). Equilibrium studies (*1–4*) have shown that two protons are released on complexation so that the $-CO_2^-$ and $-S^-$ groups are coordinated to iron(III) in the blue species. Cysteine (**I**) and penicillamine (**II**) also form blue complexes, and the observations of Stadtherr and Martin (*5*) show that $-S^-$ and $-CO_2^-$ are coordinated because S-methyl-L-cysteine and cysteine methyl and ethyl esters give no blue color.

Furthermore, the amino group is not coordinated in acidic solution because the *N*-acetyl derivatives of cysteine and penicillamine give the blue color, and penicillamine complexation (*4*) involves release of only two protons. Tomita et al. (*6*) characterized 1:1 complexes of cysteine and thioglycolic acid at −78 °C in 90% ethanol:water with absorption maxima at 620 nm ($\varepsilon \approx 500$–600 M^{-1} cm^{-1}). They also isolated violet tris complexes ($\lambda_{max} \sim 590$ nm, $\varepsilon \sim 3 \times 10^3$ M^{-1} cm^{-1}, for cysteine) and red complexes ($\lambda_{max} \sim 490$ nm, $\varepsilon \sim 1 \times 10^3$ M^{-1} cm^{-1}, for cysteine), which they assigned to $Fe(OH)(SRCO_2)_2^{2-}$.

The complexation kinetics of several $HSRCO_2H$ systems have been studied by McAuley and co-workers (*1–3*) and by Baiocchi et al. (*4*). These studies involved different conditions, with iron(III) in excess in the former and $HSRCO_2H$ in excess in the latter, and the kinetic results are in reasonable agreement and follow the usual pattern for substitution on aqueous iron(III). Cysteine was not studied, but Baiocchi et al. did study penicillamine.

More recently, Jameson and co-workers (*7, 8*) studied the reaction of aqueous iron(III) with excess cysteine at pH 2.7–4.87 and 8.5–11.68. These workers seem to have been unaware of the earlier studies of Stadtherr and Martin, Tomita et al., and Baiocchi et al., and in the lower pH range, Jameson et al. (*7*) assigned a blue species (λ_{max} 614 nm, ε 1.03 × 10^3 M^{-1} cm^{-1}) to a mono-complex of fully deprotonated cysteine with S- and N-coordination, while all previous work would indicate S- and O-coordination with an uncoordinated $-NH_3^+$. At pH > 8, the species assigned as $Fe(OH)(L)$ and $Fe(OH)(L)_2^{2-}$ by Jameson et al. appear to be $Fe(OH)(L)_2^{2-}$ and $Fe(L)_3^{3-}$, respectively, from the observations of Tomita et al. (*6*).

McAuley and co-workers (*3*) and Baiocchi et al. (*4*) studied the oxidation step for several $HSRCO_2H$ systems, although neither studied cysteine, and McAuley et al. noted only that penicillamine oxidation is much slower. With excess reductant, Baiocchi et al. proposed formation of a steady-state amount

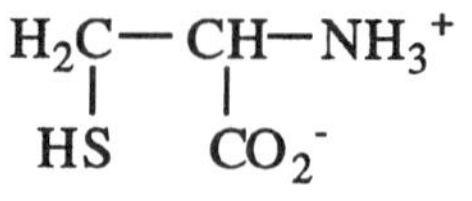

$$H_2C-CH-NH_3^+ \qquad\qquad (H_3C)_2C-CH-NH_3^+$$
$$\;\;|\qquad\;\;\;| \qquad\qquad\qquad\qquad\;\;\;|\qquad\;\;\;|$$
$$HS\quad\;\; CO_2^- \qquad\qquad\qquad\quad HS\quad\;\; CO_2^-$$

$$\mathbf{I} \qquad\qquad\qquad\qquad\qquad\qquad \mathbf{II}$$

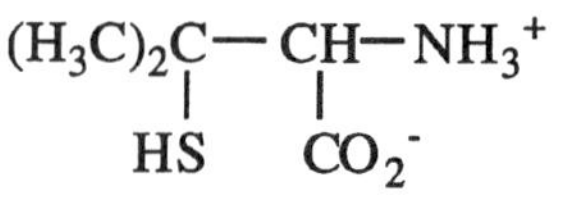

III   **IV**

of bis-complex (Fe(SRCO$_2$)$^{2-}$) as the reactive species to explain the second-order dependence on [HSRCO$_2$H]. Our analysis of the published observations for 2-mercaptosuccinic acid confirms the second-order dependence, but there is some problem with the quantitative fitting of the data. McAuley and co-workers found that the redox reactions were effectively second-order in both iron(III) and [HSRCO$_2$H] under conditions of [iron(III)] > [HSRCO$_2$H] in 0.2–0.4 M H$^+$. The detailed analysis indicated that the reaction was second-order in the complex (H$_2$O)$_4$Fe(SRCO$_2$)$^+$. McAuley et al. suggested that the oxidation might proceed through a sulfur bridged di-iron(III) species (**III**). One purpose of the present study is to determine if the structurally analogous μ-dihydroxy dimer (**IV**) might show complexation and oxidation reactivity.

The oxidation rate laws show the general feature that two mercaptocarboxylic acids are necessary to reach the transition state, and two iron(III) units are used if iron(III) is in excess. It is remarkable that these systems use this high-order reaction pathway rather than simple one-electron oxidation by iron(III) to give an unstable organic radical that is rapidly oxidized further.

Jameson et al. (7) did propose that oxidation of cysteine by iron(III) at pH 2.7–4.8 involves simple intramolecular electron transfer within the mono-cysteine–iron(III) complex. However, if this is the case, then it is difficult to explain why such a pathway was not found for the mercaptocarboxylates (3, 4), or found in the present study on cysteine. It seems more probable that Jameson et al. were observing decomposition of the bis-cysteine–iron(III) complex, since this makes their observations more compatible with those of Baiocchi et al.

The energetics of the oxidation of thiol functions provides some indication of why these systems prefer routes more complex than simple intramolecular electron transfer. Armstrong and co-workers (9) have provided values for reduction potentials of a number of species of interest, including mercaptoacetate, cysteine, and penicillamine. The potentials are rather insensitive to the substituents on the HS– group, and generally may be summarized as follows:

$$\bullet S\text{\small$\sim\!\sim$}CO_2^- + e^- = {}^-S\text{\small$\sim\!\sim$}CO_2^- \qquad\qquad E^\circ = 0.77\ \text{V}$$

$$^-O_2C\text{\small$\sim\!\sim$}S\!-\!S^{\bullet-}\text{\small$\sim\!\sim$}CO_2^- + e^- = 2\ {}^-S\text{\small$\sim\!\sim$}CO_2^- \qquad E^\circ = 0.57\ \text{V}$$

$$^-O_2C\text{\small$\sim\!\sim$}S\!-\!S\text{\small$\sim\!\sim$}CO_2^- + e^- = {}^-O_2C\text{\small$\sim\!\sim$}S\!-\!S^{\bullet-}\text{\small$\sim\!\sim$}CO_2^- \qquad E^\circ = -1.57\ \text{V}$$

$$\bullet S\text{\small$\sim\!\sim$}CO_2^- + H^+ + e^- = HS\text{\small$\sim\!\sim$}CO_2^- \qquad\qquad E^\circ = 1.33\ \text{V}$$

$$^-O_2C\text{\small$\sim\!\sim$}S\!-\!S^{\bullet-}\text{\small$\sim\!\sim$}CO_2^- + 2H^+ + e^- = 2HS\text{\small$\sim\!\sim$}CO_2^- \qquad E^\circ = 1.72\ \text{V}$$

$$^-O_2C\text{\small$\sim\!\sim$}S\!-\!S\text{\small$\sim\!\sim$}CO_2^- + 2H^+ + e^- = 2HS\text{\small$\sim\!\sim$}CO_2^- \qquad E^\circ = -0.03\ \text{V}$$

The E° for the last reaction (at 1.0 M H^+) is consistent with measured values (*10, 11*) of about –0.25 V at pH ~7 for various species. It is apparent from the first two equations that oxidation to form a radical in these systems is highly unfavorable, although overall two-electron oxidation in the last reaction is possible with quite mild oxidizing agents. The E° for the first reaction can be combined with the known complexation constants (*4*) and $E^\circ(Fe^{III}/Fe^{II}) = 0.75$ V to estimate the driving force for the following intramolecular electron transfer reaction.

$$Fe^{III}(S\text{\small$\sim\!\sim$}CO_2^-) \rightleftharpoons Fe^{II} + \bullet S\text{\small$\sim\!\sim$}CO_2^-$$

This reaction has E° of –0.54 V ($K = 9 \times 10^{-10}$ M) for cysteine and penicillamine. Clearly the intramolecular electron transfer is quite unfavorable thermodynamically. If the reverse of this reaction is assumed to have a diffusion-controlled rate constant of ~10^9 M^{-1} s^{-1}, then the upper limit for the forward rate constant is <0.9 s^{-1}.

The situation is even more unfavorable for the following reaction,

$$Fe^{III}(S\text{\small$\sim\!\sim$}CO_2^-)_2 \rightleftharpoons Fe^{II} + {}^-O_2C\text{\small$\sim\!\sim$}S\!-\!S^{\bullet-}\text{\small$\sim\!\sim$}CO_2^-$$

which has E° of –0.79 V ($K = 4 \times 10^{-14}$ M) for cysteine and penicillamine. This calculation has assumed that the formation constant of the bis complex is one-fifth that of the known value of the mono complex, but changes of a factor of 10 in this ratio just change the E° by ±0.06 V.

On the other hand, the following overall bimolecular reaction is quite favorable,

$$2Fe^{III}(S\text{\small$\sim\!\sim$}CO_2^-) \rightleftharpoons 2Fe^{II} + {}^-O_2C\text{\small$\sim\!\sim$}S\!-\!S\text{\small$\sim\!\sim$}CO_2^-$$

and has E° of 0.24 V ($K = 1 \times 10^8$ M) for cysteine and penicillamine.

The present work was undertaken to determine how the biologically important cysteine and penicillamine fit into the general picture from the mercaptocarboxylic acids. The conditions of [iron(III)] > [amino acid] and modest acidities (0.01 to 0.2 M H^+) were chosen to determine if the μ-dihydroxy dimer (**IV**) of iron(III) might show unusual reactivity. This species could bring two iron(III) and two –S– groups together in a manner somewhat analogous to the intermediate **III** proposed by McAuley et al. (*3*), and might impart unusual reactivity.

Experimental Methods

Materials. The L-cysteine hydrochloride monohydrate, L-cystine, penicillamine hydrochloride, DL-penicillamine, and D-penicillaminedisulfide were used as received (Aldrich). Stock solutions of iron(III) perchlorate were prepared by dissolving primary standard grade iron wire (Allied Chemical) in excess 3.5 M $HClO_4$, and oxidizing the iron(II) so produced with H_2O_2. The concentration of $HClO_4$ in the final solution was determined by titrating the H^+ released from Dowex 50 W–X8(H^+) cation resin and correcting for the H^+ released by adsorption of iron(III). Solutions of iron(II) perchlorate were prepared by dissolving hydrated $Fe(ClO_4)_2$ (Alfa) in aqueous perchloric acid. The iron(II) content was determined spectrophotometrically as the 1,10-phenanthroline complex, and the solutions were found to contain <4 % of the total iron as iron(III).

The lack of reproducibility of the kinetics with a deficiency of penicillamine was a source of frustration for some time. During more than 40 runs, a number of variables were tested such as the source of the penicillamine; the source and age of the iron(III) solutions; and the addition of iron(II), penicillamine disulfide, or copper(II), at the same level as the penicillamine. None of these produced a consistent kinetic pattern or reproducible results. A key observation to resolving the problem involved addition of more penicillamine to a solution after the reaction was complete. For such an experiment, a typical reaction solution was prepared by adding a 2.0 mL of penicillamine solution to 50 mL of acidic aqueous iron(III) (0.0188 M H^+, 0.010 M total iron(III) after mixing). After 70 min, the blue color had faded and a 12.5-mL aliquot of this solution was taken and mixed with 0.50 mL of the original penicillamine stock solution; the reaction was monitored spectrophotometrically. This was repeated for aliquots taken after 120 and 240 min. Qualitatively, the time for the absorbance to reach half of its initial value was 250, 110, and 180 s for runs started 70, 120, and 240 min, respectively, after the initial addition of penicillamine. At first sight these observations seem to imply that some catalytically active species is forming in the original reaction solution between 70 and 120 min and has started to disappear after 240 min. However, another explanation is that something is happening in the penicillamine stock solution, and that the age and integrity of these solutions is the source of the problem. It was finally found that reproducible results were obtained if the penicillamine stock solution was prepared and kept under anaerobic conditions.

Then, for example, if the experiment described in the preceding paragraph was repeated, the rate constants were 3 ± 0.2 M^{-1} s^{-1} for an anaerobic penicillamine solution 10, 70, 120, and 240 min after preparation. It is suspected that the problem may be trace-metal ion catalysis of the oxidation of penicillamine by O_2.

Although these reactions yield disulfide as the dominant product by thiyl radical combination, the thiyl radicals thought to be produced may be trapped competitively by O_2 under the low penicillamine concentrations of this study.

The kinetics with excess penicillamine were also fraught with reproducibility problems. It was found that the mode of mixing for these more concentrated penicillamine solutions was critical. If the stock solution of penicillamine is prepared by adding water or aqueous $HClO_4/LiClO_4$ to solid penicillamine, then highly variable results are obtained. However, if solid penicillamine hydrochloride is added to the aqueous solution, then slower, reproducible rates are observed. The only obvious difference is that the former method may yield quite concentrated penicillamine solutions during the initial stages of mixing, but it remains unclear to us why this should be a critical feature.

Kinetic Measurements. A Tritech stopped-flow system was used for the complexation kinetics and the details of the equipment have been described previously (*12*), and reported rate constants are averages of 5–8 runs. The redox reactions were mainly studied on a Cary 219 spectrophotometer, but some runs were done on a Hewlett-Packard 8451 spectrophotometer, and some were done on the stopped-flow system. To ensure internal consistency, the absorbance scale of the latter was calibrated between 0.07 and 1.40 absorbance units by comparing the absorbance of bromothymol blue solutions at 615 nm on the stopped-flow and Cary 219 systems.

For the determination of k_2, the time between mixing and start of observation (dead time) must be taken into account for a second-order analysis. For our techniques, the dead time is 5–10 s and can be quite significant, because the reaction half-times are often in the range of 20–50 s, especially for cysteine. This problem can be alleviated if one can calculate the initial absorbance on mixing. This determination can be done easily with penicillamine because the reactions are often slow enough so that the dead time is not a significant factor. For cysteine, we have used the complexation model deduced for penicillamine. The value of K_{f1} was taken from the complexation kinetics and used to determine ε_1 from slow kinetic runs at the higher acidities where the dimer and dead time are both insignificant. Then values of K_{f2} and ε_2 were determined from a number of other runs, with the initial assumption that ε_2 would be similar in magnitude to that for penicillamine and that the dead time must have a reasonable value. Overall good self-consistency is obtained with $\varepsilon_1 = 700$, $\varepsilon_2 = 1500$, and $K_{f2} = 0.07$.

Equilibrium Species Calculations. These calculations used previously determined equilibrium constants and take into account the hydrogen ion introduced with cysteine hydrochloride and that from the hydrolytic reactions of aqueous iron(III). The acid dissociation constants of cysteine and penicillamine were taken as 0.0126 M ($pK_a = 1.9$) (*13*). The hydrolysis (K_m) and dimerization constants (K_D) used were 1.62×10^{-3} M and 1.9×10^{-3} M, respectively (*14*).

Results

When mixed with aqueous iron(III), both cysteine and penicillamine give a blue color that fades with time. The fading is slower and the color appears more intense with penicillamine. The kinetics of both color formation and fad-

ing have been studied here. The rather slow loss of color with penicillamine
has allowed the initial absorbance values to be used to determine the complex
formation equilibrium constant for penicillamine with $Fe(OH_2)_6^{3+}$ and the
dimer $(H_2O)_8Fe_2(\mu\text{-}OH)_2^{4+}$. The cysteine results are consistent with an analogous equilibrium system.

Complex Formation Kinetics.   The kinetics of formation of the blue
color for cysteine and penicillamine have been studied by stopped-flow spectrophotometry at the absorbance maxima of 620 and 605 nm for cysteine and
penicillamine, respectively. Pseudo-first-order conditions were maintained
with the concentration of aqueous iron(III) (1.8–35×10^{-3} M) much larger than
that of cysteine or penicillamine (1.5–5.3×10^{-4} M), and $[H^+]$ was 0.01–0.10 M.
The absorbance–time traces are well-fitted to a first-order rate expression. For
most runs, the acid was in the iron(III) solution before mixing, and the amino
acid solutions contained $<1 \times 10^{-3}$ M H^+. In order to test for the reactivity
contribution of $Fe_2(OH_2)_8(\mu\text{-}OH)_2^{4+}$, a few runs were done with less H^+ in the
iron solution and more with the amino acid. These conditions favor formation
of the dimer before mixing, and it does not undergo dissociation on the time
scale of the complexation.

The kinetic results follow the usual pattern for complexation by aqueous
iron(III), and have been analyzed with the reactions in Scheme I, where
$Fe_2(OH)_2^{4+} = (H_2O)_8Fe_2(\mu\text{-}OH)_2^{4+}$. For both cysteine and penicillamine, the
fully protonated ligand (H_3L^+) undergoes acid dissociation at the carboxyl

$$Fe(OH_2)_6^{3+} \xrightleftharpoons{K_m} (H_2O)_5Fe(OH)^{2+} + H^+$$

$$H_3L^+ \xrightleftharpoons{K_a} H_2L + H^+$$

$$Fe(OH_2)_6^{3+} + H_3L^+ \xrightleftharpoons{K_{f1}} (H_2O)_4Fe(LH)^{2+} + 2H^+$$

$$(H_2O)_5Fe(OH)^{2+} + H_3L^+ \xrightleftharpoons{k_2} (H_2O)_4Fe(LH)^{2+} + H^+ + H_2O$$

$$Fe(OH_2)_6^{3+} + H_2L \xrightleftharpoons{k_3} (H_2O)_4Fe(LH)^{2+} + H^+$$

$$(H_2O)_5Fe(OH)^{2+} + H_2L \xrightleftharpoons{k_4} (H_2O)_4Fe(LH)^{2+} + H_2O$$

$$Fe_2(OH)_2^{4+} + H_2L \xrightarrow{k_d} Fe_2(OH)_2(LH)^{3+} + H^+$$

Scheme I.

group with $pK_a \approx 1.9$ to give the H_2L species, while ionization of the SH and NH_3^+ groups have $pK_a \geq 8$ (13) and are not included.

Scheme I actually predicts that the reaction should be biphasic, because the dimer $Fe_2(OH)_2^{4+}$ is not in equilibrium with the monomeric species on the time scale of these reactions. However, the reactions appear monophasic for two reasons: the dimer contribution is generally small, and the molar absorptivities of the monomer and dimer complex are not greatly different. Therefore, the rate constants have been fitted to eq 1, and quantitative analysis by numerical integration shows that this procedure gives surprisingly good results.

$$k_{obsd} = \left((k_2K_m + k_3K_a)[H^+] + k_4K_mK_a\right)\left\{\frac{[Fe(III)]_m}{(K_a+[H^+])(K_m+[H^+])} + \frac{1}{K_{f1}}\right\}$$
$$+ k_dK_a\left\{\frac{[Fe_2(OH)_2]}{K_a+[H^+]}\right\} \tag{1}$$

Least-squares analysis gives values of $(k_2K_m + k_3K_a)$, $k_4K_mK_a$, k_dK_a (M s^{-1}), and K_{f1} (M) for cysteine and penicillamine and the results are summarized in Table I.

Equilibrium Constant for Iron(III)–Penicillamine Complexation. In order to assess the equilibrium constant(s) for complexation, the initial absorbancies of iron(III)–penicillamine solutions were determined under various concentration conditions. Since the fading of the blue color is relatively slow, only a small linear extrapolation is necessary to obtain the initial absorbance. The time scale for these measurements (15–30 s) is such that the μ-dihydroxy–iron(III) dimer is in equilibrium with $Fe(OH_2)_6^{3+}$.

If the carboxylate and sulfhydryl groups are complexing, then monomeric complex formation can be described by the K_{f1} reaction in Scheme I, in which two protons are released from the fully protonated penicillamine cation. Since the kinetic observations indicate that the μ-dihydroxy–iron(III) dimer is forming a complex with penicillamine, the model was expanded to include $Fe_2(OH)_2(LH)^{3+}$ as described by eq 2.

$$Fe_2(OH)_2^{4+} + H_3L^+ \xrightleftharpoons{K_{f2}} Fe_2(OH)_2(LH)^{3+} + 2H^+ \tag{2}$$

Then the initial absorbance (A_0) is given by eq 3

$$A_o = \ell[L]_{tot}\frac{\varepsilon_1K_{f1}[FeOH_2] + \varepsilon_2K_{f2}[Fe_2(OH)_2]}{[H^+](K_a+[H^+] + K_{f1}[FeOH_2] + K_{f2}[Fe(OH)_2]}) \tag{3}$$

where ℓ is the path length, $[L]_{tot}$ is the total amino acid concentration, and $[FeOH_2]$ is $[Fe(OH_2)_6^{3+}]$. Since the available conditions do not establish limiting

Table I. Summary of Rate Constants for Complexation and Oxidation of Cysteine and Penicillamine by Aqueous Iron(III) at 25 °C in 1.0 M LiClO$_4$/HClO$_4$

Reactants		Rate Constant (M^{-1} s^{-1})	
		Cysteine	Penicillamine
	$k_2K_m + k_3K_a$	2.29	1.90
$(H_2O)_5Fe(OH)^{2+} + H_3L^+$	$k_2{}^a$	$\leq 1.4 \times 10^3$	$\leq 1.2 \times 10^3$ $(2.1 \times 10^3)^b$
$Fe(OH_2)_6{}^{3+} + H_2L$	$k_3{}^a$	$\leq 1.8 \times 10^2$	$\leq 1.5 \times 10^2$ $(2.5 \times 10^2)^b$
$(H_2O)_5Fe(OH)^{2+} + H_2L$	k_4	7.4×10^3	7.4×10^3 $(5.2 \times 10^3)^b$
$(H_2O)_8Fe_2(OH)_2{}^{4+} + H_2L$	k_d	4.3×10^3	8.2×10^3
$Fe(LH)^{2+} + H_3L^+$	k_1'	$<6.0^c$	$<0.08^c$
$Fe(LH)^{2+} + H_2L$	k_1''	9.2×10^1	$<0.06^c$
	$k_2'K_{f2} + k_2''K_{f2}K_D$	0.394	0.415×10^{-1}
$Fe(LH)^{2+} + Fe(LH)^{2+}$	$k_2'{}^d$	6.4×10^2	9.0
$Fe_2(OH)_2(LH)^{3+} + H_3L^+$	$k_2''{}^d$	3.0×10^3	57
$Fe_2(OH)_2(LH)^{3+} + H_2L$	k_2'''	$<1.4 \times 10^{3\,c}$	
$Fe_2(OH)_2(LH)^{3+} + Fe(LH)^{2+}$	k_3'	4.0×10^3	48.5
$Fe_2(OH)_2(LH)^{3+} + Fe_2(OH)_2(LH)^{3+}$	k_4'	$<1.2 \times 10^{3\,c}$	25.5

^{a}There is a proton ambiguity between k_2 and k_3 so that the value given is an upper limit assuming that the term is totally k_2 or k_3.

^{b}Values are from ref. 2 at 25 °C in 0.50 M NaClO$_4$.

^{c}Upper limits are determined on the basis that these terms are not detectable in the rate law and must be contributing <20% to the experimental rate constant under the conditions of their maximum possible contribution.

^{d}There is a kinetic ambiguity between k_2' and k_2''', and the former is suggested as the preferred assignment.

values of K_{fn} or ε_n for either of the complexes, we have fixed K_{f1} at 0.064 M, a value consistent with the present kinetic and previous equilibrium measurements, to obtain $K_{f2} = 0.38 \pm 0.08$ M and ε_1 and ε_2 of 974 ± 23 and 1472 ± 98 M^{-1} cm^{-1}, respectively. These parameters provide self-consistent fits of the complexation absorbance–time profiles for conditions where the dimer complex makes a significant contribution.

Oxidation–Reduction Kinetics for Cysteine.   The fading of the blue color of the iron(III) complex of cysteine generally is associated with oxidation of the sulfhydryl group to the disulfide and reduction of iron(III) to iron(II) as shown in eq 4.

$$2Fe^{III} + 2HSCH_2CH(NH_3^+)CO_2H \rightarrow 2Fe^{II} + 2H^+ + \begin{matrix} SCH_2CH(NH_3^+)CO_2H \\ | \\ SCH_2CH(NH_3^+)CO_2H \end{matrix} \quad (4)$$

The concentration conditions are the same as for the complexation studies. The stoichiometry under these conditions of excess iron(III) was investigated by determining the iron(II) produced. On the time scale of disappearance of the blue color for cysteine ($\sim$1 min), the stoichiometry is 1 iron(II) produced per amino acid, but further oxidation does occur at longer times. It also was found that the disulfide product, cystine, is oxidized by iron(III), as indicated by iron(II) production. Since the cystine solutions are colorless, the spectrophotometric observations at 620 nm do not monitor the further oxidation of cystine.

The absorbance–time curves at 620 nm for the redox reaction are not fitted satisfactorily by a first-order rate law because the absorbance decreases too slowly during the later stages of the reaction. In addition, the half-time of the reaction decreases with increases in the concentration of the deficient reagent, cysteine. These observations indicate that the reaction is higher than first-order in cysteine, and it is found that the absorbance–time profiles are well-fitted by a second-order dependence on cysteine, so that the rate is given by eq 5

$$\text{rate} = k_2[\text{cysteine}]^2 \tag{5}$$

where k_2 is a pseudo-second-order rate constant that will depend on the iron(III) and H^+ concentrations. Furthermore, it was found that addition of the products, iron(II), or cystine does not affect the rate.

The various kinetic runs have been fitted graphically by numerical integration to obtain the values for k_2. In the numerical analysis, the absorbance at the time of mixing ($t = 0$) was calculated from equilibrium constants and extinction coefficients that gave overall self-consistency for all the runs, with $K_{f1} = 0.0248$ M from the complexation kinetics. This analysis gave $\varepsilon_1 = 700$ M^{-1} cm^{-1}, $\varepsilon_2 = 1500$ M^{-1} cm^{-1}, and $K_{f2} = 0.07$ M.

The rate law can be developed from a consideration of the various reactant species as given in eq 6.

$$[LH_3^+] = [L]_{tot} \frac{[H^+]^2}{DC}$$

$$[LH_2] = [L]_{tot} \frac{K_a[H^+]}{DC}$$

$$[FeLH] = [L]_{tot} \frac{K_{f1}[FeOH_2]}{DC} \tag{6}$$

$$[Fe_2(OH)_2LH] = [L]_{tot} \frac{K_{f2}[Fe_2(OH)_2]}{DC}$$

where $DC = [H^+](K_a + [H^+]) + K_{f1}[FeOH_2] + K_{f2}[Fe_2(OH)_2]$. There are a number of reactant combinations that give a rate that is second-order in the total cysteine concentration ($[L]_{tot}$), but if termolecular terms are neglected, then the possible contributions are given by eq 7.

$$\text{rate} = k_2[\text{L}]_{\text{tot}}^2 = k_1'[\text{FeLH}][\text{H}_3\text{L}^+] + k_1''[\text{FeLH}][\text{H}_2\text{L}] + k_2'[\text{FeLH}]^2$$

$$+ k_2''[\text{Fe}_2(\text{OH})_2\text{LH}][\text{H}_3\text{L}^+] + k_2'''[\text{Fe}_2(\text{OH})_2\text{LH}][\text{H}_2\text{L}] \quad (7)$$

$$+ k_3'[\text{Fe}(\text{OH})_2\text{LH}][\text{FeLH}] + k_4'[\text{Fe}_2(\text{OH})_2\text{LH}][\text{Fe}_2(\text{OH})_2\text{LH}]$$

If the concentrations are substituted from eq 6, and the concentration of $\text{Fe}_2(\text{OH})_2$ is expressed as $[\text{Fe}_2(\text{OH})_2] = K_\text{D}[\text{FeOH}_2]^2/[\text{H}^+]^2$, then substitution and rearrangement give eq 8.

$$k_2(\text{DC})^2 = k_1'K_{\text{f1}}[\text{H}^+]^2[\text{FeOH}_2] + k_1''K_{\text{f1}}K_\text{a}[\text{H}^+][\text{FeOH}_2] + k_2'K_{\text{f2}}[\text{FeOH}_2]^2$$

$$+ k_2''K_{\text{f2}}K_\text{D}[\text{FeOH}_2]^2 + k_2'''K_{\text{f2}}K_\text{D}K_\text{a}[\text{H}^+]^{-1}[\text{FeOH}_2]^2 \quad (8)$$

$$+ k_3'K_{\text{f1}}K_{\text{f2}}K_\text{D}[\text{H}^+]^{-2}[\text{FeOH}_2]^3 + k_4'K_{\text{f2}}^2K_\text{D}^2[\text{H}^+]^{-4}[\text{FeOH}_2]^4$$

This equation predicts that $k_2(\text{DC})^2$ plotted versus $[\text{FeOH}_2]$ should be a smooth function at a particular hydrogen ion concentration. The plot in Figure 1 shows that $k_2(\text{DC})^2$ has a greater than first-order dependence on $[\text{FeOH}_2]$, and that it decreases with increasing $[\text{H}^+]$ but seems to reach a limiting value for the

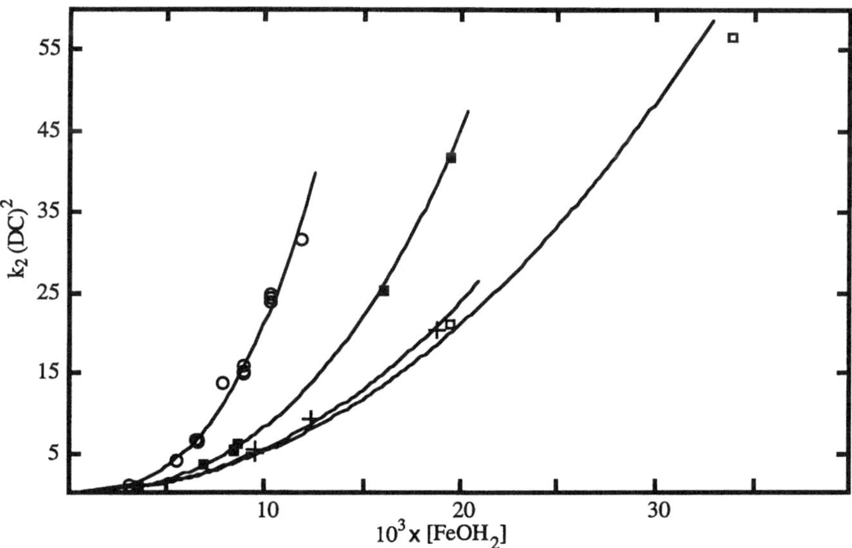

Figure 1. Variation of the second-order rate constant for oxidation of cysteine by aqueous iron(III), $k_2 \times DC\ (=[H^+](K_a + [H^+]) + K_{f1}[FeOH_2{}^{3+}] + k_{f2}[Fe_2(OH)_2{}^{4+}])$ with the concentration of $FeOH_2{}^{3+}$. The points are grouped according to the approximate values of $[H^+]$ (M) of: 0.009 (○), 0.019 (■), 0.05 (+) and 0.1 (□).

higher hydrogen ion concentrations of ~0.05 and 0.1 M. This hydrogen ion dependence shows that the k_1' and k_1'' terms are not important contributors.

Least-squares analysis reveals that the k_1', k_2''', and k_4' terms are not making significant contributions, but the standard error of the fit is improved by 22% if the k_1'' term is included, and gives values of $(k_2'K_{f2} + k_2''K_{f2}K_D) = 0.50 \pm 0.03$ and $k_3'K_{f1}K_{f2}K_D = 1.15 \pm 0.06 \times 10^{-2}$ (errors are 95% confidence limits). The best-fit values are $k_1''K_{f1}K_a = 2.9 \pm 1.7 \times 10^{-2}$, $k_2'K_{f2} + k_2''K_{f2}K_D = 0.394 \pm 0.07$, and $k_3'K_{f1}K_{f2}K_D = 1.31 \pm 0.08 \times 10^{-2}$. From these values one obtains $k_1'' = 92$, $k_2' = 6.4 \times 10^2$ or $k_2'' = 3.0 \times 10^3$, and $k_3' = 4.0 \times 10^3$ M^{-1} s^{-1}. The results are summarized in Table I.

There is a kinetic ambiguity between the k_2' and k_2'' terms. The latter involves reduction of the iron(III) dimer complex by fully protonated cysteine (H_3L^+). We suspect that H_3L^+ is a poorer reducing agent than its conjugate base (H_2L) because the k_1' term makes no significant contribution compared to k_1'', although H_3L^+ is the major form of the ligand between 0.05 and 0.1 M H^+. Therefore assignment of this term largely to k_2' appears preferable.

Oxidation–Reduction Kinetics for Penicillamine. Qualitatively, this system is analogous to cysteine except that the reaction is about 10 times slower under similar conditions. In principle, the slower reaction should make the penicillamine reaction more tractable because the initial absorbance is well-defined, and normal mixing times are small compared to the reaction time. In fact, the initial absorbance values are quite reproducible, but the kinetics were reproducible only when freshly prepared, anaerobic solutions of penicillamine were used.

The absorbance–time profiles are satisfactorily fitted by the second-order rate law described for cysteine. The addition of iron(II) under anaerobic conditions and penicillamine disulfide (or cystine) have no apparent effect on the absorbance–time profiles.

Preliminary graphical analysis (eq 8) revealed that the data are consistently analogous to those for cysteine, except that the k_4' term also is contributing. Least-squares analysis gives $k_2'K_{f2} + k_2''K_{f2}K_D = 4.15 \pm 0.3 \times 10^{-2}$, $k_3'K_{f1}K_{f2}K_D = 2.39 \pm 0.4 \times 10^{-3}$, and $k_4'K_{f2}^2K_D^2 = 3.54 \pm 0.5 \times 10^{-5}$. From these values, one obtains $k_2' = 9.0$ or $k_2'' = 57$, $k_3' = 48.5$, and $k_4' = 25.5$ M^{-1} s^{-1}. The results are summarized in Table I.

Oxidation Kinetics with Excess Penicillamine. This study was initiated to allow comparison with the similar study on cysteine of Jameson et al. (7, 8). Because of the low iron(III) concentrations, the second and higher order iron(III) contributions to the rate will be minimized, and it may be possible to establish values or upper limits for k_1' and k_1''.

Reproducibility problems also were found under these conditions, even though fresh, anaerobic penicillamine solutions were used. After many trials, it was found that the mode of mixing is critical. The slowest and reproducible rates are observed if solid penicillamine hydrochloride is added to the aqueous $HClO_4/LiClO_4$ solution, as described in the Experimental Methods section.

The kinetics were studied with a pseudo-first-order excess of penicillamine (2.6×10^{-3}–6.1×10^{-3} M) over iron(III) (2×10^{-4}–4×10^{-4} M) in ~0.021 and ~0.05 M H^+, and iron(II) perchlorate was added at the 1×10^{-3}–6×10^{-3} M level. The absorbance–time profiles appear to be normal first-order curves. However, there is some variation in the apparent pseudo-first-order rate constant with the concentration of the deficient reagent, iron(III), because of the contributions from the second-order terms discussed in the preceding section. The rate increases with increasing penicillamine concentration and decreases with increasing $[H^+]$ and iron(II) concentrations. The iron(II) inhibition is less at higher acidity. The $[H^+]$ dependence is consistent with the bis complex as the dominant reactive species. The observations can be most economically accounted for by the reactions in Scheme II.

An analysis of the initial absorbance for these conditions indicated that there is an additional absorbing species. The dimer complex is not contributing significantly at the low iron(III) concentrations used. The new species was assigned to the bis complex $Fe(LH)_2^+$, with a molar absorptivity of 2.8×10^3 M^{-1} cm^{-1} and a formation constant (K_{fl2} in Scheme II) of 1.6×10^{-2} M to predict the initial absorbance.

$$Fe(OH_2)_6^{3+} + H_3L^+ \underset{}{\overset{K_{fl}}{\rightleftharpoons}} (H_2O)_4Fe(LH)^{2+} + 2H^+$$

$$(H_2O)_4Fe(LH)^{2+} + H_3L^+ \underset{}{\overset{K_{fl2}}{\rightleftharpoons}} (H_2O)_2Fe(LH)_2^+ + 2H^+$$

$$(H_2O)_2Fe(LH)_2^+ \underset{k_{-5}}{\overset{k_5}{\rightleftharpoons}} Fe(LH)^+ + HL\bullet$$

$$Fe(LH)^+ + H^+ \underset{}{\overset{K_f'}{\rightleftharpoons}} Fe^{II} + H_2L$$

$$H_2L + HL\bullet \underset{k_{-6}}{\overset{k_6}{\rightleftharpoons}} (HL\overset{\bullet}{-}LH)^- + H^+$$

$$Fe(III) + (HL\overset{\bullet}{-}LH)^- \overset{k_7}{\longrightarrow} Fe(II) + HL-LH$$

$$HL\bullet \equiv \bullet S-C(CH_3)_2CH(NH_3^+)CO_2^-$$

$$(HL\overset{\bullet}{-}LH)^- \equiv \bullet \overset{\displaystyle S-C(CH_3)_2CH(NH_3^+)CO_2^-}{\underset{\displaystyle S-C(CH_3)_2CH(NH_3^+)CO_2^-}{|}}$$

Scheme II.

If a steady state is assumed for the radical intermediates, and the complexation reactions for Fe(III) (K_{f1}, K_{f12}) and Fe(II) (K_f') are rapidly established equilibria, then Scheme II predicts that the rate is given by eq 9

$$\text{rate} = \frac{k_5 k_6 k_7 [\text{Fe}^{III}][\text{Fe(LH)}_2]}{\dfrac{k_{-5} K_f'[\text{Fe}^{II}]_t}{[\text{H}^+] + K_f'[\text{LH}_2]}\left(k_{-6}[\text{H}^+] + k_7[\text{Fe}^{III}]\right) + k_6 k_7 [\text{Fe}^{III}]} \tag{9}$$

where $[\text{Fe}^{II}]_t = $ total Fe(II) $= [\text{Fe}^{II}_{aq}] + [\text{Fe(LH)}^+]$. A consideration of the complex formation constants for thiocarboxylate ligands (12) indicates that $K_f' < 10^{-2}$, so that $[\text{H}^+] >> K_f'[\text{LH}_2]$ for our conditions. This rate law becomes first-order in $[\text{Fe}^{III}]$ if $k_7[\text{Fe}^{III}] >> k_{-6}[\text{H}^+]$, and then eq 9 simplifies to eq 10, where the minor second-order terms described in the previous section have been added for completeness.

$$\text{rate} = \frac{k_5 [\text{Fe(LH)}_2]}{\left(\dfrac{k_{-5}}{k_6}\right)\dfrac{K_f'[\text{Fe}^{II}]_t}{[\text{H}^+]} + 1} + \text{second - order terms} \tag{10}$$

Absorbance–time profiles calculated on the basis of this rate law are shown in Figure 2. In Figure 2A, no iron(II) has been added, and these runs show the penicillamine and $[\text{H}^+]$ dependence. In Figure 2B, the effect of iron(II) is shown; the dashed curves (calculated with added $[\text{iron(II)}] = 0$) show the magnitude of the inhibition, and that the inhibition decreases at the higher acidity. All of the curves in Figure 2 have been calculated from eq 10 with $k_5 = 1.25 \times 10^{-2}$ and $k_{-5} K_f'/k_6 = 3.5$. These parameters provide a satisfactory description of all the data.

Models in which the dominant reaction is $\text{Fe(LH)}^{2+} + \text{H}_2\text{L}$ (k_1'') or $\text{Fe(LH)}^{2+} + \text{H}_3\text{L}^+$ (k_1') are also first-order in iron(III) and second-order in penicillamine, but are not consistent with the $[\text{H}^+]$ dependence. A specific rate constant that predicts the rate at 0.02 M H^+, predicts a rate that is 5 times too fast at 0.05 M H^+. This analysis leads to the upper limits for k_1'' and k_1' given in Table I. It remains something of a mystery why these pathways, especially k_1'', do not seem to contribute significantly.

Conclusions

The equilibrium constants K_{f1} and K_{f2} are both smaller for cysteine than penicillamine. This difference is typical of other metal ion systems (13) in which formation constants have been determined for both ligands. K_{f2} for complexation by the iron(III) dimer is larger than K_{f1} by 3 and 6 times for cysteine and penicillamine, respectively. For various α-mercaptocarboxylic acids, McAuley and co-workers (1–3) and Baiocchi et al. (4) are in reasonable agreement on the

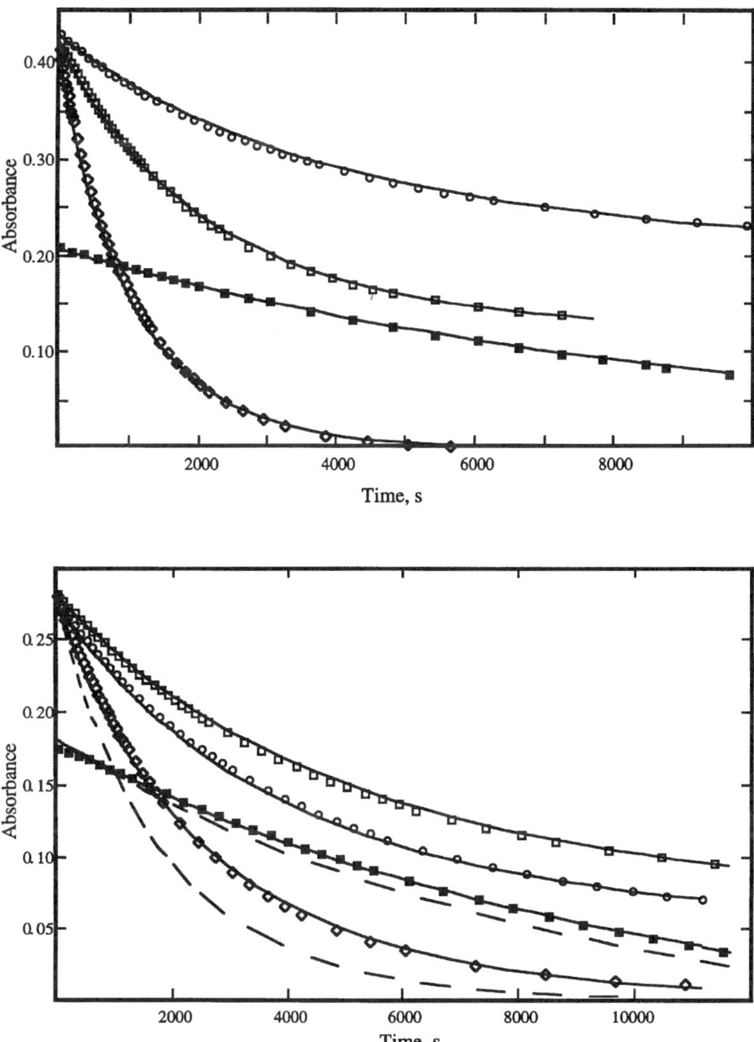

Figure 2. Absorbance–time profiles for the oxidation of excess penicillamine by aqueous iron(III). For A, the concentrations of penicillamine, total Fe(III) and H$^+$, respectively, are 2.60×10^{-3}, 2.01×10^{-4}, 2.06×10^{-2} (○); 3.80×10^{-3}, 2.01×10^{-4}, 2.11×10^{-2} (□); 6.00×10^{-3}, 2.01×10^{-4}, 2.18×10^{-2} (◇); 6.15×10^{-3}, 4.00×10^{-4}, 5.08×10^{-2} (■). For clarity, the curves are offset by 0.20, 0.12, 0, and -0.04 absorbance units , respectively. For B, the concentrations of Fe(II), penicillamine, total Fe(III) and H$^+$, respectively, are 1.14×10^{-3}, 2.60×10^{-3}, 2.04×10^{-4}, 2.07×10^{-2} (○); 2.12×10^{-3}, 2.41×10^{-3}, 1.99×10^{-4}, 2.09×10^{-2} (□); 4.00×10^{-3}, 4.09×10^{-3}, 2.02×10^{-4}, 2.17×10^{-2} (◇); 3.97×10^{-3}, 6.17×10^{-3}, 4.03×10^{-4}, 5.13×10^{-2} (■). For clarity, the curves are offset by 0.05, 0.06, 0, and -0.08 absorbance units , respectively. The curves are calculated from eq 10; dashed curves are calculated assuming [Fe(II)] = 0 for comparison to the (■) and (◇) data in B.

K_{f1} values in the general range of 0.5–3 M. For penicillamine, our value of K_{f1} = 0.068 M is in excellent agreement with that of Baiocchi et al.. These authors suggested that K_{f1} is smaller for penicillamine because it forms a six-membered chelate ring, while the α-mercaptocarboxylates form five-membered rings.

For the molar absorptivity coefficients of the iron(III)-α-mercaptocarboxylate complexes, McAuley and co-workers found values ranging from 140 to 1400 M^{-1} cm^{-1}, while Baiocchi et al. reported 700–1100 M^{-1} cm^{-1} for the same ligands. For penicillamine, our value of 9.7×10^2 is in only fair agreement with the 1200 M^{-1} cm^{-1} reported by Baiocchi et al., and we find a somewhat lower value of 7×10^2 for cysteine. These values are somewhat difficult to determine because of the transient nature of the complexes.

Previous work indicates that K_{f1} involves coordination of $-CO_2^-$ and $-S^-$ groups, and the same seems true for K_{f2} because the same number of protons are released, and the iron(III) monomer and dimer complexes have similar absorbance maxima. The dimer complex could involve chelation at just one iron(III) center, but then one might expect $K_{f1} > K_{f2}$ because of the higher charge per iron in the monomer. Alternatively, the amino acid could be bridging between the two iron centers in the dimer, in which case complexation of one end of the ligand would not adversely affect complexation of the other end at the other iron(III) center. This could explain the observation that $K_{f2} > K_{f1}$. The proton loss indicates that the μ-$(OH-)_2$ bridge is not lost, unless it is converted to an μ-$(O)_2$ bridge. The latter seems unlikely under the acidic conditions of this study.

For the complexation reaction, the specific rate constants can be calculated from the composite values determined from the least-squares analysis that are summarized in Table I. The values are quite typical for substitution on aqueous iron(III) and in reasonable agreement with the earlier work of Baiocchi et al. (4) on penicillamine. There is the usual proton ambiguity between the paths involving $Fe(OH)^{2+} + H_3L^+$ (k_2) and $Fe(OH_2)^{3+} + H_2L$ (k_3). A choice between these terms is often based on the magnitude of the calculated rate constant compared to values for other ligands. In the present cases, this criterion does not eliminate either pathway as a reasonable contributor. For the first pathway, the rate constants of $1.2–1.4 \times 10^3$ M^{-1} s^{-1} are high for reaction of a cation, and are more typical of neutral ligand complexation, but the positive charge on the $-NH_3^+$ group is somewhat removed from the reaction center. Baiocchi et al. (4) argued that assignment to k_3 leads to unreasonably high and variable values for several mercaptocarboxylic acids and penicillamine. On this basis assignment to k_2 seems more appropriate.

For the oxidation of cysteine, the dominant reaction pathways have a common feature in that they involve at least two iron(III) centers and two cysteines to reach the transition state. This appears to be a complicated way to proceed, but the disulfide product, cystine, requires that two electrons are released. This feature was also observed by Ellis et al. (1, 3) for the oxidation of several mercaptocarboxylic acids by aqueous iron(III). In the latter work, reactions of the iron(III) dimer were not observed because of the high acidity, and the only pathway identified corresponds to k_2' in the present terminology. The value of $k_2' = 6.4 \times 10^2$ M^{-1} s^{-1} for cysteine is somewhat larger than those of 48–167 M^{-1} s^{-1} (25 °C, 1.0 M NaClO$_4$) found by Ellis et al. for different reductants. However, it is still unclear how the nature of the reductant affects the reactivity, and the magnitude of k_2' seems reasonable. The alternative assignment of this kinetic term to k_2'' gives a five times larger rate constant and implies that $k_2'' > k_2'$. The latter order does not agree with the order $k_1' < k_1''$, nor with the expectation that H$_2$L should be a better reducing agent than H$_3$L$^+$.

From the earlier suggestion that the reaction of two Fe(LH)$^{2+}$ units might react though a sulfide bridged structure (**III**), one might expect that the dimer complex Fe$_2$(OH)$_2$(LH)$^{3+}$ would undergo facile reduction by cysteine as H$_2$L. In fact, the k_2''' contribution in eq 7 is too small to be detected, and only an upper limit can be given (Table I). What is revealing is that k_2''' (Fe$_2$(OH)$_2$-(LH)$^{3+}$ + H$_2$L) is smaller than k_3' (Fe$_2$(OH)$_2$(LH)$^{3+}$ + Fe(LH)$^{2+}$). This indicates that it is more favorable to have both cysteines coordinated to iron(III) centers for oxidation to occur.

The present results and the earlier work of McAuley and co-workers provide a general picture of the oxidation of mercaptocarboxylate systems under conditions of iron(III) in excess. The situation with a deficiency of iron(III) is not so clear. Baiocchi et al. studied several systems under these conditions and published details for mercaptosuccinic acid (MSAH$_3$). These data indicate a first-order dependence on [MSAH$_3$] and the proposed mechanism involves the bis complex (Fe(MSAH)$_2^-$) as a steady-state intermediate. However, the analysis assumed that formation of the mono-complex (Fe(MSAH)$^+$) is complete under all conditions, but this is certainly not the case from the formation constant also determined by Baiocchi et al.

Our study with excess penicillamine clearly shows that the reaction is second-order in penicillamine, that the bis-complex forms as an equilibrium species, and that the latter is the kinetically dominant reactant. The penicillamine reaction also shows iron(II) inhibition, and this and the other kinetic features can be accounted for by the mechanism in Scheme II. It must be acknowledged that this mechanism has some puzzling features; why, for instance, does the bis-complex undergo intramolecular electron transfer, whereas the mono-complex seems to be unreactive? The driving force, although slight in either case, would seem to favor the mono-complex. The thiyl radical, once formed, couples with free penicillamine, rather than intramolecularly with the other penicillamine ligand in the bis-complex precursor. This could be rationalized if the S– atoms are trans to each other in the bis-complex and therefore not situated to couple efficiently. Further studies on analogous systems may clarify these issues.

Acknowledgments

This work was supported by a grant from the Natural Sciences and Engineering Research Council of Canada.

References

1. Ellis, K. J.; McAuley, A. *J. Chem. Soc. Dalton Trans.* **1973**, 1533
2. Lappin, A. G.; McAuley, A. *J. Chem. Soc. Dalton Trans.* **1975**, 1560.
3. Ellis, K. J.; Lappin, A. G.; McAuley, A. *J. Chem. Soc. Dalton Trans.* **1975**, 1930.
4. Baiocchi, C.; Mentasti, E.; Arselli, P. *Trans. Met. Chem.* **1983**, *8*, 40.
5. Stadtherr, L. G.; Martin, R. B. *Inorg. Chem.* **1972**, *11*, 92.
6. Tomita, A.; Hirai, H.; Makishima, S. *Inorg. Chem.* **1968**, *7*, 760.
7. Jameson, R. F.; Linert, W.; Tschinkowitz, A.; Gutmann, V. *J. Chem. Soc. Dalton Trans.* **1988**, 943.
8. Jameson, R. F.; Linert, W.; Tschinkowitz, A. *J. Chem. Soc. Dalton Trans.* **1988**, 2109.
9. (a) Surdhar, P. S.; Armstrong, D. A. *J. Phys. Chem.* **1987**, *91*, 6532; (b) *J. Phys. Chem.* **1986**, *90*, 5915; (c) Mezyk, S. P.; Armstrong, D. A. *Can. J. Chem.* **1991**, *69*, 533.
10. Millis, K. K.; Weaver, K. H.; Rabenstein, D. L. *J. Org. Chem.* **1993**, *58*, 4144.
11. Lees, W. J.; Whitesides, G. M. *J. Org. Chem.* **1993**, *58*, 642.
12. Sisley, M. J.; Jordan, R. B. *Inorg. Chem.* **1991**, *30*, 2190.
13. Smith, R. M.; Martell, A. E.; Motekaitis, R. J. *NIST Critical Stability Constants of Metal Complexes Database;* U.S. Department of Commerce. National Institute of Standards and Technology: Washington, D.C., 1993.
14. Baes, C. F.; Messmer, R. E. *The Hydrolysis of Cations;* Wiley: New York, 1976. Milburn, R. M.; Vosburgh, W. C. *J. Am. Chem. Soc.* **1955**, *77*, 1352.

Taube's Influence on the Design of Oscillating Reactions

$ClO_2^{\bullet}$-Driven ClO_2^{-}–I^- Oscillator and Turing Structures

Irving R. Epstein, Kenneth Kustin, and István Lengyel

Department of Chemistry, Brandeis University, P.O. Box 9110, Waltham, MA 02254–9110

The deliberate design of chemical oscillators has significantly increased the number of known oscillating reactions and has facilitated the entry of chemistry into the new science of "chaos". This chapter shows how Henry Taube's research and personal contacts have influenced the design of oscillating chemical reactions. He and his co-workers have illuminated the oxidation reactions of oxychlorine species, especially the autocatalytic, substrate (iodide)-inhibited chlorite–iodide reaction, which oscillates in an open reactor. A model is presented that explains the complex dynamics that ensue when this reaction is initiated by the reaction between chlorine dioxide and iodide in a closed reactor. Coupling the starch–triiodide complex to a derivative of the chlorite–iodide oscillator has made it possible to verify experimentally the existence of Turing structures.

COMPLEXITY IS BEING UNMASKED, its secrets now being steadily revealed. A new discipline carved out of mathematics and physics provides new ideas, paradigms, and laws to understand complexity. Until now, chemistry, often the driving force for achieving understanding in other fields such as molecular biology, was poorly represented in this new scientific venture. Then oscillating reactions changed that lack of involvement. In particular, the design of dozens of new chemical oscillators paved the way for broad participation by chemists in this new field. Interestingly, Henry Taube's research and personal influence contributed to the discovery and development of the first and, in many ways, most influential of these deliberately designed oscillators.

The Need to Study Complexity

In his essay "More is Different" the physicist P. W. Anderson (*1*) takes a stand against the position articulated by Dirac, that with the development of the laws of quantum mechanics the problems of chemistry were essentially solved. "The difficulty is only that the exact application of these laws leads to equations much too complicated to be soluble," wrote Dirac (*2*). Anderson concedes that many difficult physical and chemical problems have been solved by reducing their complexity, applying fundamental laws to deduce how simpler subunits behave, and then relating the comprehensible parts to the whole. But he notes that when complex systems are involved, more often than not, the process cannot be reversed.

Even with a computer and a set of fundamental laws an engineer, a statistician, or a chemist would find it virtually impossible to predict where a steel rod will break, when the stock market will crash, or why the amino acids of life have only one chirality. The new discipline variously called nonlinear dynamics—or, more popularly, "chaos"—rejects reductionism and seeks to understand such phenomena through a different approach. Systems like the weather are recognized as being both deterministic, because individual gas molecules obey Newton's laws of motion, and random, because taken as a single entity an atmosphere of gas molecules can produce disordered as well as ordered behavior. It is the transitions from disordered to ordered behavior, and the hitherto unrecognized characteristics that emerge from the ordered state, that the new science of chaos tries to understand. This idea has achieved rapid acceptance, but not by chemists (*3*). Why has chemistry been at the margins of the development of this new science?

Part of the explanation for the chemist's reluctance to participate in the new science of chaos is that oscillating reactions, the most visible chemical version of these "emergent phenomena" (*4*) connected with complexity, at first elicited only skepticism from the chemical community. The first genuine chemical oscillator, discovered in the 1920s by William C. Bray (incidentally, Henry Taube's Ph.D. dissertation advisor), brought forth rejection because oscillations were thought to be an artifact caused by dust particles or bubbles of gas (*5*). The chemical community was more receptive to the later discovery of a second oscillator, which has become the prototypical chemical oscillator, the oxidation of malonic acid by bromate, catalyzed by a strong oxidant such as cerium(IV) and now known as the Belousov–Zhabotinsky or BZ reaction after its discoverers (*6*). Clearly, if chemical systems were to serve as models of biochemical, geological, and even astronomical oscillations and pattern formation, more examples of chemical oscillators were needed. It was the wish to fill this void that drove our program to design new chemical oscillators.

The Chlorine Dioxide–Initiated Chlorite–Iodide Reaction

To anchor the design of a chemical oscillator, we sought a system with strong nonlinear behavior. That such a strongly nonlinear system existed had been noted by chemists working in Henry Taube's laboratory in the University of Chicago. A postdoctoral fellow then, Gilbert Gordon (now professor of chemistry, Miami University, Oxford, Ohio) and David M. Kern, a member of the Dartmouth College faculty visiting there for a summer, began to work on the oxidation of iodide ion by chlorine(III). After Kern's death in 1965, Walter H. Stockmayer, chair of the Dartmouth College Department of Chemistry, found an unfinished manuscript of Kern and an undergraduate student, C.-H. Kim, which he thought was too interesting to discard. He made the considerable effort to finish the manuscript and shepherd it through the publication process, which led to its acceptance. In the article finally published, "Iodine Catalysis in the Chlorite–Iodide Reaction," Kern and Kim (7) noted that "the catalytic effect of iodine and the inhibitory effect of iodide on the chlorite–iodide reaction long ago were observed qualitatively by Bray." Thus, the personal link from Bray to Taube to Gordon, Kern, and Kim.

In terms customarily used to describe nonlinear dynamics, we would say that the chlorite–iodide reaction exhibits feedback through autocatalysis, since iodine is a product, and substrate inhibition, since iodide ion is a reactant. Not satisfied with even this degree of nonlinearity, we coupled this system to arsenite, which is oxidized in an autocatalytic process by iodate, the ultimate product of the chlorite–iodide reaction when chlorite is in excess. The iodate is reduced by arsenite to iodide, which can begin the cycle of reactions again. To keep the system far from equilibrium, we examined its dynamics in an open (flow) reactor. The system oscillated (8) and was the first of more than three dozen new chemical oscillators subsequently designed at Brandeis University (9). Soon after publication, we realized that the chlorite–iodide reaction, which is a clock reaction in a closed (batch) reactor, oscillates by itself in an open (flow) reactor (10).

Gordon has continued the area of research he started in Taube's laboratory and has developed chlorine dioxide into an interesting and useful reagent of wide applicability (11). It occurred to us that rather than add chlorite directly to iodide, the chlorite could be generated in situ in a flow or batch reactor by mixing chlorine dioxide with iodide ion. The chlorite ion, effectively an intermediate, would then react further with iodide ion producing iodine. The stoichiometries, rate laws, and dimensionless rate equations for the three concentration variables of this sequence of reactions are shown in Scheme I.

The rate equations for the three variables—$[I^-]$, $[ClO_2^-]$, and $[ClO_2^{\bullet}]$—in a flow reactor are given in eqs 1a–1c, in which subscript zero denotes inflow concentration, k_0 is the inverse reactor lifetime, $\frac{1}{2}([I^-]_0 - [I^-])$ has been substituted for $[I_2]$, and the k_{2a} term has been neglected.

$$(1) \quad ClO_2 + I^- \rightarrow ClO_2^- + \tfrac{1}{2}I_2$$

$$R_1 = k_1[ClO_2][I^-]$$

$$(2) \quad ClO_2^- + 4I^- + 4H^+ \rightarrow 2I_2 + Cl^- + 2H_2O$$

$$R_2 = k_{2a}[ClO_2^-][I^-][H^+] + 4k_{2b}\frac{[ClO_2^-][I^-][I_2]}{u+[I^-]^2}$$

Rescale, neglect k_{2a} term, and add flow.

New variables: $[I^-]$, X; $[ClO_2^-]$, Y; $[ClO_2]$, Z

$$\frac{dX}{d\tau} = -XZ - 4\frac{XY(X_0 - X)}{1+X^2} + f(X_0 - X)$$

$$\frac{dY}{d\tau} = XZ - \frac{XY(X_0 - X)}{1+X^2} - fY$$

$$\frac{dZ}{d\tau} = -aXZ + f(Z_0 - Z)$$

Scheme I. The main overall reactions of the chlorine dioxide-driven chlorite–iodide oscillator: stoichiometries, rate equations, and dimensionless rate equations: $u = 1.0 \times 10^{-14}\ M^2$. *See text for details.*

$$\frac{d[I^-]}{dt} = -k_1[ClO_2^\bullet][I^-] - 2k_{2b}\frac{[ClO_2^-]([I^-]_0 - [I^-])[I^-]}{u+[I^-]^2} + k_0([I^-]_0 - [I^-]) \quad (1a)$$

$$\frac{d[ClO_2^-]}{dt} = k_1[ClO_2^\bullet][I^-] - \tfrac{1}{2}k_{2b}\frac{[ClO_2^-]([I^-]_0 - [I^-])[I^-]}{u+[I^-]^2} - k_0[ClO_2^-] \quad (1b)$$

$$\frac{d[ClO_2^\bullet]}{dt} = -k_1[ClO_2^\bullet][I^-] + k_0([ClO_2^\bullet]_0 - [ClO_2^\bullet]) \quad (1c)$$

To convert the three differential equations (1a–1c) to the dimensionless form of Scheme I, define the following four constants $\alpha = \beta = u^{1/2}$, $\gamma = k_{2b}/(2k_1)$, $\delta = 2/k_{2b}$, and make the following substitutions to the four new variables X, Y, Z, and τ: $[I^-] = \alpha X$, $[ClO_2^-] = \beta Y$, $[ClO_2^\bullet] = \gamma Z$, and $t = \delta\tau$. It is useful to define so-called "fixed" parameters related to the inflow terms for $[I^-]$ and $[ClO_2^\bullet]$. These terms are $X_0 = [I^-]_0/u^{1/2}$, $Z_0 = 2k_1[ClO_2^\bullet]_0/k_{2b}$, and $f =$

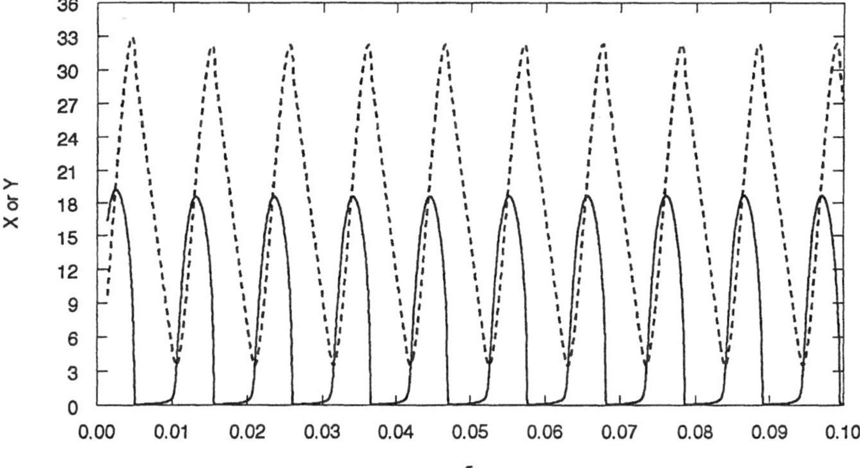

Figure 1. Computer-simulated oscillations of [I⁻] (solid line) and [ClO₂⁻] (dashed line) in the chlorine dioxide-driven chlorite–iodide oscillator in an open (flow) reactor where $\tau = t/\delta$, $X = [I^-]/\alpha$, and $Y = [ClO_2^-]/\beta$. Fixed parameters: $X_0 = 2000$, $Z_0 = 600$, f = 10, and a = 0.45.

$2k_0/k_{2b}$. To simplify the final form of the equations, the constant $a = 2k_1 u^{1/2}/k_{2b}$ is defined. Numerical integration of the dimensionless autonomous equations in the variables X and Y (Figure 1) clearly shows the oscillatory nature of the chlorine dioxide-driven chlorite–iodide reaction in a flow reactor. (The concentration of $ClO_2^{\bullet}$ in the reactor is essentially constant, and the time dependences of X and Y, proportional to $[I^-]$ and $[ClO_2^-]$, respectively, are shown.)

Modeling the $ClO_2^{\bullet}$–ClO_2^-–I^- Reaction in a Closed Reactor

To use oscillatory reactions like the chlorite–iodide oscillator to emulate the panoply of intricate behavior patterns of such complex phenomena as cardiac pacemaking, neurological bursting, and convective flow, the mechanisms of these oscillating reactions should be thoroughly understood. The approach taken is to combine mechanisms for subsets of reactions into a whole that simulates experimental results, a process that is usually carried out with the help of a computer. The danger in computer simulation is that with sufficient steps and adjustable rate constants, an apparently accurate portrayal of the dynamics may result, but one which is based on erroneous or unrealistic elementary step reactions. Therefore, the mechanism should conform to as many constraints, and contain as many experimentally determined rate constants, as possible. While constraints such as not exceeding the diffusion-controlled limit, not violating microscopic reversibility, and being consistent with the overall stoichiometry are obvious to apply, there may be further helpful constraints on the mechanism often overlooked. That is, the model should not include previously excluded mechanistic pathways. It is in this aspect of model building that we experience the almost constant presence of Taube's mechanistic research.

Within inorganic chemistry, Taube was among the first to exploit isotopic tracers to map out mechanistic pathways. Among several interesting and remarkable findings in an application of radioactive chlorine tracers to the study of the mechanisms of oxidation–reduction reactions among chlorine species, there is one result of especial importance to chlorite-based oscillators (*12*). For reactions between chlorite and hypochlorous acid (or chlorine) or in the disproportionation of chlorous acid, "…the activated complex must be unsymmetrical in every case." Therefore, any mechanism devised to explain chlorite-based oscillators, if it invokes these interactions (which of course it must), must proceed through such a pathway, or as egregious an error can be made as postulating a rate constant that exceeds diffusion control.

A mechanism consistent with these precepts, and using only experimentally determined rate laws and rate and equilibrium constants, has been devised to model the chlorine dioxide-driven chlorite–iodide reaction in a closed reactor (Scheme II) (*13*). The model was developed by studying the kinetics of the decomposition of HOI and HIO_2 and of the reactions $HOCl-I_2$, $HClO_2-I_2$, and $ClO_2^{\bullet}-I^-$ (*12*). The first reaction corresponds to initiation, or generation of chlorite. The next three reactions are related to iodine hydrolysis. Iodide inhibition arises from reversal of hydrolysis, which lowers the concentration of reactive HOI and raises the concentration of unreactive I_2. There are two autocatalytic sequences. The less effective autocatalytic route is

$$HOI \; + \; HClO_2 \rightarrow HIO_2 + HOCl$$
$$I^- + HIO_2 + H^+ \rightarrow 2HOI$$

(A)

At low pH, a more effective autocatalytic route is

$$HOI + Cl_2 + H_2O \rightarrow HIO_2 + 2H^+ + 2Cl^-$$
$$I^- + HIO_2 + H^+ \rightarrow 2HOI$$

(B)

The last three steps are largely responsible for the last stage of the batch reaction, the slow reformation of I_2, which results from the reaction between HOI and I^-. The quality of the modeling is illustrated by comparing experiment with simulation for kinetics curves at four different initial chlorine dioxide concentrations (Figure 2).

Application: The Experimental Verification of Turing Structures

The last paper written by the outstanding 20th century mathematician, Alan Turing, was titled "The Chemical Basis of Morphogenesis" (*14*). In this paper,

$$ClO_2^{\bullet} + I^- \rightarrow ClO_2^- + 0.5I_2 \qquad v_1 = 6{\times}10^3[ClO_2^{\bullet}][I^-]$$

$$I_2 + H_2O \rightleftharpoons I_2OH^- + H^+ \qquad v_2 = 3.2[I_2] \quad v_{-2} = 2{\times}10^{10}[I_2OH^-][H^+]$$

$$I_2OH^- \rightleftharpoons HOI + I^- \qquad v_3 = 1.6{\times}10^7[I_2OH^-] \quad v_{-3} = 4.8{\times}10^9[HOI][I^-]$$

$$I_2 + H_2O \rightleftharpoons I^- + H_2OI^+ \qquad v_4 = 4.4{\times}10^{-2}[I_2] \quad v_{-4} = 3.7{\times}10^9[I^-][H_2OI^+]$$

$$I^- + HClO_2 + H^+ \rightarrow HOI + HOCl \qquad v_5 = 7.5[I^-][HClO_2]$$

$$HOI + HClO_2 \rightarrow HIO_2 + HOCl \qquad v_6 = 7.2{\times}10^7[HOI][HClO_2]$$

$$HIO_2 + HClO_2 \rightarrow IO_3^- + HOCl + H^+ \qquad v_7 = 1{\times}10^6[HIO_2][HClO_2]$$

$$I^- + HOCl \rightarrow HOI + Cl^- \qquad v_8 = 4.3{\times}10^8[I^-][HOCl]$$

$$I^- + HIO_2 + H^+ \rightleftharpoons 2HOI \qquad v_9 = 1.0{\times}10^9[I^-][HIO_2][H^+] \quad v_{-9} = 22[HOI]^2$$

$$HOCl + H^+ + Cl^- \rightleftharpoons Cl_2 + H_2O \qquad v_{10} = 2.2{\times}10^4[HOCl][H^+][Cl^-] \quad v_{-10} = 22[Cl_2]$$

$$I_2 + Cl_2 + 2H_2O \rightarrow 2HOI + 2H^+ + 2Cl^- \qquad v_{11} = 1.5{\times}10^5[I_2][Cl_2]$$

$$HOI + Cl_2 + H_2O \rightarrow HIO_2 + 2H^+ + 2Cl^- \qquad v_{12} = 1.0{\times}10^5[HOI][Cl_2]$$

$$2HIO_2 \rightarrow HOI + H^+ + IO_3^- \qquad v_{13} = 25[HIO_2]^2$$

$$HIO_2 + H_2OI^+ \rightarrow IO_3^- + I^- + 3H^+ \qquad v_{14} = 140[HIO_2][H_2OI^+]$$

$$HIO_2 + HOCl \rightarrow 2IO_3 + Cl^- + H^+ \qquad v_{15} = 1.4{\times}10^3[HIO_2][HOCl]$$

Rapid equilibria

$$HClO_2 \rightleftharpoons H^+ + ClO_2^- \qquad K_{HClO_2} = [H^+][ClO_2]/[HClO_2] = 2.0{\times}10^{-2}$$

$$H_2OI^+ \rightleftharpoons HOI + H^+ \qquad K_{H_2OI^+} = [HOI][I^-]/[H_2OI^+] = 0.034$$

$$I^- + I_2 \rightleftharpoons I_3^- \qquad K_{I_3^-} = [I_3^-]/([I_2][I^-]) = 700$$

Scheme II. A model of the chlorine dioxide-driven chlorite–iodide reaction in a closed (batch) reactor.

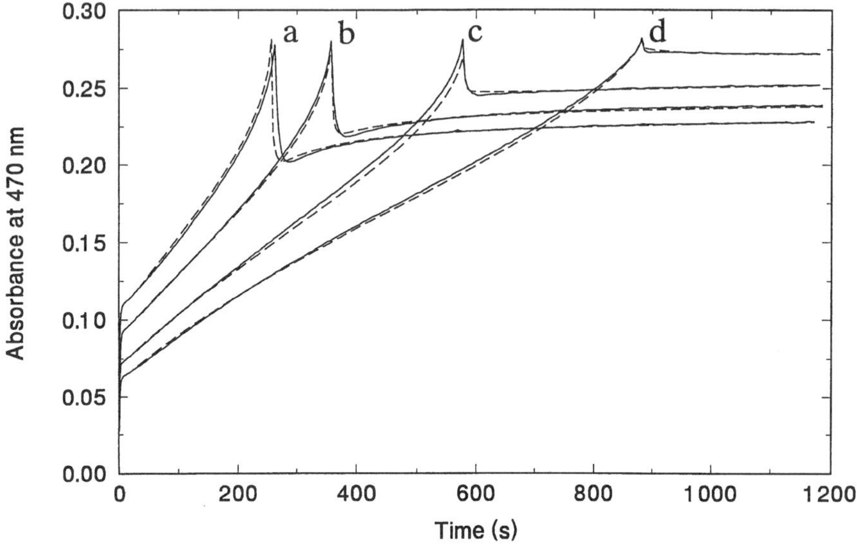

Figure 2. Computer simulation of the chlorine dioxide-driven chlorite–iodide reaction in a closed (batch) reactor. Solid lines, experimental; dashed lines, simulation. [H$^+$] = 7.14 × 10^{-3} M in each experiment (a–d). [I$^-$]$_0$: (a) 8.10 × 10^{-4} M; (b–d) 7.90 × 10^{-4} M. [ClO$_2^\bullet$]$_0$: (a) 2.90 × 10^{-4} M, (b) 2.47 × 10^{-4} M, (c) 1.97 × 10^{-4} M, and (d) 1.70 × 10^{-4} M.

he explained how biological structure and pattern could spontaneously arise from an undifferentiated, homogeneous spherical fertilized egg, a system thought to be as unlikely to break out into a pattern of stripes and spots as any analogous undifferentiated homogeneous, fully symmetrical chemical reactor. A simplified interpretation of Turing's analysis of this conundrum is that pattern formation can spontaneously result from a coupled reaction–diffusion system with two or more chemical species, significantly different diffusion coefficients for two of these species, and strong chemical interaction such as autocatalysis among the chemical species. Although almost from its first appearance, Turing's model appealed to mathematical biologists who applied it extensively to a variety of different problems of biological development, such as "How the Leopard Gets Its Spots" (*15*), there was no experimental verification of the Turing model. In particular, the requirement of significantly different diffusion rates appeared to be difficult to fulfill in a real system, because diffusion coefficients of most molecules and ions do not differ by much.

Experimental verification of Turing structures occurred when a variant of the chlorite–iodide oscillator, the chlorine dioxide–iodine–malonic acid (MA) oscillatory reaction, was studied in a flow reactor filled with a gel to avoid convection (*16*). The key to the production of a stable pattern of dots was the incorporation of starch into the gel (*17*). Starch is frequently used to monitor

Figure 3. Turing patterns in the chlorine dioxide–iodine–malonic acid (MA) reaction. Inflow concentrations: for stripes (top), $[MA]_0 = 1.0 \times 10^{-2}$ M, $[ClO_2^{\bullet}] = 1.5 \times 10^{-3}$ M at one end of the gel-filled reactor, and $[I_2]_0 = 8.0 \times 10^{-4}$ M at the other end; for hexagonally arrayed spots (bottom), $[MA]_0 = 2.0 \times 10^{-2}$ M, $[ClO_2^{\bullet}] = 1.0 \times 10^{-3}$ M at one end of the gel-filled reactor, and $[I_2]_0 = 8.0 \times 10^{-4}$ M at the other end. The wavelength of the structure in the reactor is approximately 0.2 mm in each picture. The blue color indicates the high-iodide medium where the starch–triiodide complex is formed, and the yellow color indicates the iodide-free medium.

iodine, because the intense color of the starch–I_3^- complex confers high sensitivity on this indicator. More importantly, in a gel reactor impregnated with starch, formation of the starch–iodine complex has an additional function. It acts as an impediment to the flow of iodine molecules and iodide ions through the medium. Since the chlorite ions do not experience such a retardation, the disparity in diffusion rates required by Turing's theory is fulfilled, and stable structures are produced (Figure 3).

Even an apparently simple inorganic oscillator like chlorite–iodide is astonishingly complex, as shown by the number of steps needed to model its kinetics (Scheme II). This complexity is an advantage to the scientist wishing to understand more complex systems, such as the human body, for without it no insight into the development of patterns in biology would be forthcoming. After roughly 85 years, Bray's astute observations, kept alive through the elegant experiments of the Taube group and the devotion of a colleague, and Turing's intricate theory, itself 40 years old, converged, producing a pattern of stationary spots in an open chemical reactor. We can only speculate as to what experiments will, in the future, reveal more of the underlying structure of complex systems, but the 1990s have already witnessed a dramatic surge of applications of nonlinear chemical dynamics, such as chaos control (*18*), oscillatory oxidation of CO on a platinum surface (*19*), reflection and refraction of chemical waves (*20*), and chemical computers (*21, 22*). We expect that more will be coming!

Acknowledgments

This work was supported by the National Science Foundation through research (CHE–9023294) and U.S.–Hungary cooperative (INT–9322738) grants.

References

1. (a) Anderson, P. W. *Science (Washington, D.C.)* **1972**, *177*, 393–396; (b) Anderson, P. W. *Proc. Natl. Acad. Sci. U.S.A.* **1995**, *92*, 6653–6654.
2. Dirac, P. A. M. *Proc. R. Soc. London* **1929**, *A123*, 714–733.
3. Gleick, J. *Chaos;* Viking Penguin: New York, 1987; pp 352.
4. Schweber, S. S. *Phys. Today* **1993**, *46*, 34–40.
5. Taube, H., Stanford University, personal communication, 1995.
6. Zhabotinsky, A. M. In *Oscillations and Traveling Waves in Chemical Systems;* Field, R. J.; Burger, M., Eds.; John Wiley & Sons: New York, 1985; pp 1–6.
7. Kern, D. M.; Kim, C.-H. *J. Am. Chem. Soc.* **1965**, *87*, 5309–5313.
8. De Kepper, P.; Epstein, I. R.; Kustin, K. *J. Am. Chem. Soc.* **1981**, *103*, 2133.
9. Epstein, I. R. *Chem. Eng. News* **1987**, *65*, 24–36.
10. Dateo, C. E.; Orbán, M.; De Kepper, P.; Epstein, I. R. *J. Am. Chem. Soc.* **1982**, *104*, 504–509.
11. Gordon, G. In *Chemical Oxidation: Technology for the Nineties;* Exkenfelder, W. W.; Bowers, A. R.; Roth, J., Eds.; Technomic Publishing: Lancaster, PA, 1992; pp 157–170.

12. Taube, H.; Dodgen, H. *J. Am. Chem. Soc.* **1949**, *71*, 3330–3336.
13. (a) Lengyel, I.; Li, J.; Epstein, I. R. *J. Phys. Chem.* **1992**, *96*, 7032–7037; (b) Lengyel, I.; Li, J.; Epstein, I. R.; Kustin, K. *J. Am. Chem. Soc.* **1996**, *118*, 3708–3719.
14. Turing, A. M. *Philos. Trans. R. Soc. London Ser. B* **1952**, *237*, 37–72.
15. Murray, J. D. *Sci. Am.* **1988**, *258*, 80–87.
16. Castets, V.; Dulos, E.; Boissonade, J.; De Kepper, P. *Phys. Rev. Lett.* **1990**, *64*, 2953–2956.
17. Lengyel, I.; Epstein, I. R. *Science (Washington, D.C.)* **1991**, *251*, 650–652.
18. Glanz, J. *Science (Washington, D.C.)* **1994**, *265*, 1174.
19. Jakubith, S.; Rotermund, H. H.; Engel, W.; von Oertzen, A.; Ertl, G. *Phys. Rev. Lett.* **1990**, *65*, 3013–3016.
20. Zhabotinsky, A. M.; Eager, M. D.; Epstein, I. R. *Phys Rev. Lett.* **1993**, *71*, 1526–1529.
21. Hjemfelt, A.; Weinberger, E. D.; Ross, J. *Proc. Natl. Acad. Sci. U.S.A.* **1992**, *89*, 383–387.
22. Steinbock, O.; Tóth, A.; Showalter, K. *Science (Washington, D.C.)* **1995**, *267*, 868–871.

Hydrolysis of Coordinated Nitriles and Linkage Isomerization Reactions in Ruthenium Ammine Complexes with Nitriles and Amides

Zênis Novais da Rocha[1], Glaico Chiericato, Jr., and Elia Tfouni*

Departamento de Química, Faculdade de Filosofia, Ciências e Letras de Ribeirão Preto, Universidade de São Paulo, Av. dos Bandeirantes, 3900, 14.040–901, Ribeirão Preto, São Paulo, Brazil

This chapter surveys the hydrolysis of coordinated nitriles to amides in ruthenium ammine complexes, and the subsequent afterreduction of the amido–Ru(III) complex. Nitriles coordinated to Ru(II) can undergo hydrolysis to amides after oxidation of Ru(II) to Ru(III). The hydrolysis of the Ru(II/III)-1-R-cyanopyridinium complex leads to pure amide, whereas hydrolysis of the free ligand results in a mixture of amide and pyridone. The rate constants for the hydrolysis of different coordinated nitriles to amides span several orders of magnitude. Catalytic systems can be designed. Upon reduction of the resulting Ru(III)–amide, the isonicotinamide, nicotinamide, or acrylamide complexes undergo aquation and isomerization to form the pyridyl or the olefin-bonded complex. On reduction of the Ru(III)-2-picolinamido complex, aquation and chelation occur to form $[Ru(NH_3)_5(H_2O)]^{2+}$ and picolinamide, and cis-$[Ru(NH_3)_4(2\text{-picolinamide})]^{2+}$, respectively. The spectral properties of the nitrile and amide complexes of Ru(II) and Ru(III) ammines and the preferred coordination site in ambidentate ligands are also presented.

THE HYDROLYSIS OF NITRILES TO AMIDES is well-known (*1, 2*), and the polymers, such as acrylamide and polyacrylamide, have widespread use. However, the conditions of hydrolysis of nitriles to amides are usually extreme, resulting in low yields and mixtures of products.

[1]Current address: Departamento de Química, Universidade Federal da Bahia, Salvador, Bahia, Brazil.

*Corresponding author

Coordinated nitriles can undergo hydrolysis to amides with widely varying rate constants that are often higher than those in the corresponding uncoordinated nitriles. For some coordinated nitriles, selective hydrolysis of nitriles to amides is possible, and catalytic hydrolysis systems can even be designed. In addition to this important potential application, ruthenium-coordinated nitriles and amides show interesting properties, such as the isomerization reaction that occurs after the reduction of $[Ru^{III}(NH_3)_5(amide)]$ (where the amide is bound at the nitrogen) (3–25).

This chapter is not intended as a thorough review of the literature, but rather presents some important aspects of the chemistry of Ru(II) and Ru(III) ammines with coordinated nitriles and amides (3–25). This chapter focuses, among other possible reactions, mostly on the hydrolysis of coordinated nitriles in ruthenium ammine complexes, and the reactions occurring upon the reduction of the resulting Ru(III)–amide complexes.

Ruthenium Amines with Coordinated Nitriles

A number of Ru(II)/(III) nitrile complexes with pentaammine and polypyridyl auxiliary ligands have been reported (3, 7–9, 12–38). In these complexes, Ru(II) acts as a σ-acceptor and π-donor, while Ru(III) is a σ- and π-acceptor. Nitrile ligands, such as acetonitrile (acn), benzonitrile (bzn), or cyanopyridines, which have low-lying empty π∗ orbitals of appropriate symmetry to interact with d_π-metal orbitals can coordinate to either Ru(III) or Ru(II), resulting in substitutionally inert complexes. In cationic cyanopyridinium ligands, the charge on the ligands imparts additional properties to the ruthenium complexes (12, 13–21, 30, 36). Polynitriles as ligands have also been reviewed with respect to their electric, magnetic, and spectroscopic properties (39).

Spectral and Redox Properties

In Ru(II)–nitrile complexes, as in the analogous pyridines (py–X) and pyrazine (pz) complexes, π-backbonding from the metal to the ligand plays a key role in the properties of the nitrile complexes and their derivatives. The metal-to-ligand charge-transfer (MLCT) band energies (Table I) and the Ru(III/II) redox potentials (Table II) in these $[Ru(NH_3)_5L]^{2+}$ complexes (L = substituted aromatic nitrogen heterocycles such as pyridines, pyrazines, and cyanopyridines) depend on the properties of the aromatic ring substituent of the ligand and on the solvent (26, 40, 48). The more electron-withdrawing the substituent is, the lower the MLCT absorption energy and the higher the redox potential will be. This trend in the electronegativity of the ligand substituent is related to the higher π-backbonding ability of the Ru(II) complex. However, the redox potentials (Table II) are indicative of the Ru(III)/Ru(II) affinity ratios for the nitrile ligands. Thus, they cannot be taken as an estimate of the π-backbonding ability of Ru(II), and, furthermore, the relationship of

Table I. Electronic Absorption Data of [Ru^{II}(NH$_3$)$_5$L] Complexes

L	λ_{max} in nm (log ε)		Ref.
	MLCT	*IL*	
pyridine (py)[a]	407 (3.89)	244 (3.66)	40
4-picoline (4-pic)[a]	397 (3.89)	244 (3.66)	40,41a
isonicotinamide (isn)[a]	479 (4.02)	260 (3.66)	40, 41b,41c
nicotinamide (nic)[a]	427 (3.78)	254 (3.72), 212 (3.84)	40, 41b, 41c
4-cyanopyridine (4-NCpy) (nitrile bonded)[a]	422 (4.05)	253 (4.22), 212 (4.09)	—[b]
4-NCpy (pyridyl bonded)[c]	500		28
4-NCpyH$^+$ (nitrile bonded)[d]	532 (3.91)	260 (3.93)	28
	534 (4.14)		12
1-methyl-4-cyanopyridinium (4-mcp)[a]	542		30
4-mcp[e]	545 (4.26)	267 sh, 257 sh, 242 (4.23)	30
3-NCpy (nitrile bonded)[a]	398 (3.98)	255 (4.29), 218 (4.05)	—[b]
3-NCpyH$^+$ (nitrile bonded)[d]	460 (3.67)	327 sh, 261 (4.10), 218 (3.83)	28
3-mcp[e]	462 (3.89)	258 (4.16), 210 (4.04)	30
[(NH$_3$)$_5$Ru^{II}(3-NCpy)] (nitrile bonded)[a]	422 (4.27)	254 (4.43), 214 (4.71)	—[b]
[(NH$_3$)$_5$Ru^{II}(3-NCpy)] (pyridyl bonded)[f]	438 (3.92)	256 (4.23)	34
pyrazine (pz)[a]	472 (4.03)	253 (3.78)	40,41b
pzH^{+d}	529 (4.08)		40
1-methylpyrazinium (pzCH$_3^+$)[g]	877 (2.30)	270 (3.81)	42,43
1-decyl-4-cyanopyridinium (4-decp)[e]	545 (4.35)	244 (4.22)	30
1-dodecyl-4-cyanopyridinium (4-docp)[e]	545 (4.31)	245 (4.11)	30
1-benzyl-4-cyanopyridinium (4-bcp)[e]	556 (4.31)	240 (4.22), 256 (4.26)	30
[(NH$_3$)$_5$Ru^{II}(4-NCpy)] (nitrile bonded)	518 (4.35), 416 sh	254 (4.22),206 (4.08)	—[b]
[(NH$_3$)$_5$Ru^{II}(4-NCpy)] (pyridyl bonded)[f]	492 (4.01)		34
[(NH$_3$)$_5$Rh^{III}(4-NCpy)] (pyridyl bonded)	488 (20.5)		35
(2-cyanopyridine) 2-NCpy (nitrile bonded)	406 (3.97)	256 (4.18), 222 (3.94)	—[b]
2-NCpyH$^+$ (nitrile bonded)[h]	505 (3.91)	217 (4.14)	28
1-methyl-2-cyanopyridinium (2-mcp)[e]	517 (4.18)	275 (4.36), 219 (4.08)	30
[(NH$_3$)$_5$Ru^{II}(pz)][a]	547 (4.48)	254 (3.43)	44
[(NH$_3$)$_5$Ru^{III}(pz)][a]	565 (4.32)	270 sh, 252 (3.75)	44
[(NH$_3$)$_5$Rh^{III}(pz)][a]	528 (4.25)	263 (3.81)	44
benzonitrile (bzn)	376 (3.93), 347 (3.84) sh	249 (4.21), 226 (4.17)	27
4-toluenenitrile (4-tln)	367 (3.86), 347 (3.8) sh		27
1,4-dicyanobenzene (1,4-dcb)	462 (21.6)		27
acetonitrile (acn)	229 (4.19)	350 (2.40) (LF)	27
propionitrile (prn)	262 (4.18)	350 (2.38) (LF)	23
2-cyanoethyldiphenylphosphine (2-cedp)	334 sh (2.58) (LF)	307 (3.46) (IL)	31

NOTE: λ_{max} is the wavelength of maximum absorption in nanometers; ε is the molar absorptivity; sh means shoulder; MLCT is metal-to-ligand charge transfer band; IL is internal-ligand band; LF is ligand-field band.

[a]Dilute aqueous solution. [b]This work. [c]Spectrum from [(NH$_3$)$_5$Ru^{II}OH$_2$)] (pH = 6) with excess 4-NCpy. [d]1 M HCl solution. [e]Acetonitrile solution. [f]1 M H$_2$SO$_4$ solution. [g]The spectrum of this complex displays an absorption band at 538 nm (log ε = 3.20) (ref. 43), to which an associated covalent character was assigned (λ_{max} = 540 nm; log ε = 4.20) (ref. 42) [h]2M HCl.

Table II. Electrode Potentials of Some Ruthenium Nitrile
Complexes, $[Ru(NH_3)_5L]^{3+/2+}$

L	$E^{\circ\prime}$ (mV vs. NHE)	Ref.
CH_3CN (acn)	426[a]	45
C_6H_5CN (bzn)	485[a]	45
	510[a]	46
$CH_3C_6H_5CN$ (4-tln)	475[a]	45
1,4-dicyanobenzene (1,4-dcb)	609[a]	45
4-NCpy (nitrile bonded)	592[b]	—[c]
	573[d]	47
4-NCpyH (nitrile bonded)	692[b]	—[c]
	637[e]	12
4-CH_3NCpy (4-mcp)	476[f]	20
4-NCpyRu(NH_3)_5	390, 657[b]	—[c]
	409, 684	34
4-NCpyRh(NH_3)_5	682[e]	47
3-NCpy (nitrile bonded)	327[d]	—[c]
3-NCpyH (nitrile bonded)	575[f]	12
3-CH_3NCpy (3-mcp)	414[d]	—[c]
3-NCpyRu(NH_3)_5	160, 385[d]	—[c]
	424, 649[g]	34
2-NCpy (nitrile bonded)	340[d]	—[c]

NOTE: NHE means normal hydrogen electrode.

[a] 0.1M toluenesulfonic acid/0.1 M potassium tolueneosulfonate. [b] 1 M HAc/NaAc. [c] This work. [d] 1 M KCl or 0.1 M KCl/0.1 M HCl. [e] 1 M $HClO_4$. [f] 0.1 M CF_3CO_2H/NaCF_3CO_2. [g] 1 M CF_3SO_3H.

the redox potentials and MLCT energies to backbonding is not simple or linear (45–49).

The ruthenium nitrile complexes show MLCT, internal-ligand (IL), and ligand-field (LF) absorption bands (Table I). The LF bands are usually obscured by the much more intense MLCT bands and are therefore not reported for most of the Ru(II) nitrile complexes (50). But for $[Ru(NH_3)_5$-(acn)]$^{2+}$, a ligand field band is observed at 350 nm ($\varepsilon = 163$ M^{-1} cm^{-1}) (27). In the propionitrile (prn) complexes $[Ru(NH_3)_5(prn)]^{2+}$ and trans-$[Ru(NH_3)_4(prn)-(H_2O)]^{2+}$, the LF bands are at 350 and 378 nm, respectively (23), and in the $[Ru(NH_3)_5(2\text{-cedp})]^{2+}$ (2-cedp = 2-cyanoethyldiphenylphosphine), the LF band is a shoulder at 334 nm (31).

The paradigmatic example of Ru(II) π-backbonding is the $[Ru(NH_3)_5$-(pz)]$^{2+}$ (40). The increased basicity (higher pK_a of coordinated pyrazine compared to that of free pyrazine) is attributed to Ru(II) π-backbonding. Upon protonation to $[Ru(pzH)(NH_3)_5]^{3+}$, the MLCT band is red shifted from 474 to 529 nm. The presence of an electron-donating substituent increases the MLCT energy. When CH_3^+ replaces H$^+$ on the free pyrazine nitrogen ($[Ru(pzCH_3^+)$-

$(NH_3)_5]^{3+}$), the MLCT band is red shifted to 877 nm (*42*) compared to the 529 nm band for the $[Ru(pzH^+)(NH_3)_5]^{3+}$ complex (*40*). When CH_3 replaces H in position 4 of the pyridine ring, the MLCT band at 407 nm ($[Ru(NH_3)_5py]^{2+}$) is shifted to 397 nm ($[Ru(NH_3)_5pyCH_3]^{2+}$). A similar shift is observed for the $[Ru(NH_3)_5L]^{2+}$ (L = bzn or 4-toluenenitrile (4-tln)) complexes (Table I). For the cyanopyridine complexes $[Ru(NH_3)_5(NCpy)]^{2+}$, the MLCT bands for the pairs of complexes (4-NCpyH$^+$)/(4-mcp) and (2-NCpyH$^+$)/(2-mcp) show a smaller red shift, whereas for the (3-NCpyH$^+$)/(3-mcp) a shift is barely observable. The energy splitting of the d_π orbitals is such that two absorption bands are observed for $[Ru(pzCH_3^+)(NH_3)_5]^{3+}$. The energy of the band at 540 nm (v_2) observed in this complex is solvent-insensitive and is associated with a transition involving a bonding and an antibonding orbital with high contributions from both the metal d_{yz} and the ligand π^* orbitals, thus suggesting a double-bond character between Ru(II) and $pzCH_3^+$. The energy of the band at 877 nm (v_1) is solvent-dependent and reported to be MLCT in nature (*42*). The energy difference between these two bands is indicative of the extent of backbonding.

The protonated $[Ru(4\text{-}NCpyH^+)(NH_3)_5]^{3+}$ complex shows a MLCT band at 532 nm (*28*). The $[Ru(4\text{-}mcp)(NH_3)_5]^{3+}$ has a MLCT band at 545 nm, similar to other substituted R-4-cyanopyridinium (4-rcp) [where R = methyl (4-mcp), decyl (4-decp), benzyl (4-bcp), or dodecyl (docp)] complexes (*30*), which is coincident with the v_2 band of $pzCH_3^+$. Backbonding in the cyanopyridinium complexes is less than in the $pzCH_3^+$ complex, since the v_1 band does not occur in the $[Ru(4\text{-}rcp)(NH_3)_5]^{3+}$ complexes (*30*). Thus, d_π orbital splitting and the spectroscopic resolution are smaller. In all cyanopyridinium complexes, the v_1 band is not observed and may be obscured in the v_2 lower energy part of the MLCT band.

In the cyanopyridine ligands the presence of two coordination sites, the pyridine nitrogen and the nitrile nitrogen, leads to linkage isomers, with $[Ru(NH_3)_5(OH_2)]^{2+}$, as shown in Scheme I (*28*). In strong acidic medium, the protonated nitrile bonded complex (C) is formed as a result of kinetic control. The deprotonated 4-NCpy complex (B) [λ_{max} = 425 nm (v = 23,500 cm^{-1})] can be isolated from C. In basic or slightly acidic medium, the nitrile bonded 4-NCpy complex (B) and a small amount of the complex coordinated through the pyridine nitrogen, pyCN, (A) [λ_{max} = 500 nm (v =20,000 cm^{-1})] are formed. The 4-NCpy (B) complex can be reversibly protonated, forming NCpyH$^+$ (C) [λ_{max} = 532 nm (v = 18,800 cm^{-1})]. The pK_a of the pyCN (A) is very low, while those of the nitrile-bonded complexes are higher and even higher than those of the free ligands, indicating that the nitrile-coordinated cyanopyridines are more basic then the free ligands, independent of the ring substituent position. The strong backbonding interaction of the ruthenium d_π orbitals with the π^* orbitals of the nitrile results in changes in the cyanopyridine basicity (*28*). The MLCT band energies of the Ru(II)–nitrile complexes are dependent on the nature of the nitrile, (Table I) and may not be observed, such as in $[Ru(NH_3)_5(2\text{-}cedp)]^{2+}$ (*31*), where it may lie below 200 nm.

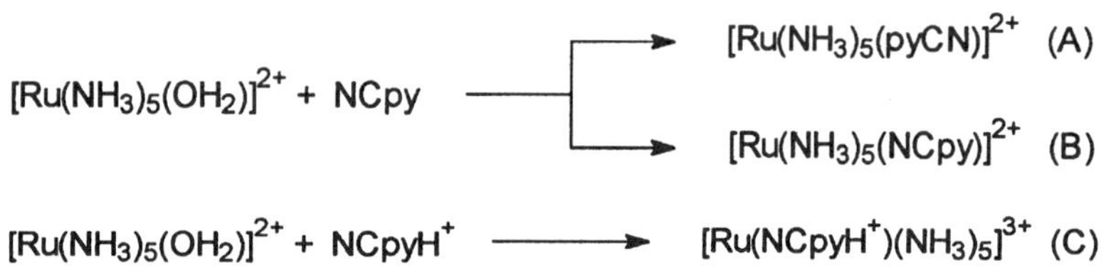

$$[Ru(NH_3)_5(OH_2)]^{2+} + NCpy \longrightarrow \begin{cases} [Ru(NH_3)_5(pyCN)]^{2+} \quad (A) \\ [Ru(NH_3)_5(NCpy)]^{2+} \quad (B) \end{cases}$$

$$[Ru(NH_3)_5(OH_2)]^{2+} + NCpyH^+ \longrightarrow [Ru(NCpyH^+)(NH_3)_5]^{3+} \quad (C)$$

Scheme I.

Cyanopyridines, similar to pyrazine, can act as bridging ligands in electron-transfer studies in binuclear species, such as $[(NH_3)_5RuLM]$ (e.g., L = pz, 4-NCpy, or 3-NCpy; and M = $Ru(NH_3)_5$; $Rh(NH_3)_5$ or $Fe(CN)_5$) (12, 32–34, 44, 47, 51–53) and in supramolecular photochemistry (54, 55). The binuclear complexes, such as $[(NH_3)_5Ru(4\text{-}NCpy)Ru(NH_3)_5]$, can exist as Ru(II)/Ru(II), Ru(III)/Ru(III), or the mixed-valence species Ru(II)/Ru(III), for which an intervalence (IT) band is present in the spectrum (Table I).

Hydrolysis of Nitriles

Hydrolysis of Uncoordinated Nitriles. Free uncharged nitriles undergo acid or base hydrolysis to form amides, but the hydrolysis reaction may result in a mixture of products, despite high yields (1, 2). In basic medium, the resulting amide undergoes further hydrolysis to carboxylic acid and ammonia (1, 2). Free 4-mcp and some 4-rcp ions undergo basic hydrolysis to a mixture of amide and pyridone derivatives, with rate constants and pyridone/amide ratios dependent on the pH and on the medium (56–59). The hydrolysis of the 4-mcp cation in organized media has also been thoroughly studied (56, 60, 61).

Hydrolysis of Coordinated Nitriles. The use of $[(NH_3)_5Ru(II)/(III)]$ complexes that coordinate the nitrile can have important advantages over kinetically labile complexes (22) and metal alloys (11) in the hydrolysis of nitriles to amides, and the study of these kinetically inert complexes is important in the design of catalytic systems.

The Ru(II) complexes, $[Ru^{II}(NH_3)_5(NCR)]^{n+}$, such as those with NCR = cyanopyridinium, undergo hydrolysis only at high-pH conditions (13, 20, 30). In these complexes, the Ru(II) inhibits the hydrolysis by donating electron density to the nitrile via π-backbonding. Upon oxidation to Ru(III) the π-backbonding is eliminated and the charge is increased, increasing the positive charge density on the nitrile carbon and favoring the nucleophilic attack. After oxidation of the Ru(II) to Ru(III), the coordinated nitriles undergo hydrolysis to amides in aqueous media, with the resulting amide complex coordinated through the amide nitrogen (eqs 1 and 2).

$$[(NH_3)_5Ru^{II}(NCR)] \rightarrow [(NH_3)_5Ru^{III}(NCR)] + e^- \tag{1}$$

$$[(NH_3)_5Ru^{III}(NCR)] + H_2O \rightarrow [(NH_3)_5Ru^{III}(NHC(O)R)] + H^+ \tag{2}$$

The hydrolysis rates of coordinated nitriles are greater than the rates of the corresponding uncoordinated nitrile and are dependent on the charge and nature of the metal center (3, 5, 6, 22, 62), nature of the nitrile, and the pH. The hydrolysis rates of Ru(III) nitrile complexes and those for the free nitriles and related species are listed in Table III.

The hydroxide-catalyzed hydrolyses of $[Rh(NH_3)_5L]^{3+}$, $[Ru(NH_3)_5L]^{2+}$, and $[Ru(NH_3)_5L]^{3+}$ (L = acn or bzn) to an amide complex have been studied by Zanella and Ford (3). The $[Ru(NH_3)_5L]^{3+}$ complexes (L = acn or bzn) have rate constants that are 10^8 higher than those of the free nitriles and 10^2 higher than for the analogous Co(III) and Rh(III) complexes (3). For the Ru(II)–acetonitrile complex, where the rate constant is 10^6 smaller (Table III), a less dramatic change is seen. The decrease in the rate constants for the Ru(II) complexes relative to Ru(III) is attributed to both the lower charge of the cation and the Ru(II) π-backbonding to the nitrile, which increases the electronic density on the nitrile carbon and makes it less prone to nucleophilic attack (3).

A mild selective conversion of nitriles to amides was proposed by Taube for $[Ru(NH_3)_5(NCR)]^{2/3+}$ (7–9). The $[Ru(NH_3)_5NCR]^{2+}$ complex is oxidized to Ru(III), which undergoes hydrolysis to Ru(III)–amide complex. After reduction of Ru(III)–amide to the Ru(II) complex, amide aquation occurs, resulting in free amide in very high yields and $[Ru(NH_3)_5(H_2O)]^{2+}$, which can be recycled by the addition of more nitrile to make the reaction catalytic in ruthenium.

The hydrolysis of the coordinated nitriles in the binuclear $[(CN)_5Fe-(pyCN)Ru(NH_3)_5]$ (4- and 3-NCpy isomers) and $[(NH_3)_5Ru(4-NCpy)Ru-(NH_3)_5]^{n+}$ complexes have been investigated (12, 64). The chemically oxidized binuclear complexes hydrolyze at a faster rate than the corresponding mononuclear Ru(III) complexes.

Recently, the hydrolysis of $[Ru^{II}(NH_3)_5(rcp)]^{3+}$ complexes with 1-R-cyanopyridinium cations (rcp) (R = methyl, decyl, dodecyl, or benzyl) has been studied (30). These $[Ru^{II}(NH_3)_5(rcp)]^{3+}$ complexes differ from the binuclear complexes $(NH_3)_5RuLM^{n+}$ by the presence of R rather than M bonded to the pyridine nitrogen, which hinders intramolecular electron-transfer to another metal center. These complexes are rather stable in acidic medium as expected, but in basic medium the hydrolysis of the coordinated nitrile results exclusively in $[Ru^{III}(NH_3)_5(amide)]^{2+}$, with no pyridone being formed (13). In basic solution the rate of reaction is pseudo-first-order in hydroxide (13). This may be a good example of inhibition of hydrolysis upon coordination to a metal center in comparison to the free nitrile and illustrates the importance of backbonding. In alkaline medium, it is difficult to know if the hydrolysis occurred before, after, or simultaneously with the oxidation of the Ru center (13). The

Table III. Rate Constants of the Hydrolysis (k_{fl}) of $[Ru^{III}(NH_3)_5(L)]^{n+}$ and Some Related Species

Species	k_{fl} (s^{-1})	pH	Ref.
$[Ru(NH_3)_5(4\text{-}NCpy)]^{3+}$ (NC bonded)	$(2.85 \pm 0.02) \times 10^{-3}$ [a]	5.0	12
	3.9×10^{-3} [a]	4.65	63
	$(4.23 \pm 0.04) \times 10^{-3}$ [b]	4.65	63
$[Ru(NH_3)_5(3\text{-}NCpy)]^{3+}$ (NC bonded)	$(0.667 \pm 0.007) \times 10^{-3}$ [a]	5.0	12
	1.7×10^{-3} [a]	4.65	63
	$(2.7 \pm 0.1) \times 10^{-3}$ [b]	4.65	63
$[Ru(NH_3)_5(2\text{-}NCpy)]^{3+}$ (NC bonded)	2.0×10^{-3} [a]	4.65	63
	$(1.38 \pm 0.01) \times 10^{-3}$ [b]	4.65	63
$[Ru(4\text{-}mcp)(NH_3)_5]^{4+}$	$(160) \times 10^{-3}$ [b]	3	20
	$(210 \pm 1) \times 10^{-3}$	2.0	12
	$(235 \pm 1) \times 10^{-3}$	4.0	12
	$(283) \times 10^{-3}$ [a]	5.0	12
	$(388 \pm 8) \times 10^{-3}$	8.0	12
$[Ru(4\text{-}bcp)(NH_3)_5]^{4+}$	710×10^{-3} [b]	3	20
$[Ru(4\text{-}decp)(NH_3)_5]^{4+}$	530×10^{-3} [b]	3	20
$[Ru(4\text{-}docp)(NH_3)_5]^{4+}$	330×10^{-3} [b]	3	20
$[Ru(3\text{-}mcp)(NH_3)_5]^{4+}$	163×10^{-3} [b]	3	63
$[(NH_3)_5Ru(4\text{-}NCpy)Ru(NH_3)_5]^{4+}$	$(19 \pm 3) \times 10^{-3}$ [a]	4.65	63
	17.5×10^{-3} [b]	4.65	63
	25×10^{-3} [a]	3	64
$[(NH_3)_5Ru(3\text{-}NCpy)Ru(NH_3)_5]^{4+}$	$(5.49 \pm 0.03) \times 10^{-3}$ [a]	4.65	63
	$(2.77 \pm 0.01) \times 10^{-3}$ [b]	4.65	63
$[Ru(4\text{-}NCpyH)(NH_3)_5]^{4+}$ (NC bonded)	$(10.4 \pm 0.1) \times 10^{-3}$ [a]	—[c]	12
$[Ru(3\text{-}NCpyH)(NH_3)_5]^{4+}$ (NC bonded)	$(3.41 \pm 0.03) \times 10^{-3}$ [a]	—[c]	12
$[Ru(NH_3)_5(acn)]^{3+}$	220	—[d]	3
$[Ru(NH_3)_5(acn)]^{2+}$	$<6 \times 10^{-5}$	—[d]	3
$[Rh(NH_3)_5(acn)]^{3+}$	1.0	—[d]	3
$[Co(NH_3)_5(acn)]^{3+}$	3.40	—[e]	5
acetonitrile	1.60×10^{-6}		—[f]
$[Ru(NH_3)_5(bzn)]^{3+}$	2.0×10^{3}	—[d]	3
	2.01×10^{-4} [a]	5	63
	3.39×10^{-5} [a]	2.3	63
benzonitrile	7.2×10^{-6}		—[f]
$[Ru(NH_3)_5(ecf)]^{3+}$ [g]	10 ± 1	—[h]	8
$[Ru(NH_3)_5(prn)]^{3+}$	$(1.9 \pm 0.1) \times 10^{2}$	—[i]	23
$[(CN)_5Fe(4\text{-}pyCN)Ru(NH_3)_5]^{-}$	$(13.1 \pm 0.1) \times 10^{-3}$	5.0	12
$[(CN)_5Fe(3\text{-}pyCN)Ru(NH_3)_5]^{-}$	$(4.39 \pm 0.06) \times 10^{-3}$	5.0	12

[a]From chemical oxidation of the Ru(II) species. [b]From electrochemical oxidation of the Ru(II) species. [c]In 1 M $HClO_4$. [d]In slightly basic medium. [e]$0.10 < [OH^-] > 0.0050$. [f]Ref. 3 and references therein. [g]ecf = ethylcyanoformate. [h]In 1.0 M H_2SO_4. [i]In basic medium.

origin of the Ru oxidation is uncertain, and although leakage of air into the solution cannot be ruled out, the rate constants were only slightly affected by saturation of the solution with oxygen (*13*). In order to avoid parallel chemical reactions, this system was studied electrochemically in acidic medium, where the chemical hydrolysis of the starting reactant does not occur. The elucidation of the mechanism for the electrochemical oxidation and reduction of these compounds is important for developing an electrochemical method for the conversion of nitriles to amides.

The selective hydrolysis of nitriles to amides has been demonstrated electrochemically for the oxidation of $[Ru^{II}(NH_3)_5(4\text{-rcp})]^{3+}$ (R = methyl, decyl, dodecyl, or benzyl) in aqueous acidic solution by cyclic voltammetry and controlled-potential electrolysis (*20*). The hydrolysis of the coordinated nitrile exclusively to coordinated amide is coupled with the reversible one-electron oxidation of Ru(II) (E_p ca. 0.51 V/SCE) (CF_3COOH/CF_3COONa, μ =0.1 M, pH 3). The overall rate constant (k_f) ranges from 160×10^{-3} s^{-1}, for 4-mcp, to 710×10^{-3} s^{-1}, for 4-bcp. After reduction of the resulting Ru(III)–amide to the Ru(II)–amide, rapid aquation to $[Ru^{II}(NH_3)_5(H_2O)]^{2+}$ and free amide occurs. This selective behavior is markedly different from that of the uncoordinated nitriles, which results in a mixture of pyridone and amide. This is an example of the selective nitrile–amide conversion catalyzed by $[Ru^{II}(NH_3)_5(H_2O)]^{2+}$, as for other nitriles (*7–9*).

The $[Ru(4\text{-rcp})(NH_3)_5]^{3+}$ (4-rcp = 4-bcp or 4-mcp) complexes sensitize the production of singlet oxygen through irradiation of the $[Ru(4\text{-rcp})(NH_3)_5)]^{3+}$ complexes in aqueous solutions (*36*). Singlet oxygen oxidizes the Ru(II) to Ru(III), and a 1-R-4-carboxoamidopyridinium complex is formed.

The $[Ru^{II}(NH_3)_5(NCR)]^{n+}$ (NCR = 4-NCpy, 3-NCpy, or 4-mcp) undergo hydrolysis to $[(NH_3)_5Ru^{III}\text{–NHC(O)R}]^{n+}$ after chemical oxidation of Ru(II) to Ru(III) (*12, 24, 25*).

In electrochemical and spectroelectrochemical studies on $[Ru^{II}(NH_3)_5\text{-}(NCR)]^{n+}$ (NCR = 2-NCpy (*21, 63*), 3-mcp (*63*), 2-mcp (*65*), 4-NCpy (*63*), or 3-NCpy (*63*)), $[(NH_3)_5Ru(4\text{-NCpy})Ru(NH_3)_5]^{n+}$ (*63, 64*) and $[(NH_3)_5Ru(3\text{-NCpy})\text{-}Ru(NH_3)_5]^{n+}$ (*63*), oxidation of Ru(II) is always followed by the hydrolysis of the coordinated nitrile to amide in acidic solution.

The hydrolysis rates of the $[Ru(NH_3)_5]$ complexes with 4-, 3-, or 2-cyanopyridines are very similar (Table III). In the rcp complexes the rate constant increases because of the added positive charge (from the CH_3^+ group on the pyridine nitrogen), as seen in comparing the rates of $[Ru(NH_3)_5(4\text{-NCpy})]^{3+}$ with $[Ru(NH_3)_5(4\text{-mcp})]^{4+}$, and $[Ru(NH_3)_5(3\text{-NCpy})]^{3+}$ with $[(NH_3)_5Ru(3\text{-mcp})]^{4+}$ (Table III). The effect of the positive charge is less prominent in the binuclear species with bridging NCpy.

The hydrolysis rates of the coordinated nitriles generally follow the electron-withdrawing ability of R in NCR. The higher this ability is, the higher the backbonding in Ru(II) complexes will be. Thus, in Ru(II)–nitrile complexes, increase in the electron-withdrawing ability of R decreases the rates, but in

Ru(III), with the elimination of backbonding, this trend is reversed and an increase in the electron-withdrawing ability of R increases the hydrolysis rates.

Ruthenium Ammines with Coordinated Amides

The Ru(III)–amide complexes are prepared by reaction of the Ru(III) precursor with the amide or through the hydrolysis of the coordinated nitrile after oxidation of the Ru(II)–nitrile complex (3, 7–9, 62). The hydrolysis of coordinated nitriles leads to an amide complex bonded to the Ru center through the amide nitrogen.

The electronic spectral properties of the Ru(III)–amide complexes (Table IV) include pH-dependent absorption bands in the 300–400-nm range assigned to ligand-to-metal charge transfer (LMCT) transitions (3, 25). Upon protonation (3), the band at 393 nm in $[Ru(NH_3)_5(NHC(O)C_6H_5)]^{2+}$ is shifted to 385 nm in $[Ru(NH_3)_5(NH_2C(O)C_6H_5)]^{3+}$, and the band at 383 nm in $[Ru(NH_3)_5(NHC(O)CH_3)]^{2+}$ is shifted to 322 nm in $[Ru(NH_3)_5(NH_2C(O)CH_3)]^{3+}$. In the $[Ru^{III}(NH_3)_5(NHC(O)pyX]^{3+}$ complexes (X = 4-NC or 3-NC) at pH 4.5 the bands at 350 nm were also assigned to a LMCT transition from the lone pair of the deprotonated amido nitrogen to the vacant d_π orbital of Ru(III) similar to the 402 nm band observed for deprotonated $Ru(NH_3)_6^{3+}$ (3, 25). These bands are blue shifted at pHs below 0.2 (20). Protonation was assumed to occur at the amide oxygen (25), since protonation in the amide nitrogen would be reflected in higher energy shifts.

Table IV. Electronic Absorption Data of $[Ru^{III}(NH_3)_5L]$ Complexes

L	λ_{max} in nm (log ε)	Ref.
CH$_3$CONH	383 (3.54); 249 (3.36)	3
CH$_3$COHNH	322 (3.19)	3
C$_6$H$_5$CONH	393 (3.61); 314 (3.57); 270 sh; 219 (3.97)	3
C$_6$H$_5$CONH$_2$	385 (3.16); 320 (3.54); 270 sh; 228 (3.97)	3
NHCO-4-py	386 (3.51)	12
	384 (3.57); 262 (3.71); 228 (3.61)	63
NHCO-4-pyH	358 (3.67)	12
NHCO-4-pyCH$_3$	358 (3.48)	12
	350	20
$[(NH_3)_5Ru(NHC(O)-4-py)]^{2+}$	346 (3.67)[a]; 276 (3.77)	63
NHC(O)-2-py	390 (3.41)[a]; 354 (3.42); 326 (3.50); 258 (3.60)	63
NHC(O)-3-py	386 (3.56)[a]; 306 (3.54); 262 (3.65)	63
	387 (3.57)	12
NHC(O)-3-pyH	357 (3.45)	12
NHC(O)-3-pyCH$_3$	360 (3.46)[a]; 314 (3.43); 266 (3.66)	63
$[(NH_3)_5Ru(NHC(O)-3-py)]^{2+}$	388 (3.56)[a]; 258 (3.80)	63

[a]HAc/Ac$^-$ solution, pH = 4.65.

Reactions Following Reduction of Ru(III)–Amides

The nature of the reaction following the reduction of the Ru(III) N-bound amide complexes, formed from the oxidized Ru(II)–nitrile complex, depends on the nature of the ligand and on whether the complex is mononuclear or binuclear. The rate constants for the reactions that follow the Ru(III) reduction in $[(NH_3)_5Ru^{III}(amide)]$ are listed in Table V.

Mononuclear complexes such as $[Ru^{III}(NH_3)_5(NHC(O)\text{-}3\text{-}pyCH_3)]^{3+}$ (63) or $[Ru^{III}(NH_3)_5(NHC(O)R)]^{3+}$ [e.g., R = 1-methyl-4-pyridine, 1-decyl-4-pyridine, 1-dodecyl-4-pyridine, or 1-benzyl-4-pyridine (20); phenyl, ethylformate, methyl, or benzoallyl (7–9)], undergo aquation upon reduction, resulting in $Ru(NH_3)_5(OH_2)^{2+}$ and the corresponding free amide, in very high yields. This aquation is a result of the lability of the amide group relative to $[Ru^{II}(NH_3)_5]$ (7–9, 20, 24). In the presence of excess nitrile, more Ru(II)–NCR complex can be formed, oxidized, hydrolyzed, and reduced to give more amide, resulting in a catalytic system.

Similarly rapid aquation occurs when binuclear complexes, such as $[(NH_3)_5Ru^{III}(NHC(O)X)Ru^{III}(NH_3)_5]^{5+}$ (X = 3-py or 4-py) are reduced to Ru(II)–Ru(II), forming $[Ru(NH_3)_5(OH_2)]^{2+}$ and the $[Ru(NH_3)_5(nic)]^{2+}$ (nic = nicotinamide) or $[Ru(NH_3)_5(isn)]^{2+}$ (isn = isonicotinamide) complex, respectively. The aquation rate constants following reduction are listed in Table V.

For the mononuclear species $[Ru^{III}(NH_3)_5(NHC(O)x\text{-}py)]^{2+}$ (x = 4 or x = 3), rather than only aquation to give $[Ru(NH_3)_5(OH_2)]^{2+}$ and the free amide (isonicotinamide or nicotinamide), after reduction, isomerization also occurs to give the pyridyl-bonded complex, $[Ru(NH_3)_5(isn)]^{2+}$ (x = 4) and $[Ru(NH_3)_5(nic)]^{2+}$ (x = 3) (24). The overall rate constant for the isn complex is 24 s^{-1} (24),

Table V. Overall Rate Constants for Reactions Following $[(NH_3)_5Ru^{III}(amide)]$ Reduction

Amide	Products	k_{f2} (s^{-1})	Ref.
4-NHC(O)pyCH$_3$	4-NH$_2$C(O)pyCH$_3$	6	24
4-NHC(O)pyC$_6$H$_5$	4-NH$_2$C(O)pyC$_6$H$_5$	10	20
4-NHC(O)py	isonicotinamide and	26	63
	$[Ru(NH_3)_5(isn)]^{2+}$	24	24
3-NHC(O)py	nicotinamide and $[Ru(NH_3)_5(nic)]^{2+}$	113	63
3-NHC(O)pyCH$_3$	3-NH$_2$C(O)pyCH$_3$	10.2	63
NHC(O)C$_6$H$_5$	benzamide	34	24
$[(NH_3)_5Ru^{II}(NHC(O)\text{-}4\text{-}py]^+$	$[Ru(NH_3)_5(isn)]^{2+}$	71	63
$[(NH_3)_5Ru^{II}(NHC(O)\text{-}3\text{-}py]^+$	$[Ru(NH_3)_5(nic)]^{2+}$	132	63

NOTE: Rate constants are for reactions at 25 °C obtained by reducing the Ru(III) analog. Products listed are in addition to $[Ru(NH_3)_5(H_2O)]^{2+}$.

SOURCE: Rate constants are from electrochemical data in references 20 and 63, and from chemical reduction and spectrophotometrical data in reference 24.

resulting in 60% aqua and 40% isn complexes, from which a calculated rate constant of isomerization of 9.6 s^{-1} was obtained (24). The isomerization of the isn complex was studied by cyclic voltammetry, controlled-potential electrolysis, and products analysis, with similar results to give an overall rate constant of 26 s^{-1}, an isomerization rate constant of 10.4 s^{-1}, and an aquation rate of 15.6 s^{-1} (63). For the reactions following the reduction of [Ru(NH$_3$)$_5$(NHC(O)-3-py)]$^{2+}$ complex, the overall rate constant is 113 s^{-1} and the isomerization rate constant is 54 s^{-1}, resulting in 48% of nic and 52% of aqua complexes (63). The linkage isomerization was followed spectrophotometrically during the controlled-potential electrolysis of [Ru^{III}(NH$_3$)$_5$(NHC(O)-3-py)]$^{2+}$ at –500 mV vs. Ag/AgCl (Figure 1). In Figure 1, the decrease of the Ru(III) species (I) is followed by an increase of absorbance (F) with the formation of the linkage isomer [Ru^{II}(NH$_3$)$_5$(nic)]$^{2+}$. This isomerization is proposed to occur via successive π-bonded complexes based on kinetic arguments and analogy to related Os(II) and Ru(II) complexes [i.e., Ru(II) bonds to the aromatic ring via η^2 in the intermediates] (24).

Only a few (NH$_3$)$_5$Ru(II) complexes of η^2-bonded to aromatic rings have been described, presumably due to the relatively low stability of the Ru(II)–aromatic bond (66). However, an isomerization reaction from amide to olefinic η^2-bonded [(acrylamido)(NH$_3$)$_5$Ru(II)] complex was reported to occur upon reduction of the amido bonded [(NH$_3$)$_5$Ru^{III}(acrylamido)] (67). The stability of the Ru(II)–olefin bond allowed the identification of η^2-bonded [(acrylamido)-

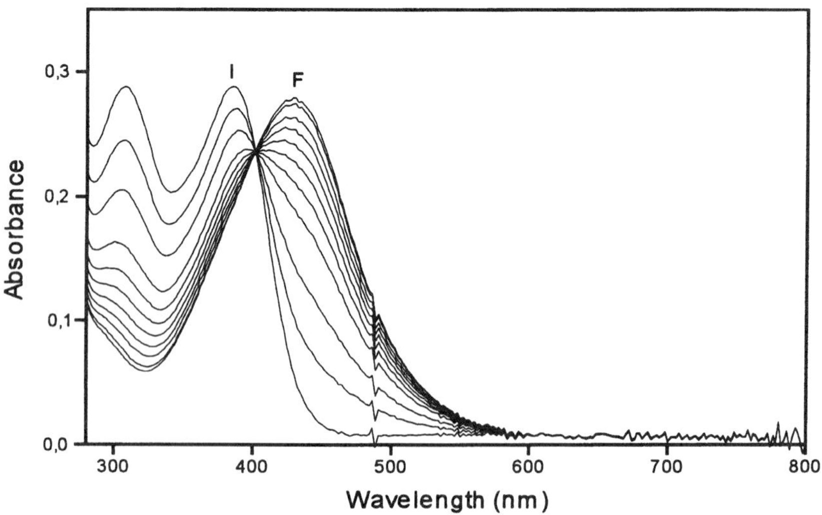

Figure 1. *Successive spectra during the reduction of [Ru^{III}(NH$_3$)$_5$(NHC(O)3-py)]$^{2+}$ by controlled-potential electrolysis at –500 mV (vs. Ag/AgCl) in HAc/Ac$^-$ solution; pH = 4.65; μ = 1 M at 25 °C. [Ru(III)]$_i$ = 1.05 × 10^{-3} M. I = initial spectrum; F = final spectrum.*

$(NH_3)_5Ru(II)]$ as a product, and an isomerization rate constant of 8 s^{-1}, at 22 °C, was determined from electrochemical data (67). In contrast, the ambidentate acrylonitrile preferentially bonds to Ru(II) through the nitrile, rather than through the olefin (29, 68). Preferential bonding to nitrile in Ru(II) was also reported for the ambidentate 2-cyanoethyldiphenylphosphine ligand in [Ru-$(NH_3)_5(2$-cedp)]$^{2+}$ (31).

The chemical reduction of the amide complexes $[(NH_3)_5Ru^{III}(NHC(O)R)]$ (R = C_6H_5, 4-C_5H_4N, or 4-C_5H_4N–CH_3^+) results in highly colored solutions. For R = C_6H_5 or 4-C_5H_4N-CH_3^+, the colors bleach with aquation (24). The one-electron reduction of an aqueous solution of $[(NH_3)_5Ru^{III}(NHC(O)R)]$ (R = 1-methyl-4-pyridyl or 1-benzyl-4-pyridyl) by controlled-potential electrolysis results in a blue color that fades quickly (20). A transient blue color was also observed when $[Ru(NH_3)_5(OH_2)]^{2+}$ was added to a solution of excess carboxamidepyridinium ion or benzamide (20, 25). The spectra of the reduced form of the Ru(III)–amide (benzamide, isonicotinamide, or 1-methyl-4-carboxamidopyridinium) have been reported recently (25), and the visible bands (Table VI) were assigned to a Ru(II)–amide MLCT transition. The wide range of MLCT band maxima observed may indicate that the Ru(II)–amide MLCT transitions are more purely CT in nature than those in the analogous $[(NH_3)_5Ru(II)$–] complexes bonded to an aromatic nitrogen (25).

With the 2-cyanopyridine complex $[(NH_3)_5Ru^{II}(2$-NCpy] an additional feature is observed. Oxidation of the nitrile-bonded $[(NH_3)_5Ru^{II}(2$-NCpy] in aqueous solution is followed by hydrolysis to the corresponding amide complex $[(NH_3)_5Ru^{III}(NHC(O)2$-py)] (21). Reduction of this Ru(III)–amide to Ru(II)–amide complex results in addition to aquation, in a chelation reaction, to give *cis*-$[(NH_3)_4Ru^{II}(2$-pca)] (2-pca = 2-picolinamide). Electrochemical and spectroelectrochemical results indicate that reduction of the Ru(III)–amide complex is followed by amide aquation (30%) and chelation (70%) with displacement of one coordinated *cis*-ammonia as depicted in Scheme II (21).

The reactions described in this chapter can be summarized in Scheme III, using the 4-NCpy ligand as an example.

The richness of the chemistry derived from the nitrile complexes of ruthenium, especially the participation of backbonding and the ability to design cat-

**Table VI. MLCT Maxima
of $[(NH_3)_5Ru^{II}(NHC(O)$–R)] Complexes**

R	λ_{max} (nm)
phenyl-	380
4-pyridyl-	475
1-methyl-4-pyridyl-	695
1-methyl-3-pyridyl-	472

NOTE: Complexes are in aqueous solution at room temperature (25).

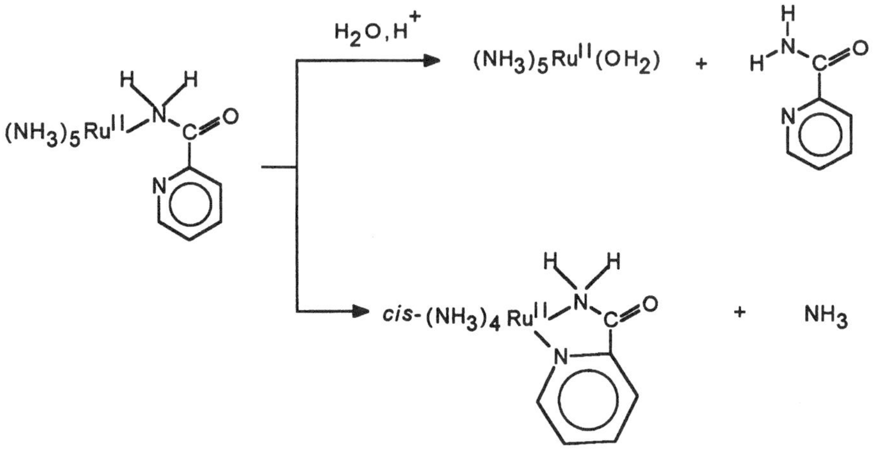

Scheme II.

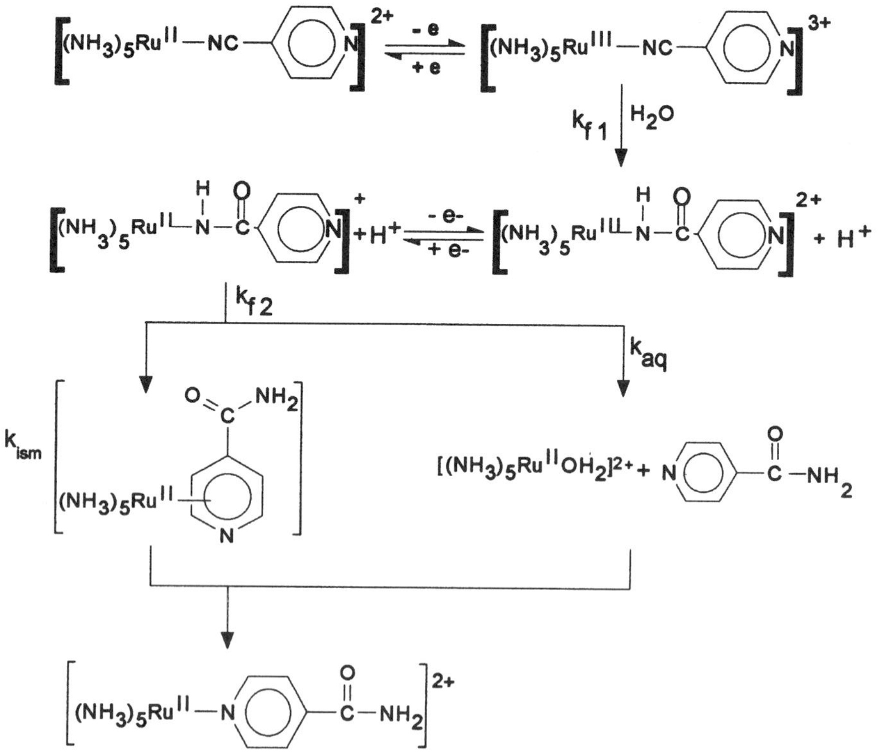

Scheme III.

alytic systems, has been demonstrated. We are currently investigating several systems with other nitriles and dinitriles such as fumaronitrile, succinonitrile, dicyanobenzene, and their amide derivatives. Ruthenium–cyclam complexes are also being studied. The amide N/O linkage isomerism is also being studied including dimethylformamide.

Acknowledgments

We thank the Brazilian agencies FAPESP, CNPq, and CAPES, and the PADCT program for grants and fellowships.

We also thank Prof. A. B. P. Lever for helpful comments in reading the manuscript.

Some of the information presented in this chapter is based on the Master's Dissertation and D. Sc. Thesis of Z. N. da Rocha (*13, 63*).

References

1. Streitwieser, A., Jr.; Heathcock, C. H. *Introduction to Organic Chemistry,* 3rd ed.; Macmillan: New York, 1985.
2. Vogel, A. I. *Vogel's Textbook of Practical Organic Chemistry;* Furniss, B. S.; Hannaford, A. J.; Smith, P. W. G.; Tatchell, A. R., Eds.; Longman Scientific and Technical: England, 1989; p 1228.
3. Zanella, A. W.; Ford, P. C. *Inorg. Chem.* **1975,** *14,* 42.
4. Pinnell, D.; Wright, G. B.; Jordan, R. B. *J. Am. Chem. Soc.* **1972,** *94,* 6104.
5. Buckingham, F. R.; Keene, F. R.; Sargeson, A. M.. *J. Am. Chem. Soc.* **1973,** 95, 5649.
6. Balahura, R. J.; Cock, P.; Purcell, W. L. *J. Am. Chem. Soc.* **1974,** 96, 2739.
7. Diamond, S. E.; Grant, B.; Tom, G. M.; Taube, H. *Tetrahedron Lett.* **1974,** *46,* 4025.
8. Diamond, S. E. Ph. D. Thesis, Stanford University, **1975.**
9. Diamond, S. E.; Taube, H. *J. Chem. Soc. Chem. Commum.* **1974,** 622.
10. Creaser, I. I.; Harrowfield, J. M.; Keene, F. R.; Sargeson, A. M. *J. Am. Chem. Soc.* **1981,** *103,* 3559.
11. Sugiyawa, K.; Miuka, H.; Watanabe, Y.; Ukai, Y.; Matsuda, T. *Bull. Chem. Soc. Jpn.* **1987,** 60, 1579.
12. Huang, H. Y.; Chen, W. J.; Yang, C. C.; Yeh, A. *Inorg. Chem.* **1991,** *30,* 1862.
13. Rocha, Z. N. M.Sc. Dissertation, Instituto de Química de Araraquara da UNESP Araraquara, 1987.
14. Naal, Z.; Tfouni, E.; Rocha, Z. N.; Benedetti, A. V. *Anais do VI Simpósio Brasileiro de Eletroquímica e Eletroanalítica;* Comissão de Organização, São Paulo, Brazil, 1988; p 3.
15. Tfouni, E.; Rocha, Z. N.; Benedetti, A. V.; Naal, Z. *VIII Encontro Latinoamericano de Eletroquímica e Corrosão.* Córdoba, Argentina, 1988.
16. Naal, Z.; Benedetti, A. V.; Tfouni, E. *Anais VII Simpósio Brasileiro de Eletroquímica e Eletroanalítica;* Comissão de Organização: Ribeirão Preto, Brazil, 1990; Vol. 1, p 66.
17. Naal, Z.; Benedetti, A. V.; Tfouni, E. *IX Congreso Iberoamericano de Eletroquímica;* La Laguna: Tenerife, Spain, 1990; pp 16-22.
18. Naal, Z. M.Sc. Dissertation, Instituto de Química de Araraquara da UNESP Araraquara, 1991.
19. Naal, Z.; Benedetti, A. V.; Tfouni, E.; Chiericato, G., Jr. *Anais VIII Simpósio Brasileiro de Eletroquímica e Eletroanalítica;* Comissão de Organização: Campinas, Brazil, 1992; p 286.

20. Naal, Z. ; Tfouni E.; Benedetti, A. V. *Polyhedron* **1993**, *13*, 133.

21. Rocha, Z. N.; Chiericato, G. Jr.; Tfouni, E. *Inorg. Chem.* **1994**, *33*, 4619.

22. Storhoff, B. N.; Lewis, H. C., Jr. *Coord.Chem.Rev.* **1977**, *23*, 1.

23. Alves, J. F. F.; Plepis, A. M. G.; Davanzo, C. U.; Franco, D. W. *Polyhedron* **1993**, *12*, 2215.

24. Chou, M. H.; Brunschwig, B. S.; Creutz, C.; Sutin, N.; Yeh, A.; Chang, R. C.; Lin, C.-T. *Inorg. Chem.* **1992**, *31*, 5347.

25. Chou, M. H.; Szalda, D. J.; Creutz, C.; Sutin, N. *Inorg. Chem.* **1994**, *33*, 1674.

26. Ford, P. C. *Coord. Chem. Rev.* **1970**, *5*, 75.

27. Clarke , R. E.; Ford, P. C. *Inorg. Chem.* **1970**, *9*, 227.

28. Clarke, R. E.; Ford, P. C. *Inorg. Chem.* **1970**, *9*, 495.

29. Ford, P. C.; Foust, R. D., Jr.; Clarke, R. E. *Inorg. Chem.* **1970**, *9*, 1933.

30. Rocha, Z. N.; Tfouni, E. *Polyhedron* **1992**, *11*, 2375.

31. Caetano, W.; Alves, J. J. F.; Lima Neto, B. S.; Franco, D. W. *Polyhedron* **1995**, *14* 1295.

32. Katz, N. E.; Creutz, C.; Sutin, N. *Inorg. Chem.* **1988**, *27*, 1687.

33. Cutin, E.; Katz, N. E. *Polyhedron* **1987**, *6*, 159.

34. Richardson, D. E.; Taube, H. *J. Am. Chem. Soc.* **1983**. *105*, 40.

35. Gelroth, J. A.; Figard, J. E.; Petersen, J. D. *J. Am. Chem. Soc.* **1979**, *103*, 3649.

36. Machado, A. E. H.; Rocha, Z. N.; Tfouni, E. *J. Photochem. Photobiol. A.* **1995**, *88*, 85.

37. Krentzien, H.; Taube, H. *Inorg. Chem.* **1982**, 21, 4001.

38. Amer, S. I.; Dasgupta, T. P.; Henry, P. M. *Inorg. Chem.* **1983**, 22, 1970.

39. Kaim, W.; Moscherosch, M. *Coord. Chem. Rev.* **1994**, *129*, 157.

40. Ford, P. C.; Rudd, F. P.; Gaunder, R.; Taube, H. *J. Am. Chem. Soc.* **1968**, *90*, 1187.

41. (a) Chaisson, D. A.; Hintze, R. E.; Stuermer, D. H.; Petersen, J. D.; McDonald, D. P.; Ford, P. C. *J. Am. Chem. Soc.* **1972**, *94*, 6665; (b) Gaunder, R. G. Ph.D. Thesis, Stanford University, 1969; (c) Malouf, G. Ph. D. Thesis, University of California, 1977.

42. Winkler, J. R.; Netzel, T. L.; Creutz, C.; Sutin, N. *J. Am. Chem. Soc.* **1987**, *109*, 2381.

43. Toma, H. E. D. M.Sc. Thesis, Universidade de São Paulo, 1974.

44. Creutz, C.; Taube, H. *J. Am. Chem. Soc.* **1973**, *95*, 1086.

45. Matsubara, T.; Ford, P. C. *Inorg. Chem.* **1976**, *15*, 1107.

46. Diamond, S. E.; Tom, G. M.; Taube, H. *J. Am. Chem. Soc.* **1975**, *97*, 2661.

47. Moore, K. J.; Lee, L.; Mabbott, G. A.; Petersen, J. D. *Inorg. Chem.* **1983**, *22*, 1108.

48. Curtis, J. F.; Sullivan, B. P.; Meyer, T. J. *Inorg. Chem.* **1983**, *22*, 224.

49. (a) Pavanin, L. A.; Giesbrecht, E.; Tfouni, E. *Inorg. Chem.* **1985**, *24*, 4444; (b) Marchant, J. A.; Matsubara, T.; Ford, P. C. *Inorg. Chem.* **1977**, *16*, 2160; (c) Taube, H. *Comm. Inorg. Chem.* **1981**, *1*, 17; (d) Johnson, C. R.; Shepherd, R. E. *Inorg. Chem.* **1983**, *22*, 2439; (e) Zwickel, A.; Creutz, C. *Inorg. Chem.* **1971**, *10*, 2395; (f) Bento, M. L.; Tfouni, E. *Inorg. Chem.* **1988**, *27*, 3410; (g) Vlecek, A. A. *Electrochim. Acta* **1968**, *13*, 1063.

50. Lever, A. B. P. *Inorganic Electronic Spectroscopy*, 2nd ed.; Elsevier: Amsterdam, Netherlands, 1984.

51. Creutz, C.; Taube, H. *J. Am. Chem. Soc.* **1969**, *91*, 3988.

52. Creutz, C. *Prog. Inorg.Chem.* **1983**, *31*, 73.

53. Yeh, A.; Haim, A. *J. Am. Chem. Soc.* **1985**, *107*, 369.

54. Balzani, V., Ed. *Supramolecular Photochemistry: Proceedings NATO Advances Workshop on Photoinduced Charge Separation and Energy Migration in Supramolecular Species*; D. Reidel: Dordrecht, Netherlands, 1987.

55. Balzani, V.; Scandola, F. *Supramolecular Photochemistry*; Ellis Horwood: Chichester, England, 1991.

56. Bonilha, J. B. S.; Chiericato, G., Jr.; Martins-Franchetti, S. M.; Ribaldo, E. J.; Quina, F. H. *J. Phys. Chem.* **1982**, *86*, 4941.

57. Kosower, E. M.; Patton, J. W. *J. Am. Chem. Soc.* **1960,** *82,* 2188
58. Kosower, E. M.; Patton, J. W. *Tetrahedron* **1966,** *22,* 2081.
59. Politi, M. J. M.Sc. Dissertation, Universidade de São Paulo, 1980.
60. Politi, M. J.; Chaimovich, H. *J. Phys. Org. Chem.* **1991,** *4,* 207.
61. Zanette, D.; Chaimovich, H. *J. Phys. Org. Chem.* **1991,** *4,* 643.
62. Zanella, A. W.; Ford, P. C. *Inorg. Chem.* **1975,** *14,* 700.
63. Rocha, Z. N., D.Sc. Thesis, Universidade Estadual Paulista, 1995.
64. Chou, M. H.; Creutz, C.; Sutin, N. *Inorg. Chem.* **1992,** *31,* 2318.
65. Oliveira, E. C.; Rocha, Z. N.; Tfouni, E., unpublished.
66. Harman. W. D.; Taube, H. *J. Am. Chem. Soc.* **1988,** *110,* 7555.
67. Katz, N. E.; Fagalde, E. *Inorg. Chem.* **1993,** *32,* 5391.
68. Foust, R. D., Jr.; Ford, P. C. *J. Am. Chem. Soc.* **1972,** *94,* 5686.

PART III:
ELECTRON TRANSFER AND TRANSITION METALS IN BIOLOGY

19

Electron Transfer in Bioinorganic Chemistry

Role of Electronic Structure and the Entatic State

Edward I. Solomon*, Michael D. Lowery, Jeffrey A. Guckert, and Louis B. LaCroix

Department of Chemistry, Stanford University, Stanford, CA 94305

The unique spectral features of oxidized blue copper proteins have often been used to support the concept that the reduced geometry is imposed on the oxidized copper site by the protein. This has been called an "entatic" or "rack" state and is thought to make a significant contribution to rapid electron transfer in biology. In this presentation, the unique spectral features of the oxidized d^9 state are shown to reflect a novel ground state wavefunction that plays a key role in defining electron transfer pathways. Further, the electronic structure of the reduced d^{10} blue copper site has been determined using a combination of variable-energy photoelectron spectroscopy and electronic structure calculations. These studies determine the change in electronic structure that occurs on oxidation and allow an evaluation of whether the reduced geometry is, in fact, imposed on the oxidized site.

HENRY TAUBE'S EXTENSIVE CONTRIBUTIONS TO INORGANIC CHEMISTRY have laid the foundations for many important areas in bioinorganic chemistry. Taube has clearly had a major impact in how we describe long-range electronic transfer in biology, which is mostly accomplished by three classes of metalloproteins: heme, iron–sulfur, and blue copper. The blue copper proteins have been generally used as the example that demonstrates the presence of an "entatic" or "rack" state in bioinorganic chemistry (1–3). In the entatic state, the protein is thought to impose an unusual geometry on a metal center that activates it for reactivity. The blue copper site that is found in proteins such as

*Corresponding author.

the plastocyanins and azurins is thought to exist in an entatic state because the oxidized site exhibits unique spectral features compared to those of normal cupric complexes (4–6). In the plastocyanin absorption spectrum, there is an extremely intense band at ~600 nm ($\varepsilon_{max} \approx 5000$ M^{-1} cm^{-1}), where normal Cu(II) complexes have weak d → d transitions ($\varepsilon_{max} \approx 40$ M^{-1} cm^{-1}), and the parallel hyperfine coupling ($A_{\parallel}$) in the electron paramagnetic resonance (EPR) spectrum is reduced by more than a factor of 2 relative to that of normal Cu(II) complexes. These unique spectral features are thought to reflect the presence of an unusual geometric structure, which is imposed on the Cu(II) site by the protein (2–3). In particular, reduced copper complexes are often tetrahedrally coordinated, whereas oxidized complexes are frequently found in tetragonal geometries due to the Jahn–Teller effect. Thus, in the entatic or rack state, the protein would oppose the Jahn–Teller distortion of the oxidized site leading to little geometric change upon oxidation and thus rapid electron transfer.

As first predicted from the unique spectral features (7), the oxidized site (Figure 1A) does, in fact, have a very different geometric structure from that of normal Cu(II) complexes (which are tetragonal) in that the blue copper site has a distorted tetrahedral geometry with two unusual ligand–metal bonds: a short thiolate S–Cu bond at ~2.1 Å from cysteine (Cys) and a long thioether S–Cu bond at ~2.8 Å from methionine (Met) (8). The remaining two ligands are fairly normal imidazole N–Cu bonds to histidine (His) residues. Also, as shown in Figure 1B, little change in geometric structure occurs on reduction (8, 9).

The unique spectral features of the oxidized d^9 site are now well understood and reflect a novel ground-state wavefunction (5, 6, 10). Recent experiments that have defined the nature of this ground state are described in the next section. It is important to emphasize that this is the highest-energy half-occupied redox-active orbital and thus plays a key role in the electron transfer function of this active site. We have also developed a new method of inorganic spectroscopy to probe the reduced d^{10} site, variable-energy photoelectron spectroscopy using synchrotron radiation (11). These experiments combined with self-consistent field–Xα-scattered wave (SCF–Xα-SW) calculations define, for the first time, the electronic structure of the reduced blue copper site and determine the change in electronic (and associated geometric) structure that occurs on oxidation. These studies are presented in this chapter and are used to evaluate whether the reduced geometry is in fact imposed on the oxidized blue copper site by the protein.

Electronic Structure of the Oxidized Blue Copper Site: Contributions to Electron Transfer Pathways

In earlier studies we used SCF–Xα-SW calculations adjusted to the ground state *g* values to develop a description of the electronic structure of the oxidized blue copper site in plastocyanin (5, 6). The ground state shown in Figure 2 (the contour is in the *xy* plane, which contains the Cu, cysteine S, and two N-

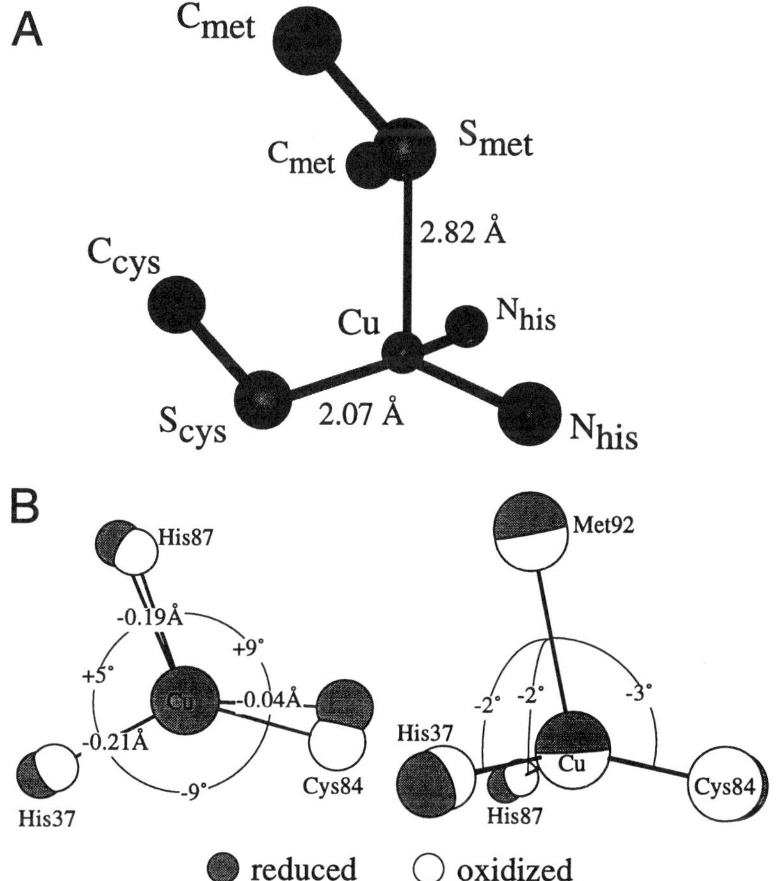

Figure 1. Plastocyanin crystallography: (A) X-ray structure of the Cu coordination environment in oxidized poplar plastocyanin (based on data from reference 8); (B) crystallographically determined structural changes in the bond lengths and angles between the oxidized and reduced sites in the NNS equatorial plane (left) and in the S(Met)CuL angles (right) (based on data from references 8 and 9). (Reproduced with permission from reference 11. Copyright 1995 American Chemical Society.)

atoms of histidine ligands in Figure 1) is quite unusual in that the $d_{x^2-y^2}$ orbital is highly covalent and the delocalization is strongly anisotropic with the dominant interaction involving the pπ orbital of the cysteine sulfur. This description of the ground state is extremely important for defining possible electronic structure contributions to function in that this wavefunction describes the redox-active orbital that is involved in long-range rapid electron transfer. As this chapter will describe, we have now experimentally confirmed the key features of this novel ground state electronic structure.

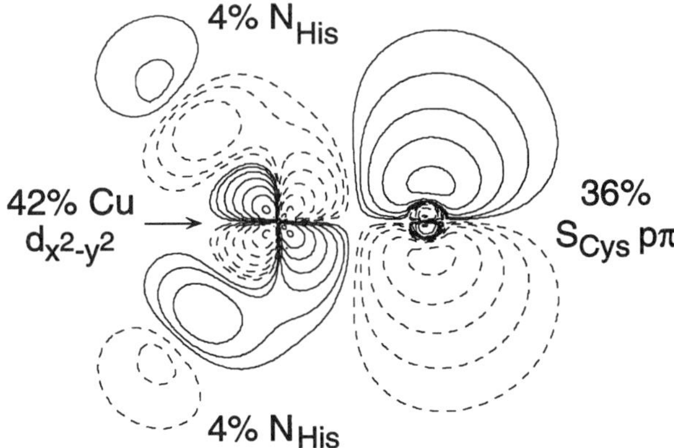

Figure 2. Contour of plastocyanin ground-state wavefunction (HOMO). Contour lines are at ±0.64, ±0.32, ±0.16, ±0.08, ±0.04, ±0.02, and ±0.01 (electrons/bohr³)¹ᐟ². The outermost contour encompasses 90% of the electron density.

The highest occupied molecular orbital (HOMO) in Figure 2 would indicate that the unusually small parallel hyperfine splitting ($A_{||}$) exhibited in the EPR spectrum of the blue copper site already described is due to the high covalency. However, the small $A_{||}$ had generally been attributed to a geometric origin where the distorted tetrahedral structure results in the Cu $4p_z$ orbital mixing into the Cu $d_{x^2-y^2}$ orbital (*12*). Approximately 12% Cu $4p_z$ mixing would account for the small $A_{||}$ observed for the blue copper site in plastocyanin. This mixing was probed experimentally through an analysis of the polarized single-crystal X-ray absorption spectral (XAS) data taken at the Cu K-edge of plastocyanin (*13*). The absorption peak observed at ~8979 eV in cupric complexes is assigned as the Cu 1s → Cu $d_{x^2-y^2}$ transition. In square planar $CuCl_4^{2-}$, there is a small amount of intensity in this transition due to its quadrupole moment (*14*). In plastocyanin the intensity increases by a factor of 2 due to 4p mixing with $d_{x^2-y^2}$, which results in the presence of electric dipole intensity for a transition from the Cu 1s orbital. Our earlier single-crystal EPR studies on plastocyanin showed that the $d_{x^2-y^2}$ orbital is perpendicular to the long thioether S–Cu bond present at the blue copper site and within 15° of the plane defined by the remaining three strong ligands (*15*). Therefore, polarized single-crystal XAS data were taken with the **E** vector of light || and ⊥ to the *z* axis (the thioether S–Cu bond) (*16*). As can be seen from the data in Figure 3, all the 8979 eV intensity is observed with **E** ⊥ *z* [i.e., **E** || (*x,y*)], which indicates that Cu $4p_{x,y}$ is mixing with the $d_{x^2-y^2}$ orbital ($4p_{x,y}$ mixing would increase $A_{||}$). This suggests that the origin of the small $A_{||}$ is not $4p_z$ mixing but instead would reflect a highly covalent site (*13*). This possibility was confirmed experimentally through XAS studies at the Cu L-edge.

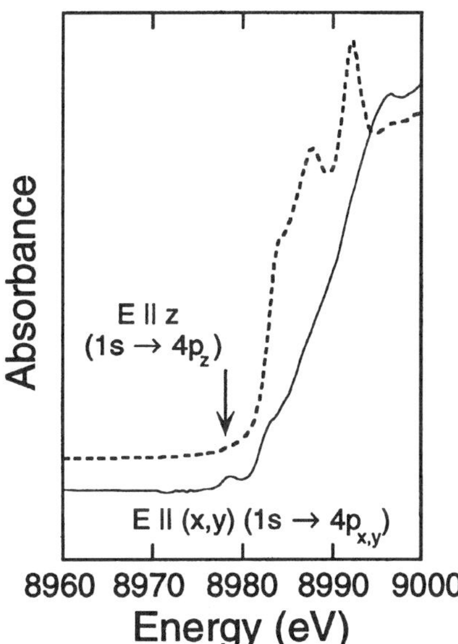

Figure 3. Polarized single-crystal X-ray absorption spectroscopy at the Cu K-edge for poplar plastocyanin (based on data from reference 16).

The Cu 2p to half-occupied HOMO transition occurs at ~930 eV. The Cu 2p → 3d transition ($L_{2,3}$-edge) is electric-dipole-allowed; thus the intensity of the 930 eV transition reflects the amount of Cu $d_{x^2-y^2}$ character in this HOMO. The L_3-edge of plastocyanin exhibits 60% of the intensity of the L_3-edge of square planar D_{4h}-CuCl$_4^{2-}$ (Figure 4) (*17*). We have studied D_{4h}-CuCl$_4^{2-}$ in great detail and found that the ground state has 61 ± 5% Cu $d_{x^2-y^2}$ character (*18*). The intensity ratio in Figure 4 thus indicates that the blue copper site has 38% Cu $d_{x^2-y^2}$ character. This value is in good agreement with the results from electronic structure calculations in Figure 2 (*17*). The fact that the delocalization dominantly involves the thiolate S was demonstrated by S K-edge XAS studies. [With Hodgson and Hedman we have been developing (*13, 19–21*) ligand K-edge X-ray absorption spectroscopy as a new method for determining the covalency of transition metal complexes.] The S 1s → half-occupied HOMO transition occurs at ~2470 eV. The S 1s → 3p transition is electric-dipole-allowed; therefore, absorption intensity at the S K-edge should reflect the amount of S 3p character mixed into the half-occupied HOMO. The K-edge of plastocyanin is a factor of 2.5 times more intense than tet *b*, a model complex prepared by Schugar (*13*) that has a fairly normal thiolate S–Cu bond with ~15% covalency (Figure 5). Thus the intensity ratio in Figure 5 indicates that the blue copper ground state has ~38% S character, which is also in reasonable agreement with the adjusted SCF–Xα -SW calculations (Figure 2).

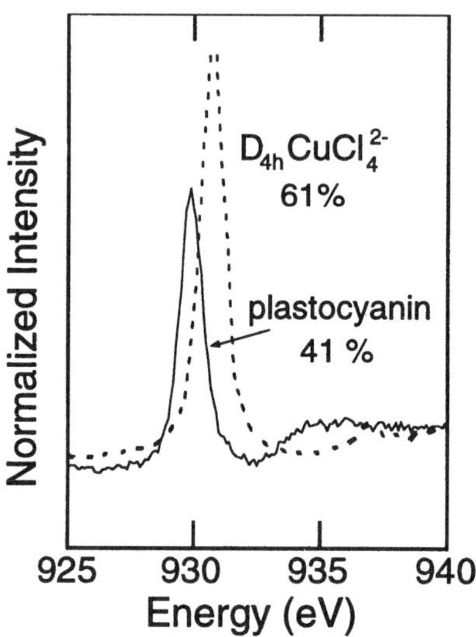

Figure 4. Cu L-edge XAS as a probe of ligand–metal covalency: XAS spectra for D$_{4h}$-CuCl$_4^{2-}$ and plastocyanin (based on data from reference 17). Values listed are the amount of Cu d character in the HOMO.

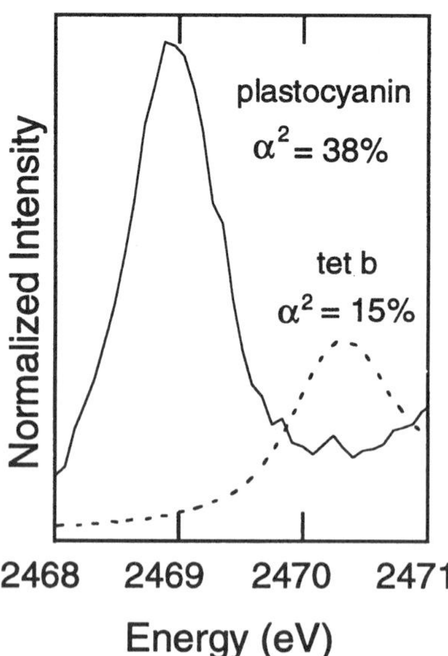

Figure 5. S K-edge XAS as a probe of ligand–metal covalency: orientation-averaged XAS spectra for tet b and plastocyanin (based on data from reference 13). α^2 is the amount of S p character in the HOMO.

A final key feature of the ground-state wavefunction depicted in Figure 2 was confirmed experimentally from our assignments of the unique absorption spectral features of the blue copper site (5). These assignments were made utilizing low-temperature absorption, polarized single-crystal absorption, and circular (CD) and magnetic circular dichroism (MCD) data, each method having different selection rules. There are a minimum of eight bands required to simultaneously fit the spectra. From the assignments indicated in the low-temperature absorption spectrum in Figure 6, Cys → Cu charge transfer dominates the intensity but with the lower-energy π charge-transfer transition being more intense than the higher-energy Cys (pseudo)-σ charge-transfer transition. This is inverted from what is observed for a normal ligand–metal bond and requires that the $d_{x^2-y^2}$ orbital be rotated 45° such that its lobes are bisected by the Cys S–Cu axis as in Figure 2. This rotation of the $d_{x^2-y^2}$ orbital occurs due to the strong π antibonding interaction with the thiolate at the short Cys S–Cu bond length.

We have correlated the ground-state wavefunction with the crystal structure of several blue copper proteins and obtained significant insight into its contributions to long-range electron transfer pathways. In plastocyanin, the cysteine is adjacent to a tyrosine that is at the "remote patch" on the protein (Figure 7) (22). In ascorbate oxidase (23) and nitrite reductase (24), it is flanked

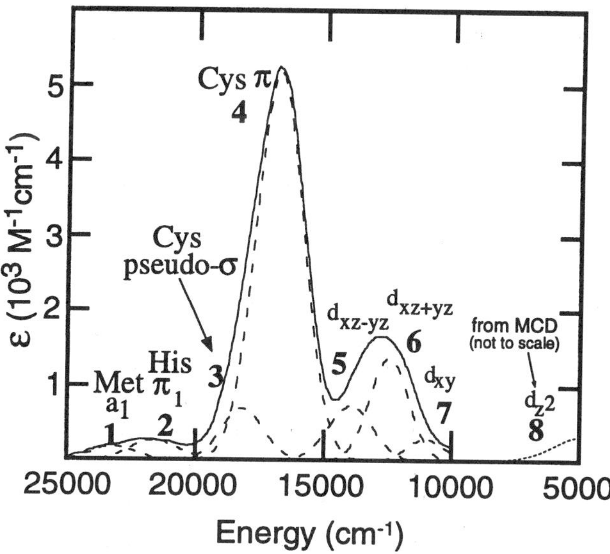

Figure 6. Blue copper excited-state spectral features: low-temperature absorption spectrum of plastocyanin. The position of band 8 has been determined from near-IR MCD data. The dashed lines indicate Gaussian resolution into the component bands (5).

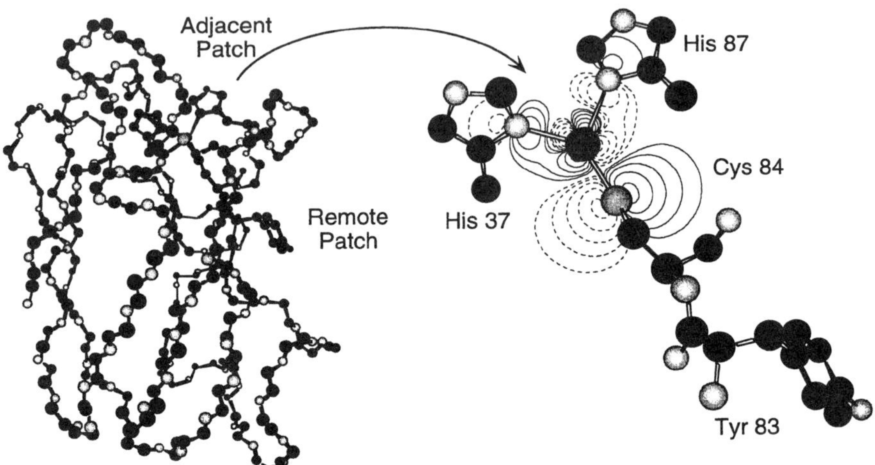

Figure 7. Proposed long-range electron transfer pathway in plastocyanin (5, 6, 10.) Left: polypeptide backbone for plastocyanin. The residues that form the proposed path are darkened. Side groups for the blue copper ligated residues and the adjacent tyrosine are included (8). Right: the plastocyanin wavefunction contours superimposed on the crystallographically defined site including the adjacent tyrosine residue that, along with the cysteine, form the path to the "remote" patch (22). The contour shows the substantial electron delocalization onto the cysteine S pπ orbital that activates electron transfer to the remote patch ~12.5 Å from the blue copper site. This low-energy, intense Cys Sπ → Cu charge transfer transition provides an effective hole superexchange mechanism for rapid long-range electron transfer.

by histidine, which is a ligand at an additional copper center. In all three proteins rapid electron transfer occurs over ~13 Å, to the remote patch in plastocyanin and the additional copper center in ascorbate oxidase and nitrite reductase. The highly anisotropic covalency involving the thiolate activates this group for directional electron transfer, while the low-energy intense Cys π → Cu charge transfer transition (Figure 6) provides a very efficient hole superexchange mechanism for rapid electron transfer (*10*). Thus the unique spectral features of the oxidized blue copper site reflect a ground-state wavefunction that is activated for rapid electron transfer to a specific site on or in the protein.

Electronic Structure of the Reduced Blue Copper Site: Contributions to Reduction Potentials and Geometry

Over recent years, we have been strongly involved in the development of variable-energy photoelectron spectroscopy (PES) using synchrotron radiation as a new method of defining the bonding and its change with ionization in inorganic complexes (*18, 25–28*). We have now used this method combined with

SCF–Xα-SW calculations to determine the electronic structure of the reduced d^{10} blue copper site and thus the change in electronic structure with oxidation (*11*). We have also been interested in whether the reduced geometry is, in fact, imposed by the protein on the oxidized site (i.e., the entatic or rack state). Since PES detects valence electrons, these studies were performed on model complexes, and as we are interested in understanding unconstrained ligand–metal bonding and how it relates to the protein, the models we studied using variable-energy PES involved the blue copper relevant imidazole, $(CH_3)_2S$, and CH_3S^- ligands bound to coordinatively unsaturated Cu(I) sites on oxide and chloride single-crystal surfaces in ultra-high vacuum (*11*). The PES spectrum of methyl thiolate on cuprous oxide is presented as an example of the data obtained (Figure 8A). The surface oxide was used to drive the deprotonation of methanethiol (which was confirmed by chemical shifts of core levels). The dashed line is the valence band spectrum of clean cuprous oxide, while the solid spectrum is that of the thiolate-bound Cu(I) site. The difference spectrum on the bottom thus gives the PES of the valence orbitals of the thiolate involved in bonding to the Cu(I). Varying the photon energy changes the relative intensities of the peaks and allows for the specific assignments indicated in Figure 8A. There are three thiolate valence orbitals involved in bonding: π, pseudo-σ, and σ (Figure 8B). The π and pseudo-σ dominate based on cross-section effects, and these split in energy indicating that the thiolate is bound to the surface copper site with an R–S–Cu angle $\phi << 180°$. Thus, the energy splitting and intensities of the valence orbital peaks in the difference spectrum in Figure 8A define the geometric and electronic structure of the thiolate S–Cu(I) bond.

Our approach was first to use variable-energy PES to experimentally estimate the geometric and electronic structure of each ligand–metal surface complex unconstrained by the protein matrix. We then used these data to evaluate and calibrate transition-state SCF–Xα-SW calculations of the surface complexes. These calculations were then extended to generate an electronic structure description of the reduced blue copper site and to determine the changes in the electronic structure that occur upon oxidation. These are described in detail in reference 11. The key points of these studies are summarized as follows.

(1) The bonding is dominated by ligand donor interactions with the unoccupied Cu 4p orbitals. Thus, even though it is a Cu(I) site there is no backbonding with the blue copper ligand set.

(2) *The long thioether S–Cu(I) (2.9 Å) bond is imposed on the copper site by the protein.* Thus, this a clear example of a protein-constrained entatic state. This long thioether bond reduces its donor interaction with the copper, which is compensated for by the thiolate leading to its short 2.1 Å bond. The long thioether S–Cu(I) bond also preferentially destabilizes the oxidized site and is thus a major contribution to the generally high reduction potentials of many of the blue copper proteins.

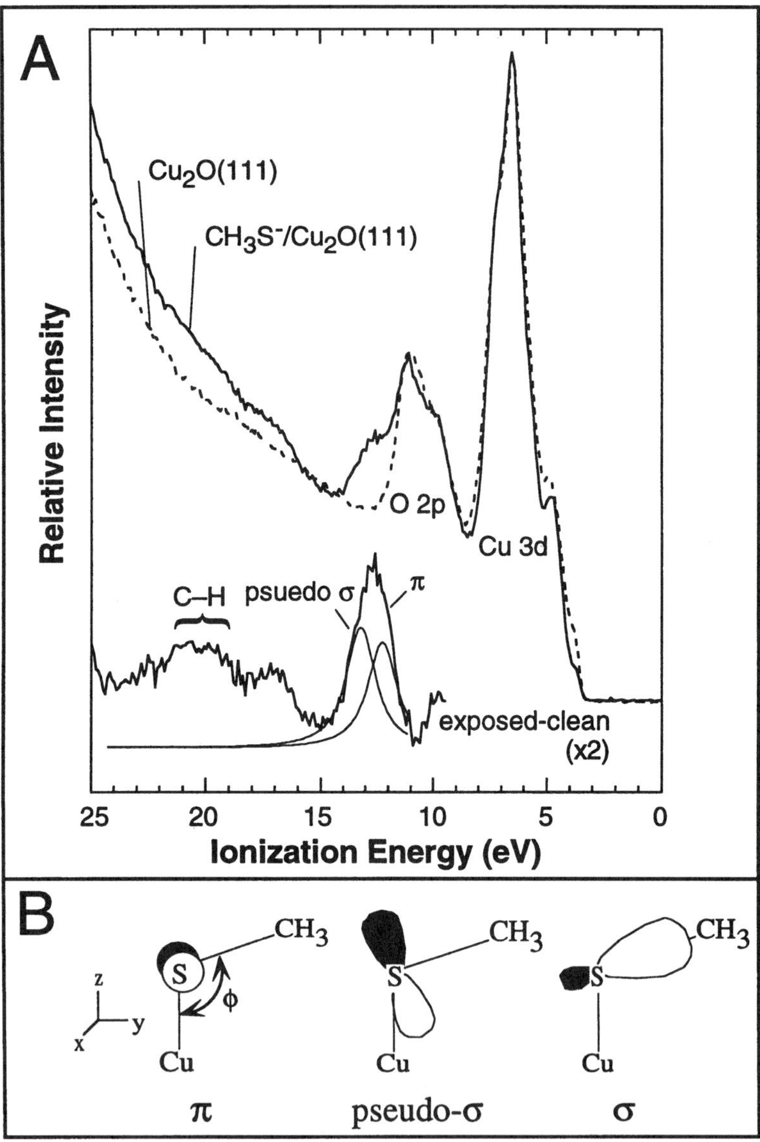

Figure 8. CH$_3$S–Cu(I) bonding. (A): valence band PES of clean Cu$_2$O(111), Cu$_2$O(111) exposed to methanethiol, and their difference spectrum with Gaussian/Lorenzian resolution of the low-energy region. (B): valence orbitals of methanethiolate. (Reproduced with permission from reference 11. Copyright 1995 American Chemical Society.)

(3) The major change in electronic structure that occurs on oxidation is the hole produced in the $d_{x^2-y^2}$-derived molecular orbital pictured in Figure 2, which is strongly antibonding with the thiolate and more weakly antibonding with the two histidine ligands.

(4) Having obtained an experimentally calibrated description of the change in electronic structure on oxidation, we could evaluate the associated change in geometric structure that would occur for a blue copper site unconstrained by the protein. This was accomplished by evaluating the electron–nuclear linear-coupling terms of the oxidized site in the reduced geometry along all the normal modes of the blue copper structure. A nonzero slope in Figure 9 corresponds to a distorting force along a specific normal mode. The geometric structural changes we predict on oxidation are consistent with the change in electronic structure described already in this chapter. There is a large distorting force to contract the thiolate S–Cu bond. This, however, is opposed by a large force constant associated with the short strong Cys S–Cu

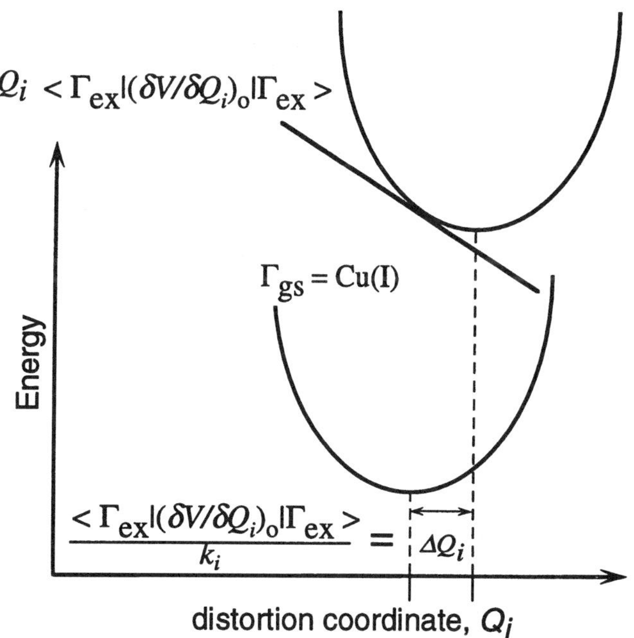

Figure 9. Distorting forces in the blue copper site on oxidation. Configuration coordinate diagram of the linear coupling term for the distorting force along the ith normal mode of vibration, Q_i, on the oxidized Cu(II) site relative to the reduced Cu(I) ground state (11). Γ_{gs} and Γ_{ex} are the ground and excited state wave functions, respectively. δV is the energy change calculated for a distortion δQ_i of the molecule. ΔQ_i is the expected distortion along the normal mode for the calculated linear coupling term.

bond. There are also moderate distorting forces along the His N–Cu bonds. These predicted distortions are in general agreement with those observed experimentally in the crystal structure differences between the reduced and oxidized blue copper protein sites (Figure 1B) (8, 9). The S–Cu bond is found to contract by 0.04 Å, while the N–Cu bonds contract by ~0.2 Å. The Cu–N contraction determined from extended X-ray absorption fine structure (EXAFS) is significantly smaller (0.07 Å) (29). Importantly, there are no significant distorting forces along bending modes, and only small angular changes are observed experimentally in the blue copper structure on oxidation.

If the protein imposed a reduced geometry on the oxidized site, this would correspond to a nonzero Jahn–Teller distorting force in the oxidized site along a bending mode. Thus the protein does not oppose a Jahn–Teller distorting force in the oxidized site, which would be the usually invoked mechanism for the entatic or rack state. The reason for this is that the combination of long thioether S–Cu bond and the associated short thiolate S–Cu bond, which are already present in the reduced site, removes all the orbital degeneracy in the oxidized site. Our low-temperature MCD studies indicate that the $d_{x^2-y^2}$ and d_{xy} orbitals are split by $>10,000$ cm^{-1}; it would be an electron–nuclear coupling term between these levels that would produce a Jahn–Teller distorting force. Thus there is little geometric change that occurs on oxidation leading to a low Franck–Condon barrier to electron transfer (8, 30).

Summary

The unique spectral features of the oxidized blue copper site are now understood. These reflect high anisotropic covalency in the ground-state wavefunction involving the thiolate that activates this residue as an efficient superexchange pathway for electron transfer. Photoelectron spectroscopy has been developed as a new spectroscopic probe of the reduced blue copper site. It has defined the change in electron structure that occurs on oxidation, which is found to be consistent with the limited geometric change observed experimentally. This limited geometric change leads to a low Franck–Condon barrier to electron transfer. Thus, the reduced geometry is not imposed by the protein on the oxidized site. The protein does impose a long thioether S–Cu (i.e., weak axial) bond on the reduced site, which is compensated by the short thiolate S–Cu bond. This raises the reduction potential, quenches the possible Jahn–Teller distortion of the oxidized site, and activates the thiolate electron transfer pathway.

Acknowledgments

E. I. Solomon wishes to express sincere appreciation to all his students and collaborators who are listed as his co-authors in the Reference list, for their commitment and contribution to this science.

This work has been supported by National Science Foundation Grant CHE–9528250.

References

1. Lumry, R.; Eyring, H. *J. Phys. Chem.* **1954**, *58*, 110–120.
2. Malmström, B. G. *Eur. J. Biochem.* **1994**, *223*, 711–718.
3. Williams, R. J. P. *Eur. J. Biochem.* **1995**, *234*, 363–381.
4. Solomon, E. I.; Penfield, K. W.; Wilcox, D. E. *Struct. Bonding (Berlin)* **1983**, *53*, 1–57.
5. Solomon, E. I.; Baldwin, M. J.; Lowery, M. D. *Chem. Rev.* **1992**, *92*, 521–542.
6. Solomon, E. I.; Lowery, M. D. *Science (Washington, D.C.)* **1993**, *259*, 1575–1581.
7. Solomon, E. I.; Hare, J. W.; Gray, H. B. *Proc. Natl. Acad. Sci. U.S.A.* **1976**, *73*, 1389–1392.
8. Guss, J. M.; Bartunik, H. D.; Freeman, H. C. *Acta. Crystallogr.* **1992**, *B48*, 790–811.
9. Guss, J. M.; Freeman, H. C. *J. Mol. Biol.* **1986**, *192*, 361–381.
10. Lowery, M. D.; Guckert, J. A.; Gebhard, M. S.; Solomon, E. I. *J. Am. Chem. Soc.* **1993**, *115*, 3012–3013.
11. Guckert, J. A.; Lowery, M. D.; Solomon, E. I. *J. Am. Chem. Soc.* **1995**, *117*, 2817–2844.
12. Bates, C. A.; Moore, W. S.; Standley, K. J.; Stevens, K. W. H. *Proc. Phys. Soc. London* **1962**, *79*, 73–93.
13. Shadle, S. E.; Penner-Hahn, J. E.; Schugar, H. J.; Hedman, B.; Hodgson, K. O.; Solomon, E. I. *J. Am. Chem. Soc.* **1993**, *115*, 767–776.
14. Hahn, J. E.; Scott, R. A.; Hodgson, K. O.; Doniach, S.; Desjardins, S. R.; Solomon, E. I. *Chem. Phys. Lett.* **1982**, *88*, 595–598.
15. Penfield, K. W.; Gay, R. R.; Himmelwright, R. S.; Eickman, N. C.; Norris, V. A.; Freeman, H. C.; Solomon, E. I. *J. Am. Chem. Soc.* **1981**, *103*, 4382–4388.
16. Scott, R. A.; Hahn, J. E.; Doniach, S.; Freeman, H. C.; Hodgson, K. O. *J. Am. Chem. Soc.* **1982**, *104*, 5364–5369.
17. George, S. J.; Lowery, M. D.; Solomon, E. I. Cramer, S. P. *J. Am. Chem. Soc.* **1993**, *115*, 2968–2969.
18. Didziulis, S. V.; Cohen, S. L.; Gewirth, A. A.; Solomon, E. I. *J. Am. Chem. Soc.* **1988**, *110*, 250–268.
19. Hedman, B.; Hodgson, K. O.; Solomon, E. I. *J. Am. Chem. Soc.* **1990**, *112*, 1643–1645.
20. Shadle, S. E.; Hedman, B.; Hodgson, K. O.; Solomon, E. I. *Inorg. Chem.* **1994**, *33*, 4235–4244.
21. Shadle, S. E.; Hedman, B.; Hodgson, K. O.; Solomon, E. I. *J. Am. Chem. Soc.* **1995**, *117*, 2259–2272.
22. Guss, J. M.; Freeman, H. C. *J. Mol. Biol.* **1983**, *169*, 521–563.
23. Messerschmidt, A.; Ladenstein, R.; Huber, R.; Bolognesi, M.; Avigliano, L.; Petruzzelli, R.; Rossi, A.; Finazzi-Agro, A. *J. Mol. Biol.* **1992**, *224*, 179–205.
24. Godden, J. W.; Turley, S.; Teller, D. C.; Adman, E. T.; Liu, M. Y.; Payne, W. J.; Legall, J. *Science (Washington, D.C.)* **1991**, *253*, 438–442.
25. Didziulis, S. V.; Cohen, S. L.; Butcher, K. D.; Solomon, E. I. *Inorg. Chem.* **1988**, *27*, 2238–2250.
26. Butcher, K. D.; Gebhard, M. S.; Solomon, E. I. *Inorg. Chem.* **1990**, *29*, 2067–2074.
27. Butcher, K. D.; Didziulis, S. V.; Briat, B.; Solomon, E. I. *Inorg. Chem.* **1990**, *29*, 1626–1637.

28. Butcher, K. D.; Didziulis, S. V.; Briat, B.; Solomon, E. I. *J. Am. Chem. Soc.* **1990,** *112,* 2231–2242.
29. Murphy, L. M.; Hasnain, S. S.; Strange, R. W.; Harvey, I.; Ingledew, W. J. In *X-ray Absorption Fine Structure;* Hasnain, S. S., Ed.; Ellis Hardwood: Chichester, England, 1990; p 152.
30. Sykes, A. G. *Adv. Inorg. Chem.* **1991,** *36,* 377–408.

20

Long-Range Intramolecular Electron Transfer Reactions Across Simple Organic Bridges, Peptides, and Proteins

Stephan S. Isied

Department of Chemistry, Rutgers, The State University of New Jersey, P.O. Box 939, Piscataway, NJ 08550

Rates of intramolecular electron transfer across peptide bridging ligands have been studied by covalently attaching inorganic reagents (Co, Ru, and Os) at their terminals and side chains. In addition to the expected variation in the rates with changes in reorganization energy and driving force as predicted by theory, quantitative information on the participation of the peptide bridging groups in these long-range electron transfer reactions can now be obtained. Our results show that peptides with organized secondary structures such as (proline)$_n$ (n $\geq$ 4) have a low distance decay constant ($\beta \sim 0.2$–0.3 Å^{-1}) as compared to $\beta \sim 1$ Å^{-1} for random coil peptides and saturated hydrocarbons. Thus more rapid rates of intramolecular electron transfer can occur across organized peptides than across rigid hydrocarbons. Experiments on electron transfer across helical, rigid non-proline peptides provide preliminary information that can be compared to the helical proline peptides. This technique has been extended to study intramolecular electron transfer in cytochrome c where the heme protein is an electron donor. These results show that the rate of intramolecular electron transfer between the heme and the metal-modified protein surface site (His 33 or Met 65) cannot be predicted from the distance between the two, even when driving force and reorganization energy are corrected for. Peptide networks between the heme and the different sites on the surface of the protein play an important role that can account for more than three orders of magnitude change in the rates of electron transfer at similar distances.

T HOSE OF US WHO HAVE BEEN ASSOCIATED with Professor Henry Taube share many memories of watching him analyze and formulate research problems. He often communicates his ideas to his associates via common sense arguments (and sometimes even using test tube experiments) full of implications and further insights. As graduate students in his laboratory, we were studying redox reactions of Co(III) and Ru(II) ammine complexes under conditions in which the time required for substitution is faster than that required for electron transfer. I clearly remember a discussion during which he suggested that if we could make binuclear complexes of the type $[(NH_3)_5Ru^{III}–bridge–Co^{III}(NH_3)_5]$, then we could rely on the large disparity in the rates of reduction of Co(III) and Ru(III) ammines (*1*) to generate the desired precursor complexes $[(NH_3)_5Ru^{II}–bridge–Co^{III}(NH_3)_5]$, which could then be directly used to measure rates of intramolecular electron transfer, independent of the substitution rates of the complexes.

To accomplish this we carried out a systematic study of the kinetics and thermodynamics of substitution and redox reactions of a variety of substituted Ru(II) tetraammine complexes (*2, 3*). From the information gained about the substitution reactions of Ru(II) complexes and the redox reactions of coordinated ligands, we were successful in developing a synthetic procedure to make the $[(SO_4)(NH_3)_4Ru^{III}–bridge–Co^{III}(NH_3)_5]$ complexes (*4*). Using four closely related pyridine carboxylate bridges (nicotinic, isonicotinic, 3- and 4-pyridine acetic acid), differences in electron transfer rates and mechanism between isomers were determined and analyzed (*4, 5*). The first binuclear $Co^{III}–Ru^{III}$ donor–acceptor complex isolated with the isonicotinic acid bridge is shown in the following structure:

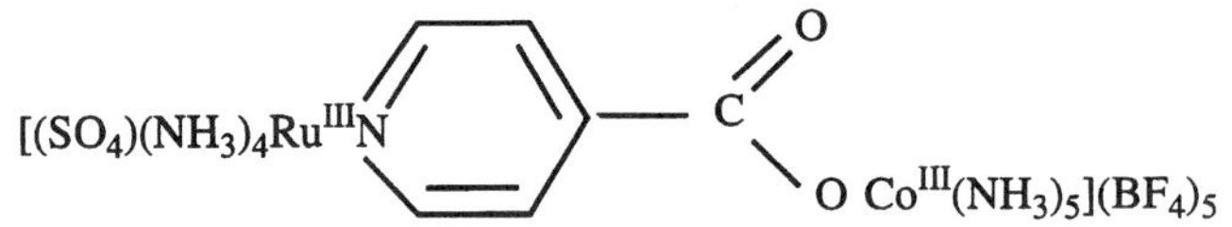

$$[(SO_4)(NH_3)_4Ru^{III}N\text{—}\langle\text{pyridine}\rangle\text{—}C(\!=\!O)\text{—}O\ Co^{III}(NH_3)_5](BF_4)_5$$

This approach has now been used for more than two decades, and newer synthetic routes (making use of $CF_3SO_3^-$ chemistry) have succeeded it (*6*). However, Henry Taube's original ideas on the importance of such bridged binuclear complexes in the study of the role of bridging ligands in mediating electron transfer continues to be a basis for studying intramolecular electron transfer in donor–acceptor complexes with bridging ligands ranging from single atoms to peptides and proteins. In this chapter I will describe how we have extended this approach to study the electron transfer mediating properties of complex bridges such as peptides and proteins.

Interest in long-range electron transfer in proteins arose from the examination of the crystal structures of redox proteins such as cytochromes, copper blue proteins, iron–sulfur proteins, and protein complexes such as cyt c–cyt c peroxidase and, more recently, the cyt c oxidase structure (*7–11*). Most of these proteins have a redox center (i.e., a heme, a copper ion, or an FeS center) constituting only a few percent of the total protein, buried within the protein.

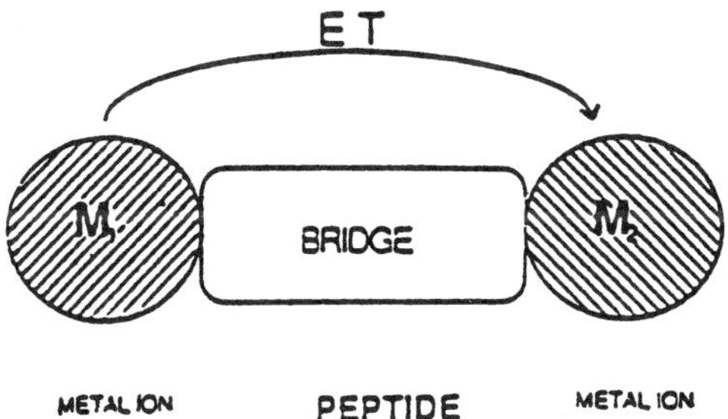

Scheme I. Donor–acceptor bridged complexes.

Crystal structures of multiredox enzyme complexes show that the redox centers are separated by peptide residues at distances of >10–20 Å. The occurrence of rapid electron transfer between redox centers separated by peptide ligands raises questions concerning the role of the polypeptide chain in providing electron transfer pathways between these centers.

Our approach for investigating the electron transfer properties of peptide and protein fragments is to study them in donor–acceptor complexes similar to that shown in Scheme I. With this strategy, electron transfer pathways in proteins can be evaluated by designing molecules that emphasize special features such as peptide bonds, hydrogen-bonding networks, polarizable amino acid side chains, and the connectivity of redox centers to the main chain or side chain of the peptide in carefully controlled experiments.

In this chapter I will discuss our studies of long-range intramolecular electron transfer in multifunctional biological bridging ligands including amino acids, peptides, and proteins and the more recently developed supramolecular peptide networks with constrained peptides, and electron transfer proteins modified with redox reagents. The goal of these systematic studies is to understand the distance dependence of long-range electron transfer reactions and its relationship to peptide conformation and specific peptide pathways necessary to achieve rapid rates of electron transfer. As will be seen in the coming sections, many interesting and unexpected results have been discovered in these investigations. These results are leading us to a better understanding of the communication between inorganic redox centers and proteins.

Oligoproline Donor–Acceptor Complexes

Early studies of energy transfer in biological molecules have made use of oligoproline peptides with organic energy donor and acceptor chromophores placed at the N- and C-terminals of the polypeptide chain. Such studies demonstrated that oligoproline peptides serve as spectroscopic rulers for distance estimation between chromophores. These energy transfer techniques were further used to establish proximity relationships in more complex biological molecules (*12*).

The advantage of the oligoproline peptide bridges as distance spacers over other naturally occurring amino acids and peptides is the very early onset of their solvent-stabilized helical secondary structure (*13–16*). A left-handed, trans-helical structure of oligoprolines begins to appear with the assembly of only three or four proline peptide units (polyproline II). This same helical secondary structure also exists in solution and in solid-state structures of tetra- and pentaproline peptides, as well as in fiber structure of longer proline polymers (*17–20*). Experimental evidence for this helical conformation comes from crystal structure in the solid state (*17–20*), from ^{13}C and two-dimensional proton NMR of $[(Pro)_5Co^{III}(NH_3)_5]$ (*21, 22*), and CD of $[(bpy)_2Ru^{II}(Pro)_n apyRu^{III}(NH_3)_5]$ ($n = 6, 7, 9$) in solution (*16*).

Redox active transition metal ion chromophores located at the N- and C-terminals of oligoproline peptides have been used to study electron transfer across bridging oligoproline peptides. The advantages of using metal ions for these studies is that substantial control over the driving force and reorganization energy of the reaction can be exercised by the choice of different transition metal donor–acceptor pairs (*13–16, 21, 23, 24*). The time scale of the intramolecular electron transfer reaction can be varied to be either slower or faster than peptide conformational changes. An example of this is seen for the diproline bridge (Table I). The intramolecular electron transfer across the same bridge can vary by more than 11 orders of magnitude, depending on the choice of metal donors and acceptors (*13–16, 21, 23, 24*). In the Co–(Pro)₂–Ru (the first entry in Table I), the rate of electron transfer is slow compared to peptide conformational changes, and hence the equilibration of many peptide conformations can occur prior to the electron transfer step (*24*). Experiments with this series, while useful for probing the existence of different peptide conformations, are not useful for determining the distance dependence of the rate of electron transfer because of complications from the multiple electron transfer

Table I. Intramolecular Rates of Electron Transfer Across the Same (Pro)₂ Bridge with Different Donor–Acceptor Metal Ion Pairs

M^{II}–M^{III}	M–M (Å)	k_{et} (s^{-1})	$-\Delta G°$
Co–Ru[a]	14.8	~10^{-4}	(–0.5)
Os–Co[b]	14.8	0.74	0.15
Os–Ru[c]	14.8	3.7×10^4	0.25
Ru–Co[d]	14.8	1.6×10^7	1.1

[a]Co–Ru is $[(H_2O)(NH_3)_4Ru^{II}-(Pro)_2-Co^{III}(NH_3)_5]$.

[b]Os–Co is $[(NH_3)_5Os^{II}i-(Pro)_2-Co^{III}(NH_3)_5]$, where i = isonicotinyl.

[c]Os–Ru is $[(NH_3)_5Os^{II}i-(Pro)_2-Ru^{III}(NH_3)_5]$.

[d]Ru–Co is $[(bpy)_2Ru^{II}L-(Pro)_2-Co^{III}(NH_3)_5]$, were bpy is 2,2′-bipyridine, and L is 4-carboxy-4′-methyl-2,2′-bipyridine.

and conformational changes of the bridging peptide that are occurring on similar time scales. In order to obtain meaningful distance dependence of rates of electron transfer, the time scale for the electron transfer reaction must be significantly faster than any large conformational changes of the bridging peptide. For oligoproline peptides the major conformational change is a trans-to-cis isomerization with a half-life of 1–2 min at room temperature [and no evidence for any other rapid, large conformational change exists to date (*12, 13*)]; thus, rates substantially faster than 1–2 min (Tables II and III) (*16*) occur prior to these conformational changes. In the $Os-(Pro)_2-Ru$ and $(bpy)_2RuL^{\bullet}-(Pro)_2-Co$ (last entries in Table I) the intramolecular electron transfer reaction across diproline is faster than trans–cis isomerization by several orders of magnitude (*16, 23*).

Table II. Rates of Intramolecular Electron Transfer Across Polyprolines $[(bpy)_2Ru^{II}L^{\bullet}-(Pro)_n-Co^{III}(NH_3)_5]^{3+}$

n	$M-M$ (Å)	k_{et} (s^{-1}) (25 °C)	$\Delta H^{\ddagger}$ (kcal mol^{-1})	$\Delta S^{\ddagger}$ (cal deg^{-1}mol^{-1})
0	9.0	—	—	—
1	12.2	$>5 \times 10^8$	—	—
2	14.8	1.6×10^7	6.0	–6
3	18.1	2.3×10^5	9.2	–3
4	21.3	5.1×10^4	9.4	–5.5
5	24.1	1.8×10^4	9.0	–9
6	27.3	8.9×10^3	8.8	–11

Intermolecular Reaction

		7.2×10^8 M^{-1} s^{-1}	3.63	–5.8

NOTE: bpy is 2,2′-bipyridine; L is 4-carboxy-4′-methyl-2,2′-bipyridine.

Table III. Rates and Activation Parameters for Intramolecular and Intermolecular Electron Transfer Across Polyprolines $[(bpy)_2Ru^{II}L^{\bullet}-(Pro)_n-apyRu^{III}(NH_3)_5]^{3+}$ ($n = 6, 7, 9$)

n	$M-M$ (Å)	k_{et} (25 °C) (s^{-1})	$\Delta H^{\ddagger}$ (kcal mol^{-1})	$\Delta S^{\ddagger}$ [cal (deg·mol)$^{-1}$]
6	31.6	1.1×10^5	5.6	–17
7	34.5	6.4×10^4	5.1	–19
8	37.5	3.8×10^4	—	—
9	40.8	2.0×10^4	4.0	–26

Intermolecular Reaction

		2.1×10^9 M^{-1}s^{-1}	3.3	–5

NOTE: bpy is 2,2′-bipyridine; L is 4-carboxy-4′-methyl-2,2′-bipyridine; apy is 4-aminopyridine.

For two series of oligoproline bridges (Tables II and III), intramolecular electron transfer reactions were studied according to eqs 1 and 2:

$$e_{aq}^- + (bpy)_2 Ru^{II}L(Pro)_n\text{–}Co^{III} \rightarrow$$

$$\{(bpy)_2 Ru^{II}L\}^\bullet\text{–}(Pro)_n\text{–}Co^{III} \xrightarrow{k_{et}} (bpy)_2 Ru^{II}L\text{–}(Pro)_n\text{–}Co^{II} \tag{1}$$
$$\downarrow$$
$$(bpy)_2 Ru^{II}L\text{–}(Pro)_n + Co^{2+}$$

where $Co^{III} = Co^{III}(NH_3)_5$, bpy = 2,2′-bipyridine, L = 4-carboxy-4′-methyl-2,2′-bipyridine, and $n = 1\text{–}6$, and

$$e_{aq}^- + (bpy)_2 Ru^{II}L(Pro)_n\text{–}apyRu^{III} \rightarrow$$

$$\{(bpy)_2 Ru^{II}L\}^\bullet\text{–}(Pro)_n\text{–}apyRu^{III} \xrightarrow{k_{et}} (bpy)_2 Ru^{II}L\text{–}(Pro)_n\text{–}apyRu^{II} \tag{2}$$

where $apyRu^{III}$ = 4-amino-pyridine–$Ru^{III}(NH_3)_5$, L = 4-carboxy-4′-methyl-2,2′-bipyridine, and $n = 6\text{–}9$. The distance dependence of the rate of intramolecular electron transfer across the bridge as a function of the number of proline residues was determined using metal donors and acceptors that have a large driving force and/or small reorganization energy. For $(Pro)_n$ bridges ($n = 1\text{–}9$), the rate of intramolecular electron transfer decreased substantially as the number of oligoproline residues increased when $\{(bpy)_2 Ru^{II}L\}^\bullet$ donor (bpy = 2,2′-bipyridine, L = 4-carboxy-4′-methyl-2,2′-bipyridine) and either $[\text{–}Co^{III}(NH_3)_5]$ or $[Ru^{III}(NH_3)_5apy]$ acceptors (apy = 4-aminopyridine) were used (*15, 16*). A schematic of one of the donor–acceptor complexes with the $(Pro)_6$ bridge for each of these series is shown in Figure 1. The rates, activation parameters, and distances for these series are summarized in Tables II and III and Figure 2.

One of the most surprising results from these studies is the presence of two different slopes in the plots of the intramolecular electron transfer rate (ln k_{et}) [or (ln $k_{et} + \Delta G^+/RT$) rate corrected for outer sphere reorganizational energy] versus the number of bridging proline residues (i.e., increasing distance) (Figure 2). For $n = 4\text{–}6$ prolines, there is a slow decrease in rate with increasing distance, while for $n = 1\text{–}3$ prolines, there is a more rapid decrease in rate with increasing distance (Figure 2). This change in rate with distance coincides with the onset of helical secondary structure adopted by the oligoproline bridge. In comparing the intramolecular electron transfer rates of oligoproline donor–acceptor complexes (where the distance of the proline bridge is the only variable), the distance dependence of the rate depends on the electronic coupling between the metal donor and acceptor after correction is made for the distance dependence of the outer sphere reorganization energy (Figure 2, Tables II and III) (*25*). At long distances the reorganization energy becomes small and the electronic factor (β) predominates. For these cases the observed rate of intramolecular electron transfer can be expressed in eq 3

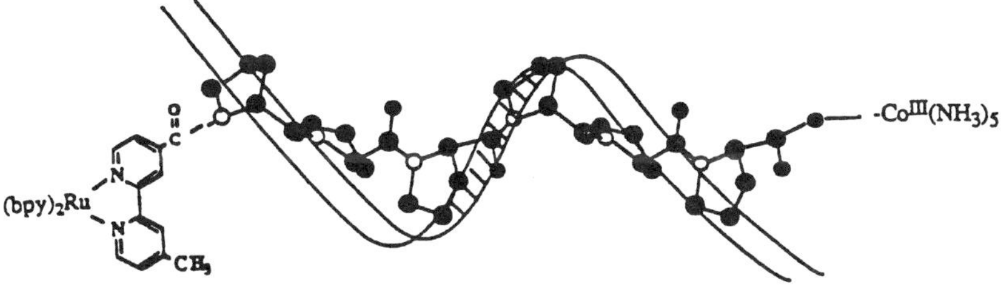

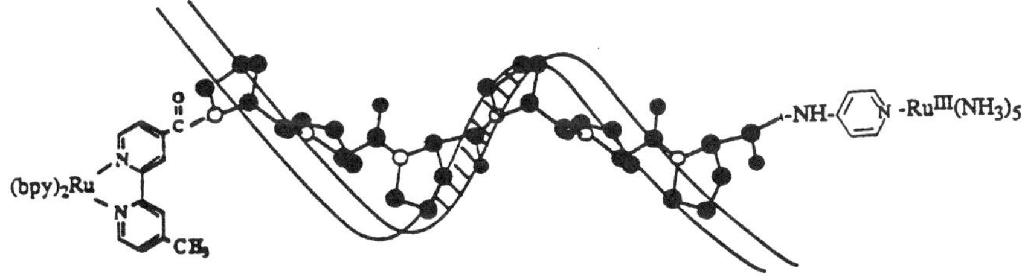

Figure 1. Helical oligoproline complexes with Ru–Co and Ru–Ru donors and
acceptors. $(Pro)_6$ is shown here.

$$k(r) = k(r_0)e^{-\beta(r-r_0)} \tag{3}$$

where $k(r_0)$ is approximated to be the rate constant at the contact distance
between the donor and acceptor (r_0), and $k(r)$ is the rate as a function of the
bridging distance in the oligoproline $(n > 4)$ complexes. A plot of ln k versus
the distance (r) will yield a slope $(-\beta)$ that is related to the attenuation of the
electron transfer rate with the distance of separation. For $n = 4$–6 prolines, a
low β value (eq 3) ~ 0.3 Å^{-1} is estimated from the data (compared to a higher β
value of 0.7–0.95 Å^{-1} determined for $n = 1$–3 prolines in different studies) (25).
This surprisingly low β value for the helical oligoproline bridges shows that the
assembled peptide bridge with a secondary structure is a better mediator of
electron transfer than can be predicted by extrapolating the distance dependence of the rate for $n = 1$–3 prolines. Such a low β for electron transfer
implies that for specific secondary structures long-range electron transfer is
not significantly impeded with distance.

For even longer helical $(Pro)_n$ bridges, $n = 6$–9, a similar low $\beta \sim 0.2$ Å^{-1} is
calculated with the $\{(bpy)_2Ru^{II}L\}^\bullet$ donor and the $[Ru^{III}(NH_3)_5apy]$ acceptor
(Figure 2, D) (16). All the helical polyproline bridges $(n > 4$ prolines) in the
different series show similar mediating properties as evidenced by the similar
change in rate (corrected and uncorrected for reorganizational energy) with
distance observed for the B and D series and for the A and C series (Figure 2).

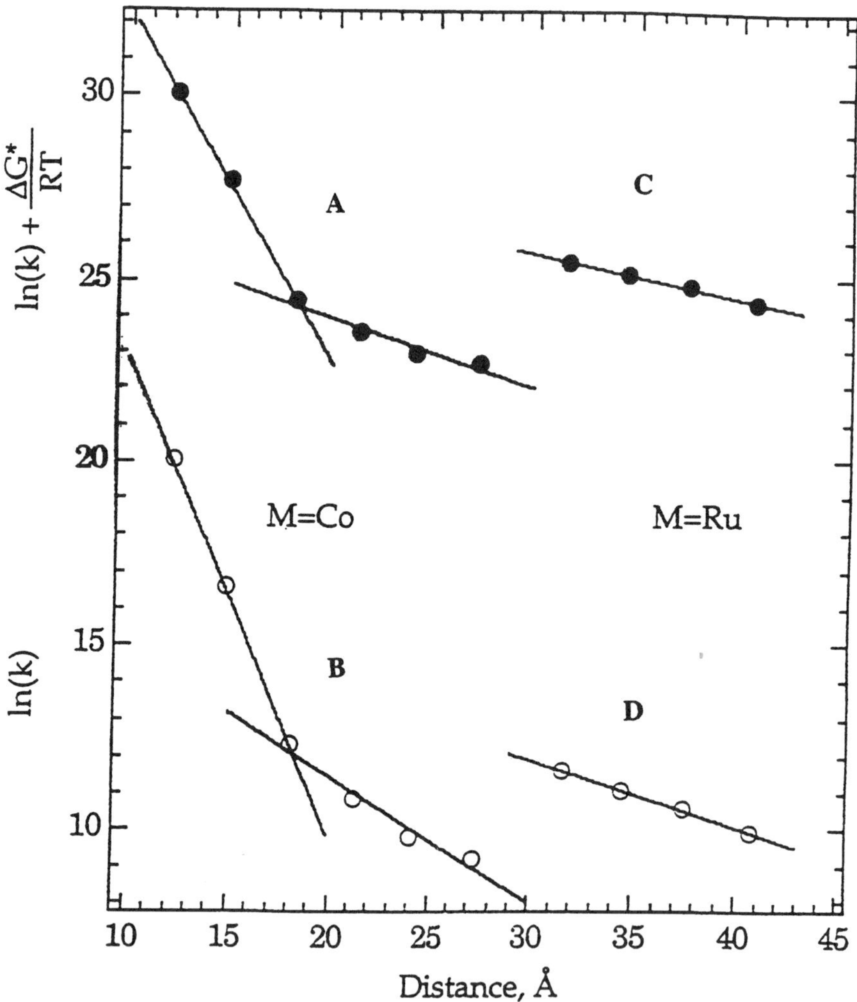

Figure 2. Intramolecular electron transfer rates for two helical oligoproline series of donor–acceptor complexes versus the bridging distance between the donor and acceptor. The rates (eqs 1 and 2) are plotted as $\ln k_{et}$ and as $\ln k_{et} + \Delta G^+/RT$, rate corrected for reorganizational energy, versus the number of bridging proline residues. The two series are $[(bpy)_2Ru^{II}L^{\bullet}-(Pro)_n-Co^{III}(NH_3)_5]$ (n = 1–6) (A and B) and $[(bpy)_2Ru^{II}L^{\bullet}-(Pro)_n-apyRu^{III}(NH_3)_5]$ (n = 6–9) (C and D) where bpy = 2,2′ bipyridine, L = 4-carboxy-4′-methyl-2,2′-bipyridine, apy = 4-aminopyridine.

Recently in a different system of porphyrin donor–acceptor complexes of helical oligoproline bridges ($n = 4$–8) a similar low β value (0.2 Å^{-1}) was estimated from fluorescence quenching studies (*26*) in ethanol solution (which, like water, is known to stabilize the helical polyproline II structure).

The longest peptide bridge across which electron transfer has been measured is the helical $(\text{Pro})_9$ bridge. In this molecule nine amino acid residues separate the metal ion donor from the acceptor (a distance of > 40 Å) (*16*). This distance of 40 Å is comparable to the diameter of a small protein such as cytochrome c. Such efficient electron transfer over a long distance (>40 Å) could not be predicted from the results for short proline bridges ($n = 1$–3).

Thus far, unexpectedly fast electron transfer at long distances has been observed in seven helical oligoproline donor–acceptor complexes. However, the origin of this efficient electron transfer is still not well-understood. The participation of the peptide chain alone and/or the solvation of the oligoproline bridge may be involved in providing this additional channel for electron transfer at long distances. Experiments in progress in deuterated media with these oligoproline complexes should provide further information on the specific electron-transfer pathway utilized in these peptide bridges.

Rigid α-Helical Peptide Donor–Acceptor Complexes

In the previous section the oligoprolines provided rigid peptide bridges with well-defined distances between metal donors and acceptors. More recently we have begun investigating non-proline α-helical peptide bridges that are cyclic and thus rigid and where the distance between the donor and acceptor is also well-defined (*14, 27*). These α-helical cyclic peptides are made rigid by cross-linking amino acid side chains with organic and inorganic reagents (*see* peptides **I**, **II**, and **III** in Chart I). Similar rigid bicyclic hexapeptides (with organic protecting groups) with two Lys side chains cross-linked to ($i + 4$) Asp residues (where i refers to the ith amino acid) have been demonstrated to be α-helical by CD and NMR analysis (*28*). Using similar methods, the helical hexapeptide donor–acceptor complex (**I**), with [(bpy)$_2$Ru–L] (L = 4-carboxy-4′-methyl-2,2′-bipyridine) at the N-terminal Lys and [(NH$_3$)$_5$Ru-apy] (apy = 4-aminopyridine) at the C-terminal Asp, has been recently synthesized and characterized (*27*). Preliminary fluorescence quenching results suggest that rapid electron transfer (on the microsecond timescale) occurs in this α-helical hexapeptide donor–acceptor complex. Electron transfer studies on this peptide donor–acceptor complex using flash photolysis and pulse radiolysis are currently in progress.

The electron transfer properties of the helical peptide donor–acceptor complex (**I**) will be compared to those of the helical $(\text{Pro})_6$ peptide bridge in the [(bpy)$_2$Ru–apy]-$(\text{Pro})_6$-[apy-Ru(NH$_3$)$_5$] donor–acceptor complex studied earlier (Table III). Both complexes have the same donor–acceptor metal ions and are separated by the same number of peptide residues. The main differ-

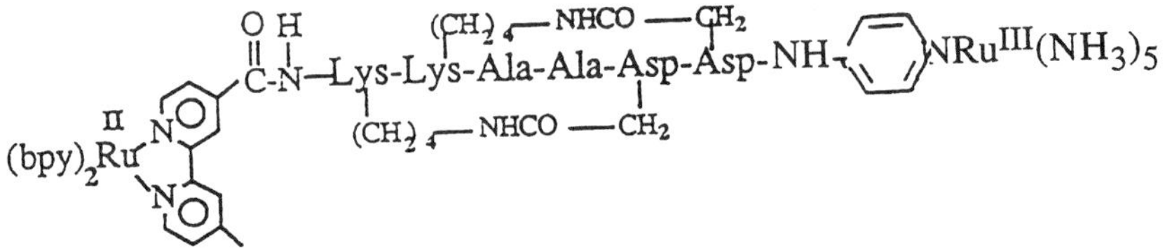

I

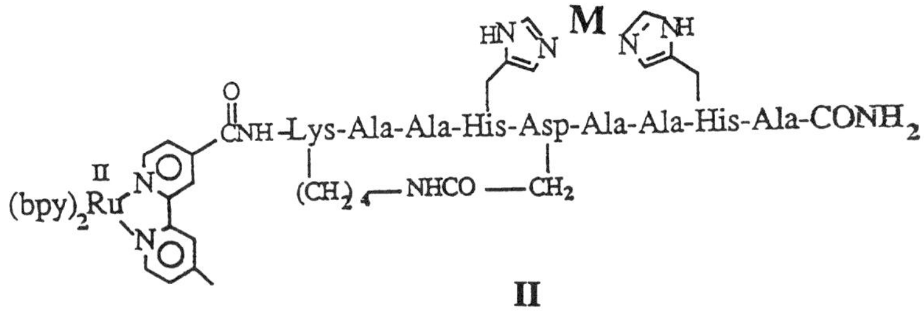

II

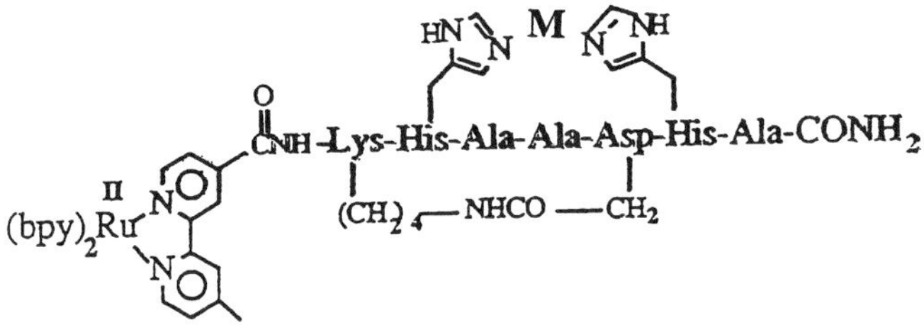

III

$$M = cis\text{-}Ru^{III}(NH_3)_4$$

Chart I. Rigid helical non-proline peptide donor–acceptor complexes.

ence between these two complexes is the different type of left-handed helical structures in the peptide bridges. In bicyclic peptide complex **I** the helicity is maintained by crosslinking of the amino acid side chains, while in $(Pro)_6$ the helicity is maintained by the specific solvation of the oligoproline peptide.

The rigid helical peptides **II** and **III** were designed to test intramolecular electron transfer pathways between the main polypeptide chain and the peptide side chain. In the donor–acceptor complexes of both bicyclic peptides **II** and **III**, one metal ion is bound at the main chain and the other is bound to the

His side chain, but with a different number of amino acids, two or four, between the donor and acceptor. Helicity in these two peptides is maintained by a lactam bridge and also by the cross-linking of two His residues to the metal acceptor, $[-Ru^{III}(NH_3)_4]$. The electron transfer properties of peptides **II** and **III** (with the metal-to-metal distance 14 and 17 Å, respectively) will provide a comparison of the rates of electron transfer between a dipeptide and a tetrapeptide in a similar rigid environment with peptide main-chain–side-chain connection to the donor and acceptor. Electron transfer in peptides **II** and **III** (side-chain–main-chain connectivity) will also be compared to similar diproline and tetraproline bridges (Table II) with main-chain–main-chain connectivity.

Preliminary fluorescence quenching studies with donor–acceptor complexes of peptides **II** and **III** indicate that rapid and measurable rates of electron transfer occur between the two metal centers ([(bpy)$_2$Ru–L] (L = 4-carboxy-4'-methyl-2,2'-bipyridine) at the N terminal and the [Ru(NH$_3$)$_4$(His)$_2$] bound to the side chains of peptides **II** and **III**) (27). Time-dependent fluorescence and absorption experiments are currently under way on these peptides.

The peptides in Chart I represent a new strategy for obtaining rigid small peptides, with amino acids different than proline, where the position of the metal donor and acceptor is well-defined. Further development of this strategy will allow one to study the effect of one or multiple polarizable amino acid side chains on the rates of long-range intramolecular transfer process (by replacing alanine with amino acids with polarizable side chains such as Phe, Tyr, Trp, or Met). The simulation of electron transfer pathways in these small, but conformationally well-defined, peptides should serve as a useful calibration for electron transfer in proteins where multiple pathways are difficult to discern.

Protein Donor–Acceptor Complexes

In the past decade electron transfer proteins have also been modified with a redox reagent (either a donor or an acceptor) that can be bound to different locations on the protein. Horse heart cyt c, modified at surface histidine residues with a variety of ruthenium ammine reagents ([Ru(NH$_3$)$_4$L]; L = NH$_3$, isn, or py), is one of the best examples of modified protein donor–acceptor complexes (29–36). These modified proteins (analogous to the donor–acceptor complex in Scheme I) have been used to study the dependence of intramolecular electron transfer on distance and the role of the protein network in facilitating the electron transfer reactions.

Rates of intramolecular electron transfer in proteins have been studied successfully by pulse radiolysis (29–34) and flash photolysis techniques (35, 36). Our studies have focused primarily on pulse radiolysis techniques because of the variety of oxidizing and reducing radicals that can be used to generate the intermediate precursor ruthenium–protein complexes. For example, pulse radiolysis using the $CO_2^{\bullet-}$ (or e_{aq}^-) radical allows one to start with the totally oxidized Ru(III)–heme(III) and generate the precursor intermediate by reduction. Using the $N_3^{\bullet}$ radical and the totally reduced protein Ru(II)–heme(II), the precursor complex can be generated by oxidation (*29–34*). The reactions summarizing the generation of the precursor complex by oxidation or reduction with pulse radiolysis and the electron transfer measurements in protein donor–acceptor complexes are shown in eqs 4 and 5.

Oxidative Method

$$N_3^{\bullet} + M_1(red)-L-M_2(red) \quad \nearrow \quad M_1(red)-L-M_2(ox) \qquad \downarrow k_{et}$$
$$(CO_3^{\bullet}) \qquad \searrow \qquad M_1(ox)-L-M_2(red) \tag{4}$$

Reductive Method

$$CO_2^{\bullet} + M_1(ox)-L-M_2(ox) \quad \nearrow \quad M_1(red)-L-M_2(ox) \qquad \downarrow k_{et}$$
$$(e_{aq}^{-}) \qquad \searrow \qquad M_1(ox)-L-M_2(red) \tag{5}$$

The azide radical $N_3^{\bullet}$ is generated by the reaction of $OH^{\bullet}$ with solutions of sodium azide, and the carbon dioxide radical anion, $CO_2^{\bullet}$, is generated by pulse radiolysis in aqueous solutions by the reaction of $OH^{\bullet}$ with formate ion.

Recently, comparative electron transfer studies were carried out in two different horse heart (hH) cytochrome c (cyt c) donor–acceptor complexes: one modified at Met 65 with $[Fe(CN)_5(H_2O)]^{2-}$ (Fe(Met 65)) and one modified at His 33 with $[Ru(NH_3)_4 isn(H_2O)]^{2+}$ (isn = isonicotinamide) (Ru(His 33)) (*14, 32*). The Met 65 and the His 33 are located on opposite sides of the heme in the protein (Figure 3). The through-space distances from Fe(heme) to the S (Met 65) ($\sim$15.3 Å) and from Fe(heme) to the N_1 (imidazole)His 33 (16 Å) are very similar. The direction of electron transfer in both complexes is from the interior of the protein to the surface- bound, transition metal redox reagent, that is, heme(II) to $[Fe^{III}(CN)_5]$(Met 65) or to $[Ru(NH_3)_4 isn]^{2+}$(His 33). Furthermore both metal acceptor complexes have comparable self-exchange rates and redox potentials. Based on these facts, similar rates of intramolecular electron transfer from the heme(II) to Fe^{III}(Met 65) or to Ru^{III}(His 33) would be expected, if the protein residues and pathways between the heme and these two metal centers are comparable. Instead, the intramolecular electron transfer rate in cyt c–(Ru(His 33)) was found to be $\geq 10^3$ times faster than that in cyt c–(Fe(Met 65)), despite the similar distance of the redox acceptor to the heme, driving force, and reorganizational energy (Table IV). For the heme(II) to Ru^{III}(His 33), intramolecular electron transfer is the preferred route

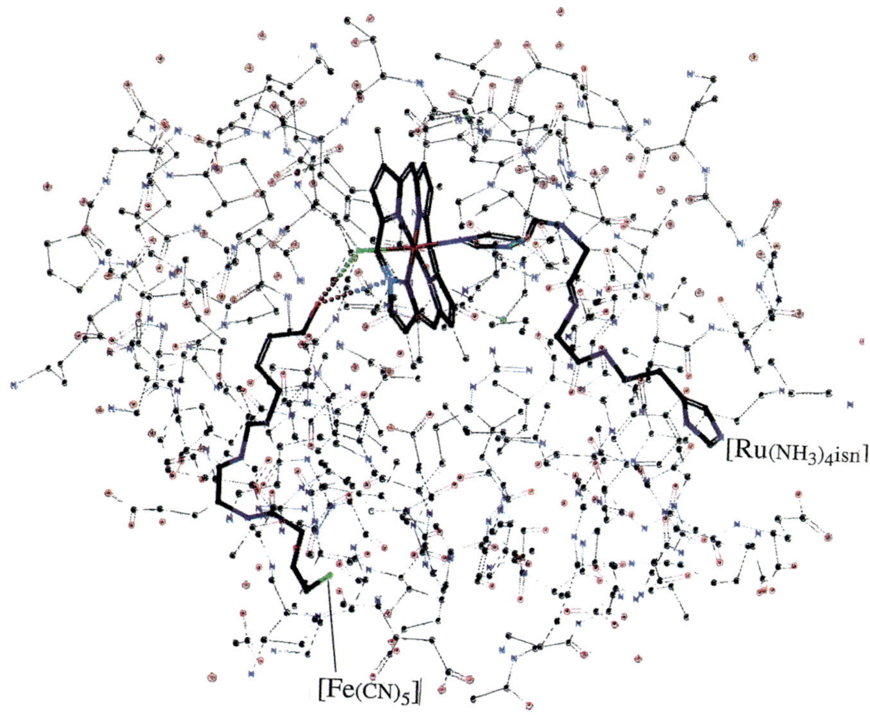

Figure 3. Two equidistant sites in horse heart cytochrome c: Met 65 and His 33. The Fe(heme) to S and to N_1 (imz) distances are 15.3 Å (Met 65) and 16.0 Å (His 33). Horse heart cytochrome c was modified with [Fe(CN)₅–] at Met 65 and [Ru(NH₃)₄(isn)–] at His 33 in two separate experiments.

($k = 440$ s⁻¹), whereas for the heme(II) to Fe^{III}(Met 65), intramolecular electron transfer is very slow ($k < 0.2$ s⁻¹). Similar results were observed with the tuna cyt c modified with [Fe(CN)₅](Met 65) (32). Thus the slow intermolecular electron transfer reaction, heme to Met 65 interaction, has now been observed for two different cyt c species. In both of these reactions the electron transfer occurs by an intermolecular pathway where electron transfer occurs from the heme(II) of one cytochrome molecule to the [Fe^{III}(CN)₅](Met 65) of another cytochrome molecule.

Calculation of the tunneling matrix element between the heme and His 33 or Met 65 is in reasonable agreement with these experimental findings (i.e., a lower tunneling matrix element is calculated for the heme → Met 65 residue than for the heme → His 33 residue) (37). [For the His 33 a proposed pathway

Table IV. Intra- and Intermolecular Electron Transfer Rates and Driving Forces for Modified Protein Complexes

Complex	k_{et} (s^{-1}) (25 °C)	k_b $(M^{-1}\,s^{-1})$ (25 °C)	$-\Delta G°$ (eV)
Hh cyt c^{II}–[Ru^{III}(NH$_3$)$_4$isn(His 33)][a] 16–4 μM	440 ± 30	$2.8 \pm 0.2 \times 10^{4}$ [b]	0.18
Hh cyt c^{II}–[Fe^{III}(CN)$_5$Met 65][c] 10.3–1.7 μM	$\leq 0.6 \pm 0.3$	$7.1 \pm 0.5 \times 10^{5}$	0.30
tuna cyt c^{II}–[Fe^{III}(CN)$_5$Met 65][c] 19.6–1.4 μM	$\leq 0.2 \pm 0.1$	$5.2 \pm 0.2 \times 10^{5}$	0.30
Ck cyt c^{II}–[Ru^{III}(NH$_3$)$_4$isn(His 33)]	220 ± 20	—	0.13
Hh cyt c^{II}–[Ru^{III}(NH$_3$)$_5$(His 33)]	53 ± 2	—	0.13

[a] 0.05 M phosphate buffer, pH 7.0, 10^{-3} M NaN$_3$.

[b] k_b is for [Ru(NH$_3$)$_5$isn]$^{3+}$ + Hh cyt c^{II}, which is an upper limit for the intermolecular electron transfer rate between two molecules of Hh cyt c^{II}–[Ru^{III}(His 33)].

[c] 0.10 M phosphate buffer, pH 7.4, 10^{-3} M NaN$_3$.

is His 33–Leu 32–Asn 31–Pro 30 with a H bond to the NH of His 18. For the Met 65 a proposed pathway is Met 65–Glu 66–Tyr 67–Tyr 67 side chain, followed by a space jump to the Met 80 or directly to the heme carbon (Figure 3). The long space jump for Met 65 is presumably mostly responsible for the large differential in rate (38)]. The difference in rates in these two modified proteins indicates that the peptide network has a marked effect on rates of electron transfer, even when the through-space distance, driving force, and reorganization energies are controlled. Furthermore, this example clearly shows that electron transfer rate from the heme can vary significantly with two similarly distant surface residues in the protein.

Another experiment illustrating this communication between the interior and surface of the protein is a study of the rate of oxidation of native cyt c compared to that of Ru(His 33)–cyt c with a common oxidant, [Co(ox)$_3$]$^{3-}$. The rate of oxidation of native cyt c(Fe^{II}) with [Co(ox)$_3$]$^{3-}$ is quite slow (39) ($k \sim 5$ M^{-1} s^{-1}, $\mu = 0.5$ M, pH 7 phosphate buffer), even though cyt c, a positively charged protein, is interacting with a negatively charged oxidant, [Co(ox)$_3$]$^{3-}$ and $\Delta G = -0.31$ eV. In contrast, the oxidation of the heme in Ru^{II}–cyt c species occurs rapidly in the presence of [Co(ox)$_3$]$^{3-}$. Thus [Co(ox)$_3$]$^{3-}$ catalyzes the rate of oxidation of cyt c(Fe^{II}) via the covalently modified ruthenium reagents with catalysis factors of $\sim 10^3$ for Hh and Ck cytochromes (Table V). This facile intramolecular electron transfer (catalysis) from the heme(Fe^{II}) $\rightarrow$ surface(Ru^{III}) in several [Ru(NH$_3$)$_4$ L](His33)-cyt c molecules is initiated by oxidation of the surface [Ru^{II}(NH$_3$)$_4$ L] reagent by the [Co(ox)$_3$]$^{3-}$oxidant, followed by rapid intramolecular electron transfer reaction between the heme(Fe^{II}) and the covalently bound [Ru^{III}(NH$_3$)$_4$ L] (eqs 6 and 7). The intramolecular reaction between the heme(Fe^{II}) and the [Ru^{III}(NH$_3$)$_4$ L] bound at His 33 (k_2) has been

Table V. Rate Constants for the Oxidation of Redox Centers by $[Co(ox)_3]^{3-}$ in Cytochrome c Derivatives

Reductant	k_1 $(M^{-1}s^{-1})$	k_2 (s^{-1})
Native cyt c(II) (Hh)	6.6	—
Native cyt c(II) (Ck)	14.3	—
$[(NH_3)_5Ru(III)]$–cyt c(II) (Hh)	45.0	0.2
trans-$[(NH_3)_4Ru(II)(isn)]$–cyt c(II) (Hh)	1.05×10^4	440
trans-$[(NH_3)_4Ru(II)(isn)]$–cyt c(II) (Ck)	5.37×10^3	220

separately determined using pulse radiolysis techniques (Table IV), and under these conditions the oxidation with $[Co(ox)_3]^{3-}$ is not the rate-limiting step.

$$L(NH_3)_4 Ru^{II}-Fe^{II}cyt\ c + Co(ox)_3^{3-} \xrightarrow{k_1}$$
$$(NH_3)_5 Ru^{III}-Fe^{II}cyt\ c + Co_{aq}^{2+} + 3ox^{2-} \tag{6}$$

$$L(NH_3)_4 Ru^{III}-Fe^{II}cyt\ c \xrightarrow{k_2} (NH_3)_5 Ru^{II}-Fe^{III}cyt\ c \tag{7}$$

The oxidation of the surface ruthenium reagent by cobalt oxalate has also been used to determine uphill electron transfer reactions where the Ru^{II} cyt c^{III} electronic isomer is the minor component (*33*).

The mediation of the electron transfer of cyt c by $[Co(ox)_3]^{3-}$ via the covalently bound ruthenium complexes has many applications where artificial redox reagents can facilitate electronic communication between the interior of a protein and the surface-modified redox reagent or a solid electrode. A similar approach has been recently used to enhance the electron exchange between covalently bound electron transfer mediators (ferrocene) and glucose oxidase, an enzyme with important medical applications (*40–42*).

The protein electron transfer experiments described here demonstrate that the location of the surface artificial redox reagent is very important for rapid communication to occur between the interior of a protein and covalently modified surface reagents. Experiments with the modification at the Met 65 (Hh cyt c) as compared to His 33 (Hh cyt c), an equidistant site from the heme, show that the two sites are not equally effective in communicating with the interior of the protein.

The examples described in this chapter provide a continuing strategy for studying various aspects of protein electron transfer reactions in well-designed peptide and protein systems. The peptide studies described here have led to the discovery of new and unexpectedly fast electron transfer channels via organized peptides over long distances. The protein experiments also show the importance of specific segments of the proteins over others in promoting the rate of electron transfer between the protein interior and surface-bound redox reagents. More systematic studies of peptide bridges and well-designed experiments with proteins are expected to lead to a better understanding of biological electron transfer in these complex molecules.

Acknowledgments

This research was supported by the U.S. Department of Energy, Division of Chemical Sciences, Office of Basic Energy Sciences, under Contracts DE-FG05-90ER1410 and DE-FG02-93ER14356 and by the Center for Advanced Food Technology (CAFT) at Rutgers University. Collaboration with J. Wishart at Brookhaven National Laboratory and the contributions of many students and colleagues at Rutgers are gratefully acknowledged.

References

1. Taube, H. *Ber. Bunsen Ges. Phys. Chem.* **1972,** *76,* 964.
2. Isied, S. S.; Taube, H. *Inorg. Chem.* **1974,** *13,* 1545–1551.
3. Isied, S. S.; Taube, H. *Inorg. Chem.* **1976,** *15,* 3070–3075.
4. Isied, S. S.; Taube, H. T. *J. Am. Chem. Soc.* **1973,** *95,* 8198.
5. Zawacky, S.; Taube, H. T. *J. Am. Chem. Soc.* **1981,** *103,* 3379.
6. Dixon, N. E.; Lawrence G. A.; Lay, P. A.; Sargeson, A. M. *Inorg. Chem.* **1983,** *22,* 846.
7. Williams, R. J. P. In *Biological Aspects of Inorganic Chemistry;* Addison, A.W.; Cullen, W. R.; Dolphin, D.; James, B. R., Eds.; J. Wiley & Sons: New York, 1977.
8. *Tunneling in Biological Systems;* Chance, B.; DeVault, D. B.; Frauenfelder, H.; Marcus, R.; Schrieffer, J. R.; Sutin, N., Eds.; Academic: New York, 1979.
9. Pelletier, H.; Kraut, J. *Science (Washington, D.C.)* **1992,** *258,* 1748.
10. Tsukihara, T.; Aoyama, H.; Yamashita, E.; Tomizaki, T.; Yamaguchi, H.; Shinzawa-itoh, K.; Nakashima, R.; Yaono, T.; Yoshikawa, S. *Science (Washington, D.C.)* **1995,** *269,* 1089.
11 Iwata, S.; Ostermeier, C.; Ludwig, B.; Michel, H. *Nature (London)* **1995,** *376,* 660.
12. Stryer, L.; Haughland, H. P. *Proc. Natl. Acad. Sci. U.S.A.* **1967,** *58,* 719.
13. Isied, S.; Ogawa, M. Y.; Wishart, J. W. *Chem. Rev.* **1992,** *92,* 381.
14. Moreira, I; Wishart, J. F.; Ogawa, M. Y.; Vassilian, A.; Arbo, B.; Sun, J.; Isied, S. S. *J. Photochem. Photobiol.* **1994,** *82,* 203.
15. Ogawa, M. Y.; Wishart, J. F.; Young, A.; Miller, J. R.; Isied, S. S. *J. Phys. Chem.* **1993,** *97,* 11456–11463.
16. Ogawa, M. Y.; Moreira, I.; Wishart, J. F.; Isied, S. S. *Chem Phys.* **1993,** *176,* 589–600.
17. Harrison, P. M.; McGavin, S. *Acta Crystallogr.* **1962,** *15,* 914.
18. Traub, W.; Shmueli, U. *Nature (London)* **1963,** *195,* 1165.
19. Matsuzaki, T. *Acta Crystallogr.* **1974,** *30,* 1029.
20. Kartha, G.; Ashida, T.; Kakudo, M. *Acta Crystallogr.* **1974,** *B30,* 1861.
21. Isied, S. S.; Vassilian, A.; Magnuson, R. H.; Schwarz, H. A. *J. Am. Chem. Soc.* **1985,** *107,* 7432–7438.
22. Xu, Y.; Sugar, I. *J. Magn. Reson.* **1993,** *101,* 145.
23. Vassilian, A.; Wishart, J. F.; van Hemelryck, B.; Schwarz, H. A.; Isied, S. S. *J. Am. Chem. Soc.* **1990,** *112,* 7278.
24. Isied, S. S.; Vassilian, A. *J. Am. Chem. Soc.* **1984,** *106,* 1732–1736.

25. Isied, S. S.; Vassilian, A.; Wishart, J. F.; Creutz, C.; Schwarz, H. A.; Sutin, N. *J. Am. Chem. Soc.* **1988,** *110,* 22.
26. Tamiaki, H.; Nomura, K.; Maruyama, K. *Bull. Chem. Soc. Jpn.* **1994,** *67,* 1863.
27. Chin, T. C. ; Brunschwig, B.; Isied, S. S., in preparation.
28. Bracken, C.; Gulyas, J; Taylor, J. W.; Baum, J. *J. Am. Chem. Soc.* **1994,** *116,* 6431.
29. Isied, S. S.; Worosila, G.; Atherton, S. J. *J. Am. Chem. Soc.* **1982,** *104,* 7659.
30. Isied, S. S.; Kuehn, C.; Worosila, G. *J. Am. Chem. Soc.* **1984,** *106,* 1722.
31. Bechtold, R.; Gardineer, M. B.; Kazmi, A.; van Hemelryck, B.; Isied, S. S. *J. Phys. Chem.* **1986,** *90,* 3800.
32. Moreira, I.; Sun, J.; Cho, M. O.-K.; Wishart, J. F.; Isied, S. S. *J. Am. Chem. Soc.* **1994,** *116,* 8396.
33. Winkler, J. R.; Gray, H. B. *Chem. Rev.* **1992,** 92, 369–379.
34. Sun, J.; Wishart, J. F.; Isied, S. S. *Inorg. Chem.* **1995,** *34,* 3998–4000.
35. Sun, J.; Wishart, J. F.; Gardineer, M. B.; Cho, M. P.; Isied, S. S. *Inorg. Chem.* **1995,** *34,* 3301–3309.
36. Sun, J. Ph.D. Thesis, Rutgers University, 1992.
37. Beratan, D. N. ; Betts, J. N.; Onuchic, J. N. *Science (Washington, D.C.)* **1991,** *252,* 1285.
38. Shin, Y.–G. K., Isied, S. S., work in progress.
39. Holwerda, R. A.; Knaff, D. B.;Gray, H. B.; Clemmer, J. D.; Crowley, R.; Simtra, J. M.; Mauk, A. G. *J. Am. Chem. Soc.* **1980,** *102,* 1142.
40. Heller, A. *Acc. Chem. Res.* **1990,** *23,* 128.
41. Badia, A.; Carlini, R.; Fernandez, A.; Battaglini, F.; Mikkelsen, S. R.; English, A. M. *J. Am. Chem. Soc.* **1996,** *118,* in press.
42. Riklin, A.; Katz, E.; Willner, I.; Stocker, A.; Buckmann, A. F. *Nature (London)* **1995,** *376,* 672.

21

Ruthenium in Biology: DNA Interactions

Michael J. Clarke

Merkert Chemistry Center, Boston College, Chestnut Hill, MA 02167

The reactivity of ruthenium complexes with nucleic acids and their constituents is reviewed. A brief survey of the antitumor activity of ruthenium complexes is followed by a discussion of ways in which ruthenium complexes coordinate to nucleic acids. The choreographed movement of ruthenium ammines on adenine and cytosine ligands is presented as a function of pH and electrochemical potential. Possible mechanisms whereby Ru^{III} coordinated at the N7 of guanine nucleosides can facilitate N-glycosidic hydrolysis and autoxidation are presented together with evidence for N-glycosidic hydrolysis following disproportionation to Ru^{IV}. The electron-spin resonance and ^1H NMR spectra of imidazole Ru^{III} complexes as models for purine complexes are discussed. The biological effects of some polynuclear ruthenium ammine species are surveyed.

HENRY TAUBE'S VIEW OF CHEMISTRY has deeply affected all his students. That his shrewd observations of nature, which began developing on a dirt farm in Saskatchewan, were turned to the study of chemistry rather than English was a serendipitous event that has affected the lives of many and the chemistry of the world. I entered the Taube group just after Carol Creutz had prepared the famous Creutz–Taube ion and was studying its spectra (*1*). Taube's application of thermodynamics, bonding theory, and structural evidence to the study of transition metal reactivity was immediately apparent—as was the role beer played in fostering the group's research efforts. One afternoon when the supply ran low, Taube funded me to replenish it. Planning to pick it up the next morning, I went back to my bench. An hour or so later, there was a soft rumbling in the lab. After a wise word from Andy Zanella regarding the future of my graduate career, the libations appeared with appropriate alacrity.

When the times inserted a hiatus into my graduate study, I met with Taube when Creutz invited him to Georgetown. We talked about new directions my research might take, and nucleic acids were mentioned in an off-hand way in conjunction with the chemistry of ruthenium. Little did I know that Taube's apparently casual remarks had led to many lifetime endeavors. When I returned to Stanford, Dick Sundberg had kindly left me a note about his work, along with a bottle of xanthine. Eventually, I picked up on Sundberg's suggestion, which led to work with other purines, pyrimidines, nucleosides (2–4), nucleotides, and the first attachment of ruthenium to a nucleic acid (5). Ruthenium interactions with nucleic acids are now widely studied by a number of groups around the world as a means of inhibiting DNA synthesis and tumor growth (6) and of probing DNA structure through photochemical and oxidative cleavage (7, 8). Other research has lead to imaging tumors by radioscintigraphy (9) and to studying what appears to be long-range electron transfer mediated by the π-stacking of DNA (10, 11), a process that Taube had thought might be feasible when ruthenium was first bound to DNA.

Antitumor Activity

A number of ammine, amine, and heterocyclic complexes of ruthenium exhibit inhibition of DNA replication (12), mutagenic activity, induction of the SOS repair mechanism (13), binding to nuclear DNA (14), and reduction of RNA synthesis (15) so that DNA is strongly suggested as the target molecule for anticancer ruthenium complexes.

   Table I summarizes the anticancer activity of a representative selection of ruthenium complexes against animal tumor models. Although the activity of

Table I. Antitumor Activity of Representative Ruthenium Complexes

Compound	Dose (mg/kg)	T/C (%)	Ref.
fac-[Cl$_3$(NH$_3$)$_3$Ru^{III}]	50	189	77
[Cl$_3$(1,5-dimethyltetrazole)$_3$Ru^{III}]	80	179	24
[CH$_3$CH$_2$COO(NH$_3$)$_5$Ru^{III}]ClO$_4$	12.5	163	77
(ImH)[Cl$_4$Im$_2$Ru^{III}]	209.3	156	78
cis-[Cl$_2$(NH$_3$)$_4$Ru^{III}]Cl	12.5	157	77
Ru^{IV}(PDTA–H$_3$)	120	152	79
cis-[I(NO)(NH$_3$)$_4$Ru^{III}]I$_2$	25	144	63
[(C$_4$O$_4$)(NH$_3$)$_5$Ru^{III}](F$_3$CSO$_3$)	21.2	140	80
(IndH)[Cl$_4$(Ind)$_2$Ru^{III}]	91.1	133	24
μ-(CH$_3$CO$_2$)$_4$Ru$_2$Cl	32	133	24
cis-[Cl$_2$(Me$_2$SO)$_4$Ru^{II}]	565	125	81
mer-[Cl$_3$(Me$_2$SO)$_2$BRu^{III}][a]		110–143	82, 83
[Cl(NH$_3$)$_5$Ru^{III}]Cl	1.5	116	84
[Ox(bipy)$_2$Ru]	3.13	101	77
[(Asc)(NH$_3$)$_5$Ru^{III}](F$_3$CSO$_3$)	10	96	80
[Cl$_2$(phen)$_2$Ru^{II}]ClO$_4$	6.25	90	77
[Cl$_3$(4Mepy)$_3$Ru^{III}]	78	83	24

NOTE: T/C values are expressed as 100 times the ratio of the lifetime of animals treated with the ruthenium drug to that for the untreated animals. Values listed are for the most common initial screens (i.e., P388 or L1210). In some cases, T/C values on other screens were considerably higher or lower. Im is imidazole; Ind is indazole; phen is 1,10-phenanthroline.
[a]B = NH$_3$, Im.

$[CH_3CH_2CO_2(NH_3)_5Ru](ClO_4)_2$ suggests that monoacido complexes can be active, generally multichloro compounds, such as *cis*-$[Cl_2(NH_3)_4Ru]Cl$, *fac*-$[Cl_3(NH_3)_3Ru]$ (*16, 17*) and Na-*trans*-$[(Im)_2Cl_4Ru]$ (Im = imidazole) (*18*) exhibit the best activity. Although *fac*-$[Cl_3(NH_3)_3Ru]$ showed good-to-excellent antitumor activity in several tumor screens, its low solubility results in poor pharmacological properties (*19, 20*); yet *mer*-$[Ru(tpy)Cl_3]$ (tpy = 2,2″:6″,2″-terpyridine) has recently exhibited good antitumor activity in the L1210 cell line (*21*). By charging with fewer nitrogen ligands and more halides, Keppler (*18, 22–26*) has increased the solubility of and activity of ruthenium complexes against colon tumors.

Multiacido ruthenium(III) complexes, particularly di- through tetrachloro complexes, appear to be transported in the blood by transferrin (*27–31*), but they also bind to other blood proteins. While some complexes, such as *trans*-$[(Im)_2Cl_4Ru]^+$, are bound more by albumin than transferrin (*32*), transferrin is likely to be the more important mode of transport to the tumor. In addition to uptake through transferrin-specific receptors, ruthenium complexes may enter through other modes of endocytosis following binding to anionic sites on the cell surface. The elevated uptake of nutrients, higher membrane permeability, and angiogenesis to increase blood flow also results in some nonspecific tumor uptake of metallopharmaceuticals. Since small ions are excreted fairly readily by the kidneys, nonspecific uptake of ruthenium ions probably occurs within a few hours, while protein-mediated uptake may extend over a longer period.

As several redox proteins are capable of reducing Ru^{III} complexes in vivo (*33*), Ru^{III} complexes may serve as prodrugs, which are activated by reduction in vivo to ligate more rapidly to biomolecules (*19*). The low O_2 content, reduced electrochemical potential (*34*), and lower pH in tumor cells (*35*) should favor the maintenance of this generally more rapidly substituting Ru^{II} state and thus provide for some degree of selective toxicity. However, in the case of multichloro complexes, binding to DNA might take place at about the same rate as cisplatin (*36*). Species capable of 1 e transfer to Ru^{III} exist in both the mitochondrial electron transfer chain and in microsomal electron transfer systems, with microsomal proteins being the more efficient (*33*). Ammin
ruthenium complexes can also be reduced by a prevalent transmembrane electron transport system, so that it may not be necessary that the complexes enter cells in order to be reduced (*37*).

Nucleic Acid Binding

Because of ion pairing, $[(H_2O)(NH_3)_5Ru^{II}]^{2+}$ binding to nucleic acids occurs fairly rapidly and is strongly ionic-strength-dependent. For tRNAs the rate for binding to guanine N7 sites (G^7) is equal to $k[Ru][P_{RNA}]$, where $k = 5.96$ M^{-1} s^{-1} at 25 °C and $\mu = 0.045$ (*38*). In DNA, binding occurs primarily at G^7 sites, which are relatively exposed in the major groove, but secondary binding occurs on the exocyclic ammines of adenine and cytosine residues (*39*). Equilibrium binding constants for guanine sites are 5100 M^{-1} and 7800 M^{-1} on helical and single-stranded CT–DNA, respectively, but only 2900 M^{-1} for RNA. In part, this is due to the additional oxygen on the sugar in ribonucleotides, which has a modest effect on the basicity of G[7].

Possibilities for interfering with DNA metabolism include (1) blockage or lack of recognition by replicating enzymes at the metallated G, thereby halting DNA synthesis; (2) intra- or interstrand DNA cross-linking by the metal; (3) protein–DNA cross-links; and (4) chemical reactions of the guanine residue induced by the metal ion. Fenton's chemistry may also occur, but the low oxygen concentration in tumors and the relative inefficiency of [Guo(NH$_3$)$_5$Ru^{III}] in cutting DNA by Fenton's chemistry (39) make this mode of damage unlikely for coordinatively saturated ruthenium ammine complexes.

Steric crowding in octahedral complexes probably minimizes intrastrand *cis*-G[7]–G[7] cross-linking, but trans cross-links to other than adjacent G sites and cross-links to exocyclic amines on A and C remain possibilities. For example, *mer*-[RuCl$_3$(Me$_2$SO)$_3$], *trans*-[(Me$_2$SO)$_2$Cl$_4$Ru]$^-$ (which both yield *mer, cis*-[RuCl$_3$(H$_2$O)$_2$(Me$_2$SO)] in solution), *trans*-[Cl(SO$_2$)(NH$_3$)$_4$Ru]$^+$ (40, 41), and *mer*-[Cl$_3$(tpy)Ru] (21) all produce DNA interstrand cross-links. Although interstrand cross-linking has been suggested for *cis*-diaqua polypyridyl complexes (42), studies with *trans*-[(Ind)$_2$Cl$_4$Ru]$^-$ revealed few intrastrand or DNA–protein cross-links (43).

Movement of Ru on Purines and Pyrimidines

While the initial complexation of adenosine by [(H$_2$O)NH$_3$)$_5$Ru^{II}] appears to be largely at N1, N7 coordination appears to occur with 1-methyladenosine (1-MeAdo) (44, 45). Equation 1, where R (nm) is the distance between the Ru^{III} and the ionization site, predicts that oxidation to Ru^{III} will generate a shift in the pK_a of the N6 amine of about 10 units. Indeed, electrochemical measurements suggest the pK_a of [(Ado-κN1)(NH$_3$)$_5$Ru^{III}] to be 8.2. Consequently, the ionized N1 presents an excellent binding site for Ru^{III}, to which this metal ion rapidly linkage isomerizes as shown in Scheme I (20).

$$\Delta pK_a = \frac{1.36}{R^2} - 2.7 \tag{1}$$

Once coordinated at the exocyclic amine, hydrogen bonding occurs between two ammine ligands and the anionic N1 of adenine or N3 of cytosine (46). Hydrogen bonding between ammine ligands and the adjacent nitrogen on the pyrimidine ring is negated upon protonation of the ring site, which causes the metal to rotate about the N$_{exo}$–C bond. pK_a values for the N1 and N3 sites for the isocytosine (ICyt) complex are 2.9 and 10.0, respectively, and the corresponding values for 6-methylisocytidine (6-MeICyt) are 3.1 and 10.2. Both rotamers are evident by electrochemistry and NMR in the pH range around

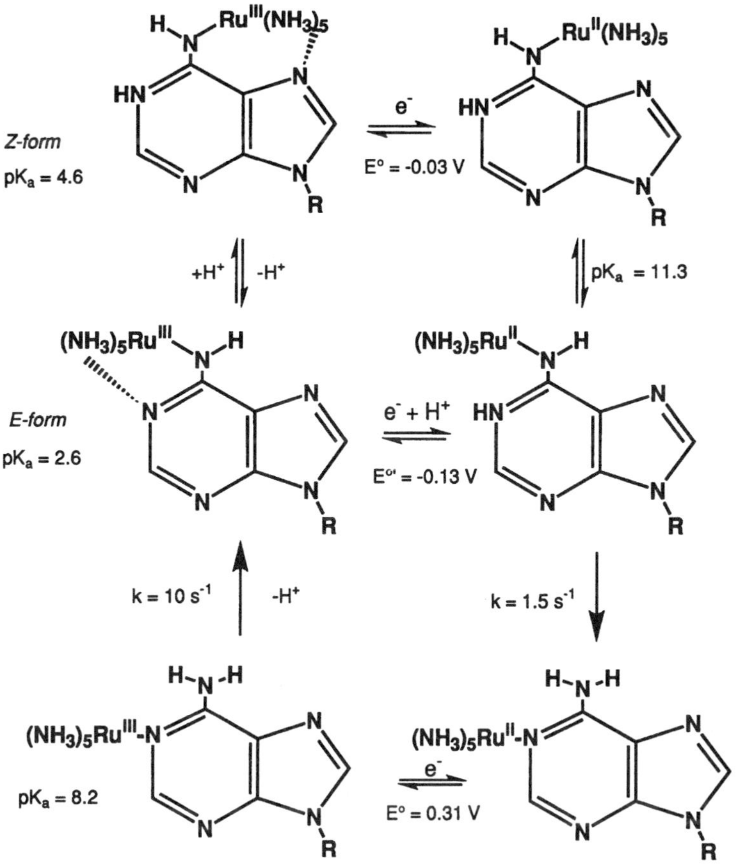

Scheme I. *Terpsichorian movements of [(NH$_3$)$_5$Ru^{II,III}] on adenosine, which can be controlled by pH and reduction potential.*

the pK_a values of the isomers (~3.3), and activation parameters for the rotamerizations have been determined by variable-temperature NMR. For [(ICyt)(NH$_3$)$_5$Ru^{III}], activation enthalpy ($\Delta H^{\ddagger}$) = 2.69 kcal/mol, activation entropy ($\Delta S^{\ddagger}$) = –33.9 cal/mol K, and activation energy (E_a) = 3.34 kcal/mol. For [(6-MeICyt)(NH$_3$)$_5$Ru^{III}], $\Delta H^{\ddagger}$ = 1.56 kcal/mol, $\Delta S^{\ddagger}$ = –37.2 cal/mol K, and E_a = 2.18 kcal/mol. The rotational activation parameters for an analogous ligand, 1,7,7-trimethylcytosine, when measured by a similar technique, are $\Delta H^{\ddagger}$ = 15.2 kcal/mol (E_a = 15.7 kcal/mol) and $\Delta S^{\ddagger}$ = 0.2 cal/mol K. The 13 kcal decrease in $\Delta H^{\ddagger}$ likely arises from the appreciable π-bonding between the Ru^{III} and the amide, so that π-interactions between the amide and the pyrimidine ring are substantially decreased (*47*).

A second motion of the metal occurs upon reduction to Ru^{II}, which rapidly linkage isomerizes to the adjacent π-acceptor ring nitrogen. Taken together the reversible pirouette (rotation) and side-step (linkage isomerization) movements, which can be choreographed by pH and electrochemical potential, constitute a true terpsichorean motion as indicated in Scheme I (*20*).

Like Ru^{III} (*2–5, 44, 48, 49*), $[(NH_3)_5Os^{III}]$ forms stable complexes by binding to the N7 of guanine, xanthine, and hypoxanthine ligands and also forms a stable complex at the C8 of 1,3,7-trimethylxanthine (1,3,7-Me₃Xan; caffeine) (*50*). While osmium complexes are known to readily bind across double bonds in aromatic systems and electrochemical anomalies may be consistent with metal ion movement on purines, no direct evidence has been obtained for this with $[(NH_3)_5Os^{III/II}]$ on the ligands indicated (*50*). Elsewhere in this volume, Shepherd (*51, 52*) describes the binding of ruthenium complexes with edta-type ligands to double bonds in uracil and provides insight into the applications of such bonding in potential chemotherapeutic agents.

Ruthenium-Induced Reactivities on Nucleic Acids and Their Constituents

Autoxidation. A reaction unique to Ru^{III} is the base-catalyzed air oxidation of N7-coordinated nucleosides to 8-oxo-nucleosides when coordinated to $[(NH_3)_5Ru^{III}]$ or similar ions (*53*). This likely proceeds through hydroxyl attack at C8 followed by passing two sequential electrons to oxygen through the ruthenium as outlined in Scheme II. At low dissolved oxygen concentrations, the reaction is first-order in oxygen. The reaction is also first-order in the ruthenium complex and first-order in hydroxide. In air, the second-order rate constants (M^{-1} $s^{-1}/10^{-2}$) are 7.7, 6.6, 3.5, and 20 for L = inosine (Ino), guanosine (Guo), deoxyguanosine (dGuo), and 1-methylguanosine (1-MeGuo), respectively. The increased rate for L = 1-MeGuo arises from its inability to ionize at N1 in basic media. Consequently, this ligand does not bear an anionic charge and so is more conducive to hydroxide attack. Similarly, the rate for L = Ino is greater than for L = Guo as the former lacks the electron-donating amine. Likewise, the rate for L = Guo is greater than that for dGuo because the added oxygen on the sugar slightly depletes the aromatic ring of electron density. The relatively low values for $\Delta H^{\ddagger}$ and negative values for $\Delta S^{\ddagger}$ for this reaction with L= Ino and dGuo as listed in Table II are also consistent with an initial hydroxide attack (Scheme II).

N-Glycosidic Hydrolysis. Ruthenium(III) functions as a general acid in promoting hydrolysis of the glycosidic bond in $[(dG^7)(NH_3)_5Ru^{III}]$, ($t_{1/2} = 1.5$ days, 56 °C, pH 7) (*54*). Consequently, hydrolysis of DNA might be induced by π-acceptor cation coordination at N7, but this has not been observed with $[(NH_3)_5Ru^{III}]$ (*39*). However, increasing the reduction potential by substitution of a single pyridine does yield strand cleavage, which is thought to involve a Ru^{IV} intermediate. In basic media, compounds of the type *trans*-[L-κN⁷(py)-$(NH_3)_4Ru^{III}]$ [py = pyridine, and L = Guo, dGuo, 1 MeGuo, and 9-methylgua-

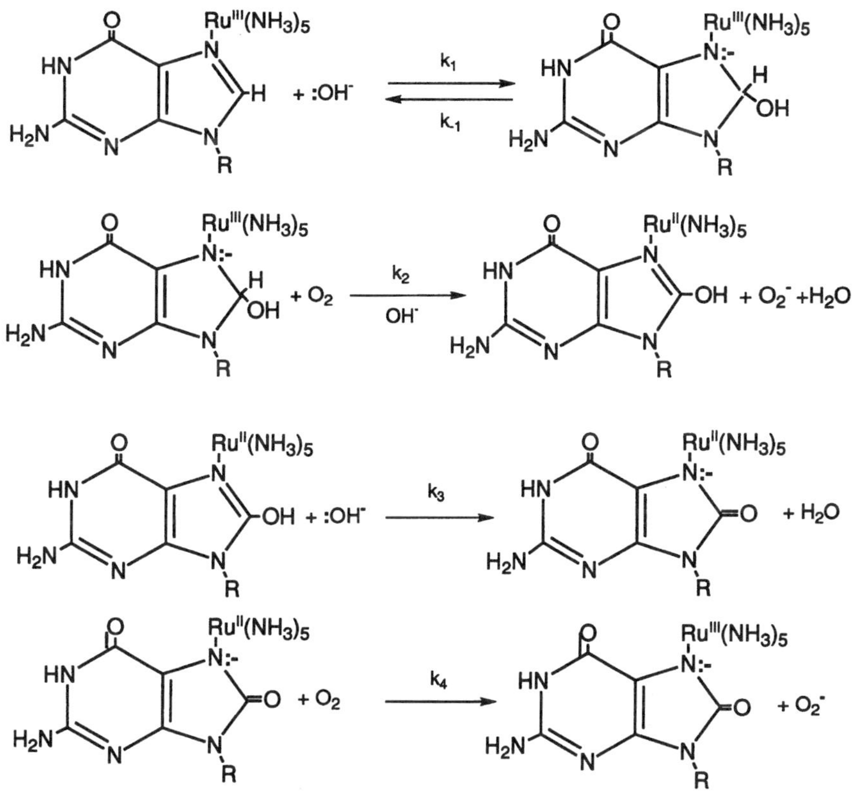

Scheme II. Likely mechanism for the autoxidation of [L(NH$_3$)$_5$Ru^{III}], where L = inosine, guanosine, deoxyguanosine, or 1-methylguanosine.

Table II. Activation Parameters for the Autoxidation of [L(NH$_3$)$_5$Ru^{III}]

L	$\Delta H^{\ddagger}$ (kcal/mol)	$\Delta S^{\ddagger}$ (kcal/mol K)	pH	Ref.
Ino	13 ± 1	−29 ± 5	10.27	53
dGuo	14 ± 1	−30 ± 5	12	76

NOTE: $\Delta H^{\ddagger}$ and $\Delta S^{\ddagger}$ are enthalpy and entropy of activation, respectively. L means ligand. Ino is inosine; dGuo is deoxyguanosine.

nine (9-MeGua)] undergo disproportionation following the rate law d[Ru^{II}]/dt $= k_o$[Ru^{III}] $+ k_1$[OH$^-$][Ru^{III}], yielding the corresponding complexes of Ru^{II} and, presumably, Ru^{IV}. Disproportionation rates are independent of [O$_2$] and, for L = Guo, are $k_o = 2.9 \times 10^{-4}$ s^{-1} and $k_1 = 6.4$ M^{-1} s^{-1}.

Since disproportionation of [py(NH$_3$)$_5$Ru^{III}] is approximately second-order in the metal ion (55), the rate law for the guanine complexes indicates that the limiting step in the dominant, hydroxide-dependent pathway is probably not electron transfer between Ru^{III}s, but rather the deprotonation of an ammine, which stabilizes Ru^{IV} while the pyridine stabilizes Ru^{II}. The relative second-order rate constants follow the series 1-MeGuo > Guo ~ dGuo > 9-MeGua >> Gua, so that ionization of the purine at N1 or N9 slows the rate. Activation parameters for k_1 (pH = 11.50) with L = Guo are $\Delta H^{\ddagger} = 17.4 \pm 0.8$ kcal/mol ($E_a = 18.0 \pm 0.8$ kcal/mol) and $\Delta S^{\ddagger} = 2.4 \pm 0.1$ cal/mol K. At pH 12, both dilute and concentrated solutions of $[Guo(Py)(NH_3)_4Ru^{III}]$ yielded approximately 50% $[Gua(Py)(NH_3)_4Ru^{III}]$ with ribose as the dominant organic product.

The appearance of $trans$-$[Gua(py)(NH_3)_4Ru^{III}]$ and free ribose indicates cleavage of the glycosidic bond, which probably occurs by general acid hydrolysis catalyzed by Ru^{IV} in a manner analogous to the Maxam–Gilbert G-reaction (56, 57). The rate of appearance of $trans$-$[Gua(py)(NH_3)_4Ru^{III}]$ following completion of the disproportionation reaction in the pH range 9.2–11.9 is complicated by purine loss, anation, and possibly redox reactions. The appearance of $trans$-$[8$-O-$Guo(py)(NH_3)_4Ru^{III}]$ as a minor product (6%), which occurred only when oxygen was present, and the lack of complexes of 8-oxoguanine indicate that purine oxidation is not likely to be responsible for the piperidine-independent strand scission. Scheme III summarizes a likely mechanism for disproportionation followed by general acid cleavage of the N-glycosidic bond.

Spectra of Ru^{III}-Imidazole Complexes

In general, Ru–N distances follow in the sequence $N_{amine} > N_{py} > N_{Im}$; however, the differences are frequently not statistically significant (58). Nevertheless, in concert with other evidence (59), this trend and the relatively low-energy ligand-to-metal charge transfers (LMCTs) evident in imidazole complexes of $[(NH_3)_5 Ru^{III}]$ suggest modest π-donor characteristics for the imidazole ligand. The energy of imidazole $\rightarrow Ru^{III}$ ($\pi \rightarrow d_{\pi}$) is strongly modulated by the π-donor ability of the ligand trans to it in $trans$-$[L(Im)(NH_3)_4 Ru^{III}]^{3+}$ as shown in Figure 1, which plots the energy of the LMCT (E_{LMCT}) versus the reduction potential ($E°$) for the complex and (inset) as a function of the Lever electrochemical parameter (E_L) (60) for the trans ligand, which reflects its π-donor ability.

The eclipsed conformation of the two imidazoles in $trans$-$[(Im)_2(NH_3)_4 Ru^{III}]^{3+}$ (and other complexes with $trans$-$[Im$–Ru^{III}–$Im]$ moieties) arises partly from steric effects (*see* Figure 2), which are minimized by having the two imidazoles in the same plane with the equatorial ligands bent away from this plane. Also favoring the eclipsed conformation is that it leads to a nondegenerate orbital ground state (d_{xz^1}, d_{yx^2}, $d_{x^2-y^2}$). The eclipsed imidazoles undergo π-interactions with a single orbital (d_{xz}), as opposed to the staggered conformation, which leads to a doubly degenerate ground state. Assuming noninteracting imidazoles, spin–orbit effects alone predict the eclipsed conformation to

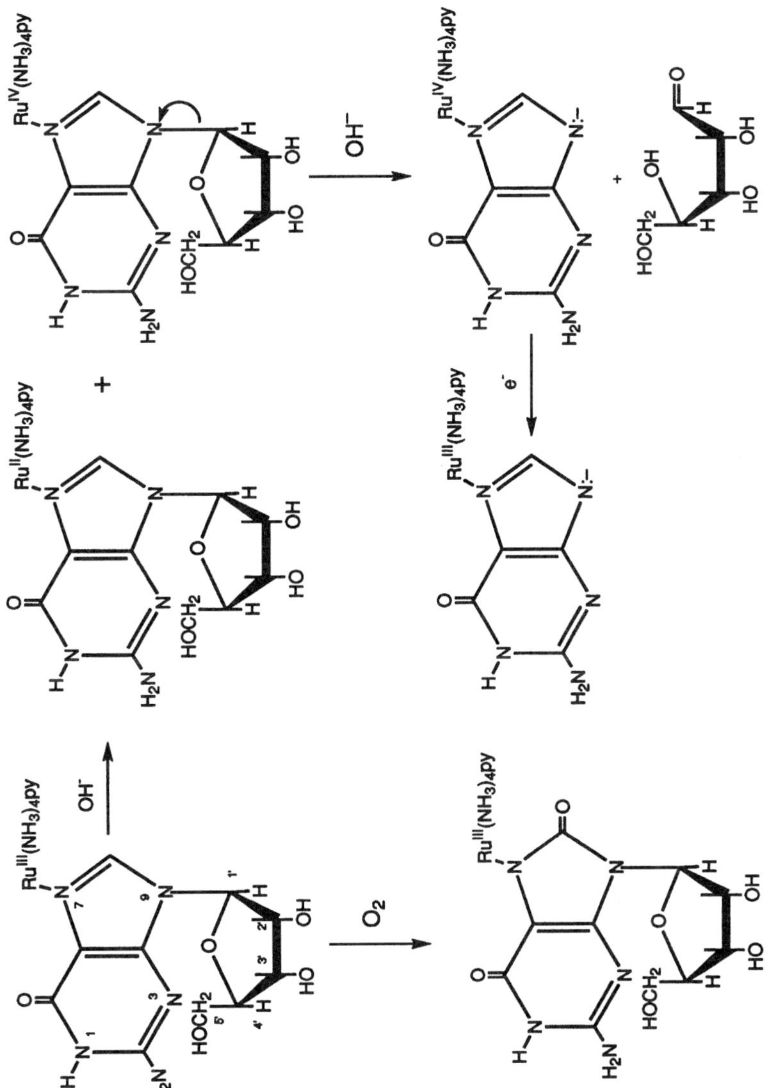

Scheme III. Possible mechanism for the disproportionation and hydrolysis of trans-[L(pyr)(NH₃)₄Ru^{III}], where L = inosine, guanosine, deoxyguanosine, or 1-methylguanosine.

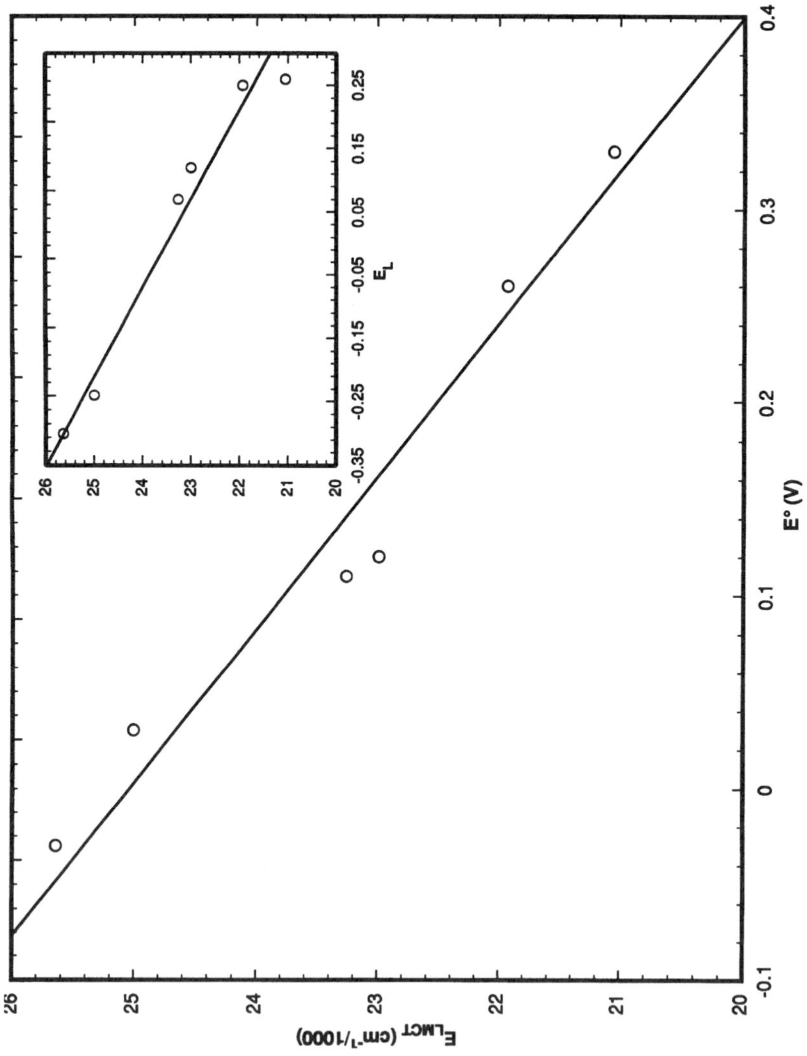

Figure 1. Plot of E_{LMCT} versus $E°$ and (inset) E_L.

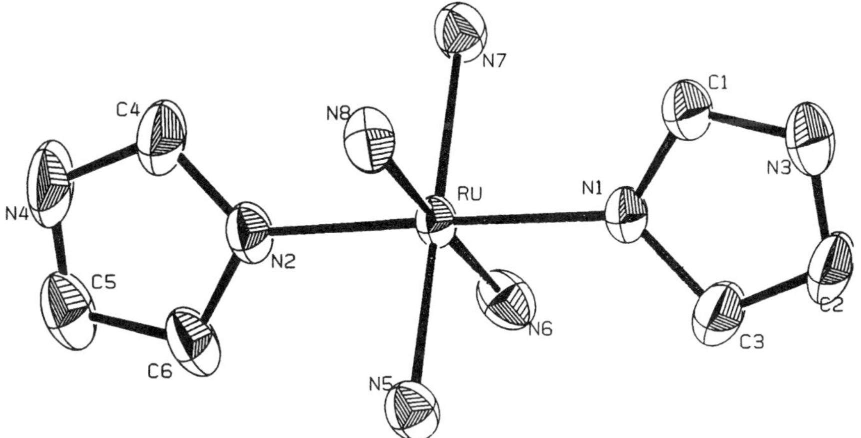

Figure 2. ORTEP (75) plot of trans-$[(Im)_2(NH_3)_4 Ru^{III}]^{3+}$.

be about 0.8 kJ mol^{-1} (70 cm^{-1}) lower in energy than the staggered conformation (58).

Spin–orbit effects generally mix the t_{2g} orbitals in ruthenium(III) complexes so that the paramagnetic field and EPR spectra are usually rhombic. Consequently, the ^{1}H NMR spectra are dramatically affected, but dipolar shifts can be estimated by the standard paramagnetic shift equation, provided the ligands are not rotating (58). For the monotonic series *trans*-[L(Im)(NH$_3$)$_4$-Ru^{III}], where L = isonicotinamide (Isn), py, NH$_3$, Im, Cl$^-$, and SO$_4^{2-}$, the NMR contact shift was shown to vary linearly with $E°$ for the complexes. Figure 3 shows a linear correlation between %d-character in the ground state and $E°$, which arises from π-ligand affects altering the relative energies of d$_{xz}$ and d$_{yz}$ such that their relative participations in the ground state are inverse to one another (58). Since each of the three g values is proportional to the fraction of d-character for each of the t_{2g} orbitals, respectively, there is a linear relationship between Δg_{12}, the difference between the two largest g values (proportional to %d$_{xz}$ and %d$_{yz}$, respectively) and %d$_{xz}$ (Figure 3, top inset) and also between Δg_{12} and $E°$ (Figure 3, bottom inset).

The UV–vis, electron-spin resonance, and ^{1}H NMR studies on this series of complexes have implications for long-range electron transfer studies in proteins. Ruthenium complexes are often bound to histidyl imidazole sites in electron transfer proteins as artificial redox sites. In studying the long-range electron transfer through the protein, the electrochemical potential of the ruthenium center is often modulated by varying the π-donor/acceptor abilities of its ligands. However, in doing so, the overlap of π-molecular orbitals involving the Ru and C5, which couples the imidazole into the protein backbone, should also be changed. Consequently, while it is not generally acknowledged,

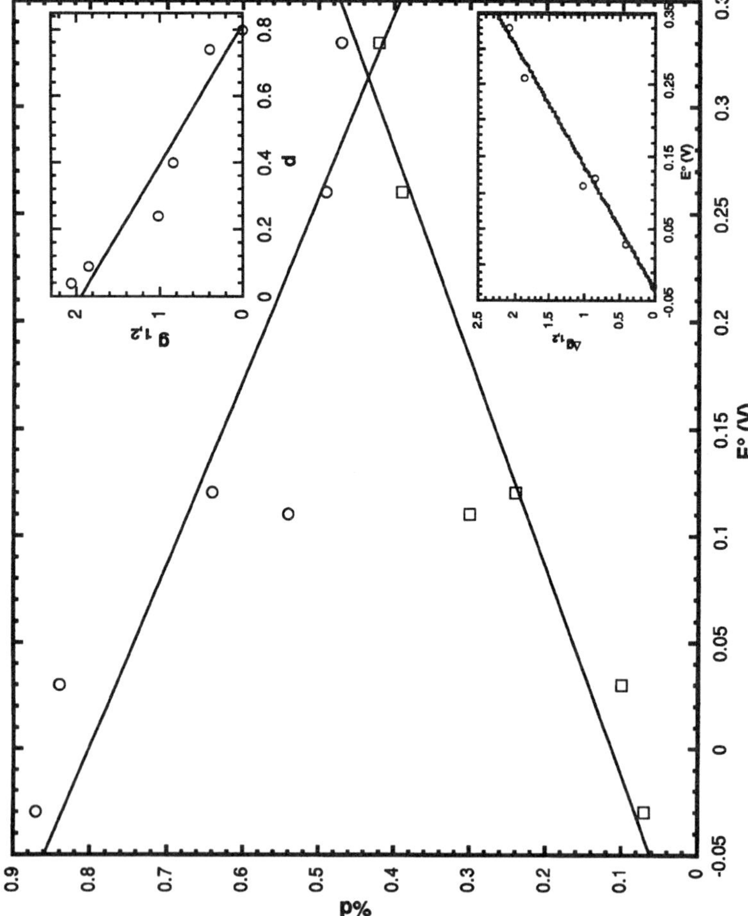

Figure 3. For the series trans-$[L(Im)(NH_3)_4 Ru^{III}]^{3+}$, where $L = Isn$, Py, NH_3, Im, Cl^- and SO_4^{2-}, plot of fraction of d-orbital character (open circles, d_{xz}; open squares, d_{yz}) in the ground state versus $E°$. Inset (top): plot of Δg_{12}, the difference between the two largest g values for each complex, versus $\%d_{xz}$. Inset (bottom): Δg_{12} versus $E°$.

two terms ($\Delta G°$ and the electronic coupling term, $\mathbf{H}_{AB}$) in the Marcus equation (eq 2) are simultaneously varied.

$$k_{ET} = \sqrt{\frac{\pi}{h^2\lambda k_B T}}\mathbf{H}_{AB}^2 \exp\left[-\frac{(\Delta G° + \lambda)^2}{4\lambda k_B T}\right] \tag{2}$$

For the homologous series of complexes, *trans*-[L(Im)(NH$_3$)$_4$ Ru^{III}]$^{3+}$, there is about 12% change in participation of the π-bonding d$_{xz}$ in the ground state for every 100-mV change in $E°$, and this participation varies over almost a factor of 2 from 87% for the anionic, π-donor ligands (Cl$^-$ and SO$_4^{2-}$) to 46% for the π-acceptor, isonicotinamide ligand. If the ordering of orbitals remains the same over the series, this translates into a factor of approximately 4 ($\mathbf{H}_{AB}^2$) in k_{ET} with a maximum variation of 6–8. The range of $\mathbf{H}_{AB}^2$ might be expected to increase as more π-donor or π-acceptor ligands are added.

Biological Activity of Di- and Trinuclear Ruthenium Complexes

Mixed-valent, μ-carboxylato complexes of the type μ-(RCO$_2$)$_4$ClRu$_2$ (R = CH$_3$, CH$_3$CH$_2$) have shown activity against P388 lymphocytic leukemia (*24*), presumably by a mechanism analogous to the structurally similar rhodium complexes. Dunbar et al. (*61*) have shown that 9-ethylguanine (9 EtGua) reacts with μ-(F$_3$CCO$_2$)$_4$(F$_3$CCO$_2$)Ru$_2$ to form the complex *cis*-[μ-(F$_3$CCO$_2$)$_4\mu$-(EtGua)Ru$_2$] in which the guanines bridge between the two Ru^{II} atoms in a N7–O6 head-to-tail fashion.

Mixtures containing the mixed-valent complex ruthenium red, [(NH$_3$)$_5$Ru^{III}ORu^{IV}(NH$_3$)$_4$ORu^{III}(NH$_3$)$_5$]$^{6+}$ (*62*), have been used for over a century as a cytological stain. Consequently "ruthenium red" (Ru-red) has been observed to exert a number of biological and biochemical effects (*63*). Crude preparations of this complex mixed with other di- and trinuclear species are easily prepared by refluxing RuCl$_3$ in ammonia (*64, 65*). For the past three decades these preparations have been used as a selective stain for mitochondria and muscle fibrils in both visible and electron microscopy (*66*). Ru-red and other complexes in the mixture bind to polyanions such as plant pectins and the protective mucopolysaccharide (mainly hyaluronic acid) coat that surrounds some tumor cells (*67*). Such binding probably accounts for the radioimaging of tumors with ^{103}Ru-red (*68*). Ru-red also inhibits tumor growth, which may arise from its ability to interfere with calcium metabolism (*69*). While it has been reported to interact with DNA (*70*) and would be expected to do so on the basis of charge interactions alone, no definitive studies have been performed. A dimeric complex related to Ru-red, μ-O-[(H$_2$O)$_2$(bpy)$_2$Ru^{III}]$_2^{4+}$ (bpy = 2,2'-bipyridyl), does bind to DNA at relatively low levels ([Ru]$_{DNA}$/[P]$_{DNA}$ = 0.02) and with low stereoselectivity, which may favor the $\Lambda\Lambda$ isomer. Association at this level also stabilizes the thermal melting of DNA by about 8 °C.

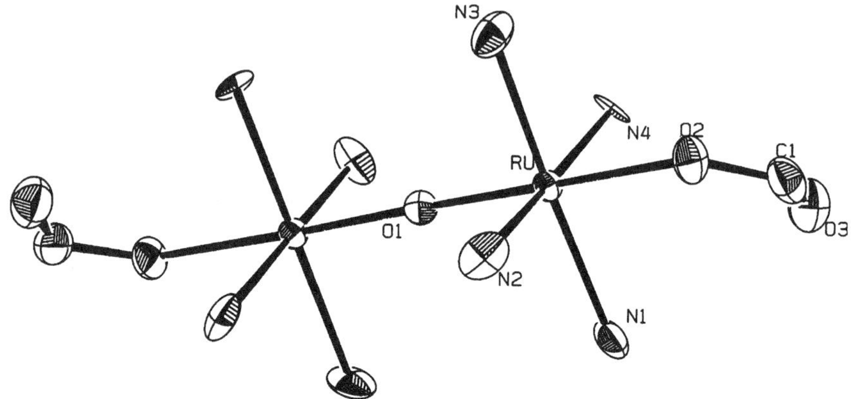

Figure 4. ORTEP (75) diagram of μ-O-[(HCO$_2$)(NH$_3$)$_4$Ru]$_2^{3+}$.

Irreversible thermal denaturation of DNA with this complex covalently bound has been taken as an indication of interstrand cross-links (*42*).

Staining with Ru-red blocks Ca^{2+} transport in a number of biological systems (*71*). However, it is an impurity in Ru-red preparations that is responsible for most of the inhibition of Ca^{2+} uptake in mitochondria (*72*). As shown in Figure 4, this has been identified as an oxygen-bridged, dimeric species, μ-O-[X(NH$_3$)$_4$Ru]$_2^{3+}$, where X = Cl$^-$ or OH$^-$ (*73, 74*). The tripositive charge of these complexes probably allows them to ion-pair and hydrogen-bond with anionic, carboxylate sites on the surface of proteins and possibly at Ca^{2+} binding sites as well. Substitution of the apical ligands may also result in covalent binding to carboxylate residues (*73*). No significant spectroscopic change occurs in the presence of DNA, and the dimeric complexes have not exhibited anticancer activity.

The ability to readily deform by bending along and rotation about the Ru–O–Ru axes probably allows these complexes to optimize their electrostatic and hydrogen-bonding interactions with biopolymers. The instability of these dimeric and trimeric complexes at neutral and high pH probably also gives rise to decomposition products in vivo, which may involve ammine loss and dissociation to dimers and possibly monomers (*73*).

Conclusion

While no complexes of ruthenium have yet reached clinical trials, ruthenium is probably second only to platinum in the number of its complexes that exhibit anticancer activity. An advantage over platinum is that it is generally less toxic. Moreover, the work of Henry Taube and his students has provided a broad range of ways to vary substitution rates, reduction potential, and other properties through changes in its ligands. The ability to control the movements of the ruthenium between binding sites on biological ligands, particularly DNA, may provide a unique way of modulating its biological activity.

Acknowledgments

Funding for this work was provided by National Institutes of Health Grant GM–26390.

References

1. Creutz, C.; Taube, H. *J. Am. Chem. Soc.* **1969**, *91*, 3988–3989.
2. Clarke, M. J.; Taube, H. *J. Am. Chem. Soc.* **1974**, *96*, 5413–5419.
3. Krentzien, H.; Clarke, M. J.; Taube, H. *Bioinorg. Chem.* **1975**, *4*, 143–151.
4. Clarke, M. J.; Taube, H. *J. Am. Chem. Soc.* **1975**, *97*, 1397–1403.
5. Clarke, M. J. Ph.D. Thesis, Stanford University, 1974.
6. Clarke, M. J.; Stubbs, M. *Met. Ions Biol. Syst.* **1995**, *32*, in press.
7. Barton, J. K. In *Bioinorganic Chemistry*; Bertini, I.; Gray, H.; Lippard, S. J.; Valentine, J. S., Ed.; University Science Books: Mill Valley, CA, 1994; pp 455–504.
8. Thorp, H. H.; Cheng, C.-C.; Goll, J. G.; Neyhart, G. A.; Welch, T. W.; Singh, P.; Thorp, H. H. *J. Am. Chem. Soc.* **1995**, *117*, 2970–2980.
9. Srivastava, S. C.; Mausner, L. F.; Clarke, M. J. In *Progress in Clinical Biochemistry and Medicine*; Clarke, M. J., Ed.; Springer-Verlag: Berlin, Germany, 1989; Vol. 10, pp 111–149.
10. Stemp, E. D. A.; Arkin, M. R.; Barton, J. K. *J. Am. Chem. Soc.* **1995**, *117*, 2375–2376.
11. Meade, T. J.; Kayyem, J. F. *Angew. Chem. Intl. Ed. Engl.* **1995**, *34*, 352.
12. Kelman, A. D.; Clarke, M. J.; Edmonds, S. D.; Peresie, H. J. *J. Clin. Hematol. Oncol.* **1977**, *7*, 274–288.
13. Yasbin, R. E.; Matthews, C. R.; Clarke, M. J. *Chem. Biol. Interact.* **1980**, *30*, 355.
14. Marx, K. A.; Kruger, R.; Clarke, M. J. *Mol. Cell. Biochem.* **1989**, *86*, 155–162.
15. Marx, K. A.; Seery, C.; Malloy, P. *Mol. Cell. Biochem.* **1989**, *90*, 37–95.
16. Bottomley, F. *Can. J. Chem.* **1977**, *55*, 2788.
17. Durig, J. R.; Danneman, J.; Behnke, W. D.; Mercer, E. E. *Chem. Biol. Interact.* **1976**, *13*, 287–294.
18. Keppler, B. K.; Rupp, W.; Juhl, U. M.; Endres, H.; Niebel, R.; Balzer, W. *Inorg. Chem.* **1987**, *26*, 4366–70.
19. Clarke, M. J. In *Metal Complexes in Cancer Chemotherapy*; Keppler, B. K., Ed.; VCH: Weinheim, 1993; pp 129–157.
20. Clarke, M. J. *Prog. Clin. Biochem. Med.* **1989**, *10*, 25–39.
21. van Vliet, P. M.; Sarinten, M. S.; Toekimin, S. M. S.; Haasnoot, J. G.; Reedijk, J.; Nováková, O.; Vrána, O.; Brabec, V. *Inorg. Chim. Acta* **1995**, *231*, 57.
22. Seelig, M. H.; Berger, M. R.; Keppler, B. K. *J. Cancer Res. Clin. Oncol.* **1992**, *118*, 195.
23. Berger, M. R.; Garzon, F. T.; Schmal, D.; Keppler, B. K. *Anticancer Res.* **1989**, *9*, 761–765.
24. Keppler, B. K.; Henn, M.; Juhl, U. M.; Berger, M. R.; Niebl, R.; Wagner, F. E. In *Ruthenium and Other Non-Platinum Metal Complexes in Cancer Chemotherapy*; Clarke, M. J., Ed.; Springer-Verlag: Heidelberg, Germany,1989; Vol. 14, pp 41–70.
25. Keppler, B. K.; Balzer, W.; Seifried, V. *Arzneim. Forsch.* **1987**, *37*, 770–1.
26. Keppler, B. K.; Hartman, M. *Metal Based Drugs* **1994**, *1*, 145–149.

27. Srivastava, S. C.; Richards, P.; Meinken, G. E.; Larson, S. M.; Grunbaum, Z. In *Radiopharmaceuticals*; Spencer, R. P., Ed.; Grune & Stratton, Inc.: New York, 1981; p 207.

28. Som, P.; Oster, Z. H.; Matsui, K.; Gugliemi, G.; Persson, B.; Pellettieri, M. L.; Srivastava, S. C.; Richards, P.; Atkins, H. L.; Brill, A. B. *Eur. J. Nucl. Med.* **1983,** *8,* 491.

29. Srivastava, S. C.; Mausner, L. F.; Clarke, M. J. In *Ruthenium and Other Non-Platinum Metal Complexes in Cancer Chemotherapy*; Clarke, M. J., Ed.; Springer-Verlag: Heidleberg, Germany, 1989; Vol. 10, pp 111–150.

30. Kratz, F.; Keppler, B. K.; Messori, L.; Smith, C.; Baker, E. N. *Metal Based Drugs* **1994,** *1,* 169–173.

31. Kratz, F.; Hartmann, M.; Keppler, B.; Messori, L. *J. Biol. Chem.* **1994,** *269,* 2581–2588.

32. Keppler, B. K. In *Metal Complexes in Cancer Chemotherapy*; Keppler, B. K., Ed.; VCH: Weinheim, Germany, 1993; pp 187–220.

33. Clarke, M. J.; Bitler, S.; Rennert, D.; Buchbinder, M.; Kelman, A. D. *J. Inorg. Biochem.* **1980,** *12,* 79–87.

34. Miklavcic, D.; Sersa, G.; Novakovic, S. *J. Bioelectr.* **1990,** *9,* 133.

35. Vaupel, P.; Kallinowski, F.; Okunleff, P. *Cancer Res.* **1989,** *49,* 6449–6465.

36. Dhubhghaill, O. M. N.; Hagen, W.; Keppler, B. K.; Sadler, J. *J. Chem. Soc. Dalton Trans.* **1994,** 3305–3311.

37. Laliberté, J. F.; Sun, I. L.; Crane, F. L.; Clarke, M. J. *J. Bioenerg. Biomembr.* **1987,** *19,* 69–81.

38. McNamara, M.; Clarke, M. J. *Inorg. Chim. Acta* **1992,** *195,* 175–185.

39. Clarke, M. J.; Jansen, B.; Marx, K. A.; Kruger, R. *Inorg. Chim. Acta* **1986,** *124,* 13–28.

40. Wlotzka, B.; Starck, D.; Qu, Y.; Farrell, N.; Orgel, L. E., Virginia Commonwealth University, personal communication, 1995.

41. Alessio, E.; Balducci, G.; Calligaris, M.; Costa, G.; Attia, W. M. *Inorg. Chem.* **1991,** *30,* 609–618.

42. Grover, N.; Welch, T. W.; Fairley, T. A.; Cory, M.; Thorp, H. H. *Inorg. Chem.* **1994,** *33,* 3544–3549.

43. Fruhauf, S.; Zeller, W. J. *Cancer Res.* **1991,** *51,* 2943.

44. Clarke, M. J. *J. Am. Chem. Soc.* **1978,** *100,* 5068–5075.

45. Clarke, M. J. *Inorg. Chem.* **1980,** *19,* 1103–1104.

46. Graves, D. J.; Hodgson, D. K. *J. Am. Chem. Soc.* **1979,** *101,* 5608–5612.

47. La Chance-Galang, K.; Clarke, M. J., unpublished.

48. Clarke, M. J. *Inorg. Chem.* **1977,** *16,* 738–744.

49. Kastner, M. E.; Coffey, K. F.; Clarke, M. J.; Edmonds, S. E.; Eriks, K. *J. Am. Chem. Soc.* **1981,** *103,* 5747–5752.

50. Johnson, A.; O'Connell, L. A.; Clarke, M. J. *Inorg. Chim. Acta* **1993,** *210,* 151–157.

51. Shepherd, R. E.; Zhang, S.; Lin, F.-T.; Kortes, R. A. *Inorg. Chem.* **1992,** *31,* 1457.

52. Shepherd, R. E.; Zhang, S. *Inorg. Chim. Acta.* **1992,** *191,* 271.

53. Gariepy, K. C.; Curtin, M. A.; Clarke, M. J. *J. Am. Chem. Soc.* **1989,** *111,* 4947–52.

54. Clarke, M. J.; Morrissey, P. E. *Inorg. Chim. Acta* **1984,** *80,* 69–70.

55. Rudd, D. P.; Taube, H. *Inorg. Chem.* **1971,** *10,* 1543–1544.

56. LaChance-Galang, K. J.; Zhao, M.; Clarke, M. J. *Inorg. Chem.* **1996,** *35,* 6021–6026.

57. Maxam, A. M.; Gilbert, W. *Methods Enzymol.* **1980,** *65,* 499–560.

58. LaChance-Galang, K. J.; Doan, P. E.; Clarke, M. J.; Rao, U.; Yamano, A.; Hoffman, B. *J. Am. Chem. Soc.* **1995,** *117,* 3529–3538.

59. Elliott, M. G.; Shepherd, R. E. *Inorg. Chem.* **1987,** *26,* 2067–73.

60. Lever, A. B. P. *Inorg. Chem.* **1990,** *29,* 1271–1285.

61. Dunbar, K. R.; Matonic, J. H.; Saharan, V. P.; Crawford, C. A.; Christou, G. *J. Am. Chem. Soc.* **1994**, *116*, 2201–2202.
62. Carrondo, M. A.; Griffith, W. P.; Hall, J. P.; Skapski, A. C. *Biochim. Biophys. Acta* **1980**, *627*, 332–334.
63. Clarke, M. J. *Met. Ions Biol. Syst.* **1980**, *11*, 231–283.
64. Fletcher, J. M.; Greenfield, B. F.; Hardy, C. J.; Scargill, D.; Woodhead, J. L. *J. Chem. Soc. A* **1961**, 2000–2007.
65. Luft, J. H. *Anat. Rec.* **1971**, *171*, 347–368.
66. Hirabayashi, Y.; Sakagami, T.; Yamada, K. *Acta Histochem. Cytochem.* **1990**, *23*, 165–175.
67. Oberc-Greenwood, M. A.; Muul, L. M.; Gately, M. K.; Kornblith, P. L.; Smith, B. W. *J. Neuro Oncol.* **1986**, *3*, 387–396.
68. Anghileri, L. H. *Strahlentherapie* **1975**, *149*, 173–175.
69. Anghileri, L. J.; Marchal, C.; Matrat, M.; Crone-Escanya, M. C. *Neoplasma* **1986**, *33*, 603–608.
70. Murano, E.; Paoletti, S.; Cesaro, A. *Analyt. Biochem.* **1990**, *187*, 120–124.
71. Hurley, T. W. *Am. J. Physiol.* **1988**, *23*, 621–627.
72. Reed, K. C.; Bygrave, F. L. *FEBS Lett.* **1974**, *46*, 109–114.
73. Emerson, J.; Clarke, M. J.; Ying, W.-L.; Sanadi, D. R. *J. Am. Chem. Soc.* **1993**, *115*, 11799–11805.
74. Ying, W.–L.; Emerson, J.; Clarke, M. J.; Sanadi, D. R. *Biochemistry* **1991**, *30*, 4949–4952.
75. Johnson, C. K. *ORTEPII: Report ORNL-5138.;* Oak Ridge National Laboratory: Oak Ridge, TN, 1976.
76. LaChance-Galang, K. Ph.D. Thesis, Boston College, 1995.
77. Clarke, M. J. In *Inorganic Chemistry in Biology and Medicine*; Martell, A. E., Ed.; American Chemical Society: Washington, DC, 1980; Vol. 190, pp 157–180.
78. Keppler, B. K.; Rupp, W. *J. Cancer Res. Clin. Oncol.* **1986**, *111*, 166–168.
79. Vilaplana, R. A.; Gonzalez-Vilchez, F.; Ruiz-Valero, C. *Inorg. Chim. Acta* **1994**, *224*, 15.
80. Bryan, D. M.; Pell, S. D.; Kumar, R.; Clarke, M. J.; Rodriguez, V.; Sherban, M.; Charkoudian, J. *J. Am. Chem. Soc.* **1988**, *110*, 1498–1506.
81. Mestroni, G.; Alessio, E.; Calligaris, M.; Attia, W. M.; Quadrifoglio, F.; Cauci, S.; Sava, G.; Zorzet, S.; Pacor, S.; Monti-Bragadin, C.; Tamaro, M.; Dolzani, L. In *Ruthenium and Other Non-Platinum Metal Complexes in Cancer Chemotherapy*; Clarke, M. J., Ed.; Springer-Verlag: Heidelberg, Germany, 1989; Vol. 10, pp 74–88.
82. Pacor, S.; Sava, G.; Ceschia, V.; Bregant, F.; Mestroni, G.; Alessio, E. *Chem Biol. Interact.* **1991**, *78*, 223.
83. Sava, G.; Pacor, S.; Mestroni, G.; Alessio, E. *Anti-Cancer Drugs* **1992**, *3*, 25–31.
84. Armor, J., Boston University, personal communication, 1970.

Ru(II)–Polyaminopolycarboxylate Complexes for Improved DNA Probes

Rex E. Shepherd*, Ya Chen, Songsheng Zhang, Fu-Tyan Lin, and Richard A. Kortes

Department of Chemistry, University of Pittsburgh, Pittsburgh, PA 15260

The potential use of Ru^{II} polyaminopolycarboxylates (Ru^{II}–pacs) to bind at the C5–C6 bonds of pyrimidine nucleobases of DNA and, as a binuclear derivative, to span the major groove of DNA is discussed. A complex that illustrates this potential chemotherapeutic use is presented in studies of $[Ru_2(ttha)(DMU)_2]^{2-}$ ($ttha^{6-}$ = triethylenetetraminehexaacetate; DMU = 1,3-dimethyluracil). Factors in promoting η^2 attachments that will dearomatize diazines are examined in studies of $[Ru^{II}(hedta)(H_2O)]^-$ in coordination with pyrimidines (pyrimidine, 4-methylpyrimidine, and 4,6-dimethylpyrimidine) and pyridazine (pyd). Rapid coordination of pyrimidines and pyridazine occurs at the N1 position. A further migration to $\eta^2(1,2)$, $\eta^2(1,6)$, and $\eta^2(5,6)$ locations occurs slowly for pyrimidines. The η^2 attachments are described by ^1H NMR assignments. Coordination at N1 of pyd is followed by an internal nucleophilic substitution of the enhanced nitrogen base at N2. The asymmetrically coordinated bidentate $[Ru(hedta)(pyd)]^-$ complex readily forms the bis derivative by disrupting the strained Ru^{II}–pyd bonding. Structural factors that optimize Ru^{II}–pac coordination for olefins or pyrimidine nucleobases are studied with $[S,S-Ru(Me_2edda)(H_2O)_2]$. The best Ru^{II}–pac for η^2 DNA coordination would have a metallo headgroup that utilizes the coordination of the asymmetric ligand N,N-ethylenediaminediacetate for the Ru^{II} center.

DURING THE PAST DECADE there has been considerable attention devoted to the development of novel small molecules that recognize DNA sites in a sequence-specific fashion (*1–20*) or that form selective covalent attachments to DNA (*1, 11–13, 17–19*). Such agents may be useful in the design of novel

---

*Corresponding author

chemotherapeutics and as tools for biotechnology (*1–7, 11, 21–25*). Stable covalent attachments of metallodrugs and coordination complexes have almost always involved Pt(II) or Pd(II), Ru(II)/(III), and Rh(III) complexes with bonding via the N7 sites of guanosine bases (and to a lesser extent N7 of adenosine units) (*1, 11, 18, 19, 23, 24, 26*). The N7 sites of G and A project into the major groove of B-DNA, making these positions the most available nitrogen bases to exterior reagents. Shape-selective agents that utilize noncovalent, intercalative binding in the major groove of B-DNA by large, planar aromatic moieties including phen (1,10-phenanthroline), phi (9,10-phenanthrenequinone diimine), or dppz (dipyridophenazine) donor ligands attached to Ru(II), Rh(III), and Co(III) chromosphores have been heavily studied by Barton and colleagues (*27–31*) and by Thorp and colleagues (*12, 14, 20*) as selective DNA binding and cleavage agents. The selectivity of these intercalative associations have been improved by attachments of molecular-recognition-enhancing side chains of oligonucleotide sequences and groove-binding peptide sequences (*28, 29*). Binding to DNA can also be enhanced by H-bonding and structural features of ancillary ligands at the metal chromophore, particularly with aliphatic amines (*27, 30*).

Covalent attachment of metallodrugs at sites other than N7 loci on the DNA helix offers some exciting promise for chemotherapeutic agents. The distortions of the DNA helix by derivatization at positions other than N7 of G bases, and the inherent difference in electronic disruptions that differ from N7 attachment, may lead to an altered chemotherapeutic response. Ideally, some tumor cell lines that are cisplatin-resistant might be responsive to reagents that produce a different extent of kinking, DNA base-pair disruption, or release of a regional DNA binding protein upon coordination than is induced by N7 derivatizations with typical Pt(II) antitumor agents. Consequently, reagents that can attach at sites other than N7 of G or A bases in the major groove are desirable. Also, highly hydrophobic metallo headgroups containing polypyridine or donors such as phi and dppz, useful in designing tools of molecular biology, may have limitations as metallodrugs due to transport problems for these complexes, either in the circulatory system or across cell walls. Therefore, there may be advantages in obtaining agents that lack the necessity of intercalative interactions with DNA as it pertains to potential chemotherapeutic use. Indeed, cisplatin and its most active antitumor derivatives as drugs generally lack large hydrophobic ligands, which reduce solubility in the cellular medium.

Our group has discovered an alternative binding association for Ru(II) agents with pyrimidine nucleobases that provides a novel potential site of metallation on a DNA helix other than N7 of G bases (*32*). Using small nucleosides and related pyrimidine bases, we observed that the ruthenium(II) polyamino-polycarboxylate (Ru^{II}–pac), [Ru^{II}(hedta)]$^-$ (hedta^{3-} = *N*-hydroxyethylethylene-diaminetriacetate), will bind at either the normal N3 coordination position of pyrimidines related to cytidine and uridine, or as an η^2 unit to the C5–C6 olefinic site of pyrimidines (*32–34*). The same C5–C6 bond region, which projects into the major groove of B-DNA for C and T base residues, is the site of OsO_4 addition reactions for DNA staining and heavy atom labeling purposes (*35–38*). But these adducts are unstable in the absence of 2 M pyridine or 0.2 M 2,2′-bipyridine (bpy) (*35–38*), a condition unsuitable for chemotherapies or most molecular biology procedures. The η^2 complexation of pyrimidine nucleobases at the C5–C6 bond by [Ru^{II}(hedta)]$^-$ occurs with all the properties attributed to the partial dearomatization of the N-heterocyclic ring, similar to the dearomatizing influence of (NH$_3$)$_5$Os^{II} with arenes (*39–42*).

In this chapter we discuss several factors that are involved in promoting η^2 attachment of the Ru^{II}–pac complexes with pyrimidines and other closely related diazines. During the study of [Ru^{II}(hedta)]$^-$ complexes of simple pyrimidines that lack electron-withdrawing exo oxygens of the pyrimidine nucleobases, we have observed the migration of [Ru^{II}(hedta)]$^-$ from the initial N1 ring attachment to η^2 coordination (*43, 44*) by means of both ^{1}H NMR and electrochemical methods. Three different η^2 complexes are detected for pyrimidine by ^{1}H NMR: $\eta^2(1,2)$, $\eta^2(1,6)$, and $\eta^2(5,6)$ (*43*). The $\eta^2(5,6)$ or olefinic coordination is similar to the (NH$_3$)$_5$Os^{II}L^{2+} species, L = 2,6-lutidine (*45*) and L = pyrrole (*46*), observed previously by Taube and co-workers (*32–34*), and analogous to the $\eta^2(5,6)$ coordination of [Ru^{II}(hedta)]$^-$ with pyrimidine nucleobases. The $\eta^2(1,2)$ and $\eta^2(1,6)$ modes of coordination have not been observed previously for stable Ru^{II} complexes of N-heterocycles. However, the existence of $\eta^2(1,2)$ species were proposed 16 years ago by Durante and Ford (*47*) as flash photolysis intermediates in the photochemistry of [(NH$_3$)$_5$Ru(pyridine)]$^{2+}$ upon metal-to-ligand charge transfer (MLCT) excitation. Another metastable η^2-bonded N-heterocyclic intermediate has been proposed by Creutz and co-workers (*48*) to account for the amido-N to pyridine-N isomerism of [(NH$_3$)$_5$Ru^{II}(isonicotinamide)]$^{2+}$ upon electrochemical reduction of the amido-bound Ru^{III} analogue.

The design of a possible binuclear Ru(II) major groove-spanning chelate that opts for η^2 attachments as interstrand cross-links is modeled in this chapter by the [Ru$_2$^{II}(ttha)(DMU)$_2$]$^{2-}$ complex (ttha^{6-} = triethylenetetraminehexaacetate; DMU = 1,3-dimethyluracil). In this complex, the DMU ligands represent the potential sites of η^2 derivatization along the parallel strands of the B-DNA helix, and the [Ru$_2$^{II}(ttha)]$^{2-}$ chromophore provides the interstrand cross-link. The concept of the major groove-spanning chelates is an extension of the findings of Farrell and co-workers (*49–53*) that bis-Pt(II) diamine chelates, which are tethered with a distance of four to six methylene spacers, provide N7-coordinating Pt(II) antitumor agents that do not exhibit acquired platinum resistance overtime, and which are also active against some cisplatin-resistant tumor cell lines. Extensions of Farrell's theme have already been carried out for Pt(II) and Pd(II) complexes by ourselves (*54, 55*) and by studies of Taylor (*56*) and Beck (*57*).

Ru$^{II/III}$–pacs are known to enter and localize in some tumor cells with resultant antitumor activity (*22, 58–60*). Ruthenium complexes often have exhibited comparable activity to the platinum-based drugs, yet with fewer toxic side effects (*61, 62*). These ruthenium-based agents include several ruthenium ammines (*59–62*), *cis*- or *trans*-[Ru(dmso)$_4$Cl$_2$] (*19, 63–65*), and *trans*-[Ru(imidazole)$_2$Cl$_4$]$^-$, the latter known as ICR (*66–71*). ICR has shown inhibition of colorectal tumors (unaffected by cisplatin) and action on P-388 leukemia that is superior to cisplatin or 5-fluorouracil (*66–71*). Thus Ru^{II}–pac reagents that exploit η^2 coordination may provide a novel, active class of chemotherapeutic agents with toxicity factors that are lower than Pt(II) drugs. Recently [Ru(H$_2$cydta)Cl$_2$]·2 H$_2$O (cydta^{4-} = cyclohexanediamine-*N,N,N′,N′*-tetracetic acid) was synthesized and shown to possess antitumor activity against Ehrlich ascitic tumors, P-388 leukemia cells, and MX-1 transplanted mammary carcinomas (*72a*). The mode of interaction of [Ru(H$_2$cydta)Cl$_2$], a Ru^{IV} complex, with the tumor targets is not yet known. The complex may serve as a prodrug for lower oxidation states of Ru^{II} or Ru^{III} within the cell. [Ru(H$_2$cydta)Cl$_2$]·2H$_2$O is, however, a well-defined example of a Ru–pac with antitumor action (*72a*).

Finally, in the development of what factors are involved in the dearomatization of N-heterocycles, we have observed that although pyridazine should have a barrier comparable to the pyrimidines for an η^2 rearrangement from a bound N1 pyridazine ligand to another η^2 site in the molecule, such a migration fails to occur. Instead, an enhanced nucleophilic displacement of a cis carboxylato functionality of the hedta^{3-} ligand chelate occurs. This is facilitated by means of the improved basicity of the second nitrogen, adjacent to the coordinated N1 position. The substitution process generates a bidentate pyridazine that does not undergo further migration to form η^2 derivatives. This points to the very special set of circumstances that promotes η^2 complexation with pyrimidines and pyrimidine nucleobases with Ru^{II}–pacs and sets limits on which nucleobases of DNA will be dearomatized by Ru^{II}–pacs. This may prove to be a fortunate aspect of selectivity in the action of Ru^{II}–pacs toward DNA.

Experimental Methods

Reagents and Transfer Methods. The ligands pyrimidine (pym), 4-methylpyrimidine (4-CH$_3$pym), 4,6-dimethylpyrimidine (Me$_2$pym), and pyridazine (pyd) were obtained from Aldrich and used as supplied. ^{1}H NMR spectra of the free ligands confirmed high-purity. Ar and N$_2$ gases were passed through Cr(II) scrubbing towers followed by a deionized water rinse tower to remove traces of O$_2$ from the inert gas stream and to saturate the blanketing gases with H$_2$O vapor to avoid dehydration of aqueous samples. When samples were prepared in D$_2$O, the tank Ar gas was used as delivered from Air Products. This avoids water vapor enriching the D$_2$O solvent in unwanted amounts of HOD.

Na[Ru^{II}(hedta)(H$_2$O)]·4H$_2$O was prepared as described previously (*72b, 73*) following modifications of Shimizu's procedure (*74*). The starting ruthenium complex in D$_2$O was treated with Zn/Hg in Ar-purged 10-mL glass round bottom flasks, sealed with rubber septa. Stirring was maintained with rice-sized magnetic stir bars. Reduced solutions were transferred via syringe needles and syringe tubing into a separate 10-mL flask along the Ar stream. The reaction flask contained weighed amounts of the ligand to achieve desired 0.50:1.00, 0.80:1.00, 1.00:1.00, or 2.00:1.00 ligand:[Ru^{II}(hedta)]$^-$ stoichimetries. For ^{1}H NMR experiments the solvent was D_2O, initially adjusted to pD ~ 3 with DCl solution. After mixing and reacting for an appropriate time, depending on the experiment, the Ru^{II}(hedta)–pyrimidine or Ru^{II}(hedta)–pyridazine solution was further transferred by syringe tubing into an Ar-purged NMR tube. After filling with a solution, the septum-sealed NMR tubes were further wrapped with parafilm to provide an additional barrier against air diffusion into the tube.

Instrumentation. Spectra on samples prepared as just described were then obtained as a function of time on Bruker AF300 NMRs following standard procedures reported previously (*32–34, 72, 73*).^{1}H NMR spectra were obtained at 70.46 kG at 300.13 MHz in D_2O referenced against HOD (4.80 ppm) and DSS (0.00 ppm). Assignments were assisted for ^{1}H spectra by standard integration and decoupling procedures. Assignments were further checked by obtaining spectra on fresh samples studies by ^{1}H–^{1}H and ^{1}H–^{13}C correlation experiments on a Bruker 500 NMR–LD12 instrument. Confirmation of the given assignments and additional details are reported elsewhere (*43*). The pD of samples is known to rise to ca. pD 6 over Zn/Hg. For samples at lower pD values, adjustment was made by adding Ar-purged 1.00 M DCl via a syringe to the NMR tube, calculating the pD value from the volume dilution with the sample.

Electrochemical studies were performed on solutions of Ar- or N_2-purged 0.100 M NaCl plus the desired complex at 22 °C. Measurements were made using an IBM 225 electrochemical analyzer for cyclic voltammetry and differential-pulse polarography modes. The standard three-electrode configuration was utilized with a glass-carbon working electrode, a Pt-wire auxiliary, and a saturated sodium chloride calomel electrode (SSCE) reference (*33, 75*). The sweep rates were 50 mV/s for cyclic voltammetry (CV) and 40 mV/s for differential pulse polarography (DPP). Confirmation of the percentages of the various N1 and η^2 forms (*see* text) were made by means of comparison of the ratios of DPP areas and the respective integration ratios of ^{1}H and ^{13}C NMR lines for the several species present. A flowing N_2 stream was maintained above the solution for electrochemistry to protect the samples against air oxidation for samples maintained for up to 24 h. For longer times, samples were prepared in Ar-purged glass bulbs sealed with septa as described. Aliquots of Ar-protected samples were taken at convenient times and diluted in Ar-purged 0.100 M NaCl for electroanalytical procedures on freshly transferred samples.

Results and Discussion

Toward a Major-Groove-Spanning, η^2-Coordinated DNA Cross-Link. Prior studies of Ru^{II}(pacs) by Matsubara and Creutz (*76*) and by Diamantis and Dubrawski (*77, 78*) have shown that [Ru^{II}(edta)(H_2O)]$^{2-}$ forms complexes with a wide range of common π-acceptor ligands (L = N_2, CO, RCN, pyridines, and pyrazines) as shown in Figure 1. N-heterocycles stabilize the Ru^{II} form such that typical Ru$^{II/III}$(edta)L$_n$ (n = 1 or 2) half-wave potentials

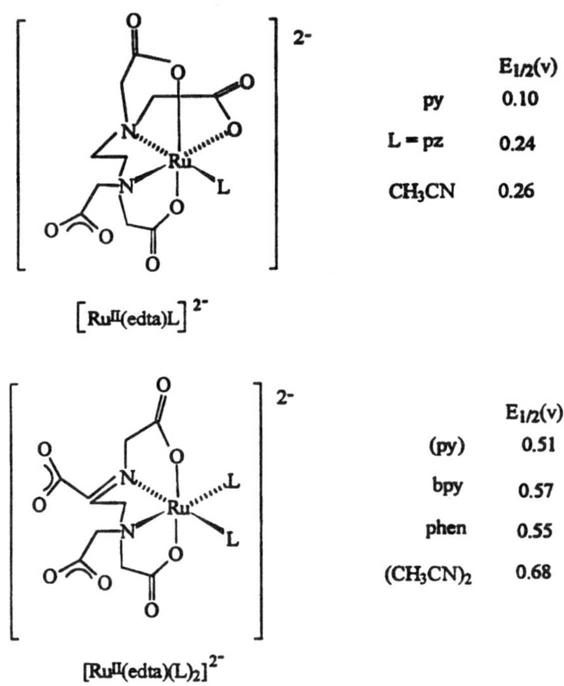

Figure 1. Mono and bis π-acceptor complexes of $[Ru^{II}(edta)]^{2-}$ (edta^{4-} = ethylenediaminetetraacetate). [Data are from references 76 (top) and 77 (bottom).]

$(E_{1/2})$ occur at ca. 0.10–0.25 for pyridine donors ($n = 1$) (76) and at ca. 0.51–0.55 for the bis complexes of pyridine donors (77). Diamantis and Dubrawski also established that the bis-substituted N-heterocyclic complexes of $[Ru^{II}(edta)]^{2-}$ form with the pyridine donors in the same plane as the ethylenediamine backbone N-donors of edta^{4-}. Thus, displacement of a more-strained in-plane glycinato donor (79, 80) occurs in forming the bis derivatives, $[Ru^{II}(edta)L_2]^{2-}$, which leave carboxylato donors in the axial positions. This substitution pattern is also followed by $[Ru^{II}(hedta)(H_2O)]^-$ of this chapter, and for S,S-$[Ru^{II}(Me_2edda)(H_2O)_2]$ (81).

When $[Ru^{II}(hedta)(H_2O)]^-$ is combined with pyrimidine nucleosides, the η² isomers, with coordination at the C5–C6 olefinic site, are observed for uridines and cytidines in addition to the complexation of $[Ru^{II}(hedta)]^-$ at the N3 positions (32–34) as shown in Figure 2. A more positive $E_{1/2}$ value at ca. 0.60–0.80 is observed for the η²-bound species, with the value dependent on the number and nature of the electron-withdrawing groups that are α or attached to the double bond (32–34). Electron-withdrawing halogen (X) units favor the η² coordination mode, inducing up to 88% of the $[Ru^{II}(hedta)]^-$ complexes of the halouracils bound in the η² isomer (12% bound as the N3 isomer)

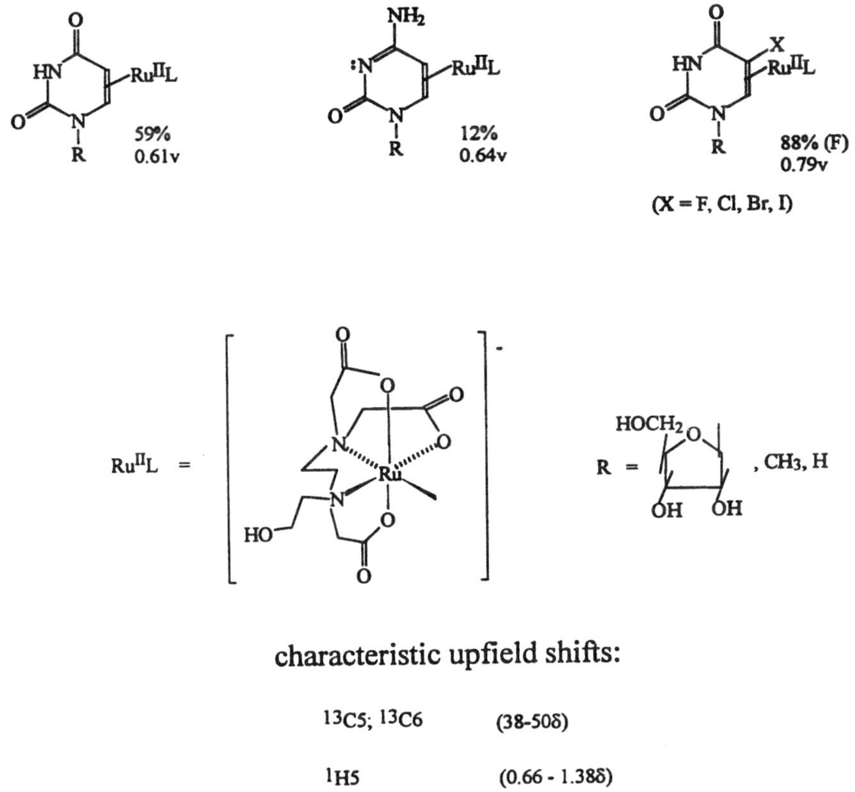

characteristic upfield shifts:

Figure 2. η^2-[Ru^{II}(hedta)]$^-$–nucleobase complexes ((hedta^{3-} = N-hydroxyethylethylenediaminetriacetate). (Data are from references 32 and 33.)

(33). The η^2 complexes exhibit significant ^{13}C upfield shifts of 38–50 ppm relative to the free ligand values of the C5 and C6 carbons; likewise upfield ^{1}H NMR shifts of 0.66–1.38 ppm for the H5 proton of ligands related to uridine and cytidine are observed for the η^2 isomers. These influences are similar to the ^{13}C and ^{1}H NMR shifts relative to free ligand values when olefins are coordinated to [Ru^{II}(hedta)]$^-$ or [Ru(NH$_3$)$_5$]$^{2+}$ (73, 82a) and have served as a signature of η^2 coordination at olefinic portions of dearomatized N-heterocycles. We have recently shown, however, that the presence of an alpha NR(R') functionality, adjacent to the olefin bond, will promote a downfield ^{1}H NMR shift upon formation of an η^2 linkage with [Ru^{II}(hedta)]$^-$ (82b). In the cases we have studied to date, the ^{13}C NMR reveals upfield shifts of +40 to +80 ppm for the same carbons whose hydrogens experience downfield shifts. Thus the local electronic environment may produce effects that have not been observed previously for derivative of more simple olefins. Therefore, assignments based on one parameter such as proton shifts alone must be viewed with caution (43, 82b).

If one wishes to exploit the η^2 coordination mode of Ru^{II}–pacs with pyrimidine nucleobases for an eventual chemotherapeutic role that avoids the acquired metallodrug resistance problem, it would be advantageous to utilize the multiple-site coordination offered by a binuclear system. An example using N7 coordination is shown by the prior studies of Farrell and co-workers (49–53) with major-groove-spanning Pt(II)-diamine binuclear complexes. A view of the interstrand cross-links of these derivatives with Pt(II) binding at N7 of adjacent G bases along opposite strands of a DNA helix is shown pictorially in Figure 3.

Some related binuclear Pt(II) complexes that have aminocarboxylate ligand headgroups tethered at the equivalent distance of 6 or 8 methylene units are shown in Figure 4 for the $[Pt^{II}_2(hdta)X_2]$ and $[M_2(egta)X_2]$ (M = Pt^{II} or Pd^{II}; X = Cl^- or inosine) chelates that have been prepared recently in our laboratories (54, 55). These serve as aminocarboxylate analogues of Farrell's diamine-based complexes (49–53) and potential antitumor agents themselves. However, we have used them as models of what might be achieved using Ru^{II}–pacs that are tethered at suitable distances to afford interstrand DNA cross-links.

The binuclear Ru^{II}–pac, $[Ru^{II}_2(ttha)(H_2O)_2]^{2-}$, has been studied recently in our laboratories. It forms the gamut of doubly substituted π-acceptor complexes, $[Ru_2^{II}(L)_2(ttha)]^{2-}$ (L = py, pz, CO, RCN, and olefins) (83). The $ttha^{6-}$ ligand provides two adjacent binding sites for the Ru centers that are nearly equivalent of side-by-side tethered $hedta^{3-}$ binding pockets. The short interior $(CH_2)_2$ bridge between the ethylenediamine backbone units favors the formation of intrabridged binuclear species. We have studied the pyrimidine-bridged $Ru^{II}Ru^{II}$ and $Ru^{II}Ru^{III}$ system previously (84) and the hydroxy- or oxo-bridged $Ru^{II}Ru^{III}$, $Ru^{III}Ru^{III}$, and $Ru^{III}Ru^{IV}$ derivatives recently (85). The $[Ru^{II}Ru^{III}(OH)(ttha)]^{2-}$ complex is related to the mixed-valence $Fe^{II}(OH)Fe^{III}$ core of purple acid phosphatases.

Figure 5 presents the data obtained when 1,3-dimethyluracil (DMU) is combined with $[Ru^{II}_2(ttha)(H_2O)_2]^{2-}$ at a 4 DMU:$[Ru_2(ttha)(H_2O)_2]^{2-}$ ratio. The

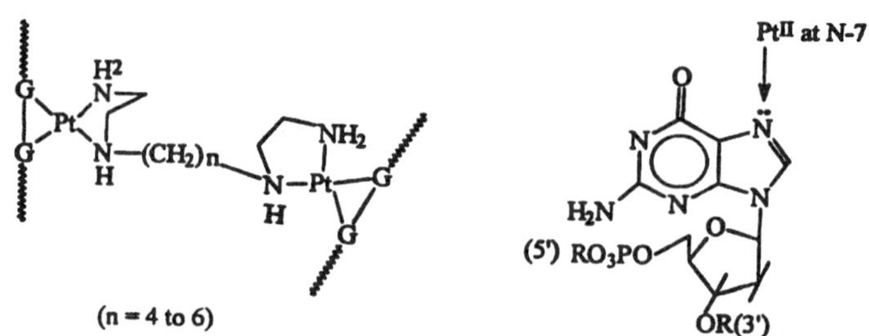

Figure 3. Bis(Platinium–diamine) major-groove cross-link on DNA: a stylized view with 1,2-GG intrastrand coordination (as discussed in reference 49).

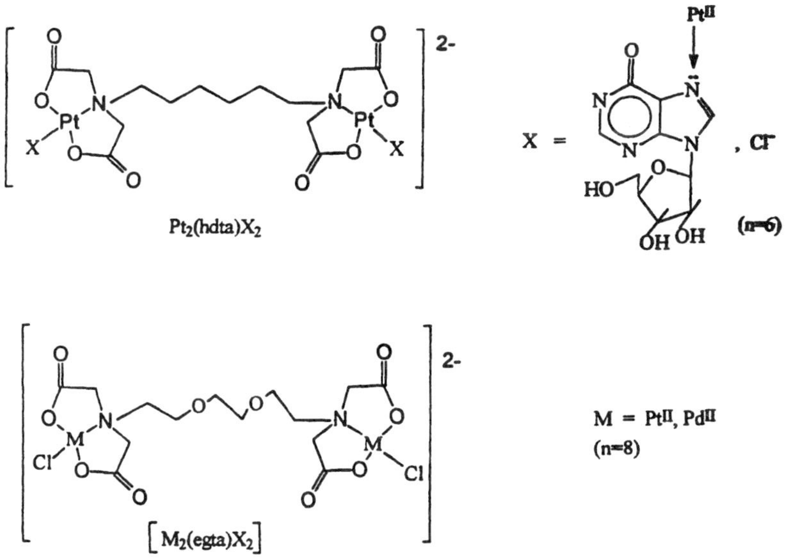

Pt$_2$(hdta)X$_2$

X = , Cl$^-$ HO OH OH   (n=6)

[M$_2$(egta)X$_2$]

M = Pt^{II}, Pd^{II}
(n=8)

Figure 4. Bis(Platinium–PAC) major-groove-spanning chelates (as discussed in references 54 and 55.)

^{1}H NMR of the aromatic ring region shows a 0.67 ppm upfield shift of the H5 proton and –0.27 ppm (downfield) shift of the coordinated H6 proton of DMU. The $E_{1/2}$ value (CV, DPP data) of the Ru$^{II/III}$ sites of the bis complex is shifted from 0.05 V of the diaqua derivative to 0.68 V for [Ru$_2$(ttha)(DMU)$_2$]$^{2-}$.

Studies at incremental increases in the DMU:Ru^{II} ratio have shown sequential addition of two DMU ligands at the separate Ru^{II} binding sites (Y. Chen, unpublished data). The influence of the Ru^{II}–pac ligand environment on ^{1}H NMR shifts of the H5 and H6 protons of DMU compared to [(NH$_3$)$_5$Ru-(DMU)]$^{2+}$ and [(NH$_3$)$_5$Os(DMU)]$^{2+}$ (86) is presented in Table I. For the pentaammine Ru(II) and Os(II) complexes, large upfield shifts (1.16–1.98 ppm) are observed for both H5 and H6 protons upon forming η^2 complexes with DMU. However, both the Ru^{II}–pac systems of [Ru^{II}(hedta)]$^-$ and each headgroup of [Ru$^{II}_2$(ttha)]$^{2-}$ promotes the downfield shift of the H6 proton of DMU (ca. –0.28 ppm) and less significant upfield shifts for the H5 proton (0.84 or 0.67 ppm vs. 1.98 for Ru(NH$_3$)$_5$DMU^{2+}). This is an example of the α N(R)R′ influence on ^{1}H NMR shifts mentioned previously. Models also show the H6 proton of DMU is placed above a coordinated glycinato functionality, which may contribute to deshielding of the H6 proton.

Clearly, [Ru$^{II}_2$(ttha)(DMU)$_2$]$^{2-}$ serves as a useful model that Ru^{II}–pacs may be prepared which could ultimately bridge two strands of a DNA helix. The methyl groups of DMU serve two roles in this model. The CH$_3$ at N1 is the same attachment as the N1-to-deoxyribose linkage in DNA. The methylation at

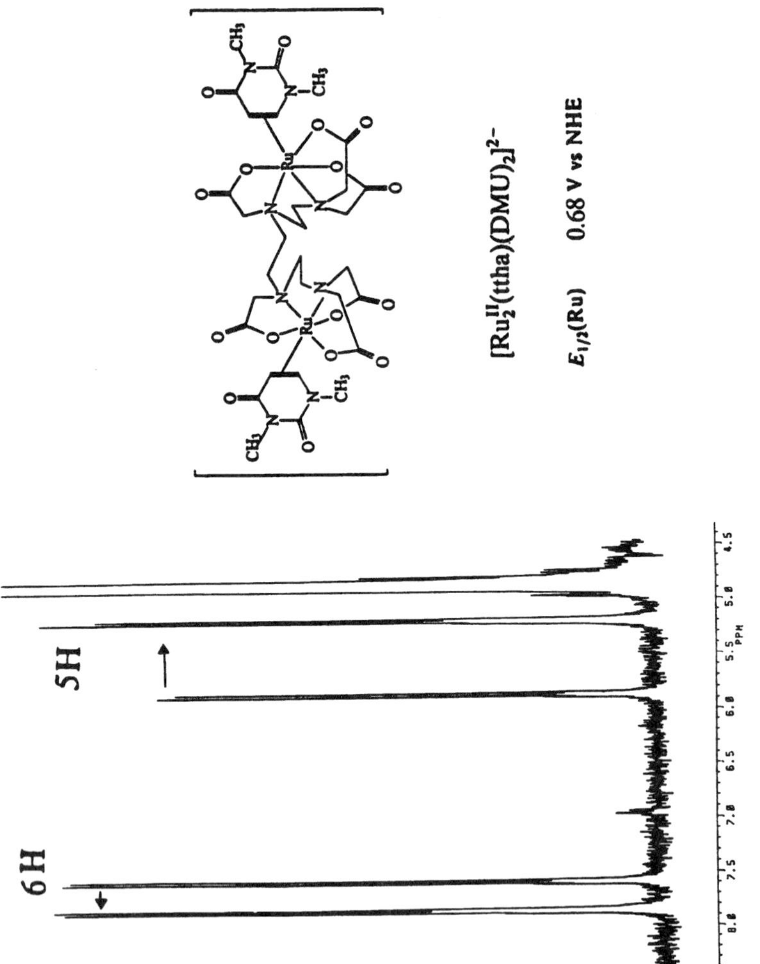

Figure 5. The [Ru₂(ttha)(DMU)₂]²⁻ complex and its aromatic ¹H NMR shifts relative to free DMU.

Table I. DMU ^{1}H NMR Data

	6H ($\Delta\delta$)	5H ($\Delta\delta$)	$CH_3(1)$	$CH_3(3)$	Ref.
Free DMU	7.58	5.85	3.40	3.29	
$[(NH_3)_5Ru^{II}]$	6.42 (1.16)	3.87 (1.98)	3.48 (−0.08)	3.16 (0.13)	86
$[(NH_3)_5Os^{II}]$	5.89 (1.69)	4.32 (1.53)	3.35 (0.05)	3.11 (0.18)	86
$[Ru^{II}(hedta)]^-$	7.79 (−0.29)	5.01 (0.84)	3.45 (−0.05)	3.03 (0.26)	32
$[Ru^{II}_2(ttha)]^{2-}_{1/2}$	7.85 (−0.27)	5.18 (0.67)	3.45 (−0.05)	3.03 (0.26)	This study

NOTE: DMU is 1,3-dimethyluracil. $\Delta\delta = (\delta_{\text{free L}} - \delta_{\text{comp}})$. A positive $\Delta\delta$ is an upfield shift.

N3 in this model represents the protection at N3 that is provided by the H-bonding of the cross-strand base-pair partner inside a B-DNA helix. Thus, upon approach of a Ru^{II}–pac headgroup toward the C5–C6 region of a pyrimidine base that projects toward the major groove, the Ru^{II} center will encounter a potential η^2 bonding partner. The ^{1}H NMR studies were performed such that the spectrum shown in Figure 5 was obtained within 30 min after mixing under second-order conditions. For complete reaction within this time, $k \sim 3$ M^{-1} s^{-1} for DMU substitution. This is comparable with other substitution rates of π-acceptor ligands (13–30 M^{-1} s^{-1}) measured with $[Ru^{II}(edta)]^{2-}$ (76, 87). Therefore η^2 coordination of pyrimidine nucleobases can compete with bonding at N-donors as long as H-bonding between the helical strands is maintained.

Two improvements to the model are necessary to obtain the best potential major groove-spanning Ru^{II} binuclear chelate. The internal methylene tether would need to be lengthened from $(CH_2)_2$ to $(CH_2)_{4-6}$, sufficient to span the major groove, but to allow flexibility during coordination as shown by the studies of Farrell (49–53). Also, the major groove is lined by phosphate linkages such that a 1− charge per base pair is established along the DNA helix. This repels negatively charged species. Since $[Ru_2(ttha)(H_2O)_2]^{2-}$ is a 2− ion, reduction in the number of anionic donors by at least one per site is suggested for an active Ru^{II}–pac groove-spanning chelate. In tumor cells the environment is hypoxic and reducing (88). Clarke (22, 59, 62, 89) has presented the case that ruthenium ammines, whose redox potentials are similar to Ru^{II}–pacs, would be transported through cells as the Ru^{III} form, and then reduced at or near the DNA groove to promote binding at DNA in the Ru^{II} form. A ligand with the chelation of uedda^{2-} (*N,N*-ethylenediaminediacetate; two saturated N donors and two carboxylates per Ru site) would either be neutral in the Ru^{II} form or even cationic as the bis Ru^{III} derivative. Work is in progress toward the synthesis of a Ru^{II} major groove-spanning chelate designed around $[Ru^{II}(uedda)]$ headgroups for this purpose as an extension of the $[Ru_2(ttha)(DMU)_2][^{2-}$ model presented in this chapter.

Factors Favoring N-base to η^2 Migrations. Other than the $[Ru^{II}(hedta)(pyrimidine\ nucleobase)]$ complexes (32–34), there are only a few

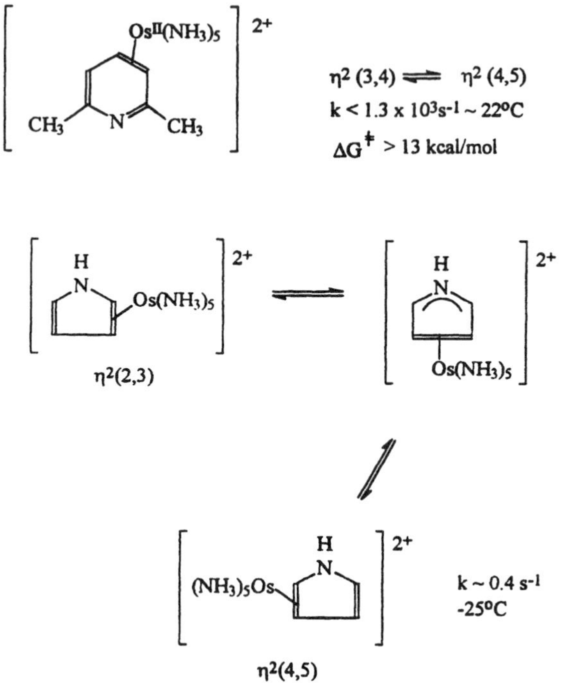

Figure 6. η^2-Coordinated N-heterocycles described previously by Taube and co-workers (see references 45 and 46.)

literature reports of dearomatized N-heterocycles promoted by metallation. The $(NH_3)_5Os^{II}$ chromophore has been observed to form olefinic η^2 coordination with 2,6-lutidine (45) and pyrrole (46) as illustrated in Figure 6. Taube and co-workers observed a slow fluxional migration of $[(NH_3)_5Os]^{2+}$ between $\eta^2(3,4)$ and $\eta^2(4,5)$ positions of 2,6 lutidine and a rapid $\eta^2(2,3)$ to $\eta^2(4,5)$ migration with pyrrole. Only chemical intermediates have been proposed for η^2 interactions of $[(NH_3)_5Ru]^{2+}$ with pyridines as illustrated in Figure 7 and discussed in the introduction to this chapter (47, 48).

The energetic factors for the process of migration from an N-bound Ru^{II} nitrogen heterocycle to a position of η^2 coordination that promotes dearomatization are shown in Figure 8. The activation barrier involves two primary factors: (1) losses due to breaking of the $Ru^{II}-N$ bond and (2) energy losses associated with dearomatization of the N-heterocycle. The losses due to σ bond breaking will parallel the ability of the nitrogen base to donate a σ electron pair to Ru^{II} or H^+. Hence, the ligand pK_a is one parameter that needs to be considered along with the energy estimates to dearomatize the N-heterocycle. The resonance energy and pK_a data for pyridine, pyrrole, pyrazine, pyrimidine, and pyridazine are shown at the bottom of Figure 8. Data are collected

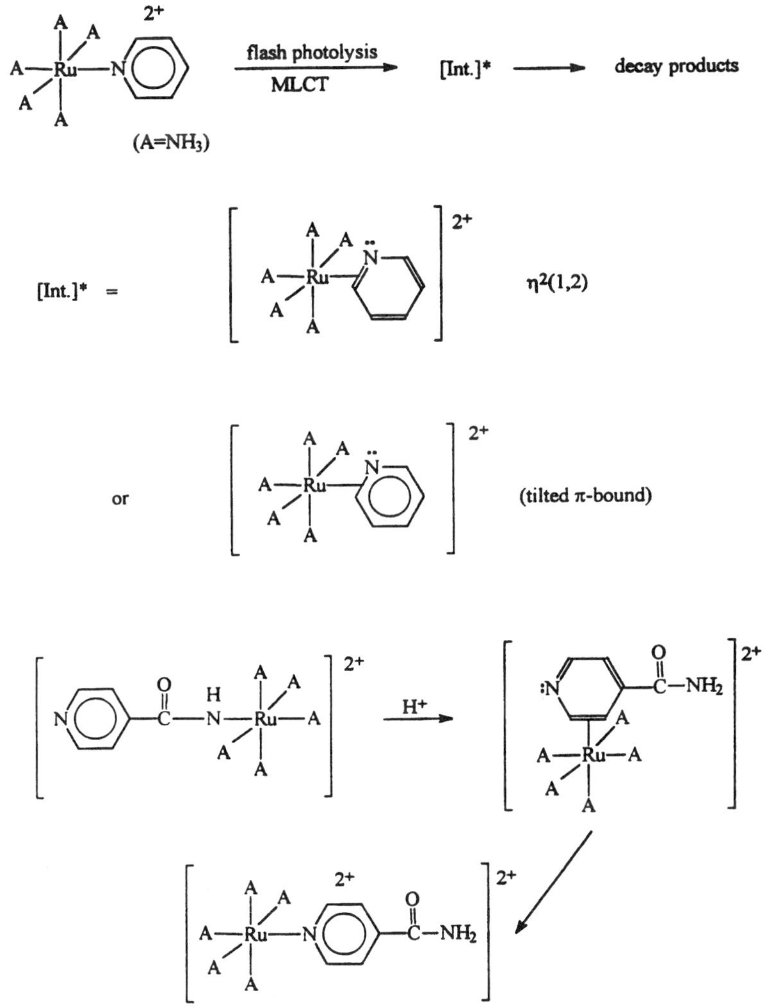

Figure 7. Previously proposed $(NH_3)_5Ru^{II}$–pyridine ligand η^2 intermediates. [Adapted from references 47 (top) and 48 (bottom).]

from references 90–94. The high pK_a of pyridine and its high resonance energy of 27 kcal/mol provide ample reason why η^2 complexes will be very slow to form whenever substitution occurs first at the σ lone pair of the N of a pyridine. We have estimated the activation barrier for an N1 to $\eta^2(1,2)$ migration to be ≥ 36.5 kcal/mol (*43*). The thermodynamic data suggest that the lowest N-base to η^2 reorganization barrier will be for pyrimidines (resonance energy $\sim$ 14 kcal/mol) and pyridazine (resonance energy $\sim$ 10–12 kcal/mol). Of these two, pyrimidine with a weaker σ donor N-base ($pK_a = 1.31$) compared to pyri-

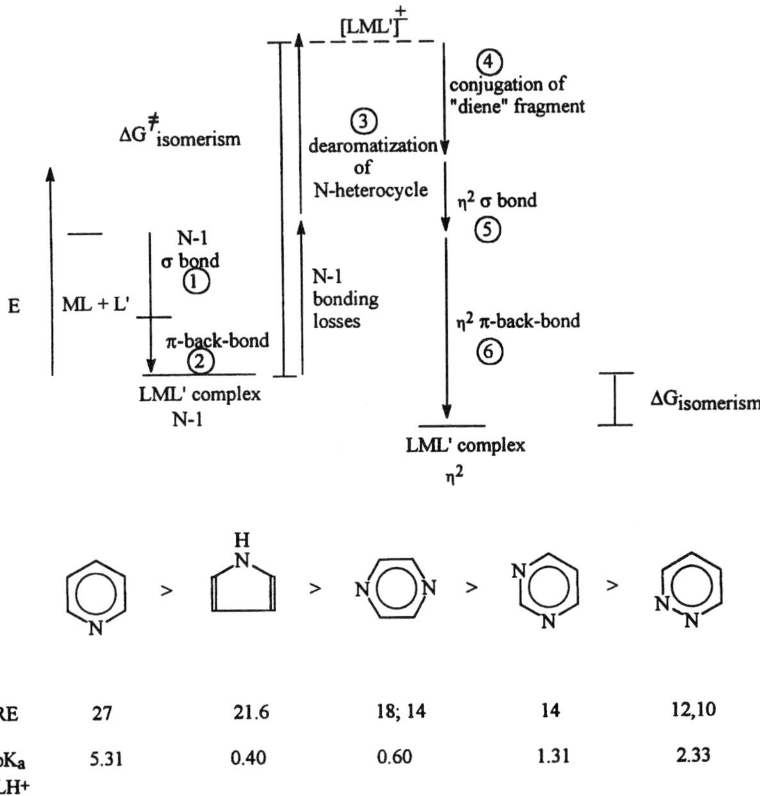

RE	27	21.6	18; 14	14	12,10
pK$_a$ **LH+**	5.31	0.40	0.60	1.31	2.33

Figure 8. Steps in N-bound to η^2-bound isomerism rearrangements for Ru^{II} and Os^{II} complexes, and resonance energy and pK$_a$ data for N-heterocyclic ligands.

dazine (pK$_a$ = 2.33) will be the most likely to produce N-base to η^2 migrations. Pyridazine is the next most likely prospect. Since pyridazine is a somewhat better π-acceptor than pyrimidine (95–98), reorganization factors for N-base to η^2 migrations may be estimated to be similar for pyrimidines and pyridazines, overall.

We have only recently observed the anticipated N1-to-η^2 migrations for pyrimidines as shown in Figure 9 with pyrimidine and methylated pyrimidines (43). Substitution occurs rapidly at N1 with rate constants of about 31 M^{-1} s^{-1} for pym as judged by the absorbance changes and subsequent ^{1}H NMR spectral measurements.

When electrochemical studies are performed, the initial solutions of the pym, 4-CH$_3$pym, or Me$_2$pym complexes exhibit only one CV or DPP wave with $E_{1/2}$ values near 0.14 V for N1-bound complexes. Data for all complexes are presented in Figures 10–13 along with the measured ^{1}H NMR chemical

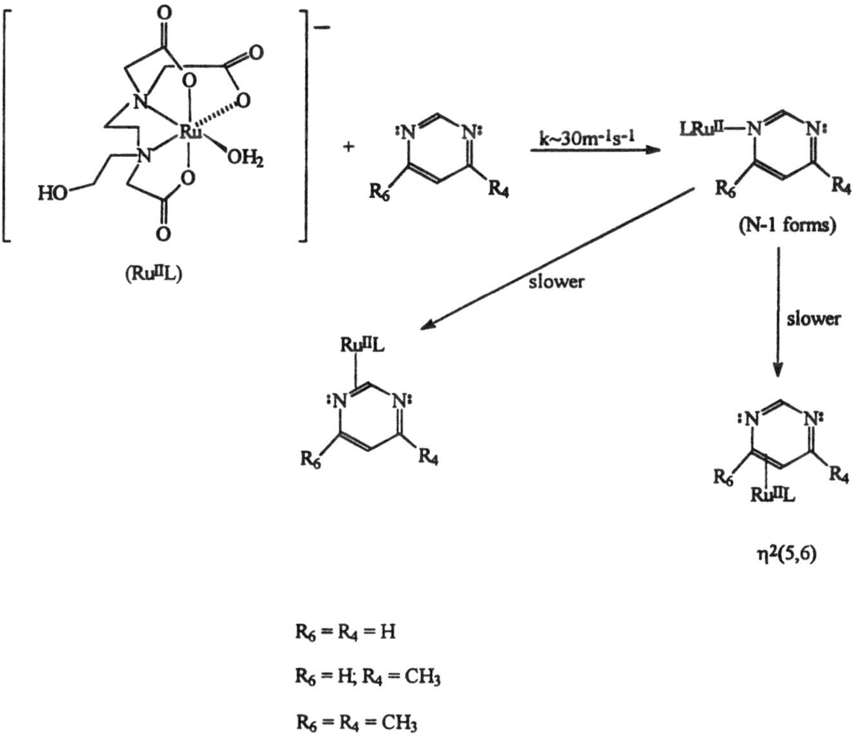

Figure 9. *Sequential steps in forming [Ru^{II}(hedta)]$^-$ N-bound and η^2-bound isomers.*

shifts and $\Delta\delta = (\delta_{freeL} - \delta_{complex})$ values. The CV and DPP data after 4 days are shown for the pym and 4-CH$_3$ pym complexes in Figure 14 (*43*).

The development of the η^2 isomers and the loss of the N1-bound isomers is clearly observed. The extent of migration to $\eta^2(5,6)$ is retarded by electronic releasing methyl substituents which favor N1 coordination, but presence of ring methyls also stabilizes $\eta^2(5,6)$ relative to $\eta^2(1,2)$ species as detected by ^{1}H NMR (*43*). During an electrochemical sweep in the DPP mode, oxidation of the η^2 complex is accompanied by aquation as well as an apparent migration back to N1 by the Ru^{III} complex. This is detected by an increase in the CV reduction wave for the N1 species. The fraction of the η^2 complex that undergoes aquation gives rise to an Ru$^{III/IV}$ oxidation wave at 1.14 V. The extent of migration back to N1 is retarded for 4-CH$_3$pym. We have not studied these processes in detail as we have focused our attention on further identification of the properties of the Ru^{II} η^2 complexes. These studies have proved to be complicated as described in subsequent sections. The results do, however, reveal an electrochemical catalysis for reformation of the N1 form by the overall sequence of steps for the pyrimidine case, shown in Scheme I.

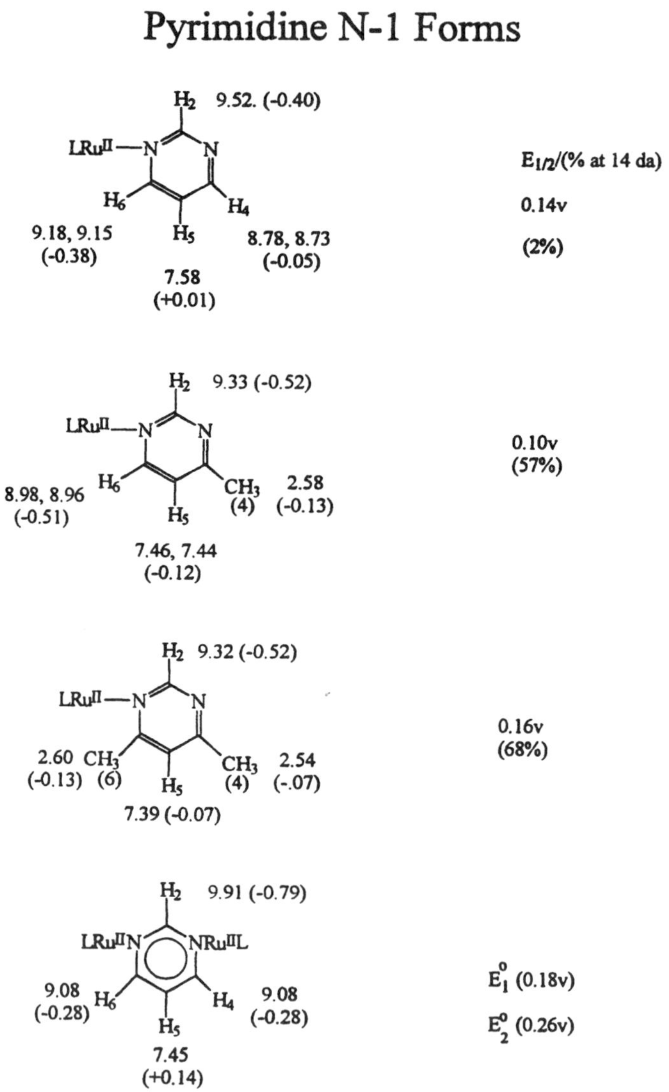

Pyrimidine N-1 Forms

Figure 10. ^1H *NMR assignments for proton chemical shifts, $Ru^{II/III}$ $E_{1/2}$ values, and isomer distribution after 14 days for $[Ru^{II}(hedta)L]^-$ complexes (N1 forms); L = pyrimidine, 4-methylpyrimidine, and 4,6-dimethylpyrimidine.*

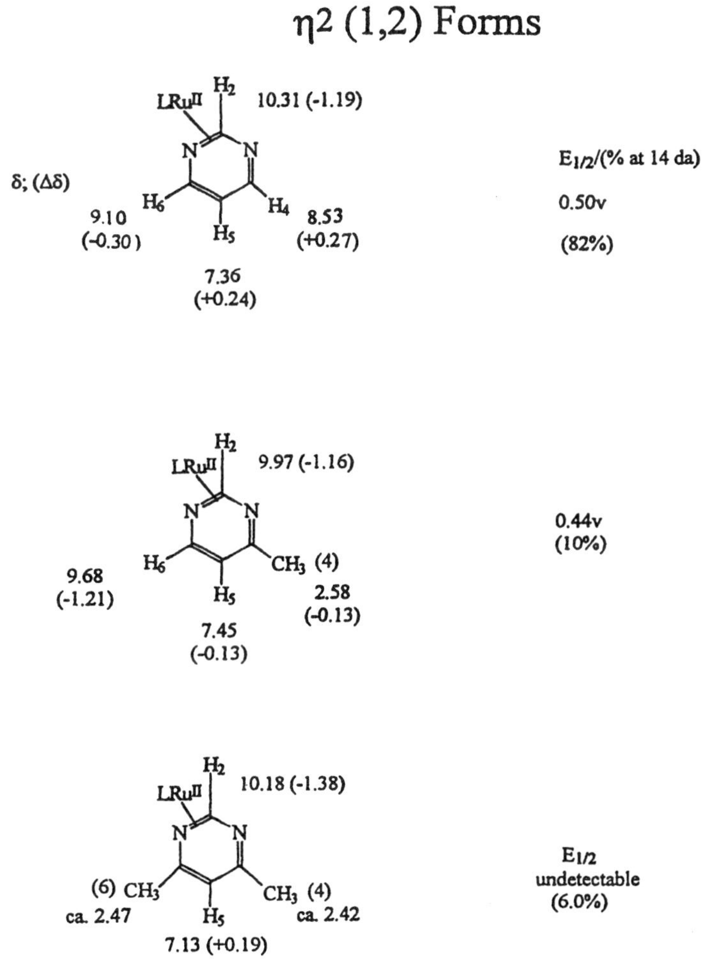

η2 (1,2) Forms

δ; (Δδ)

LRu^{II} H$_2$ 10.31 (-1.19)

H$_6$ 9.10 (-0.30)

H$_4$ 8.53 (+0.27)

H$_5$ 7.36 (+0.24)

E$_{1/2}$/(% at 14 da)

0.50v

(82%)

LRu^{II} H$_2$ 9.97 (-1.16)

H$_6$ 9.68 (-1.21)

CH$_3$ (4) 2.58 (-0.13)

H$_5$ 7.45 (-0.13)

0.44v
(10%)

LRu^{II} H$_2$ 10.18 (-1.38)

(6) CH$_3$ ca. 2.47

CH$_3$ (4) ca. 2.42

H$_5$ 7.13 (+0.19)

E$_{1/2}$
undetectable
(6.0%)

Figure 11. ^{1}H NMR assignments for proton chemical shifts, Ru$^{II/III}$ E$_{1/2}$ values, and isomer distribution after 14 days for [Ru^{II}(hedta)L]$^-$ complexes (η^2(1,2) forms); L = pyrimidine, 4-methylpyrimidine, and 4,6-dimethylpyrimidine.

η2 (5,6) Forms

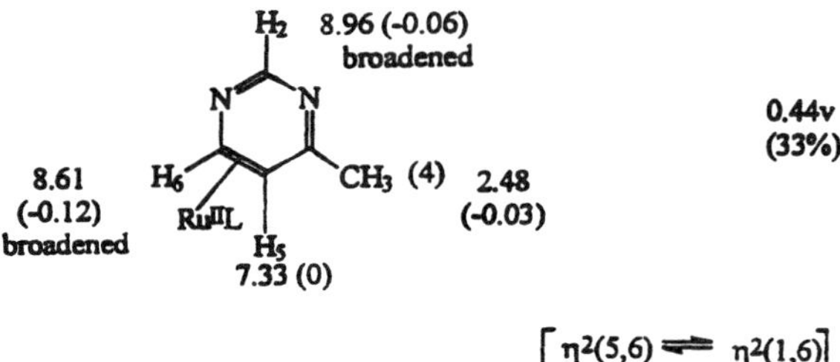

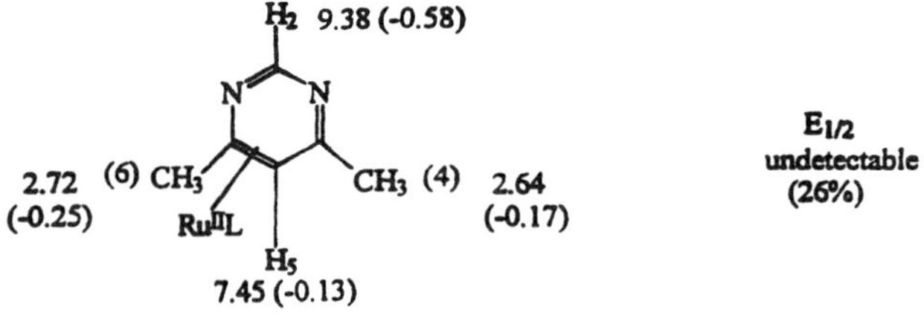

No. CH3 groups	O(pym)	1(4Me)	2(4,6 Me2)
% Total η2 forms	98	46	32
Δδ 1H5	+0.42	0.0	-0.13

Figure 12. ^{1}H NMR assignments for proton chemical shifts, $Ru^{II/III}$ $E_{1/2}$ values, and isomer distribution after 14 days for [Ru^{II}(hedta)L]⁻ complexes (η^2(5,6) forms); L = pyrimidine, 4-methylpyrimidine, and 4,6-dimethylpyrimidine.

$\eta2$ (1,6) Form

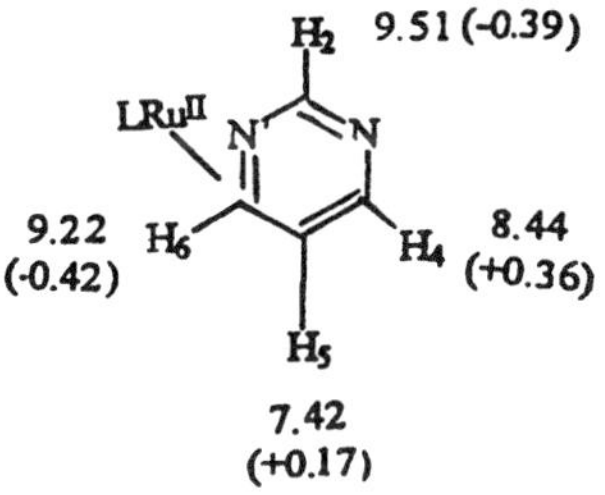

Figure 13. ^1H NMR assignments for proton chemical shifts, $Ru^{II/III}$ $E_{1/2}$ values, and isomer distribution after 14 days for [Ru^{II}(hedta)L]$^-$ complexes; L = pyrimidine, 4-methylpyrimidine, and 4,6-dimethylpyrimidine. $\eta^2(1,6)$ species.

In order to obtain accurate values for the chemical shifts of the N1 isomer of [Ru^{II}(hedta)(pym)]$^-$, it was found to be most convenient to form the bridged binuclear complex {[Ru^{II}(hedta)$_2$(pym)]}$^{2-}$, which is an analogue of the [Ru_2(ttha)(pym)]$^{2-}$ complex (84) studied previously. The small amount of [Ru(hedta)(pym)]$^-$ monomer was clearly identifiable in the presence of the binuclear because these differ in splitting pattern and chemical shifts as shown in Figure 15 wherein the binuclear pyrimidine protons are labeled 2B–6B for the H2–H6 protons and the monomer complex is labeled 2M–6M in the same manner.

At 0.80:1.00 stoichiometry, the amount of the binuclear species can be minimized. This allows the η^2 migration products to be followed with time by ^1H NMR. The observed spectrum after 14 days is shown in Figure 16. A combination of integration, COSY, NOESY, and COLOC, and decoupling techniques, as well as the different rates of growth of several species, allow the assignment of the H2–H6 protons for three new species. Similar time-dependent ^1H NMR, ^{13}C, and COSY studies were carried out for 4-CH_3pym and Me$_2$pym complexes of [Ru^{II}(hedta)]$^-$. The assignments of the ^1H NMR chemical shifts and estimates of isomer abundance and results of parallel CV and DPP studies are given in Figures 10–13.

The details of the ratios of N1 to η^2 migration products have been recently presented elsewhere (43). However, several ^1H NMR shift trends are noteworthy. The $\Delta\delta$ values for H2 of N1-bound species are all near –0.50 ppm, a downfield shift. ^1H NMR shift and $E_{1/2}$ values for the binuclear pym complex are presented in Figure 10 for the first time. The ^1H NMR parameters show only a modest increase in downfield shift of H2 and H4 protons compared to the monomer, when two Ru^{II} centers act in concert on the pym ring. The percentage of N1 isomer that persists at equilibrium after 14 days is also provided in Figure 10. Only 2% of the N1 isomer remains for the pym complex, almost all converting to one of the η^2 isomers. The $\eta^2(1,2)$ isomers exhibit a strong down-

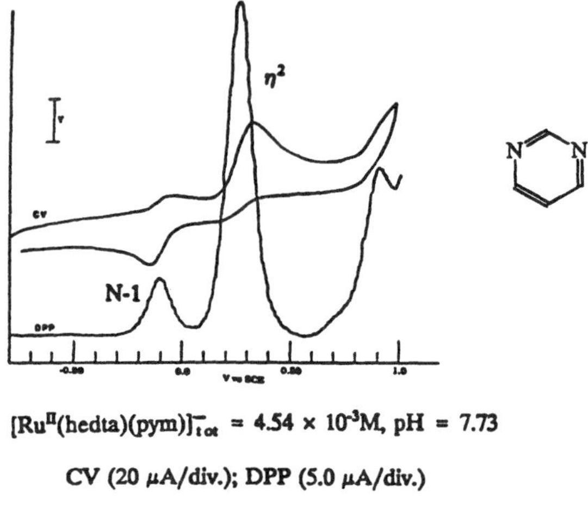

[Ru^{II}(hedta)(pym)]$^-_{Tot}$ = 4.54 × 10^{-3}M, pH = 7.73

CV (20 μA/div.); DPP (5.0 μA/div.)

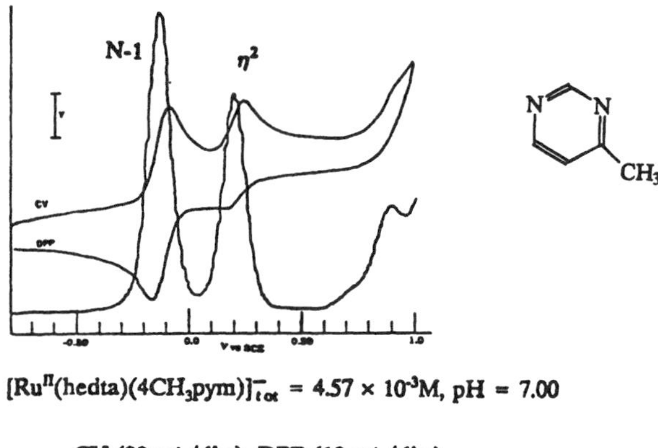

[Ru^{II}(hedta)(4CH$_3$pym)]$^-_{Tot}$ = 4.57 × 10^{-3}M, pH = 7.00

CV (20 μA/div.), DPP (10 μA/div.).

Figure 14. Cyclic voltammograms and differential pulse polarograms for pyrimidine and 4-methylpyrimidine complexes of [Ru^{II}(hedta)]$^-$ after 4 days: presence of N1 and η^2 isomers.

field shift influence of the Ru^{II} center attached η^2 adjacent to C2. This carbon also has a neighboring pair of N atoms that serve as withdrawing groups. A net downfield shift at −1.16 to −1.38 ppm is observed for all three pyrimidines. Upfield shifts of protons across the ring for $\eta^2(1,2)$ derivatives as a possible bond-localization influence are observed for pym, but the effect is canceled with 4-CH$_3$pym.

The $\eta^2(5,6)$ isomer is assigned by the upfield shifts of the H5 and H6 protons in the pym complex and the more positive $E_{1/2}$ values, reminiscent of [Ru^{II}(hedta)(olefin)]$^-$ complexes (73, 34). The H6 protons of the $\eta^2(5,6)$ isomers

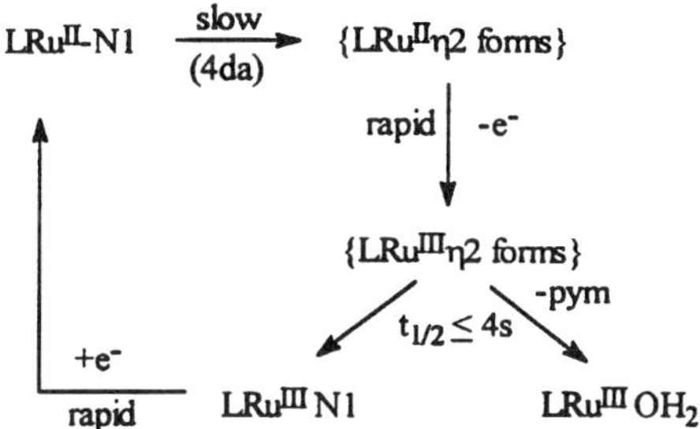

Scheme I. Electrochemically catalyzed isomerisms and aquation of [Ru^II(hedta)-(pym)]⁻.

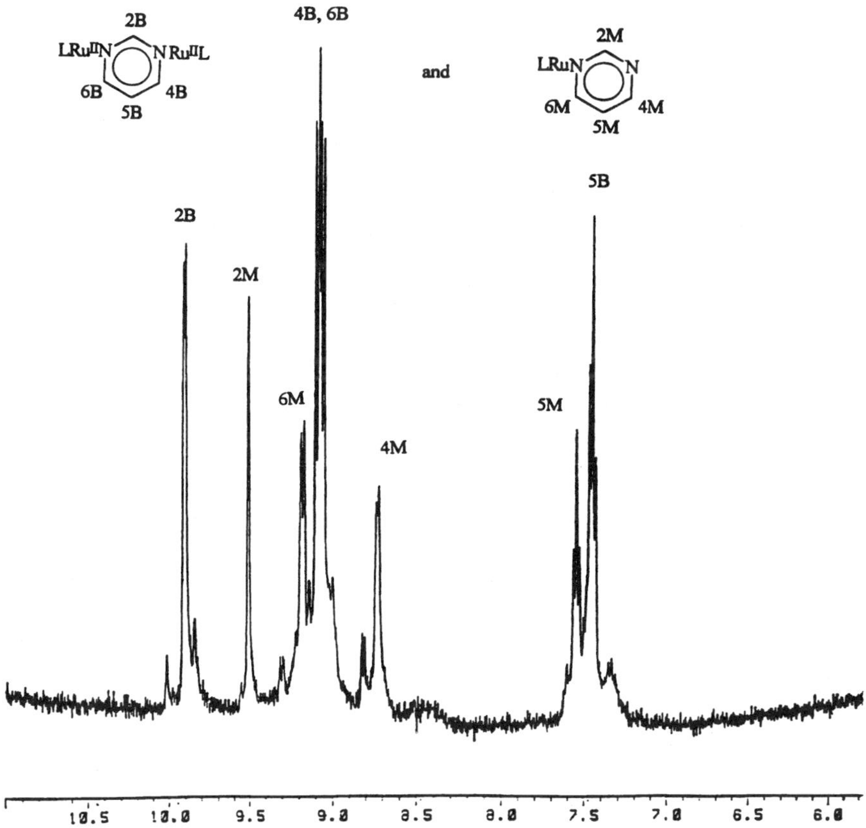

Figure 15. ¹H NMR spectra of monomeric N1 and bridged-binuclear pyrimidine complexes of [Ru^II(hedta)]⁻.

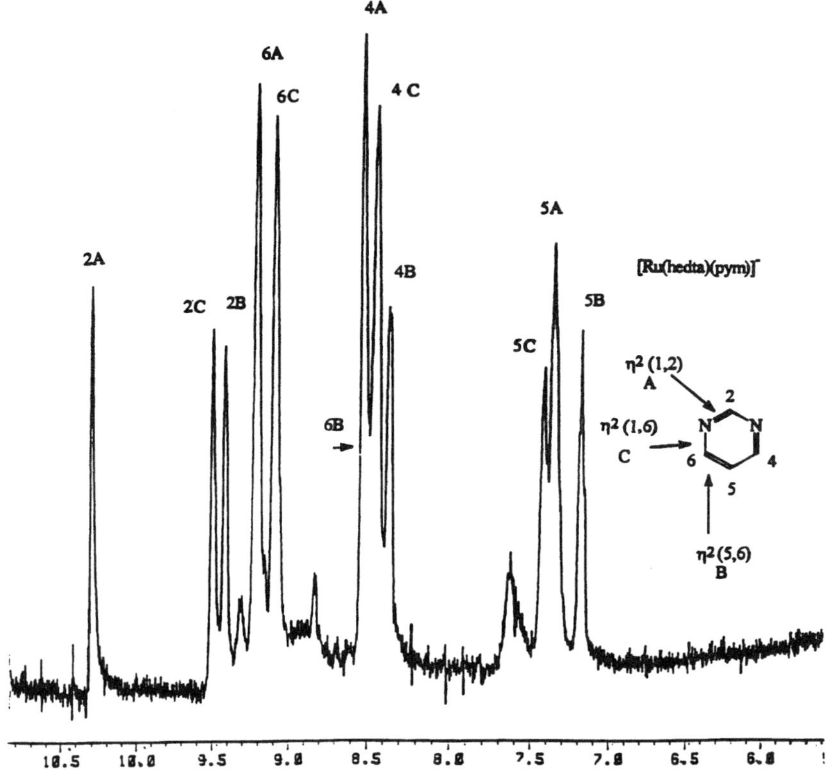

Figure 16. ^1H NMR spectra for the equilibrium distribution of η^2 isomers of $[Ru^{II}(hedta)(pym)]^-$. (The trace species represent components of the bridged binuclear complex and 2% of residual N1 isomers.)

are either upfield of the free ligand or not as far downfield as the N1 isomers, indicating the back-donation influence of Ru(II). An influence on the H5 proton shift is seen upon varying the extent of ring methylation. In the pym complex, H5 is shifted 0.42 ppm upfield of the free ligand value, but substitution of one methyl group negates the effect, and two methyl groups result in a downfield shift. It is not clear whether this represents an electronic influence of methyl or whether the favored isomer avoids steric hindrance by controlling the face of η^2 coordination to the pym rings, and in doing so, places the H5 above the glycinato carboxylate. This influence was already noted for DMU. Broadening of the resonance lines for the H6 and H2 protons of the assigned $\eta^2(5,6)$ isomer of the 4 CH_3pym complex of $[Ru^{II}(hedta)]^-$ may be reconciled by a fluxional process involving a shift between $\eta^2(5,6)$ and the near position $\eta^2(1,6)$ (43). The $\eta^2(1,6)$ isomer was detected as a separate stable isomer for only the pym system. This isomer exhibits less of a downfield shift fo the H6 proton than for its $\eta^2(1,2)$ counterpart, but also the cross-ring upfield shifts of H4 and H5 like those observed for the $\eta^2(1,2)$ species. The lesser downfield shift of H6 for the $\eta^2(1,6)$ species can be reconciled by the presence of a carbon neighbor atom at C5 instead of a nitrogen neighbor for C2 in the case of the $\eta^2(1,2)$ derivatives. Methylation of the ring causes an absence of the $\eta^2(1,6)$ isomer (and also disfavors $\eta^2(1,2)$ isomers) both by increasing the basicity of the N1 nitrogen and the enhancing stability of the N1 forms of these complexes, but also because of some steric effects for Me_2pym of coordination at an olefinic group with substituents that cannot move far from coplanar with the ring by bending back as is the common effect with η^2 coordination on simple olefins.

The Pyradizine Complex.   In several studies of various $[Ru^{II}(hedta)$-$(pyrazines)]^-$ and $[Ru^{II}(hedta)(pyrimidines)]^-$ under conditions of a large excess of free ligands, we have observed a second slow substitution reaction forming $[Ru^{II}(hedta)L_2]^-$ species. These reactions occur slowly with rate constants of ca. 4.28×10^{-5} s^{-1} at 25 °C (2-methylpyrazine, $t_{1/2} \sim 4.5$ h). Therefore bis additions are not normally a problem in the studies of the pyrimidine and pyrazine complexes of $[Ru^{II}(hedta)]^-$ at 1:1 stoichiometry. When $[Ru^{II}(hedta)(D_2O)]^-$ and pyridazine was combined at pD $\sim$ 6 in a 1:1 stoichiometry, the ^{1}H NMR pattern shown in Figure 17 is observed. There are two sets of four distinguishable singlet patterns. The more abundant species has chemical shifts at 9.54, 9.06, 7.55, and 7.63 δ and the lesser abundant species has chemical shifts at 9.11, 8.72, 7.63 and 7.45 δ (the 7.63 ppm resonances are overlapped). Upon adjustment of the solution pD to $\sim$1.0, the major component has a line coalescence of the more upfield pair of lines in a sharp resonance at 7.48 ppm. The downfield pair of resonances become very broadened with a maximum near 8.0 ppm (inset in Figure 17). The less abundant species is unchanged except that the most upfield resonance is obscured by the 7.48 ppm line of the other species. Also noted in the spectrum at pD 6 are the appearance of high-field ethylenediamine backbone proton peaks indicative of the less rigid backbone of a $[Ru^{II}(hedta)]^-$ chromophore, which has one glycinato fragment displaced from the in-plane coordination. These lines appear at 2.66 and 2.87 ppm, which have moved out from under the complex pattern of ethylene backbone and coordinated glycinato splitting patterns that cover the range of 3.10–4.30 ppm.

When a second equivalent of pyridazine is added to the sample at pD 6, a rapid conversion occurs as indicated by the less complex pattern in Figure 18. Electrochemical studies implicate formation of the bis complex $[Ru^{II}(hedta)$-$(pyd)_2]^{2-}$, consistent with virtually equivalent in-plane N1-coordinated ligands with ^{1}H NMR shifts as assigned on Figure 18. A separate sample prepared at pD $\sim$ 1.0 does not rapidly form the bis chelate upon addition of a second equivalent of pyridazine.

These observations are all consistent with the sequence presented in Scheme II. When the 1:1 complex is prepared, the rapid substitution of the

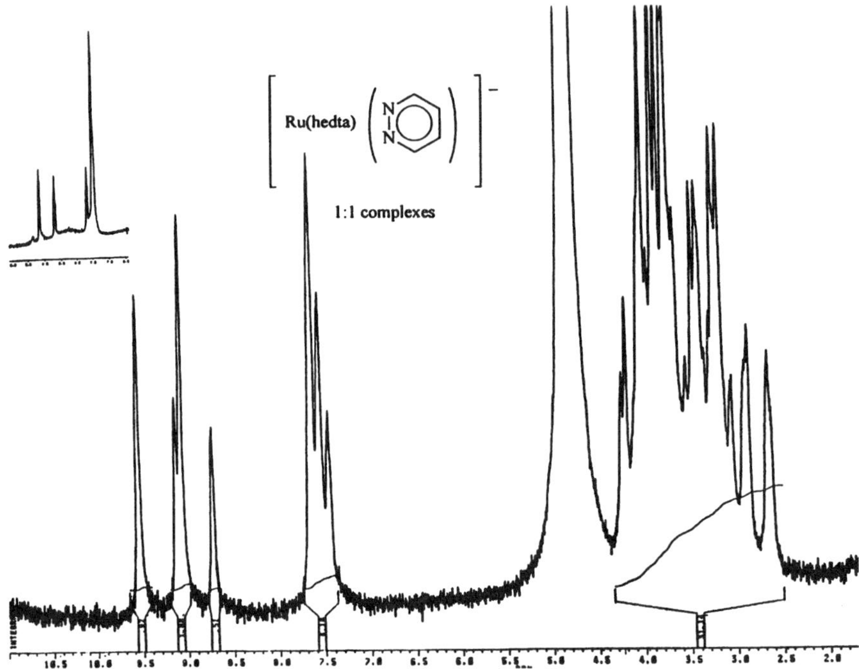

Figure 17. ^1H *NMR spectrum showing two species formed from [Ru^{II}(hedta)-(H$_2$O)]$^-$ and pyridazine at pD of 6. Inset: spectrum at pD ~ 1.0.*

aqua site occurs, forming the normal N1-bound derivative. However, unlike all previously studied pyrimidines and pyrazines, the pyridazine ligand has a second potential nitrogen base at N2. Upon coordination to the back-donating [Ru^{II}(hedta)] fragment, the basicity is enhanced by several orders of magnitude. This nitrogen may act as an internal nucleophile to generate the less-abundant, equilibrium species in which a carboxylate donor is displaced. This frees the rigidity of the ethylenediamine backbone for the bidentate pyridazine complex as observed by ^1H NMR. The coordination is asymmetric as indicated by the four distinguishable singlets of the bidentate complex. This factor implies that the chelate form is rather strained. It should be amenable to displacements by better nitrogen base donors. The presence of a second equivalent of pyridazine shifts the equilibrium through the bidentate (1:1) complex to the bis (2:1) complex very rapidly ($t_{1/2} < 1$ min) as indicated by the first ^1H NMR spectrum that can be obtained within 10 min of mixing, which shows only the bis complex.

The influence of protonation supports this mechanism. The protonation of the electron pair at N2 promotes a linkage isomerism for the site of Ru^{II} attachment to pyridazine. As the Ru^{II} center migrates between N1 and N2, the pro-

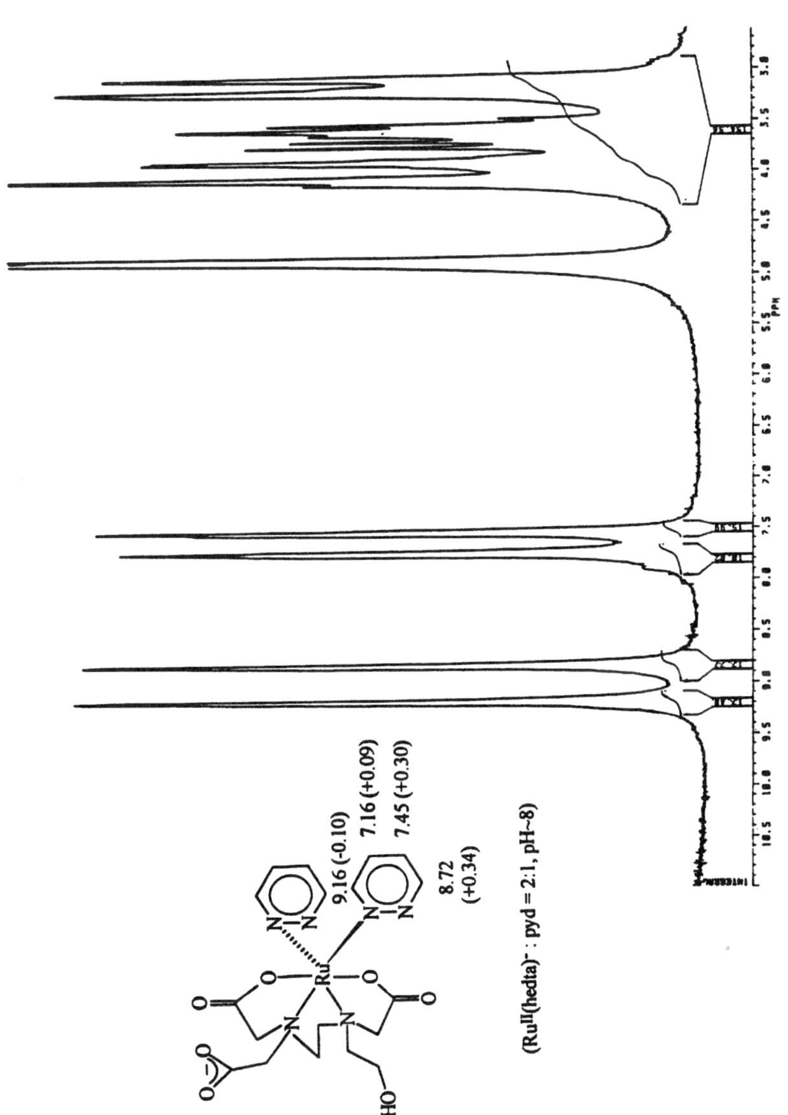

Figure 18. ^{1}H NMR spectrum of [Ru^{II}(hedta)(pyridazine)$_2$]$^-$ at pD ~ 6.

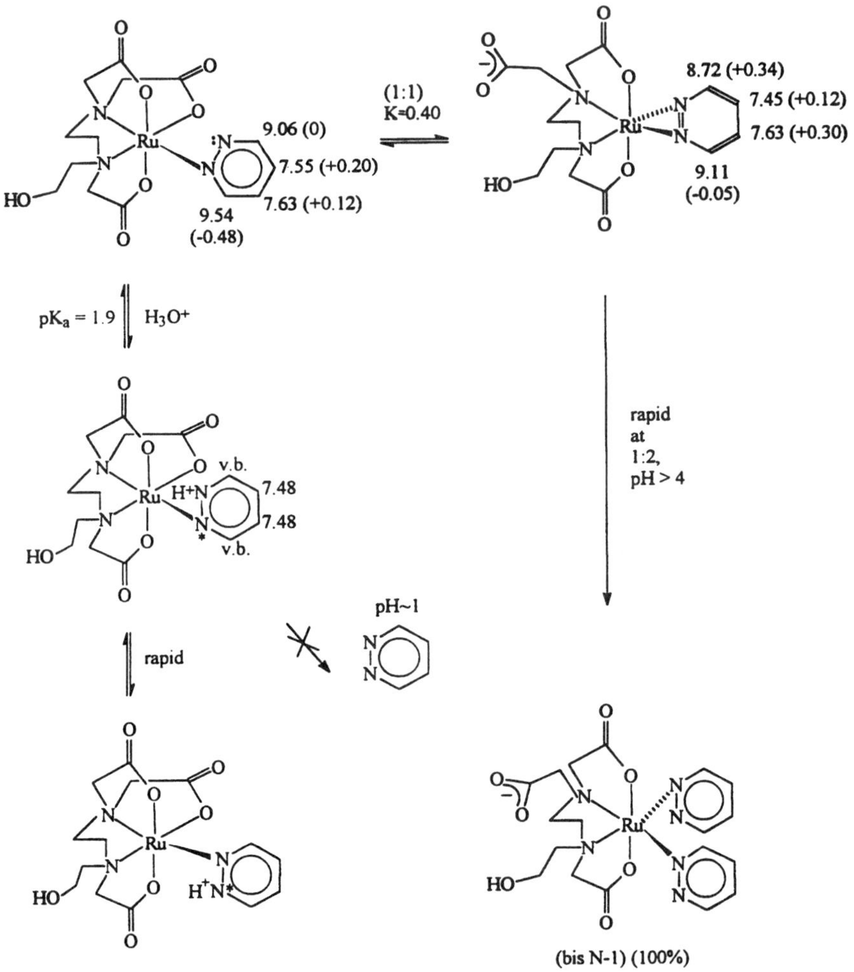

9.06 (0)
7.55 (+0.20)
7.63 (+0.12)
9.54 (−0.48)

(1:1) K=0.40

8.72 (+0.34)
7.45 (+0.12)
7.63 (+0.30)
9.11 (−0.05)

$pK_a = 1.9$　H_3O^+

rapid at 1:2, pH > 4

v.b.
7.48
7.48
v.b.

rapid

pH~1

(bis N-1) (100%)

Scheme II. Substitution reaction scheme and ^1H NMR chemical shifts for pyridazine mono- and bis-complexes of $[Ru^{II}(hedta)]^-$.

ton migrates to the other nitrogen base. This process broadens the ^1H NMR resonances of H3 and H6 because of the magnetic differences caused by the fluxional process. Protons H4 and H5 are made equivalent by the linkage isomerism process, and since they are more remote from the Ru^{II} center, these protons are not severely broadened. The coalescence into a single resonance at 7.48 ppm is consistent with the protonation and fluxional processes between N1 and N2 of the coordinated pyridazine. If a sample at 1:1 pyridazine: $[Ru^{II}(hedta)(H_2O)]^-$ is prepared at pD ~ 1.0, the bidentate species cannot be formed. Addition of a second equivalent of pyridazine to make the bis species is blocked. Somewhat surprisingly, once the bidentate species is formed by the initial mixing at 1:1, a rapid shift to pD ~ 1.0 does not rapidly protonate the bidentate species (cf. inset ^{1}H NMR of Figure 17). Thus forming even the weak second bond to Ru(II) protects this lone pair from rapid protonation by H_3O^+.

A time-dependent study of the ^{1}H NMR spectrum shown in Figure 17 showed no change indicative of any further N-base to η^2 type of migrations. This suggests that the barrier to form η^2-bound species pyridazine is, in fact, greater than for pyrimidine. Evidence from the pK_a data of $(NH_3)_5Os^{II}LH^{3+}$ complexes (L = pyrimidine, pyridazine, and pyrazine) (95) indicates that coordinated pyridazine is a stronger base by 1.6 log units than the pyrimidine complex (pK_a of 3.7 vs. 2.1). The enhancement in basicity over the free ligand values are 0.79 and 1.37 log units for pyrimidine and pyridazine, respectively. Even though the site of protonation is much closer to the Os^{II} center, which would present an electrostatic disadvantage, the pyridazine acts as the better coordinated base. This implies that pyridazine benefits more by the π-back donation from Os^{II}. The electrostatic effect is larger for the $[(NH_3)_5RuLH]^{3+}$ series (97), but the trend is the same. Since $[Ru^{II}(hedta)]^-$ is an anionic center, the influence of the electrostatic repulsion of the adjacent proton in $[Ru(hedta)(pydH)]^{3+}$ should follow the trend of the Os^{II} series more closely. Electrochemical studies indicate a pK_a of ~ 2.0 for pyd compared to <1.0 for pym. This implies that migrations from N1 coordination to an η^2 ring location will have a larger π-backbonding component to overcome in the activation step (Figure 8). The barrier for N-base to η^2 migration was estimated to be 24.5 kcal/mol for pyrimidine from time-dependent NMR data. One can then estimate 14 kcal/mol for loss of aromaticity and another 10 kcal/mol from bonding factors (43). The absence of an observed N1 to η^2 migration for pyridazine implies the bonding contribution must be ≥15 kcal/mol, or a total free energy barrier of ≥27 kcal/mol. This translates into a slower rate of isomerism of ca. ≥70-fold.

Structural Controls on η^2 Coordination. We recently prepared $[Ru^{II}(Me_2edda)(H_2O)_2]$ with the goal of lowering the negative charge of the Ru^{II}–pac by one unit for purposes of enhancing the attraction of Ru^{II}–pacs for the major groove of DNA (81). Subsequent studies of the affinity of potential η^2-bound ligands revealed aspects of molecular recognition in the coordination of olefinic groups with S,S-$[Ru^{II}(Me_2edda)]$. The crowded nature of two in-plane glycinato chelate arms permits only small olefins to bind as axial ligands in the cleft afforded by S,S-$[Ru^{II}(Me_2edda)]$ (*see* Scheme III). The signature electrochemical wave of complexes that form η^2 derivatives (ethylene or *cis*-2-butene) is observed near 0.68 V. Larger ligands such as branched olefins and DMU are rejected on steric grounds. 1,3-Butadiene exhibits the capacity to associate initially as a mono-η^2 adduct. The second olefinic donor of 1,3-butadiene is then poised to displace an in-plane glycinato donor. A subsequent rearrangement promotes a "walk" of the 1,3-butadiene to provide the more sta-

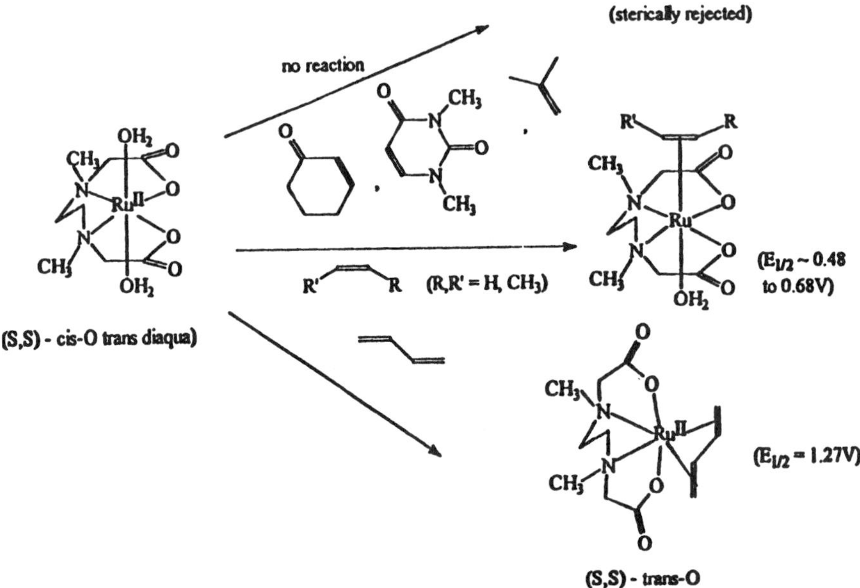

Scheme III. Molecular recognition for olefinic ligands by S,S-[Ru(Me₂edda)-(H₂O)₂] (discussed in reference 81).

ble η^4 bidentately coordinated 1,3-butadiene. This promotes a shift of the glycinato donors into the available axial sites. The ^1H NMR spectrum is consistent with s-trans coordination of the bound 1,3-butadiene. Although s-trans stereochemistry is favored by 3 kcal/mol, coordination to metals occurs most commonly as the s-cis isomer (*99–101*). Ru^{II} complexes with harder N and O donors appear to stabilize the s-trans stereochemistry. Two other Ru^{II}(1,3-butadiene) complexes are also s-trans: [Ru^{II}(2,3-dimethyl-1,3-butadiene)(acac)₂] (*102*) and [Ru(NH₃)₄(1,3-butadiene)](FP₆)₂ (*103*).

The selectivity shown by S,S-[Ru^{II}(Me₂edda)(H₂O)₂] toward larger olefins, and in particular the C5–C6 region of DMU, indicates that a more open face is required if a Ru^{II}–pac headgroup is to be used for η^2 coordination along the major groove of DNA. Models show that the unsymmetrical isomer of Me₂edda, having both glycinato donors attached at the same nitrogen, provides an accessible Ru^{II} center similar to [Ru^{II}(hedta)]⁻ in spatial freedom for entering ligands. Work is in progress to design a binuclear groove spanning chelates of Ru^{II} that utilizes this structural feature and related Ru^{II}(pac) systems.

Acknowledgements

We gratefully acknowledge the support of the Research Corporation for these studies.

References

1. Pyle, A. M.; Barton, J. K. *Prog. Inorg. Chem.* **1990**, *38*, 413.
2. *Metal–DNA Chemistry;* Tullius, T. D., Ed; ACS Symposium Series 402; American Chemical Society: Washington, DC, 1989.
3. Nielsen, P. E. *J. Mol. Recognit.* **1990**, *3*, 1.
4. Dervan, P. B. *Science (Washington, D.C.)* **1986**, *232*, 464.
5. Moser, H. E.; Dervan, P. B. *Science (Washington, D.C.)* **1987**, *238*, 645.
6. Hecht, S. M. *Acc. Chem. Res.* **1986**, *19*, 383.
7. Stubbe, J.; Kozarich, J. W. *Chem. Rev.* **1987**, *87*, 1107.
8. Reid, T.; Baldini, A.; Rand, T. C.; Ward, D. C. *Proc. Natl. Acad. Sci. U.S.A.* **1992**, *89*, 1388.
9. Tizard, R.; Cate, R. L.; Ramachandran, K. L.; Wysk, M.; Voyta, J. C.; Murphy, O. J.; Bronstein, I. *Proc. Natl. Acad. Sci. U.S.A.* **1990**, *87*, 4514.
10. Sigman, D. S. *Acc. Chem. Res.* **1986**, *19*, 180.
11. Bruhn, S. L.; Toney, J. H.; Lippard, S. J. *Prog. Inorg. Chem.* **1990**, *38*, 477.
12. Grover, N.; Gupta, N.; Thorp, H. H. *J. Am. Chem. Soc.* **1992**, *114*, 3390.
13. Barton, J. K.; Lolis, E. *J. Am. Chem. Soc.* **1985**, *107*, 708.
14. Neyart, G. A.; Grover, N.; Smith, S. R.; Kalsbeck, W. A.; Fairley, T. A.; Cory, M.; Thorp, H. H. *J. Am. Chem. Soc.* **1993**, *115*, 4423.
15. Muller, J. G.; Chen, X.; Dadiz, A. C.; Rokita, S. E.; Burrows, C. J. *J. Am. Chem. Soc.* **1992**, *114*, 6407.
16. Chen, X.; Burrows, C. A. A.; Rokita, S. E. *J. Am. Chem. Soc.* **1992**, *114*, 322.
17. Hartwig, J. F.; Pil, P. P.; Lippard, S. J. *J. Am. Chem. Soc.* **1992**, *114*, 8292.
18. Clarke, M. J.; Jansen, B.; Marx, K. A.; Kruger, R. *Inorg. Chim. Acta* **1986**, *124*, 13.
19. Mestroni, G.; Zassinovich, G.; Allesso, E.; Bontempie, A. *Inorg. Chim. Acta* **1987**, *137*, 63.
20. Welch, T. W.; Neyhart, G. A.; Goll, J. G.; Ciftan, S. A.; Thorp, H. H. *J. Am. Chem. Soc.* **1993**, *115*, 9311.
21. *Platinum and Other Metal Coordination Compounds in Cancer Chemotherapy;* Nicolini, M., Ed.; Martinus Nijhoff: Boston, MA, 1987.
22. Clarke, M. J. In *Platinum, Gold, and Other Chemotherapeutic Agents;* Lippard, S. J., Ed.; American Chemical Society: Washington, DC, 1983; Vol. 209, p 335.
23. Lippard, S. J. In *Platinum and Other Metal Coordination Compounds in Cancer Chemotherapy;* Howell, S. B., Ed.; Plenum: New York, 1991; p 112.
24. Reedjik, J.; Richtinger-Schepman, A. M.; van Oosterom, A. J.; van de Putte, P. *Struct. Bonding (Berlin)* **1987**, *67*, 53–89.
25. Lippert, B.; Arpalahti, J.; Krizanovic, O.; Micklitz, W.; Schwartz, F.; Trotscher, G. In *Platinum and Other Metal Coordination Compounds in Cancer Chemotherapy;* Nicolini, M., Ed.; Martinum Nijhoff: Boston, MA, 1987, pp 563–581.
26. Sherman, S. E.; Gibson, D.; Wang, A. H.-J.; Lippard, S. J. *Science (Washington, D.C.)* **1988**, *230*, 412.
27. Sitlani, A.; Long, E. C.; Pyle, A. M.; Barton, J. K. *J. Am. Chem. Soc.* **1992**, *114*, 2303.
28. Jenkins, Y.; Barton, J. K. *J. Am. Chem. Soc.* **1992**, *114*, 8736.
29. Sardesai, N. Y.; Zimmermann, K.; Barton, J. K. *J. Am. Chem. Soc.* **1994**, *116*, 7502.
30. Krotz, A. H.; Hudson, B. P.; Barton, J. K. *J. Am. Chem. Soc.* **1993**, *115*, 12577.
31. Sitlani, A.; Dupureur, C. M.; Barton, J. K. *J. Am. Chem. Soc.* **1993**, *115*, 12589.
32. Zhang, S.; Holl, L. A.; Shepherd, R. E. *Inorg. Chem.* **1990**, *29*, 1012.
33. Shepherd, R. E.; Zhang, S.; Lin, F.-T.; Kortes, R. A. *Inorg. Chem.* **1992**, *31*, 1457.
34. Shepherd, R. E.; Zhang, S. *Inorg. Chim. Acta* **1992**, *191*, 271.
35. Chang, C. H.; Beer, M.; Marzilli, L. G. *Biochemistry* **1977**, *16*, 33.
36. Daniel, F. B.; Behrman, E. *J. Am. Chem. Soc.* **1975**, *97*, 7352.

37. Gilkin, G. C.; Vojiskova, M.; Reva-Descalzi, L.; Palacek, E. *Nucleic Acids Res.* **1984,** *12,* 1725.
38. Palacek, E.; Rosovska, E.; Boublikova, P. *Biochem. Biophys. Res. Commun.* **1988,** *150,* 731.
39. Harman, W. D.; Taube, H. *J. Am. Chem. Soc.* **1987,** *109,* 1883.
40. Harman, W. D.; Taube, H. *Inorg. Chem.* **1987,** *26,* 2917.
41. Harman, W. D.; Taube, H. *J. Am. Chem. Soc.* **1988,** *110,* 5725.
42. Harman, W. D.; Taube, H. *J. Am. Chem. Soc.* **1988,** *110,* 7555.
43. Chen, Y.; Lin, F.-T.; Shepherd, R. E. *Inorg. Chem.* **1997,** in press.
44. Shepherd, R. E.; Zhang, S. *Transition Met. Chem. (London)* **1994,** *19,* 146.
45. Cardone, R.; Taube, H. *J. Am. Chem. Soc.* **1987,** *109,* 8101.
46. Cardone, R.; Harman, W. D.; Taube, H. *J. Am. Chem. Soc.* **1989,** *111,* 5969.
47. Durante, V. A.; Ford, P. C. *Inorg. Chem.* **1979,** *18,* 588.
48. Chow, M. H.; Brunschweig, B. S.; Creutz, C.; Sutin, N.; Yeh, A.; Chang, R. C.; Lin, C.-T. *Inorg. Chem.* **1992,** *31,* 5347.
49. Roberts, J. D.; Van Houten, B.; Qu, Y.; Farrell, N. P. *Nucleic Acids Res.* **1989,** *17,* 9719.
50. Farrell, N.; Qu., Y. *Inorg. Chem.* **1989,** *28,* 3416.
51. Farrell, N.; Qu, Y.; Hacker, M. P. *J. Med. Chem.* **1990,** *33,* 2179.
52. Farrell, N.; Qu, Y.; Feng, L.; Van Houten, B. *Biochemistry* **1990,** *29,* 9522.
53. Farrell, N. P.; de Almeida, S. G.; Skov, K. A. *J. Am. Chem. Soc.* **1988,** *110,* 5018.
54. Shepherd, R. E. In *Trends in Inorganic Chemistry;* Goyathri, A., Ed.; Research Trends: Trivandrum, India, 1993; Vol. 3, pp 503–530.
55. (a) Kortes, R. A.; Shepherd, R. E. *Inorg. Chim. Acta* **1997,** in press; (b) Kortes, R. A.; Shepherd, R. E. *Trans. Met. Chem. (London)* **1997,** in press.
56. Alul, R.; Cleaver, M. B.; Taylor, J.-S. *Inorg. Chem.* **1992,** *31,* 3636.
57. Shuhmann, E.; Altman, J.; Karaghiosoff, K.; Beck, W. *Inorg. Chem.* **1995,** *34,* 2316.
58. Dwyer, F. P.; Mayhew, E.; Roe, E. M. F.; Shulman, A. *Brit. J. Cancer* **1965,** *19,* 195.
59. Clarke, M. J. In *Metal Ions in Biological Systems;* Sigel, H., Ed; Dekker: New York, **1980,** *11,* 231.
60. Margalit, R.; Gray, H. B.; Podbielski, L.; Clarke, M. J. *Chemico-Biological Interactions* **1986,** *59,* 231.
61. Cauci, S.; Viglino, P.; Esposito, G.; Quadrifoglio, F. *J. Inorg. Biochem.* **1991,** *43,* 739.
62. Clarke, M. J.; Galang, R. D.; Rodriguez, V. M.; Kumar, R.; Pell, S.; Bryan, D. M. In *Platinum and Other Metal Coordination Compounds in Cancer Chemotherapy;* Nicolini, M., Ed.; Martinus Nijhoff: Boston, MA, 1987; pp 582-600.
63. Allesso, E.; Mestroni, G.; Mardin, G.; Attia, W. M.; Calligaris, M.; Sava, G.; Zorzet, S. *Inorg. Chem.* **1988,** *27,* 4909.
64. Cauci, S.; Allesso, E.; Mestroni, G.; Quadrafoglio, F. *Inorg. Chem.* **1987,** *137,* 19.
65. Sava, G.; Zorzet, S.; Giraldi, T.; Mestroni, G.; Zassinovich, G. *Eur. J. Cancer Clin. Oncol.* **1984,** *20,* 841.
66. Keppler, B. K.; Rupp., W.; Juhl, U. M.; Endres, H.; Niebel, R.; Balger, W. *Inorg. Chem.* **1987,** *26,* 4366.
67. Keppler, B. K.; Weke, D; Endres, H.; Rupp, W. *Inorg Chem.* **1987,** *26,* 844.
68. Keppler, B. K.; Rupp, W. *J. Cancer Res. Clin. Oncol.* **1986,** *111,* 166.
69. Keppler, B. K.; Henn, M.; Juhl, U. M.; Berger, M. R.; Niebel, R.; Wagner, F. E. *Prog. Clin. Biochem. Med.* **1989,** *10,* 41.
70. Garzon, F. T.; Berger, M. R.; Keppler, B. K.; Schmahl, D. *Cancer Chemother. Pharmacol.* **1987,** *19,* 347.
71. Keppler, B. K.; Schmahl, D. *Arzneim-Forsch.* **1986,** *36,* 1822.
72. (a) Vilaplana, R. A.; Gonzalez-Vilchez, F.; Ruiz-Valero, C. *Inorg. Chim. Acta* **1994,** *224,* 15; (b) Zhang, S.; Shepherd, R. E. *Inorg. Chem.* **1988,** *27,* 4712.

73. Elliott, M. G.; Zhang, S.; Shepherd, R. E. *Inorg. Chem.* **1989**, *28*, 3036.
74. Shimizu, K. *Bull. Chem. Soc. Jpn.* **1977**, *50*, 2921.
75. Shepherd, R. E.; Zhang. S.; Dowd, P.; Choi, G.; Wilk, B.; Choi, S.-C. *Inorg. Chim. Acta* **1990**, *174*, 249.
76. Matsubara, T.; Creutz, C. *Inorg. Chem.*, *18*, 1956.
77. Diamantis, A. A.; Dubrawski, J. W. *Inorg. Chem.* **1983**, *22*, 1934.
78. Diamantis, A. A.; Dubrawski, J. V. *Inorg. Chem.* **1981**, *20*, 1142.
79. Hoard, J. L.; Kennard, C. H. L.; Smith G. S. *Inorg. Chem.* **1963**, *2*, 316.
80. Terrill, J. B.; Reilley, C. N. *Inorg. Chem.* **1966**, *5*, 1988.
81. Zhang, S.; Chen, Y.; Shepherd, R. E. *Inorg. Chim. Acta* **1995**, *230*, 77.
82. (a) Elliott, M. G.; Shepherd, R. E. *Inorg. Chem.* **1988**, *27*, 3322; (b) Shepherd, R. E.; Zhang, S.; Chen, Y. *Inorg. Chim. Acta* **1996**, in press.
83. Zhang, S.; Shepherd, R. E. *Transition Met. Chem. (London)* **1992**, *17*, 190.
84. Zhang, S.; Shepherd, R. E. *Transition Met. Chem. (London)* **1992**, *17*, 97.
85. Zhang, S.; Shepherd, R. E. *Inorg. Chem.* **1994**, *33*, 5262
86. Zhang, S.; Shepherd, R. E. *Inorg. Chim. Acta* **1989**, *163*, 237.
87. Matsubara, T.; Creutz, C. *J. Am. Chem. Soc.* **1987**, *100*, 6255.
88. Gullino, P. M. *Adv. Exp. Biol. Med.* **1976**, *75*, 521.
89. Clarke, M. J. In *Met. Complexes Cancer Chemother.*; Keppler, B. K., Ed.; VCH: Weinheim, 1993; pp 129–156.
90. Brown, D. J. In *Comprehensive Heterocyclic Chemistry*; Katrintzky, A. R.; Rees, C. W., Eds.; Pergamon: New York, 1984; Vol. 3, p 59.
91. Lenhert, A. G.; Castle, R. N. In *Pyridazines. The Chemistry of Heterocyclic Compounds*; Castle, R. N., Ed.; Wiley: New York, 1973. **PAGE NUMBER(S)?**
92. Barlin, G. G. *The Pyrazines. The Chemistry of Heterocyclic Compounds*; Wiley: New York, 1982; p 7.
93. Tjebbes, J. *Acta. Chem. Scand.* **1962**, *16*, 916.
94. Cox, J. P. *Tetrahedron* **1963**, *19*, 1175.
95. Sen, J.; Taube, H. *Acta Chem. Scand. Ser. A.* **1979**, *33*, 125.
96. Lay, R. A.; Magnuson, R. H.; Sen, J.; Taube, H. *J. Am. Chem. Soc.* **1982**, *104*, 7658.
97. Ford, P.; Rudd, D. F. P.; Gaunder, R. G; Taube, H. *J. Am. Chem. Soc.* **1968**, *90*, 1187.
98. Wiberg, R. B.; Lewis, T. P. *J. Am. Chem. Soc.* **1970**, *92*, 7154.
99. Kreiter, C. G. *Adv. Organomet. Chem.* **1986**, *26*, 297.
100. Erker, G.; Kruger, C.; Muller, G. *Adv. Organomet. Chem.* **1985**, *24*, 1.
101. Bennett, M. A.; Bruice, M. I.; Matheson, T. W. *Comprehensive Organometallic Chemistry*; Wilkinson, G.; Stone, F. G. A.; Abel, E. W., Eds.; Pergamon: New York, 1982; Vol. 4, pp 741–754.
102. Ernst, R.; Melendez, E.; Stahl, L.; Ziegler, M. L. *Organometallics* **1991**, *10*, 3635.
103. Sugaya, T.; Atsuko, T.; Sago, H.; Sano, M. *Inorg. Chem.* **1996**, *35*, 2692.

The Role of Inorganic Chemistry in Cellular Mechanisms of Host Resistance to Disease

James K. Hurst

Department of Chemistry, Washington State University, Pullman, WA 99164–4630

Phagocytic cells associated with host resistance to disease appear to be capable of generating a variety of inorganic oxidants that function as microbicidal agents. In neutrophils and related cells containing myeloperoxidase, hypochlorous acid is a primary microbicide; the nature of oxidants produced by other types of phagocytic cells are less well characterized, but may involve metal-mediated reactions of H_2O_2 or intermediary formation of peroxonitrite (ONO_2^-) ion. The biochemical role of these oxidants is reviewed from the perspective of mechanistic inorganic chemistry. Results from recent studies suggesting unique roles for HCO_3^- and CO_2 in these processes are also described.

$\mathbf{U}$PON ENCOUNTERING A BACTERIUM, phagocytic white blood cells undergo a progression of biochemical transformations that lead to isolation of the organism within a highly inimical environment (1). Prominent among these transformations is activation of a respiratory chain that catalyzes the one-electron reduction of oxygen to superoxide ion (2). As discussed herein, the subsequent fate of $\cdot O_2^-$ ion is the subject of considerable debate and experimental investigation. Remarkably, the conceptual basis of almost all current discussions of the identities of the ultimate toxins and their modes of action is the seminal mechanistic work (3–7) on the redox chemistry of main group elements carried out by Henry Taube and associates four decades ago. Not only do the general mechanistic principles laid out in these studies serve as a design model for investigating similar reactions in the more complex biological arena, but his early research on the chemistry of oxygen, hydrogen peroxide, and the oxides of chlorine and nitrogen, in particular, remains directly relevant to current mechanistic issues concerning cellular disinfection processes. Notable among the latter are his studies on the reaction between HOCl and H_2O_2 to form electronically excited $(^1\Delta)O_2$ (4–5) and reactions between HNO_2 and H_2O_2 to form peroxonitrous acid (ONO_2H) and its subsequent isomerization to nitric acid (6).

Phagocytosis: A Closer Look

The events comprising bacterial phagocytosis are illustrated stylistically in Figure 1. Binding at receptor sites on the cellular membranes of phagocytic cells elicits a secondary-messenger cascade that leads ultimately to phosphorylation of specific cytosolic proteins (8). This phosphorylation triggers assembly of cytosolic and membrane-localized proteins to form a functioning electron transport chain. The respiratory chain appears to be unusually simple, containing only FAD and b-type hemes as redox components (8–9). It is vectorially organized across the plasma membrane, so that electrons from cytosolic donors (NADPH) are translocated to the external environment where the O_2 reductase site is located (10). Simultaneously with respiratory activation, the plasma membrane invaginates, surrounding the bacterium and eventually pinching off

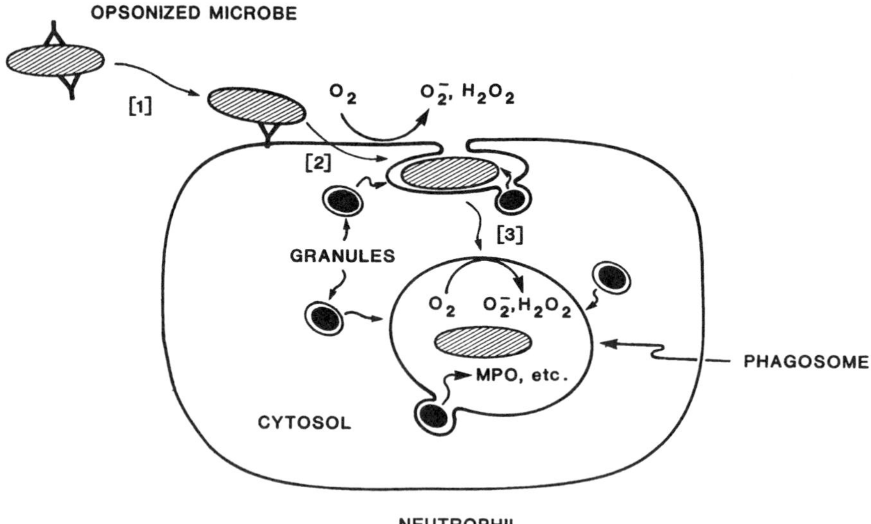

Figure 1. Diagram of phagocytosis by neutrophils. [1] Binding of the opsonized bacterium at receptor sites initiates a secondary-messenger cascade that evokes the respiratory burst; [2] phagocytosis proceeds by membrane invagination, which ultimately pinches off [3] to form the new vacuole (phagosome). Simultaneous degranulation leads to both extracellular secretion and intraphagosomal accumulation of granule components. MPO means myeloperoxidase. (Reproduced with permission from reference 10. Copyright 1989 CRC Press.)

to form an internalized vacuole (phagosome) containing the entrapped bacterium. Because the membrane everts in this process, the respiratory chain is now oriented to generate $\cdot O_2^-$ within the phagosome, the bulk of which appears to undergo nonenzymatic disproportionation to form H_2O_2. In neutrophils (the predominant type of phagocytic white blood cells), these events are accompanied by simultaneous migration of lysosomal granules containing enzymes and related biopolymers from the cytosol to the phagosomal membrane, where the membranes fuse and the lysosomal contents are discharged into the phagosomal volume (*1*). Among the lysosomal enzymes is a unique peroxidase (myeloperoxidase, MPO) that is capable of catalyzing the two-electron oxidation of Cl^- to HOCl (*11*). This enzyme is present in astonishingly high concentrations, comprising 2–5% of the total weight of the neutrophil (*12*). The entire process from binding to lysosomal degranulation and killing of the bacterium requires only a few minutes; in many cases the bacterial cell morphology is indistinguishable from normal viable cells, indicating that gross physical disruption of the bacterial envelope is not the cause of death (*13–14*).

Although the immediate source of electrons for O_2 reduction is NADPH, this reductant is cyclically regenerated by glucose oxidation via the hexose monophosphate pathway (*1*). Consequently, the oxidized respiratory end product is CO_2, which can be expected to rise severalfold above the normal physiological levels of ~25 mM during the course of stimulated respiration. The O_2 reductase site is thought to be the heme prosthetic group of the cytochrome (*8, 9*). Although structural characterization is incomplete (e.g., evidence suggesting that the cytochrome contains two hemes per FAD is accumulating), resonance Raman spectral analyses indicate that the environment of the heme (or both hemes) is 6-coordinate low spin in both Fe^{II} and Fe^{III} oxidation states (*14, 15*). This strong axial ligation precludes direct O_2 binding to the heme iron, forcing the reaction to proceed by an "outer-sphere" mechanism (*16*), hence, in one-electron steps. Thus, formation of $\cdot O_2^-$ as the immediate product is ensured.

Two lines of evidence indicate that oxidative mechanisms are important to bacterial killing. One is that many bacteria are killed much less efficiently by neutrophils under anaerobic conditions than in the presence of oxygen (*17, 18*). The other is that individuals with a congenital defect known as chronic granulomatous disease, characterized by a defective $\cdot O_2^-$-generating oxidase but otherwise normal phagocytic capabilities, suffer chronic life-threatening bacterial infections (*1*). Thus, formation of O_2-derived oxidants is essential to cellular defense mechanisms. In contrast, the role of MPO has been controversial. Individuals with hereditary MPO deficiency, characterized by the absence of a functional peroxidase but an otherwise normal phagocytic response, exhibit only minor clinical manifestations of this disease (*19, 20*). Nonetheless, MPO-deficient neutrophils are by several criteria considerably less effective than normal neutrophils in in vitro studies of bactericidal potency (*19*).

In addition to its unique capacity to efficiently catalyze Cl^- oxidation, MPO catalyzes two-electron oxidation of other halides and pseudohalides (*21,*

22), as well as one-electron oxidation of organic substrates (23). This diversity of catalytic capability calls into question its true physiological function. However, based upon competition studies, the preferred substrate in normal physiological environments appears to be Cl⁻ (22). Furthermore, evidence consistent with direct chlorination within the phagosome has been obtained using a recoverable fluorescent probe (Q. Jiang, unpublished observations). In the probe studies (Figure 2), fluorescein was attached to 0.5–2.0-μm carboxy-derivatized polyacrylamide spheres via cystamine linker groups. Following opsonization, that is, targeting with serum-derived antibodies and complement, the particles were avidly phagocytosed by isolated neutrophils. Fluorescence changes consistent with chlorination of the probe were observed during and immediately following phagocytosis. The dye was subsequently recovered in near-quantitative yield by cell lysis, followed by its release from the particle by cleavage of the cystamine disulfide bond upon addition of a thiol. The only reaction products detected by HPLC analysis co-chromatographed with and had spectroscopic features analogous to authentic samples of mono- and dichlorofluorescein thiols; these structures have now been confirmed by ion electrospray mass spectrometry (D. F. Barofsky and D. A. Griffin, unpublished observations). Thus, MPO-catalyzed intraphagosomal chlorination undoubtedly occurs. Since HOCl is freely diffusible from the enzyme active site (11) and the same ring-chlorinated products are obtained from reaction between HOCl and fluorescein (24), one infers that HOCl can be formed within the phagosome.

To summarize, the most probable primary set of reactions leading to formation of bactericidal agents in normal neutrophils is enzyme-catalyzed one-electron reduction of O_2 by glucose, followed by disproportionation of the respiratory end-product $^{\bullet}O_2^-$ to H_2O_2 and its MPO-catalyzed oxidation of Cl⁻ to HOCl (Figure 3). Alternatively, HOCl could be formed by direct reaction of $^{\bullet}O_2^-$ with MPO, yielding compound III (the $Fe^{III}-{}^{\bullet}O_2^-$ adduct) as an intermediary species (25, 26)

$$\text{MPO}(Fe^{III}) + {}^{\bullet}O_2^- + H^+ \rightarrow \text{compound III } (Fe^{III}-{}^{\bullet}O_2H)$$

followed by its one-electron reduction by a second O_2^- to give compound I (a ferryl π-cation):

$$\text{compound III} + {}^{\bullet}O_2^- + H^+ \rightarrow \text{compound I } (Fe^{IV}=O\ \pi\text{-cation}) + H_2O + O_2$$

which is the active form of the catalyst (27)

$$\text{compound I} + Cl^- + H^+ \rightarrow \text{MPO} + \text{HOCl}$$

and the same product formed by direct reaction of MPO with H_2O_2. An additional function of $^{\bullet}O_2^-$ may be to prevent accumulation of compound II, a ferryl

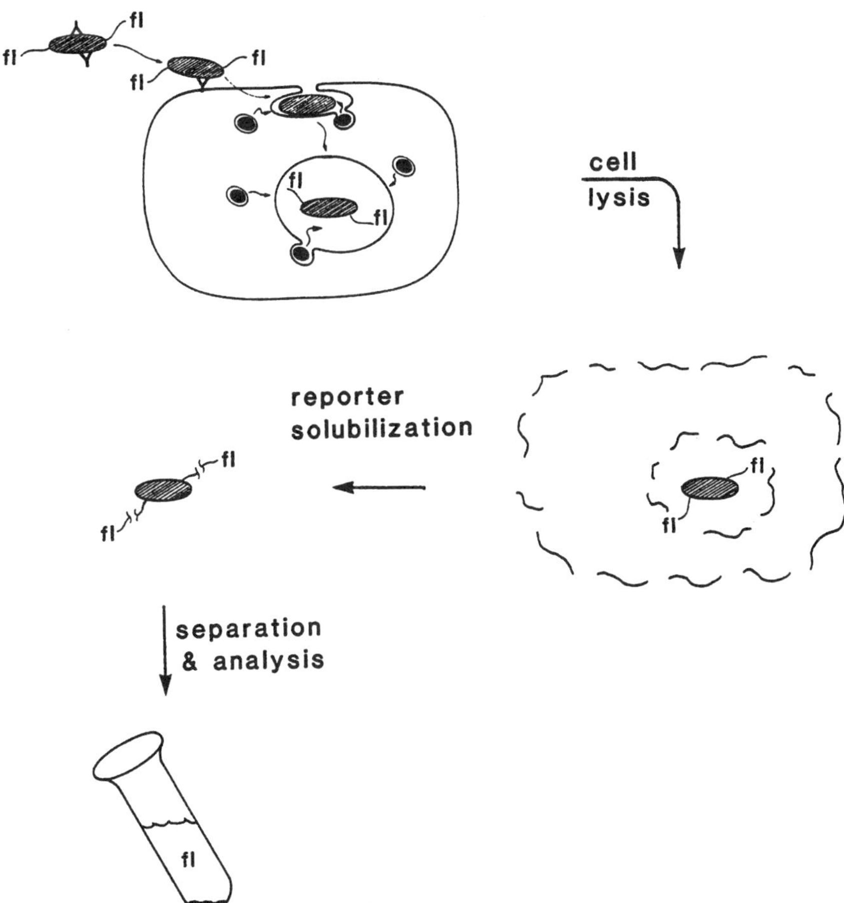

Figure 2. Recoverable probes for detecting chlorination and other postphagocytic events; the symbol fl refers to the fluorescein moiety. In a typical experiment, the fluorescein-labeled particles are vortex-mixed with neutrophils to elicit binding, and the subsequent changes in fluorescence properties of the dye are monitored. The dye is then recovered for chemical analysis by (1) homogenization of the neutrophils to release the particles, (2) cleavage of the linker disulfide bond with dithiothreitol to release the dye, and (3) centrifugation to remove cell debris and the now-unlabeled polyacrylamide beads. (Reproduced with permission from reference 78. Copyright 1993 Plenum Press.)

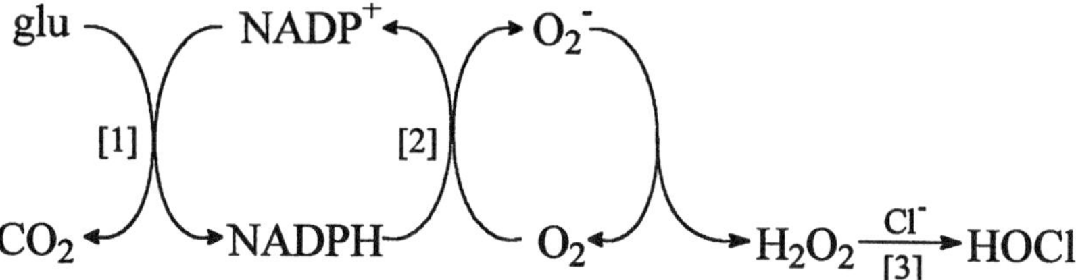

Figure 3. Flow diagram for respiratory generation of HOCl and CO$_2$ from glucose and O$_2$. Catalysis by [1] the enzymes involved in the hexose monophosphate shunt, [2] the NADPH oxidase, and [3] myeloperoxidase. The symbol glu refers to glucose.

form of the catalyst that is incapable of oxidizing Cl$^-$, by reducing it to the native ferric MPO state (*28*) and thereby returning it to the catalytic cycle, viz.:

$$\text{compound II (Fe}^{IV}\text{=O)} + {}^{\bullet}O_2^- + 2H^+ \rightarrow MPO + O_2 + H_2O$$

In any event, directly or indirectly, ${}^{\bullet}O_2^-$ is an appropriate substrate for MPO-catalyzed formation of HOCl. Hypochlorous acid might itself be the bactericidal agent produced by these cells or, alternatively, might react with endogenous amines to form the corresponding N-chloramines, which are also potent bactericides. This issue of the identity of the ultimate toxin is presently unresolved (*10, 29*) but will not affect our discussion of bactericidal mechanisms because the chemical principles governing reactivities are the same for the two oxidants.

MPO-Dependent Mechanisms: The Chemical Basis for HOCl Toxicity

Hypochlorous acid is highly toxic to prokaryotic cells (*1, 19*). A comparison of in vitro bactericidal assays for a prototypic bacterium, *Escherichia coli*, using various oxidants under roughly comparable conditions is given in Table I. In this comparison, HOCl is at least 10^3-fold more toxic than the more strongly oxidizing H_2O_2, ONO_2^- ion, and ${}^{\bullet}OH$ radical (*30*). Approximately 10^8 molecules of HOCl are required to kill one *E. coli* cell (*31*); this means that 1 mL of a properly distributed commercial bleach solution (5% HOCl) would be sufficient to kill 5 g of cells! Other oxidants, for example, H_2O_2 in the presence of Cu^{2+} and a reducing agent (*32*), and $HCO_3^{\bullet}$ radical (*33*), also exhibit high toxicity, approaching that of HOCl (Table I). How can we understand these wide variations in toxicity? Since the oxidant quantities required to inflict lethal damage in the more effective bactericidal systems are remarkably small, cellular death must be associated with destruction of a limited number of vulnerable sites within the bacterium. Correspondingly, toxicity is associated with oxidant

Table I. LD$_{90}$ for Inorganic Toxins Against *Escherichia coli* in Suspension

Toxin	LD$_{90}$ (molecules/ cell)	Relative Toxicity	Cell Density (viable cells/mL)	E$_7^o$ (V)	Refs.
HOCl	0.4–$5.0 \times 10^{8\,a,b}$	1–12	5×10^8	0.17 (HOCl/Cl$^{\bullet}$); 1.08 (HOCl/Cl$^-$)	31, 72
H$_2$O$_2$	$>3 \times 10^{11}$	$<2 \times 10^{-3}$	10^7–10^9	0.33 (H$_2$O$_2$/$^{\bullet}$OH); 1.36 (H$_2$O$_2$/H$_2$O)	32, 73
H$_2$O$_2$/Cu/ ascorbate	$10^{8-9\,c}$	~1	10^7–10^9		32
HCO$_3^{\bullet}$	~$10^{10\,d}$	0.05	10^5–10^6	~1.7 (HCO$_3^{\bullet}$/HCO$_3^-$)^f	33
$^{\bullet}$OH	$>2 \times 10^{11\,a}$	$<2 \times 10^{-3}$	10^6	~2.3 ($^{\bullet}$OH/H$_2$O)	33, 74
ONO$_2^-$	$2.4 \times 10^{11\,a}$	2×10^{-3}	10^6	1.4 (ONO$_2^-$/$^{\bullet}$NO$_2$)	56, 66
$^{\bullet}$NO	$>6 \times 10^{12\,a}$	2×10^{-5}	10^6	0.4 ($^{\bullet}$NO/^{3}NO$^-$)	56, 66
CuCl	$3 \times 10^{6\,e}$	200	10^7–10^9	0.4 (Cu^{2+}/CuCl$_2^-$)^g	73

NOTE: Values are for strain ATCC 25922 in 0.05–0.10 M phosphate, pH 7.4, containing 0.15 M NaCl, unless otherwise specified. LD$_{90}$ is dose level required to kill 90% of the cells. Relative toxicity values are scaled to HOCl at pH 7.4 (LD$_{90}$ = 5×10^8 HOCl/*E. coli*). E$_7^o$ values are standard reduction potentials (vs. NHE) at pH 7.0.

^{a}NaCl absent.

^{b}pH dependent (pH 5.0–7.4).

^{c}[Cu^{2+}] and [ascorbate] dependent.

^{d}0.1 M carbonate buffer, pH 6.5–7.4.

^{e}Anaerobic media.

^{f}Calculated from E°($^{\bullet}$CO$_3^-$/CO$_3^{2-}$) = 1.6 V (75) and the proton dissociation constants (76).

^{g}Calculated from E°(Cu^{2+}/Cu$^+$) = 0.16 V (73) and the cuprous chloride association constant (77).

selectivity for these target sites, rather than its capacity to inflict massive, non-specific oxidative damage to the cells. The key to understanding the toxicity of an oxidant therefore lies in understanding its inherent chemical reactivity.

E. coli are killed within 100 ms after exposure to lethal doses of HOCl (*31*). For this oxidant, the vulnerable sites are clearly among the more reactive biomolecules. A useful model system to explore the reactivity of HOCl is its oxidation of H$_2$O$_2$:

$$HOCl + H_2O_2 \rightarrow O_2 + H_2O + H^+ + Cl^- \tag{1}$$

A crucial early observation by Cahill and Taube (*4*) was that both O atoms in O$_2$ were obtained from H$_2$O$_2$, which precludes any reasonable radical reaction mechanism. One-electron oxidations by HOCl are also unlikely on thermodynamic grounds since they require formation of high-energy $^{\bullet}$OH or $^{\bullet}$Cl radicals as reaction products (*34*). As illustrated by Taube's early mechanistic studies (*3–7*) on atom transfer reactions, two-electron oxidation requires some form of incipient bond formation. Rate measurements provided indirect evidence for this type of association. Specifically, reaction 1 was dominated by a pathway whose rate law is $-d[HOCl]_o/dt = k[H_2O_2]_o[HOCl]_o$ where the subscript o refers to total reactant concentration. The bimolecular rate constant (k) exhibited a bell-shaped pH–rate profile indicating that the true reactant pairs are either HOCl and HO_2^- or OCl^- and H_2O_2 (35). The reactions of H_2O_2 with a series of chlorine(+1) compounds for which the dissociable proton was replaced by a nondissociable electron-withdrawing group (X) gave rate laws of the form $d[O_2]/dt = k[HO_2^-][X–Cl]$, with the rate constant for reaction of *tert*-butyl hypochlorite approaching the value calculated for reaction between HOCl and HO_2^- (36).

These observations implicate HOCl and HO_2^- as the true reactants in the HOCl–H_2O_2 reaction. What is remarkable about this reaction is that the other possible pairs, namely, H_2O_2 and HOCl, H_2O_2 and OCl^-, or HO_2^- and OCl^-, are unreactive, despite there existing a relatively large thermodynamic driving force for reaction between them (35). [A slow reaction occurs between HOCl and H_2O_2, the rate law for which is $d[O_2]/dt = k[HOCl][H^+][Cl^-]$, independent of the H_2O_2 concentration; the rate is enhanced in acetate and phthalate buffers, possibly by general acid catalysis (35). In physiological environments, HCO_3^- might cause similar effects (D. T. Sawyer, personal communication), although this possibility has not been examined.]

The unique feature of the HOCl and HO_2^- reactant pair is that it combines a relatively electrophilic chlorine atom with a strongly nucleophilic HO_2^- ion, which favors electrophile–nucleophile interactions of the type shown in Figure 4. In this model, incipient bond formation is thought to lead to two-electron transfer from the electronegative hydroperoxide oxygen atom to chlorine, leading to net oxidation–reduction. Additional supporting evidence consistent with this transition state structure are the observations that (1) di-*tert*-butyl hydroperoxide, which has no dissociable proton, is unreactive toward HOCl (D. T. Sawyer, personal communication) and (2) the rate constant for reaction of HO_2^- decreases proportionately with decreasing electron-withdrawing character of the chlorine substituent group (X), that is, with decreasing electrophilic character of the chlorine atom (Figure 4). Whether a discrete ClOOH intermediate is formed, as has been proposed for the reaction between Cl_2 and H_2O_2 in acidic media (4, 37), cannot be established from the kinetic data.

Based upon these kinetic properties, we would expect HOCl to react preferentially with nucleophilic centers in biomolecules, and, indeed, this is what is observed. Hypochlorous acid displays an exceptionally wide range of reactivity toward prototypic biological partners, rapidly oxidizing electron-rich π-delocalized centers such as nitrogen heterocycles (hemes or nucleotide bases), iron–sulfur clusters, and conjugated polyenes (e.g., carotenes), as well as amino acids containing highly polarizable sulfur atoms and amines, while being virtually unreactive toward compounds not possessing nucleophilic sites (38).

This same selectivity has been shown to extend to biomolecules within bacteria. Specifically, oxidation of sulfhydryl substituents and N-chlorination of

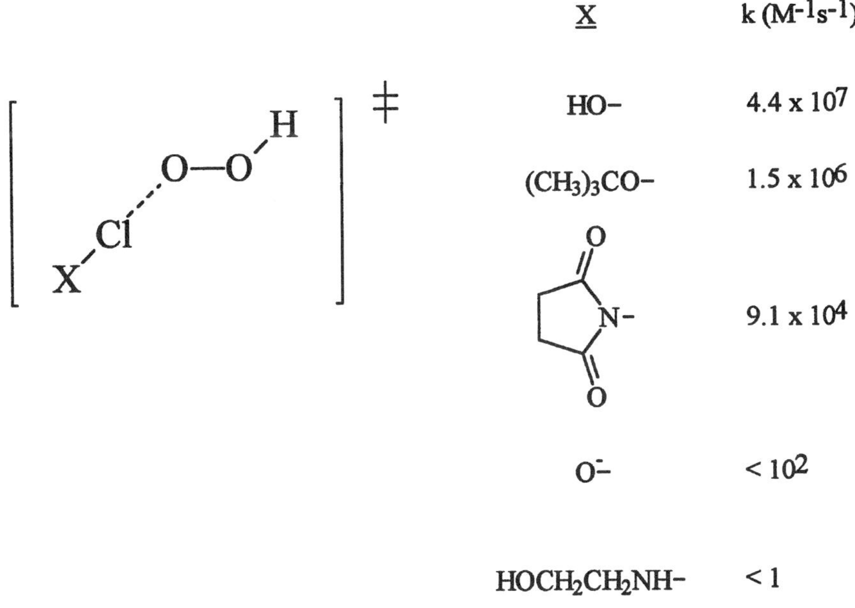

$\underline{X}$	$k\ (M^{-1}s^{-1})$
HO–	4.4×10^7
$(CH_3)_3CO–$	1.5×10^6
(succinimido) N–	9.1×10^4
O^-	$< 10^2$
$HOCH_2CH_2NH–$	< 1

Figure 4. Hypothetical transition-state structure and second-order rate constants for the reaction of hydroperoxide ion with chlorine (+1) compounds. Rate constants are defined by the rate law $d[O_2]/dt = k[HO_2^-][X–Cl]$; the critical parameter controlling reactivity is the electrophilic character of the chlorine atom. Kinetic data are from references 35 and 36.

amines on the bacterial envelope (39, 40) and oxidation of plasma membrane-localized respiratory components [cytochromes (41) and iron–sulfur cluster-containing dehydrogenases (42)] and carotene pigments (38) are early events in the progressive oxidation of bacteria by HOCl. Selectivity for these sites is sufficiently high that titration curves for oxidation of particular sites (or loss of biological function) can be constructed against the amount of added HOCl. Comparison of these titrimetric curves with analogous survival curves for the organism provides a means of assessing relationships between oxidation at these sites and cellular death. The pattern that has emerged from numerous studies of this type (reviewed in reference 10) is that HOCl kills bacteria by selectively inactivating proteins located in their plasma membranes that are associated with energy transduction, including active transport of metabolites, ATP synthesis, and (in respiring organisms) electron transport (43). Without these capabilities, the cell is unable to generate metabolic energy and is functionally dead.

The molecular sites of oxidative attack by HOCl have not yet been identified for any of these dysfunctions (42). However, we have recently found in *E. coli* that inactivation of F_1-ATPase, an $\alpha_3\beta_3\gamma$-multimeric component of the

F_oF_1-ATP synthase, involves cumulative inactivation of each (α, β, and γ) of the protein subunits (*44*). (As the central element for chemiosmotic coupling of ATP hydrolysis/synthesis to all other forms of cellular energy, the F_oF_1-ATP synthase is critical to cell survival.) The extent of oxidative damage to the individual subunits was apparently very minor because it could not be detected by two-dimensional gel electrophoresis, a technique that under favorable circumstances can detect modifications of a single amino acid within a protein. Thus, dramatic loss of activity was not accompanied by any substantial changes in protein structure, indicating that only a very limited number of target sites were attacked. Consequently, HOCl must have exhibited high selectivity toward amino acid groups that are essential for enzymatic activity.

MPO-Independent Toxicity: Alternate Oxidants

The Role of H_2O_2. Despite the evidence that HOCl is a primary microbicide generated by the respiratory burst (Figure 3), the clinical manifestations of hereditary MPO deficiency are very minor (*19, 20*). Since oxidative mechanisms are central to host cellular defense mechanisms against invading pathogens (*17, 18*), this observation implies that phagocytic cells are also capable of generating other oxidants that are effective microbicides. The identities of these oxidants are presently unknown.

Historically, the search for the alternate oxidant has focused on H_2O_2, which is known to accumulate during the respiratory burst (*45*). Many bacteria, particularly pathogenic ones (*46*), are resistant to relatively large amounts of H_2O_2 under most medium conditions. For example, dilute cellular suspensions of *E. coli* can survive for extended periods in media containing as much as 0.1 M H_2O_2 (Table I). However, in the presence of trace amounts of cupric ion and a reducing agent, the toxicity of H_2O_2 is markedly potentiated, approaching that of HOCl (Table I). This enhanced toxicity has generally been ascribed to the ability of the metal ion to catalyze formation of hydroxyl radical through Fenton-type reactions (*47, 48*); alternatively, hypervalent cupryl or copper–peroxo complexes might be formed by reaction between Cu^I and H_2O_2 (*49, 50*). [Interestingly, we have found that ferric ion is totally ineffective in promoting H_2O_2 toxicity under a wide variety of medium conditions, despite apparently being capable of very similar chemistry (*32*).]

Hydroxyl radical, when generated randomly by pulse radiolysis in the external medium, is not toxic to bacteria in dilute suspension (Table I). This is thought to be a consequence of its extreme oxidizing potential, which precludes selectivity in reactions with biomolecules and also leads to scavenging by buffer components, generating considerably less reactive secondary radicals. In recognition of this problem, a "site-specific" mechanism has been proposed (*51*) in which the catalytic copper ions confer selectivity for bacterial targets by binding at specific coordination sites on the bacterial envelope. Cuprous ion-catalyzed generation of •OH would then lead to destruction of biomolecules in the immediate vicinity of the coordination site. This reaction model has received widespread acceptance and is supported indirectly by data from various in vitro studies (*51*).

If •OH is nontoxic because it is too short-lived and too nonselective to inflict oxidative damage at vulnerable bacterial sites with a significant degree of probability, might not a secondary radical formed from •OH that is less reactive—and, consequently, longer lived and more selective toward potential reductants—show increased bactericidal capabilities? We examined this possibility by performing radiolysis experiments of bacteria suspended in bicarbonate-containing buffers under conditions where nearly all the •OH formed was converted to bicarbonate radical (*33*). Under these conditions, radiolytic killing was markedly enhanced. The effect was shown to be due to externally generated radicals by addition of •OH scavengers to the aqueous medium, which completely protected the bacteria from the enhanced killing.

A typical result is illustrated in Figure 5. Physiological fluids generally contain ~25 mM HCO_3^- ion; this level could be severalfold higher within

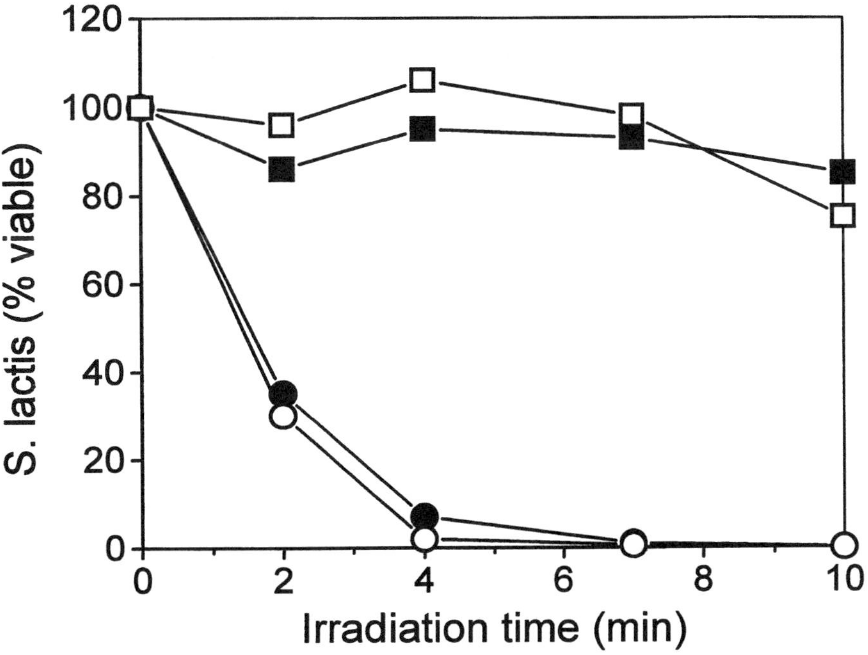

Figure 5. γ-Irradiation of Streptococcus lactis *(2 x 10⁵ cells/mL) (adapted from reference 33). Squares: 0.1 M NaH₂PO₄/Na₂HPO₄, pH 7.4; circles: 0.1 M NaHCO₃/ CO₂, pH 7.4. Open symbols refer to media in which 0.15 M NaCl has been added and solid symbols refer to media containing no added chloride ion. The ordinate is the percent survival based upon unirradiated controls.*

phagocytic cells undergoing stimulated respiration (Figure 3). Therefore, formation of $HCO_3^{\bullet}$ by reaction of HCO_3^- with $^{\bullet}OH$ generated in copper-catalyzed reactions, or possibly by direct reaction of HCO_3^- with hypervalent copper species, could be an important physiological bactericidal mechanism. If so, this set of reactions constitutes an alternative to the "site-specific" model. The fundamental difference between the two is that in the latter specificity is dictated by the bacterial copper-binding sites, but in the former the most reactive sites within the bacterium are selected for oxidation according to the inherent reactivity of the oxidant. Both concepts are illustrated in the scheme given in Figure 6.

One conceptual problem with alternatives to the "site-specific" hypothesis is that the biological milieu contains relatively high concentrations of reactive molecules that might be expected to compete effectively with the bacterium for phagocyte-generated oxidative toxins. For example, in dilute bacterial suspensions, addition of relatively low concentrations of oxidant scavengers completely protected bacteria from the lethal effects of $HCO_3^{\bullet}$ (33). However, spatial confinement within phagosomes will promote reaction of short-lived oxidants with the bacterium. Specifically, when the physical dimensions of the reaction compartment are small relative to the mean diffusion length of the oxidant (i.e., the distance it travels before reacting with an aqueous-phase partner), scavengers will no longer efficiently protect the entrapped particle from oxidation. The critical dimensions of the compartment are dictated by the intrinsic reactivity of the oxidant and the composition of the aqueous medium.

To evaluate the influence of compartmentation upon reactivity, we developed (30) the diffusional model shown in Figure 7. Here, oxidants that are randomly generated throughout the intraphagosomal fluid react by parallel pathways with reductants in the fluid (k_1), with the phagosomal membrane (k_o), or with the bacterial envelope (k_i). Sufficient data are available on the biological composition of physiological fluids and the reactivities of both proteins and fluid components with $^{\bullet}OH$ and $HCO_3^{\bullet}$ to allow reasonable estimates of the homogeneous (k_1) and surface (k_o, k_i) rate constants. The mathematics can be solved to determine the dependence of the fraction of oxidant generated that reacts with the bacterium upon the thickness (d) of the intraphagosomal fluid layer. [Appropriate values of d, from electron microscopy (52), are 0.1–0.5 μm.] Representative calculations are given in Figure 8.

The results predict that, as expected, $^{\bullet}OH$ will be effectively scavenged by solution-phase reactants unless it is generated immediately adjacent to the bacterial surface. In contrast, a significant fraction of $HCO_3^{\bullet}$ is predicted to react with the bacterium at normal phagosomal dimensions, despite the presence of relatively high concentrations of oxidant scavengers in the intraphagosomal fluid. This condition arises because the reactivity of $HCO_3^{\bullet}$ in this environment gives a diffusion length that is approximately the same as the reaction zone thickness (d). Thus, $HCO_3^{\bullet}$ appears to be a plausible phagocyte-generated bactericidal agent in physiological systems, provided that the bacterium is con-

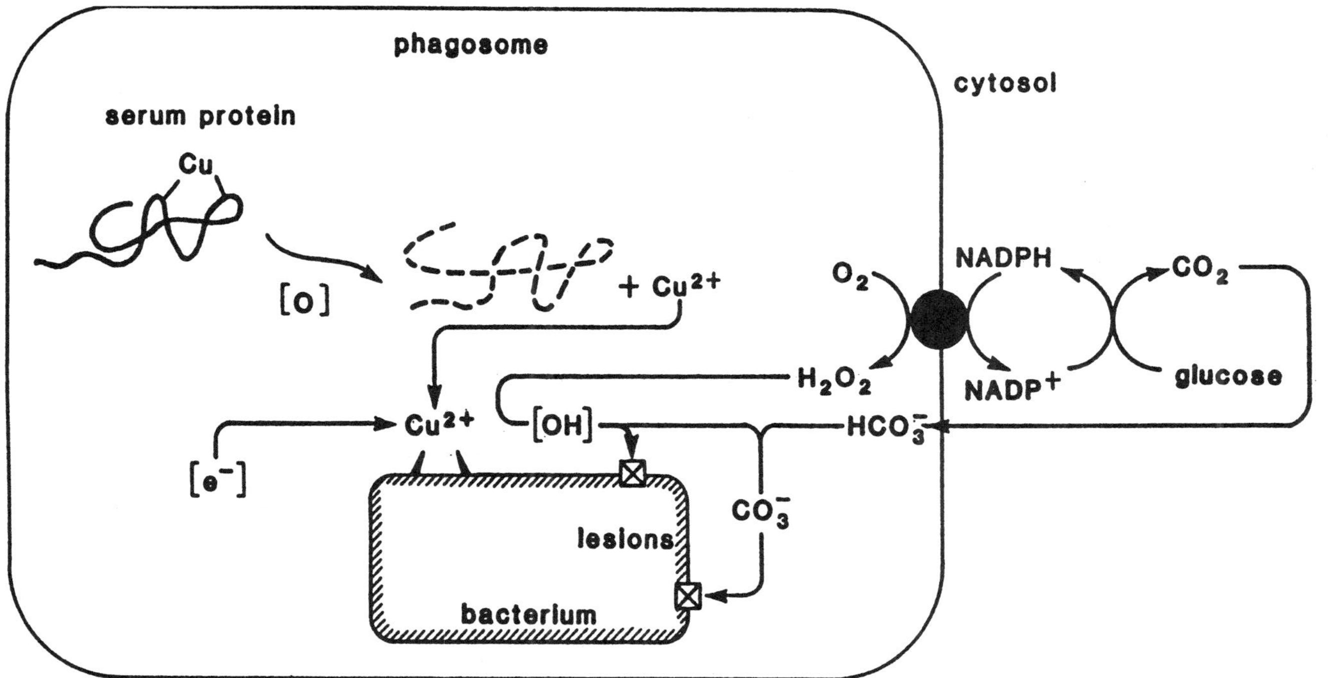

Figure 6. *Speculative model (32) for enhanced intraphagosomal toxicity. Serum Cu-binding protein undergoes oxidative [O] or proteolytic damage, releasing its Cu ion, which relocates to the bacterial envelope, where it either catalyzes formation of lethal oxidants [Cu(III) or •OH] from H_2O_2 and endogenous reductants [e⁻], causing damage at that site, or catalyzes formation of $HCO_3^{\bullet}$ radical, which selectively oxidizes vulnerable sites on the envelope. (Reproduced with permission from reference 78. Copyright 1993 Plenum Press.)*

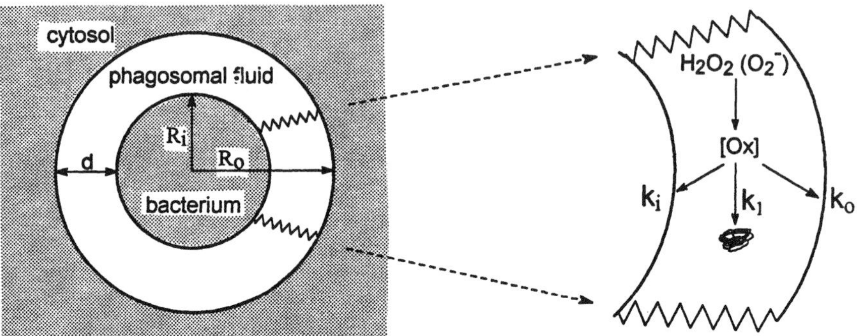

Figure 7. Kinetic model for intraphagosomal reactions of leukocyte-generated oxidants. The reaction zone is defined as the volume between the bacterium and the phagosomal membrane. Oxidants obtained from secondary reactions involving products of the respiratory burst are assumed to be generated uniformly throughout the reaction zone, which then either react with reductants in the phagosomal fluid (k_1) or at target sites on the bacterial envelope (k_i) or phagosomal membrane (k_o). (Reproduced from reference 30. Copyright 1995 American Chemical Society.)

fined in a subcellular reaction compartment. Under these conditions, "site-directed" mechanisms are not required for bacterial killing.

The biological origin of the catalytic cupric ions that are necessary to potentiate bacterial killing by H_2O_2 is problematic. The insensitivity of aerobic organisms such as *E. coli* to H_2O_2 in metal-free environments (32) suggests that their metal ions are ineffective in promoting toxicity. Consequently, if redox-active metal ions are involved in schemes such as depicted in Figure 6, they must be host-derived. This arrangement is conceptually appealing because it suggests a mechanism for controlling H_2O_2 toxicity wherein metal-binding proteins from serum that are carried into the phagosome release their metals, triggering the lethal reactions. However, several studies involving addition of serum or specific serum-derived metal complexes to model reaction systems (reviewed in reference 32) have failed to demonstrate potentiation of H_2O_2 toxicity. Thus, a biological metal-releasing mechanism has not yet been found. Until one can be demonstrated, the whole issue of metal-mediated H_2O_2 toxicity must be considered unsubstantiated. Otherwise, it would seem that either H_2O_2 concentrations within the phagosomes of peroxidase-deficient phagocytic cells must reach the very high levels necessary to overwhelm the bacterial defense systems (32) or that H_2O_2 is not the alternate oxidant. What else is there?

The Role of $\cdot O_2^-$. Superoxide ion, the immediate product of the respiratory burst (Figure 3), is not very strongly oxidizing in aqueous physiological environments (53) and, in general, does not appear to be intrinsically bactericidal (*10*). However, recent evidence suggests (*54, 55*) that phagocytic cells have

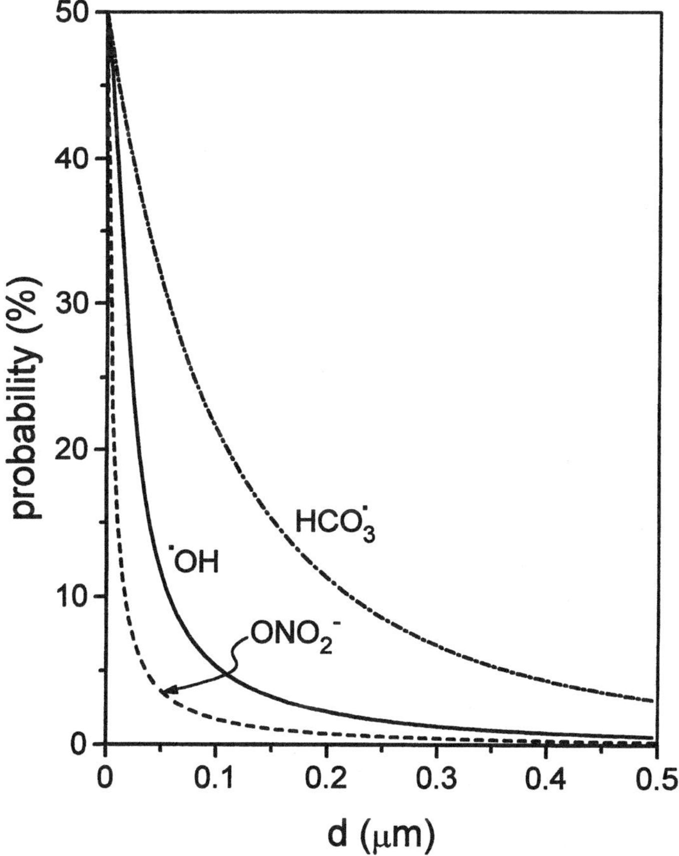

Figure 8. Dependence of the fraction of oxidants reacting with the bacterium upon the magnitude of the reaction zone thickness (d). The solid curve is for •OH with $k_i = k_o = 250$ cm/s and $k_1 = 4 \times 10^7$ s^{-1}; the dashed curve is for ONO$_2^-$ ion with $k_i = k_o = 1.2 \times 10^{-6}$ cm/s and $k_1 = 55$ s^{-1}; and the dot–dashed curve is for HCO$_3^•$ with $k_i = k_o = 2.5$ cm/s and $k_1 = 4 \times 10^5$ s^{-1}. Rate constants were estimated from published kinetic data using the procedures described in reference 30.

the capacity to enzymatically generate •NO. Nitric oxide is also not bactericidal at physiologically relevant concentration levels (56) (Table I), but reacts very rapidly with •O$_2^-$ to form the peroxonitrite ion (ONO$_2^-$). [The bimolecular rate constant for this reaction is the largest yet recorded for any reaction of •O$_2^-$ (57).] Peroxonitrite ion is a powerful, relatively long-lived oxidant (58) with bactericidal capabilities (56) (Table I). It undergoes intramolecular isomerization to nitrate ion with $t_{1/2} \sim 1$ s in neutral solutions (59). In addition to a role as phagocyte-generated microbicidal agent, it has been proposed to cause oxidative damage to other tissues that are capable of simultaneously generating •O$_2^-$ and •NO, giving rise to a variety of pathogenic conditions (leading references are given in reference 60). However, ONO_2^- is unstable in carbonate-containing media (*61*), and low concentration levels of HCO_3^- protect *E. coli* from the toxic effect of ONO_2^- in dilute bacterial suspensions (*62*). Since bicarbonate is *the* physiological buffer, we thought it important to investigate its reaction with ONO_2^-.

Stopped-flow analyses revealed that the decay of ONO_2^- in carbonate buffers was highly pH-dependent (*60*); the bimolecular rate constant gave the bell-shaped profile shown in Figure 9. Analogous to the $HOCl–H_2O_2$ reaction

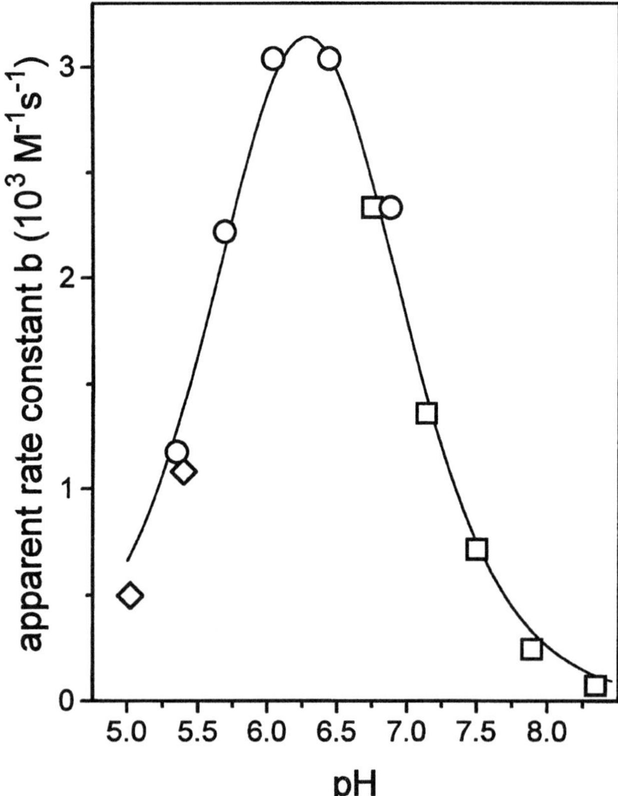

Figure 9. Dependence of the apparent rate constant (b) for reaction between per-oxonitrite and carbonate upon the medium acidity. The constant b is defined by: $-d[ONO_2]_o/dt = b[ONO_2]_o[HCO_3]_o$, *where* $[ONO_2]_o = [ONO_2^-] + [ONO_2H]$, *and* $[HCO_3]_o = [HCO_3^-] + [CO_2]$. *Conditions: 200 mM phosphate (squares); 150 mM pyrophosphate (circles); 400 mM acetate (diamonds); 23 °C. Rates have been corrected for spontaneous isomerization of* ONO_2H *to* HNO_3. *The solid line is the theoretical fit to the data using appropriate dissociation constants and a bimolecular rate constant,* $k = 3 \times 10^4$ M^{-1} s^{-1}, *for reaction between* HCO_3^- *and* CO_2. *(Reproduced from reference 60. Copyright 1995 American Chemical Society.)*

discussed already, these kinetic data indicate that the actual reactant species are either HCO_3^- and ONO_2H or CO_2 and ONO_2^-. In general, it would not be possible to identify which of these pairs are the actual reactants because the rate laws are homomorphic. However, in neutral solutions, equilibration between CO_2 and HCO_3^- (i.e., the reaction $CO_2 + H_2O \rightarrow HCO_3^- + H^+$) is slow (63) relative to the reaction between carbonate and ONO_2^-, so that it is possible by applying pH-jump methods to prepare neutral solutions that contain essentially all or none of the carbonate in the form of CO_2.

Typical results of these experiments are illustrated in Figure 10a; they show that reaction with CO_2 is extremely rapid, whereas reaction with HCO_3^- is below the detectable limit set by the ONO_2^- isomerization and HCO_3^- dehydration rates. Thus, of the possible reactant pairs (HCO_3^- and ONO_2H, HCO_3^- and ONO_2^-, or CO_2 and ONO_2^-), only CO_2 and ONO_2^- are reactive. This conclusion is supported by the additional observations that decay profiles for ONO_2^- were biphasic when mixed with solutions containing excess HCO_3^-, but limiting CO_2, yet followed simple second-order kinetics when CO_2 was in excess of ONO_2^-. The breakpoints of the biphasic curves coincided with the initial amount of CO_2 in the solutions; this indicates that, when exposed to ONO_2^-, the existing CO_2 was rapidly consumed, after which the reaction became limited by the rate of CO_2 generation by HCO_3^- dehydration. These effects are illustrated in Figure 10b, where the conditions for the right-hand trace are $[ONO_2^-] > [CO_2]$ and for the left-hand trace are $[ONO_2^-] \cong [CO_2]$. In both solutions, the total carbonate concentration exceeded that of ONO_2^-, so that the reactions proceeded to completion.

The rate constant for reaction between CO_2 and ONO_2^- ($k = 3 \times 10^4$ M^{-1} s^{-1} at 23 °C) is comparable to the largest rate constants reported for reaction of ONO_2^- with organic compounds (64–67) and raises the issue of the physiological consequences of this reaction. Although the kinetic data base is limited, we have applied our diffusional model (30) to analysis of the reactivity of ONO_2^- within the phagosome (Figure 7). In this case, the lifetime of ONO_2^- is sufficiently long that the diffusion length of the ion far exceeds the phagosomal dimensions. Consequently, the distribution of oxidizing equivalents will be dictated solely by the relative rates of reaction with oxidizable species present in the milieu. Even at 25 mM HCO_3^- [recall that intraphagosomal concentrations are probably higher (Figure 3)], the calculations predict negligible direct reaction of ONO_2^- with the bacterium (Figure 8) because the rate of reaction with CO_2 simply overwhelms all other possible reactions. Within the context of the model (Figure 7), this means that nearly all ONO_2^- formed will disappear by the k_1 pathway. Clearly, any chemistry associated with ONO_2^- formation by phagocytes or other biological tissues must proceed through the intermediacy of the product of the ONO_2^-–CO_2 reaction (presumably the $ONO_2CO_2^-$ adduct).

As with the $HOCl$–HO_2^- pair, the unique reactivity between CO_2 and ONO_2^- can be attributed to strong electrophile–nucleophile interactions between the electron-deficient central carbon atom in CO_2 and the nucle-

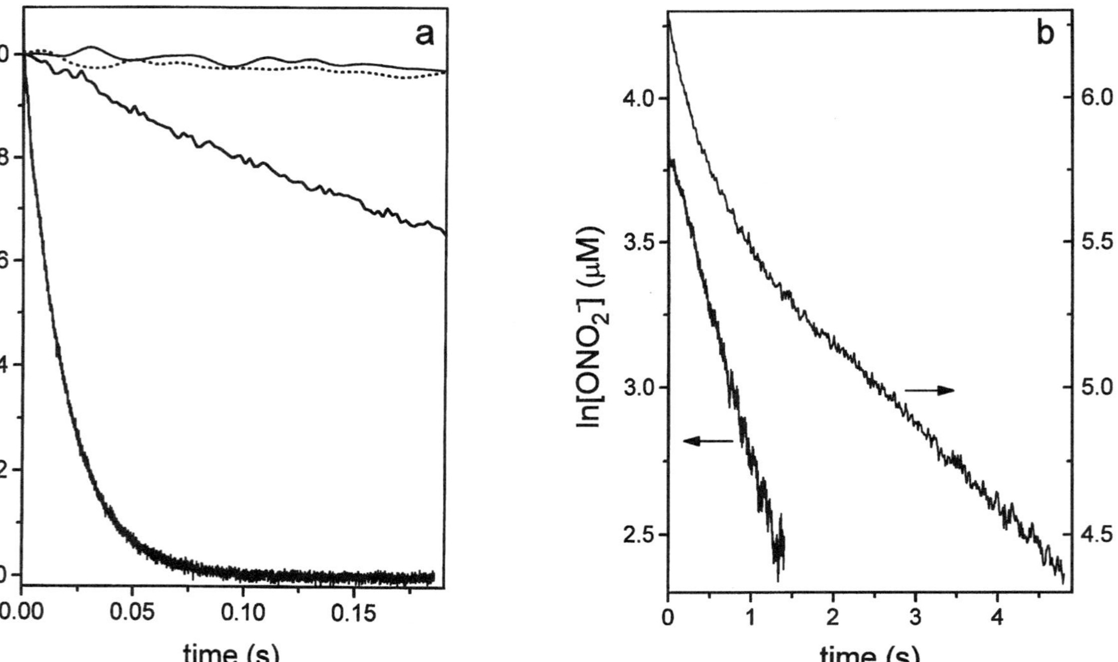

Figure 10. Stopped-flow kinetic traces of peroxonitrite decay. Changes in ONO_2^- concentrations were monitored spectrophotometrically at 302 nm following mixing of alkaline (pH ~11) solutions containing $ONO_2^- \pm HCO_3^-/CO_3^{2-}$ with acidic phosphate-buffered solutions $\pm$ CO_2. Panel a: for all reactions, buffer concentrations were chosen so that the pH after mixing was 7.6; initial reactant concentrations were $[ONO_2]_o \cong 500 \ \mu M$ and $[HCO_3]_o \cong 5 \ mM$ (60). Solid traces: ONO_2^- plus HCO_3^- (upper); ONO_2^- plus $HCO_3^-/CO_2 = 10/1$ (middle); ONO_2^- plus CO_2 (lower); dashed trace: ONO_2^- with no added carbonate. Panel b: reaction of 530 μM ONO_2^- (right-hand trace) and 48 μM ONO_2^- (left-hand trace) with 1.3 mM carbonate (~40 μM CO_2) in 250 mM phosphate (pH 7.5).

ophilic terminal oxygen atom of the ONO_2^- peroxo group. In contrast, the absence of detectable reaction between HCO_3^- and ONO_2H can be attributed to their considerably reduced electrophilic and nucleophilic character. Although the electron-withdrawing $-N=O$ substituent reduces the peroxo group nucleophilicity relative to other peroxo anions (e.g., HO_2^-), it also reduces its basicity (59), allowing the reactant pair CO_2 and ONO_2^- to be present simultaneously in neutral solutions at concentration levels that are sufficiently high to ensure a large overall reaction rate between them. In contrast, more basic peroxides that might form analogous peroxocarbonates (e.g., HO_2^-), would be considerably less reactive toward CO_2 because they exist primarily in their unreactive conjugate acid forms (H_2O_2) in neutral solutions. Formation of ONO_2^- might therefore serve the unique function of allowing entry into a pathway for generating peroxide-based toxins that is not accessible by H_2O_2 itself in physiological environments. The proposed set of reactions involving $^\bullet O_2^-$ and $^\bullet NO$ is summarized in Figure 11.

The chemistry of the putative $ONO_2CO_2^-$ adduct is largely unexplored. Preliminary kinetic studies from our laboratory (S. V. Lymar, unpublished observations) suggest that it decomposes with a $t_{1/2} \leq$ milliseconds, so that compartmentation may be required for it to be an effective bactericide. Homolytic cleavage of the weak peroxo O–O bond would yield the toxic (33) $HCO_3^\bullet$ radical by a pathway that obviates the need for metal ion catalysis. Heterolytic cleavage of the O–O bond would yield NO_2^+, which is a highly reactive, but very short-lived, nitrating agent (68). Formation of either of these products appears

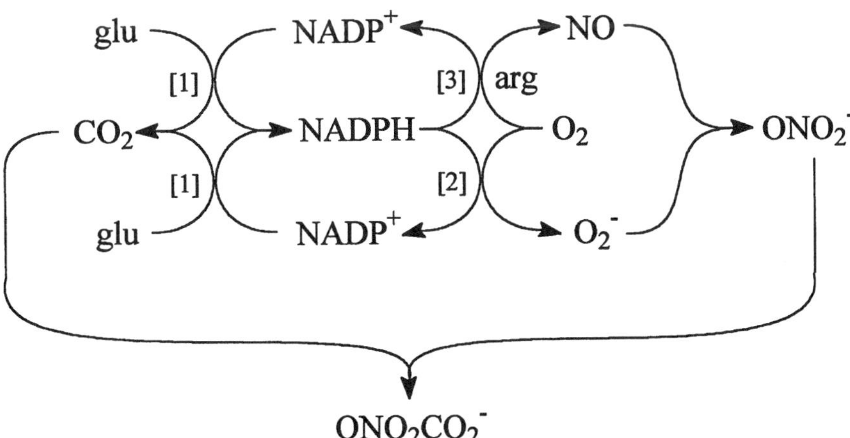

Figure 11. *Flow diagram of generation of* $ONO_2CO_2^-$ *from glucose, arginine, and* O_2. *Catalysis by [1] the enzymes involved in the hexose monophosphate shunt, [2] the NADPH oxidase, and [3] nitric oxide synthase. The symbols glu and arg refer to glucose and arginine, respectively.*

to be thermodynamically feasible. Carbonate effectively promotes ring nitration of aromatic compounds such as tyrosine by ONO_2^- in neutral aqueous media (69). The rate of this reaction is at least 100-fold slower than the estimated lifetime of NO_2^+ (68), suggesting that $ONO_2CO_2^-$ or other secondary products are the actual nitrating reagents (Q. Jiang and S. V. Lymar, unpublished observations). These observations support the notion that similar carbonate-promoted reactions of ONO_2^- may be expressed in biological systems.

Summary

A favorite quotation of Taube's students is "Nature continues to display more imagination than any observer" (70). In the case of host resistance to disease, this statement takes on almost prophetic qualities. Thus, a unique peroxidase has evolved in the neutrophil that is capable of forming bleach, and a growing body of circumstantial evidence suggests that this may be the primary oxidative agent of bactericidal action in these cells. Humans, who long ago discovered the disinfectant properties of HOCl (71), have only relatively recently discovered that it is formed in their own bodies. The identities of alternate oxidants that are generated by phagocytic cells remain obscure. In this brief review, I have emphasized the potential role of the physiological buffer, carbonate, either in the form of an oxidant radical generated by metal-catalyzed reactions with H_2O_2 or as a component of a remarkable $^\bullet O_2^- - ^\bullet NO - CO_2$ redox system. My arguments for carbonate participation are speculative and not widely accepted. They are, however, supported by our initial laboratory findings and, I believe, are based upon sound fundamental mechanistic principles in which both electrophile–nucleophile interactions and microphase compartmentation may play essential roles.

Nitric oxide stands as a further example of nature's inventiveness. Only recently discovered in biological tissues, its potential role in pathogenesis and host resistance is being vigorously examined. If the redox chemistry of ONO_2^- is involved in these processes, then it almost certainly must react with CO_2 in an intermediary step, a point which is not yet widely recognized. Certainly the chemistry that is being discovered is surprising to the participants in this endeavor, but is in keeping with Taube's gentle admonition. If one is concerned that all fundamental discoveries in this area have now been made, consider the last entry to Table I where Cu^I alone, neither a strong oxidant or reductant, is on a molar basis 10^2-fold yet more toxic to bacteria than HOCl, although Cu^{II} alone is virtually nontoxic.

Note Added in Proof

Decomposition of the $ONO_2CO_2^-$ adduct, discussed on p. 415, is more complicated than originally realized. In the presence of excess NO_2^- ion, HCO_3^- is an immediate reaction product, as indicated by the reaction dynamics (Figure

10b). However, when the NO_2^- ion concentration is kept low, decomposition occurs with immediate formation of CO_2 (S. V. Lymar, unpublished observations). In both cases, the ultimate products are the same, i.e., nitrate ion and the equilibrium distribution of HCO_3^- and CO_2. This remarkable difference in reaction pathways can be rationalized by a reaction sequence involving one-electron oxidation of NO_2^- to $^\bullet NO_2$, forming the $^\bullet ONO_2CO_2^{2-}$ radical, which decomposes to HCO_3^- and $^\bullet NO_2$, followed by $^\bullet NO_2$ radical combination yielding N_2O_4, and its subsequent hydrolysis to NO_2^- and NO_3^- ions.

Acknowledgments

I am deeply grateful to the students and associates with whom I have been engaged on this research for their dedication, enthusiasm, and insights, and to colleagues who have willingly served as expert resources in unfamiliar subject areas. I owe a special debt of gratitude to Henry Taube, who first taught me (as well as anyone could hope to) how to think about science and who has served as an inspiration and guide throughout my career.

This research was supported by National Institute of Allergy and Infectious Diseases Grant AI–15834.

References

1. Klebanoff, S. J.; Clark, R. *The Neutrophil-Function and Clinical Disorders*; North Holland: Amsterdam, Netherlands, 1978.
2. Chanock, S. J.; Benna, J. E.; Smith, R. M.; Babior, B. M. *J. Biol. Chem.* **1994**, *269*, 24519–24522.
3. Halperin, J.; Taube, H. *J. Am. Chem. Soc.* **1950**, *72*, 3319–3320.
4. Cahill, A. E.; Taube, H. *J. Am. Chem. Soc.* **1952**, *74*, 2313–2318.
5. Anbar, M.; Taube, H. *J. Am. Chem. Soc.* **1954**, *76*, 6243–6247.
6. Anbar, M.; Taube, H. *J. Am. Chem. Soc.* **1958**, *80*, 1073–1077.
7. Taube, H. *Record Chem. Prog.* **1956**, *17*, 25–33.
8. Morel, F.; Dousierre, J.; Vignais, P. V. *Eur. J. Biochem.* **1991**, *201*, 523–546.
9. Segal, A. W.; Abo, A. *Trends Biochem. Sci.* **1993**, *18*, 43–47.
10. Hurst, J. K.; Barrette, W. C., Jr. *Crit. Rev. Biochem. Mol. Biol.* **1989**, *24*, 271–328.
11. Harrison, J. E.; Schultz, J. *J. Biol. Chem.* **1976**, *251*, 1371–1374.
12. Bos, A.; Wever, R.; Roos, D. *Biochim. Biophys. Acta* **1978**, *525*, 37–44.
13. Ayoub, E. M.; White, J. G. *J. Bacteriol.* **1969**, *98*, 728–736.
14. Segal, A. W.; Geisow, M.; Garcia, R.; Harper, A.; Miller, R. *Nature (London)* **1981**, *290*, 406–409.
15. Hurst, J. K.; Loehr, T. M.; Curnutte, J. T.; Rosen, H. *J. Biol. Chem.* **1991**, *266*, 1627–1634.
16. Taube, H. *Adv. Inorg. Chem. Radiochem.* **1959**, *1*, 1–53.
17. Mandell, G. L. *Infect. Immun.* **1974**, *9*, 337–341.
18. Vel, W. A. C.; Namavar, F.; Verweig, M. J.; Pubben, A. N. B.; McLaren, D. M. *J. Med. Microbiol.* **1984**, *18*, 173–180.
19. Klebanoff, S. J. In *Peroxidases in Chemistry and Biology*; Everse, J.; Everse, K. E.; Grisham, M. B., Eds.; CRC Press: Boca Raton, FL, 1991; Vol. 1, pp 1–35.

20. Johnson, K. R.; Nauseef, W. M. In *Peroxidases in Chemistry and Biology;* Everse, J.; Everse, K. E.; Grisham, M. B., Eds.; CRC Press: Boca Raton, FL, 1991; Vol. 1, pp 63–81.

21. Weiss, S. J.; Test, S. T.; Eckmann, C. M.; Roos, D.; Regiani, S.; *Science (Washington, D.C.)* **1986,** *234,* 200–203.

22. Thomas, E. L.; Fishman, M. *J. Biol. Chem.* **1986,** *261,* 9694–9702.

23. Hurst, J. K. In *Peroxidases in Chemistry and Biology;* Everse, J.; Everse, K. E.; Grisham, M. B., Eds.; CRC Press: Boca Raton, FL, 1991; Vol. 1, pp 37–62.

24. Hurst, J. K.; Albrich, J. M.; Green, T. R.; Rosen, H.; Klebanoff, S. J. *J. Biol. Chem.* **1984,** *259,* 4812–4821.

25. Cuperus, R. A.; Muijsters, A. O.; Wever, R. *Biochim. Biophys. Acta* **1986,** *871,* 78–84.

26. Kettle, A. J.; Winterbourn, C. C. *Biochem. J.* **1988,** *252,* 529–536.

27. Harrison, J. E.; Araiso, T.; Palic, M. M.; Dunford, H. B. *Biochem. Biophys. Res. Commun.* **1980,** *94,* 34–40.

28. Kettle, A. J.; Winterbourn, C. C. *Biochem. J.* **1989,** *263,* 823–828.

29. Thomas, E. L.; Learn, D. B. In *Peroxidases in Chemistry and Biology;* Everse, J.; Everse, K. E.; Grisham, M. B., Eds.; CRC Press: Boca Raton, FL, 1991; Vol. 1, pp 84–121.

30. Lymar, S. V.; Hurst, J. K. *Chem. Res. Toxicol.* **1995,** *8,* 833–840.

31. Albrich, J. M.; Hurst, J. K. *FEBS Lett.* **1982,** *144,* 157–161.

32. Elzanowska, H.; Wolcott, R. G.; Hannum, D. M.; Hurst, J. K. *Free Radical Biol. Med.* **1995,** *18,* 437–449.

33. Wolcott, R. G.; Franks, B. S.; Hannum, D. M.; Hurst, J. K. *J. Biol. Chem.* **1993,** *269,* 9721–9728.

34. Stanbury, D. M. *Adv. Inorg. Chem.* **1989,** *33,* 69–138.

35. Held, A. M.; Halko, D. J.; Hurst, J. K. *J. Am. Chem. Soc.* **1978,** *100,* 5732–5740.

36. Hurst, J. K.; Carr, P. A. G.; Hovis, F. E.; Richardson, R. J. *Inorg. Chem.* **1981,** *20,* 2435–2438.

37. Connick, R. E. *J. Am. Chem. Soc.* **1947,** *69,* 1509–1514.

38. Albrich, J. M.; McCarthy, C. A.; Hurst, J. K. *Proc. Natl. Acad. Sci. U.S.A.* **1981,** *78,* 210–214.

39. Thomas, E. L. *Infect. Immun.* **1979,** *25,* 110–116.

40. Thomas, E. L. *Infect. Immun.* **1979,** *23,* 522–531.

41. Rosen, H.; Rakita, R. M.; Waltersdorph, A. M.; Klebanoff, S. J. *J. Biol. Chem.* **1987,** *262,* 15004–15010.

42. Hurst, J. K.; Barrette, W. C., Jr.; Michel, B.; Rosen, H. *Eur. J. Biochem.* **1991,** *202,* 1275–1282.

43. Barrette, W. C., Jr.; Hannum, D. M.; Wheeler, W. D.; Hurst, J. K. *Biochemistry* **1989,** *28,* 9172–9178.

44. Hannum, D. M.; Barrette, W. C., Jr.; Hurst, J. K. *Biochem. Biophys. Res. Commun.* **1995,** *212,* 868–874.

45. Test, S. T.; Weiss, S. J. *J. Biol. Chem.* **1984,** *259,* 399–405.

46. Beaman, L.; Beaman, B. L. *Annu. Rev. Microbiol.* **1984,** *38,* 27–48.

47. Simic, M. G.; Taylor, K. A.; Ward, J.; Sonntag, C. *Oxygen Radicals in Biology and Medicine;* Plenum: New York, 1988.

48. Halliwell, B.; Gutteridge, J. M. *Methods Enzymol.* **1990,** *186,* 1–85.

49. Sutton, H. C.; Winterbourn, C. C. *Free Radical Biol. Med.* **1989,** *6,* 53–60.

50. Sawyer, D. T.; Kang, C.; Llobet, A.; Redman, C. *J. Am. Chem. Soc.* **1993,** *115,* 5817–5818.

51. Chevion, M. *Free Radical Biol. Med.* **1988,** *5,* 27–37.

52. Rozenberg-Arska, M.; Salters, M. E. C.; van Strijp, J. A. G.; Geuze, J. J.; Verhoef, J. *Infect. Immun.* **1985,** *50,* 852–859.

53. Sawyer, D. T.; Valentine, J. S. *Acc. Chem. Res.* **1981**, *14*, 393–400.
54. Carrerras, M. C.; Pargament, G. A.; Catz, S. D.; Pederoso, J. J.; Boveris, A. *FEBS Lett.* **1994**, *341*, 65–68.
55. Ischiropoulos, H.; Zhu, L.; Beckman, J. S. *Arch. Biochem. Biophys.* **1992**, *298*, 446–451.
56. Brunelli, L.; Crow, J. P.; Beckman, J. S. *Arch. Biochem. Biophys.* **1995**, *316*, 327–334.
57. Huie, R. E.; Padmaja, S. *Free Rad. Res. Commun.* **1993**, *18*, 195–199.
58. Koppenol, W. H.; Moreno, J. J.; Pryor, W. A.; Ischiropoulos, H.; Beckman, J. S. *Chem. Res. Toxicol.* **1992**, *5*, 834–842.
59. Edwards, J. O.; Plumb, R. C. *Prog. Inorg. Chem.* **1994**, *41*, 599–635.
60. Lymar, S. J.; Hurst, J. K. *J. Am. Chem. Soc.* **1995**, *117*, 8867–8868.
61. Keith, W. G.; Powell, R. E. *J. Chem. Soc. A* **1969**, *1969*, 60.
62. Zhu, L.; Gunn, C.; Beckman, J. S. *Arch Biochem. Biophys.* **1992**, *298*, 452–457.
63. Kern, D. M. *J. Chem. Educ.* **1960**, *37*, 14–23.
64. Radi, R.; Beckman, J. S.; Bush, K. M.; Freeman, B. A. *J. Biol. Chem.* **1991**, *266*, 4244–4250.
65. Pryor, W. A.; Jin, X.; Squadrito, G. L. *Proc. Natl. Acad. Sci. U.S.A.* **1994**, *91*, 11173–11177.
66. Beckman, J. S.; Ischiropoulos, H.; Zhu, L.; van der Woerd, M.; Smith, C.; Chen, J.; Harrison, J.; Martin, J. C.; Tsai, M. *Arch Biochem. Biophys.* **1992**, *298*, 438–445.
67. Beckman, J. S.; Beckman, T. W.; Chen, J.; Marshall, P. A.; Freeman, B. A. *Proc. Natl. Acad. Sci. U.S.A.* **1990**, *87*, 1620–1624.
68. Moodie, R. B.; Schofield, K.; Taylor, P. G. *J. Chem. Soc. Perkin Trans. 2* **1979**, *1979*, 133–136.
69. van der Vliet, A.; O'Neill, C. A.; Halliwell, B.; Cross, C. E.; Kaur, H. *FEBS Lett.* **1994**, *339*, 89–92.
70. Taube, H. *J. Chem. Educ.* **1959**, *36*, 451–455.
71. Alcock, T. *An Essay on the Use of Oxide of Sodium and of Lime, as Powerful Disinfecting Agents, and of the Chloruret of Oxide of Sodium, More Especially as a Remedy of Considerable Efficacy, in the Treatment of Hospital Gangrene: Phagedenic, Syphilitic, and Ill-Conditioned Ulcers: Mortification and Various Other Diseases;* Burgess and Hill: London, 1827.
72. Koppenol, W. H. *FEBS Lett.* **1994**, *347*, 5–8.
73. *CRC Handbook of Chemistry and Physics, 70th Edition;* Weast, R. C. Ed.; CRC Press: Boca Raton, FL, 1989; pp 151–158.
74. Koppenol, W. H. In *Focus on Membrane Lipid Oxidation;* Vigo-Pelfey, C., Ed.; CRC Press: Boca Raton, FL, 1989; Vol. 1, pp 1–13.
75. Huie, R. E.; Clifton, C. L.; Neta, P. *Radiat. Phys. Chem.* **1991**, *38*, 477–481.
76. Eriksen, T. E.; Lind, J.; Merenyi, G. *Radiat. Phys. Chem.* **1985**, *26*, 197–199.
77. Sillen, L. G.; Martell, A. E. *Stability Constants of Metal-Ion Complexes;* Burlington House: London, 1964.
78. Hurst, J. K. In *The Activation of Dioxygen and Homogeneous Catalytic Oxidation;* Barton, D. H. R.; Martell, A. E.; Sawyer, D. T., Eds.; Plenum: New York, 1993; pp 267–286.

Application of Taube Insights to Nuclear Medicine

In Vivo Inorganic Chemistry

Edward Deutsch

Mallinckrodt Medical, Inc., St. Louis, MO 63134

The methodologies and tools developed by Henry Taube and his colleagues over the past several decades make it possible to rationally investigate mechanistic inorganic chemistry within a living animal or human. "In vivo inorganic chemistry" can indeed be studied. This chapter describes one such investigation in which the principles of modern inorganic chemistry are applied to the development of a new heart imaging radiopharmaceutical for use in diagnostic nuclear medicine. Based on the insights and teachings of Taube, the fundamental principles of periodicity and redox reactivity are applied within a subtle experimental design to first understand why reducible $^{99m}Tc(III)$ cations fail as heart imaging agents, and then the concepts of π backbonding, coordination chemistry, and steric control of reaction kinetics are applied to design and develop nonreducible $^{99m}Tc(III)$ cations that are successful heart imaging radiopharmaceuticals. One of these agents has already provided clinically useful information for hundreds of patients and is currently in commercial development.

WITHIN THIS CELEBRATION of Henry Taube's 80th birthday, it is especially appropriate to subtitle this review chapter "In Vivo Inorganic Chemistry". Taube has always loved the intricacies of the English language, and this subtitle utilizes these intricacies to generate an apparent oxymoron. Yet, it is the methodologies and tools developed by Taube and his colleagues over the past several decades that make it possible to rationally investigate mechanistic inorganic chemistry within a living animal or human; that is, one can indeed investigate inorganic chemistry in vivo. This chapter describes such an investigation that was conducted in our research group as part of a long-term program to apply the principles and techniques of modern inorganic chemistry to the development of new radiopharmaceuticals for use in nuclear medicine. These ultimately successful applications of modern inorganic chemistry to the practical problems of nuclear medicine stem in large part from the scientific foundations and insights provided by our mentor, Henry Taube.

Our research has focused largely on the inorganic chemistry of technetium and rhenium as relevant to nuclear medicine. For readers not familiar with this field, a complete introduction with considerable background information can be found in our 1989 review (1), which was part of the published proceedings from the 3rd International Conference on Technetium and Rhenium in Chemistry and Nuclear Medicine (2); more recent advances, additional background information, and a slightly different perspective can be found in the published proceedings of the 4th Conference on this subject (3).

Technetium

The preeminent isotope in diagnostic nuclear medicine is clearly [99m]Tc (where m stands for "metastable"). Fully 85% of the procedures in diagnostic nuclear medicine are conducted with this isotope, and procedures based on other isotopes would be converted to ^{99m}Tc if suitable technetium radiopharmaceuticals were available. This preeminence derives from several factors: the ready availability of ^{99m}Tc from an inexpensive [99m]Mo/^{99m}Tc generator obviates the need for an on-site cyclotron or nuclear reactor; the nuclear properties of ^{99m}Tc are nearly ideal for medical applications (6 h half-life, no particulate radiation, and a single γ ray emission of moderate energy); and, most importantly, a rich and diverse chemistry allows technetium to be incorporated into a variety of chemical forms that provide the basis for many different radiopharmaceuticals. [99m]Tc decays to [99g]Tc (where g stands for ground state), a very long-lived, weak β-emitter that is available in gram quantities and is used for chemical characterizations.

Rhenium

While β-emitting isotopes of many elements have been proposed for radiotherapeutic use, we have focused our research on those of rhenium. Both [186]Re (half-life of 90 h) and [188]Re (half-life of 17 h) emit β particles of therapeutically useful energies and γ rays of energies suitable for imaging (1).

Tc/Re

The periodic relationship between technetium and rhenium has direct consequences for the development of radiopharmaceuticals based on these elements. Most obviously, the gross similarities between Tc and Re chemistry might lead to the development of rhenium-based therapeutic agents derived from existing technetium-based diagnostic agents. More subtly, by understanding and elaborating the differences between Tc and Re chemistry, analogous pairs of technetium and rhenium complexes might be used to probe the in vivo mechanisms of action of Tc/Re radiopharmaceuticals.

^{99m}Technetium Phosphine Complexes as Heart Imaging Agents

Fundamental studies utilizing ^{99g}Tc and the prototypical bidentate phosphine ligand DMPE (1,2-bis(dimethylphosphino)ethane) have established the identity, structures, and properties of three stable, monocationic complexes: *trans*-$[Tc^VO_2(DMPE)_2]^+$, *trans*-$[Tc^{III}Cl_2(DMPE)_2]^+$, and $[Tc^I(DMPE)_3]^+$ (*1, 4*). By appropriate control of reaction conditions (primarily pH and temperature), the corresponding three ^{99m}Tc complexes can each be prepared in >90% radiochemical purity within formulations suitable for evaluation as imaging agents (*4*). This series of complexes nicely illustrates the variety of oxidation states available to technetium, and their successive generation by the action of the two-equivalent reductant DMPE. Further studies established that the corresponding Tc(II) complexes *trans*-$[Tc^{II}Cl_2DMPE)_2]^0$ and $[Tc^{II}(DMPE)_3]^{2+}$ are readily accessible by chemical or electrochemical means. It is this facile redox reactivity of low-valent technetium complexes that leads to the example of "in vivo inorganic chemistry" discussed in this chapter.

We had proposed in 1980 that cationic complexes of ^{99m}Tc might function as heart imaging agents, analogous to ^{201}Tl, the only such agent in use at the time (*5*); in this way, the optimal nuclear properties and ready availability of ^{99m}Tc might be advantageously used in the diagnosis of heart disease. Thus, a focus of our early work was to evaluate the three prototypical DMPE–Tc cations as potential heart imaging agents (*6*). Within this overall program, the first ^{99m}Tc complex of any type to be evaluated as a potential heart imaging agent in humans was *trans*-$[^{99m}Tc^{III}Cl_2(DMPE)_2]^+$ (*7*).

While there is definite accumulation of *trans*-$[^{99m}Tc^{III}Cl_2(DMPE)_2]^+$ in normal heart muscle, this agent tends to rapidly wash out of the heart and accumulate in the liver. Since the liver is adjacent to the heart, this transfer of activity from the heart to the liver gradually decreases the target-to-nontarget ratio of the agent and severely limits its utility. From a series of indirect experiments we suspected that the cause of this washout might be the in vivo reduction of Tc(III) to Tc(II); this reduction would convert the initial cationic complex to a neutral species that would be expected to accumulate in the liver. However, the very low concentrations of technetium (*8*) involved in these in vivo studies would make it very difficult, and probably impossible (*9*), to conduct a direct experiment that would detect the hypothesized in vivo reduction of Tc(III) to Tc(II). [The total concentration of technetium in a radiopharmaceutical preparation of *trans*-$[^{99m}Tc^{III}Cl_2(DMPE)_2]^+$ is about 10^{-8} M, and the corresponding concentration in tissues is less than about 10^{-10}M.] To establish the validity of our hypothesis would thus require an indirect, subtle approach similar to the approaches used by Taube in his pioneering studies of reaction mechanisms.

To determine whether in vivo reduction of *trans*-[$^{99m}Tc^{III}Cl_2(DMPE)_2$]$^+$ occurs, we designed an "in vivo inorganic" experiment based on the periodic relationship between technetium and rhenium (*10*). The chemistries of Tc and Re are quite similar due to their relative positions in the periodic table and the lanthanide contraction. Thus, analogous complexes of Tc and Re have identical sizes, shapes, charges, dipole moments, and lipophilicities, and the biological milieu cannot distinguish between them on the basis of any external property: to blood proteins or cell membranes, for example, *trans*-[$Tc^{III}Cl_2(DMPE)_2$]$^+$ and *trans*-[$Re^{III}Cl_2(DMPE)_2$]$^+$ appear indistinguishable. However, there is a substantive difference in the redox chemistries of analogous Tc and Re complexes; in general, Re complexes are about 200 mV more difficult to reduce than are their Tc analogs (*1*). [Differences in redox potentials determined at 25 °C are reasonably expected to be of the same magnitude at body temperature (37 °C). The in vivo redox system is generally buffered by the glutathione thiol–disulfide interchange.]

For the pair of complexes *trans*-[$M^{III}Cl_2(DMPE)_2$]$^+$ (M = Tc, Re), the Re complex is 190 mV more difficult to reduce to its M(II) form, while X-ray crystal structures confirm that the two complexes are structurally equivalent (*10*). Thus, this pair of complexes can be used to test our hypothesis of in vivo reduction; if *trans*-[$Tc^{III}Cl_2(DMPE)_2$]$^+$ undergoes in vivo reduction, the Re analog will not since its reduction potential will be out of the range accessible to biological systems. As a control, we used the pair of M(I) complexes [$M^I(DMPE)_3$]$^+$ (M = Tc and Re) since neither of these complexes undergoes reduction at biologically accessible potentials (*10*). For both the experiment (M(III) complexes) and the control (M(I) complexes), analogous ^{99m}Tc and ^{186}Re complexes were mixed in the same solution and then coinjected into rats; the ratio of $^{99m}Tc/^{186}Re$ was then determined in a variety of tissues as a function of time. This procedure allows a more accurate determination of relative biodistributions because the ratio of ^{99m}Tc and ^{186}Re activities can be determined more precisely than can the individual activities.

As expected, when [$^{186}Re^I(DMPE)_3$]$^+$ and [$^{99m}Tc^I(DMPE)_3$]$^+$ are coinjected into rats, the ratio $^{99m}Tc/^{186}Re$ is essentially 1.0 in all organs at all times after injection. In other words, the biological milieu cannot distinguish between these M(I) (M = Tc or Re) complexes. However, when the M(III) complexes *trans*-[$^{99m}Tc^{III}Cl_2(DMPE)_2$]$^+$ and *trans*-[$^{186}Re^{III}Cl_2(DMPE)_2$]$^+$ are coinjected into rats, the resulting ratio of $^{99m}Tc/^{186}Re$ in tissues varies by almost a factor of 50 (from 0.14 to 6.4; *see* Figure 1) (*10*). Thus, the biological milieu readily distinguishes between these complexes, primarily because it can reduce the Tc(III) complex, but not the Re(III) complex. Moreover, as expected, the ^{99m}Tc complex washes out of the heart faster than does the ^{186}Re analog. Thus, the biodistribution of ^{99m}Tc reflects the in vivo reduction of the

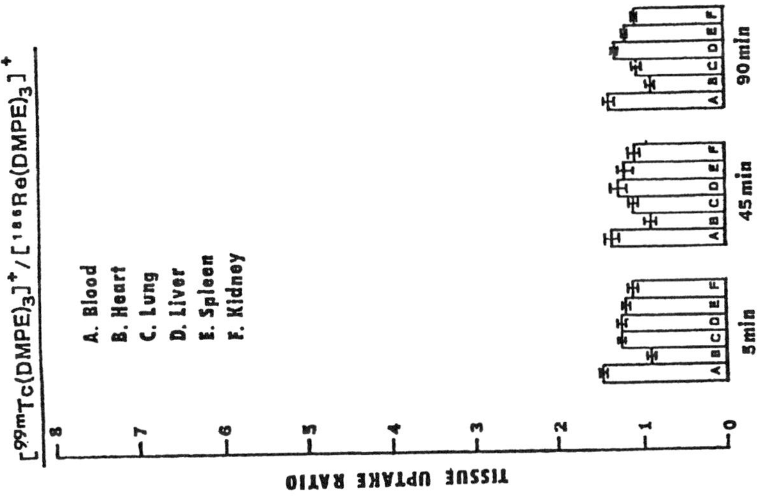

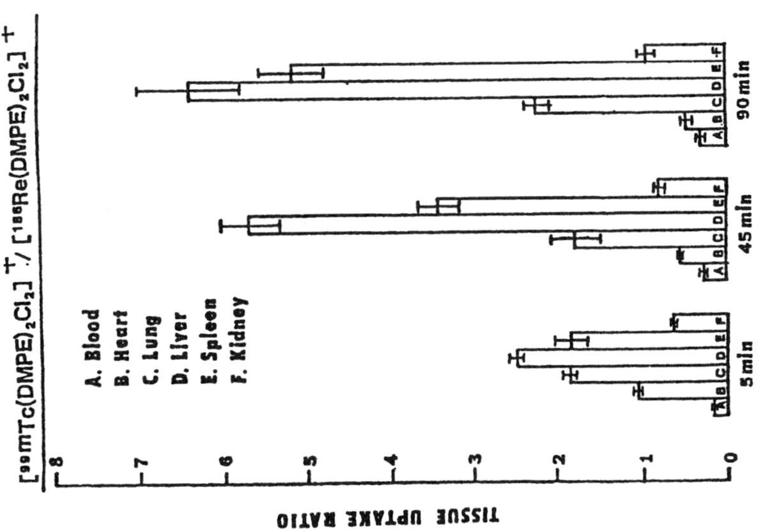

Figure 1. ^{99m}Tc/^{186}Re tissue uptake ratios as a function of time for the experiment described in the text (based on data from reference 10).

Tc(III) cation to the Tc(II) neutral, while the biodistribution of [186]Re reflects the behavior of a purely cationic species. These experiments demonstrate that *trans*-[[99m]Tc^{III}Cl$_2$(DMPE)$_2$]$^+$ undergoes in vivo reduction, and that this in vivo redox activity detracts from the utility of this particular Tc(III) complex as a potential heart imaging agent. The "in vivo inorganic chemistry" experiment briefly summarized here was reported in 1985 (*10*), and to our knowledge it contains one of the first animal biodistribution studies, and the first animal scintiphotographs, to be reported in *Inorganic Chemistry*.

Nonreducible [99m]Tc(III) Complexes as Heart Imaging Agents

The experiments just described provide strong, albeit indirect, evidence that *trans*-[[99m]Tc^{III}Cl$_2$(DMPE)$_2$]$^+$ fails as a heart imaging agent at least in part because it undergoes in vivo reduction. This implies that if the in vivo reduction of Tc(III) to Tc(II) could be avoided, a cationic Tc(III) complex would be retained in the heart and an improved heart imaging agent might result. Moreover, the retention of a nonreducible Tc(III) complex in the heart would provide direct evidence for the "in vivo inorganic chemistry" that causes washout of *trans*-[[99m]Tc^{III}Cl$_2$(DMPE)$_2$]$^+$ from the heart. [Any nonreducible ^{99m}Tc cation would provide the same test, and in fact nonreducible Tc(V) and Tc(I), cations have been developed as heart imaging agents (*4*). The *trans*-[Tc^{V}O$_2$(DMPE)^2]$^+$ and [Tc^I(DMPE)$_3$]$^+$ cations mentioned earlier in this chapter fail as heart imaging agents because they bind too tightly to blood proteins, not because they wash out of the heart (*11, 12*).]

There are a variety of ways to modify *trans*-[Tc^{III}Cl$_2$(DMPE)$_2$]$^+$ so that it becomes more difficult to reduce. One approach involves changing the axial ligand from chloride to the strongly σ-donating thiolato group; this increases the negative charge on the Tc center and makes it more resistant to reduction. The preparation of *trans*-[Tc^{III}(SR)$_2$(DMPE)$_2$]$^+$ complexes required the development of new synthetic routes, and the characterizations of these thiolato complexes led to some interesting and unexpected chemistry (*13–20*). As expected, these complexes are more difficult to reduce from Tc(III) to Tc(II) than is the prototypical chloro complex; for example, *trans*-[Tc^{III}(SCH$_3$)$_2$-(DMPE)$_2$]$^+$ is about 300 mV more difficult to reduce than is *trans*-[Tc^{III}Cl$_2$-(DMPE)$_2$]$^+$. Preliminary animal studies established that the Tc(III)-thiolato complexes are indeed taken up and retained in the heart, and it is likely this series of compounds could yield an effective ^{99m}Tc heart imaging agent. However, a parallel route to generating nonreducible Tc(III) complexes proved to be more productive, and we did not further pursue development of the Tc(III)–thiolato complexes.

Our alternate approach to rendering Tc(III) complexes resistant to one-equivalent reduction is based upon the fundamental π-acid character of phosphine ligands. The greater the number of phosphine ligands in the Tc(III) coordination shell, the greater is the stability of the Tc(II) state. Thus, by reducing the number of phosphine ligands from the four in *trans*-[$Tc^{III}Cl_2$-$(DMPE)_2$]$^+$, the stability of Tc(II) should be lessened and the Tc(III) complex should be stabilized to reduction. This principle is manifested in the synthesis and characterization of a series of Schiff base Tc(III) complexes that contain only two phosphine ligands, for example, *trans*-[Tc^{III}($acac_2$en)(PEt_3)$_2$]$^+$, where $acac_2$en is the prototypical Schiff base ligand bis(acetylacetonato)ethylenediamine (*21*). These complexes contain an $N_2O_2P_2$ donor atom set, as opposed to the Cl_2P_4 donor atom set of *trans*-[$Tc^{III}Cl_2(DMPE)_2$]$^+$, and, as expected, they are much more difficult to reduce to Tc(II); the prototypical *trans*-[Tc^{III}-($acac_2$en)(PEt_3)$_2$]$^+$ complex is about 800 mV more difficult to reduce than is *trans*-[$Tc^{III}Cl_2(DMPE)_2$]$^+$ (*22*).

The strategy of making Tc(III) resistant to in vivo reduction does in fact succeed in generating a ^{99m}Tc cation that does not wash out of the heart. The first such agent evaluated in humans was *trans*-[$^{99m}Tc^{III}$($acac_2$en)(PMe_3)$_2$]$^+$ (*23*), and we gave this agent the trivial designation Q2 to indicate that it is the second in the series of these biologically nonreducible Tc(III) complexes that contain an $N_2O_2P_2$ donor atom set. The fact that this biologically nonreducible Tc(III) agent does not wash out of the heart, whereas the biologically reducible *trans*-[$^{99m}Tc^{III}Cl_2(DMPE)_2$]$^+$ does, effectively confirms our conclusion that it is in vivo reduction to the neutral Tc(II) form that underlies the washout phenomenon.

Further chemical studies, as well as evaluations in animals and humans, led to significant improvements in this "Q series" of Tc(III) cations. The biodistributions of the Q cations can be modified by alterations in either the phosphine or the Schiff base ligands, while the kinetics of formation of the Q agent from starting materials (the free phosphine and Schiff base ligands, plus 99m-pertechnetate) are most easily controlled via modifications in the Schiff base ligand. Alterations to the pendant groups bonded to the phosphorous atoms of Q2 led to Q3, which exhibits a superior biodistribution (*24*), and subsequent incorporation of a rigid ring into the Schiff base ligand of Q3 led to Q12 (structures shown in Figure 2), which can easily be prepared in >90% yield by a one-step reaction, and which is now being commercialized as a ^{99m}Tc heart imaging agent (*25–27*).

Extensive clinical studies in hundreds of patients suffering from a variety of heart diseases and disorders have now shown that both Q3 and Q12 are effective imaging agents that rapidly accumulate in normal heart tissue and then are retained within the heart for several hours (*27*). These characteristics, combined with the relatively high heart/liver uptake ratio observed with these agents, enhance their clinical utility and efficacy. Q12 has been approved for sale in Sweden, and we expect it to be approved for sale in the rest of Europe and in the United States. The successful development of these nonreducible Tc(III) agents depended upon a synergistic interaction between the classical field of inorganic chemistry and the practical field of nuclear medicine. These two disparate disciplines were brought together to investigate "in vivo inor-

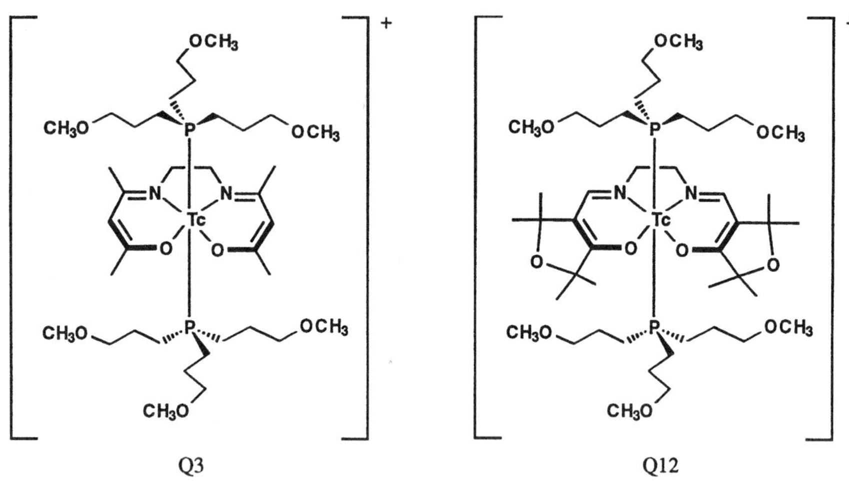

Figure 2. Structural representations of the cationic ^{99m}Tc complexes designated Q3 and Q12.

ganic chemistry" and thereby to generate coordination complexes that will contribute significantly to the health and well being of mankind. Henry Taube's insights provided the foundation for this outcome, and of this he can be justifiably proud.

References

1. Deutsch, E.; Libson, K.; Vanderheyden J.-L. In *Technetium and Rhenium in Chemistry and Nuclear Medicine 3*; Nicolini, M.; Bandoli, G.; Mazzi, U., Eds.; Raven: New York, 1989; pp 13–22.
2. *Technetium and Rhenium in Chemistry and Nuclear Medicine 3*; Nicolini, M.; Bandoli, G.; Mazzi, U., Eds.; Raven: New York, 1989.
3. *Technetium and Rhenium in Chemistry and Nuclear Medicine 4*; Nicolini, M.; Bandoli, G.; Mazzi, U., Eds; SGEditoriali: Padova, Italy, 1995.
4. Deutsch, E. *Radiochim. Acta* **1993**, *63*, 195–197.
5. Deutsch, E.; Glavan, K. A.; Ferguson, D. L.; Lukes, S. R.; Nishiyama, H.; Sodd, V. J. *J. Nucl. Med.* **1980**, *21*, 56.
6. Deutsch, E.; Ketring, A. R.; Libson, K.; Vanderheyden, J.-L.; Hirth, W. W. *Nucl. Med. Biol.* **1989**, *16*, 191–232.
7. Gerson, M. C.; Deutsch, E. A.; Nishiyama, H.; Libson, K. F.; Adolph, R. J.; Grossman, L. W.; Sodd, V. J.; Fortman, D. L.; Vanderheyden, J.-L. E.; Williams, C. C; Saenger, E. L. *Eur. J. Nucl. Med.* **1983**, *8*, 371–374.
8. Holland, M. E.; Deutsch, E.; Heineman, W. R.; Holland, E. M. *Appl. Radiat. Isot.* **1986**, *37*, 165–171.
9. Deutsch, E.; Hirth, W. *J. Nucl. Med.* **1987**, *28*, 1491–1500.
10. Vanderheyden, J.-L.; Heeg, M. J.; Deutsch, E. *Inorg. Chem.* **1985**, *24*, 1666–1673.
11. Gerson, M. C.; Deutsch, E. A.; Libson, K. F.; Adolph, R. J.; Ketring, A. R.; Vanderheyden, J.-L.; Williams, C. C.; Saenger, E. L. *Eur. J. Nucl. Med.* **1984**, *9*, 403–407.

12. Deutsch, E.; Ketring, A. R.; Libson, K.; Vanderheyden, J.-L.; Hirth, W. W. *Nucl. Med. Biol.* **1989**, *16*, 191–232.

13. Konno, T.; Heeg, M. J.; Deutsch, E. *Inorg. Chem.* **1988**, *27*, 4113–4121.

14. Konno, T.; Kirchhoff, J. R.; Heineman, W. R.; Deutsch, E. *Inorg. Chem.* **1989**, *28*, 1174–1179.

15. Konno, T.; Heeg, M. J.; Deutsch, E. *Inorg. Chem.* **1989**, *28*, 1694–1700.

16. Konno, T.; Heeg, M. J.; Stuckey, J. A.; Kirchhoff, J. R.; Heineman, W. R.; Deutsch, E. *Inorg. Chem.* **1992**, *31*, 1173–1181.

17. Konno, T.; Kirchhoff, J. R.; Heeg, M. J.; Heineman, W. R.; Deutsch, E. *J. Chem. Soc. Dalton Trans.* **1992**, 3069–3075.

18. Konno, T.; Seeber, R.; Kirchhoff, J. R.; Heineman, W. R.; Deutsch, E. *Transition Met. Chem. (London)* **1993**, *18*, 209–217.

19. Okamoto, K.-I.; Kirchhoff, J. R.; Heineman, W. R.; Deutsch, E. *Polyhedron* **1993**, *12*, 749–757.

20. Okamoto, K.-I.; Chen, B.; Kirchhoff, J. R.; Ho, D. M.; Elder, R. C.; Heineman, W. R.; Deutsch, E. *Polyhedron* **1993**, *12*, 1559–1568.

21. Jurisson, S. S.; Dancey, K.; McPartlin, M.; Tasker, P. A.; Deutsch, E. *Inorg. Chem.* **1984**, *23*, 4743–4749.

22. Ichimura, A.; Heineman, W. R.; Deutsch, E. *Inorg. Chem.* **1985**, *24*, 2134–2139.

23. Deutsch, E.; Vanderheyden, J.-L.; Gerundini, P.; Libson, K.; Hirth, W.; Colombo, F.; Savi, A.; Fazio, F. *J. Nucl. Med.* **1987**, *28*, 1870–1880.

24. Libson, K.; Messa, C.; Kwiatkowski, M.; Zito, F.; Best, T.; Colombo, F.; Matarese, M.; Wang, X.; Fragasso, G.; Fazio, F.; Deutsch, E. In *Technetium and Rhenium in Chemistry and Nuclear Medicine 3*; Nicolini, M.; Bandoli, G.; Mazzi, U., Eds.; Raven: New York, 1989; pp 365–368.

25. Rossetti, C.; Vanoli, G.; Paganelli, G.; Kwiatkowski, M.; Zito, F.; Colombo, F.; Bonino, C.; Carpinelli, A.; Casati, R.; Deutsch, K.; Marmion, M.; Woulfe, S. R.; Lunghi, F.; Deutsch, E.; Fazio, F. *J. Nucl. Med.* **1994**, *35*, 1571–1580.

26. Gerson, M. C.; Lukes, J.; Deutsch, E.; Biniakiewicz, D.; Rohe, R. C.; Washburn, L. 430C.; Fortman, C.; Walsh, R. A. *J. Nucl. Cardiol.* **1994**, *1*, 499–508.

27. Gerson, M. C.; Lukes, J.; Deutsch, E.; Biniakiewicz, D.; Washburn, L. C.; Walsh, R. A. *J. Nucl. Cardiol.* **1995**, *2*, 224–230.

Author Index

Affiliation Index

Subject Index